"This is by far THE most comprehensive book on the Internet. A must for novices, even experts will learn something new from this book. The attention to detail is first-rate, yet its presentation understandable for everyone."

Scott Yanoff
Author of the "Yanoff List" of Internet services
distributed electronically throughout the Internet

D0520301

"This book is exactly what I have needed for a long time! Ev.... .... other Internet books, *The Internet Complete Reference* is a must-read. It is at once tremendously useful and thoroughly readable. Harley Hahn's irreverent sense of humor and clear, insightful explanations will both keep you laughing and make you an Internet expert. I give it an absolute, unqualified recommendation for anyone who uses or plans to use the Internet."

Kenn Nesbitt, Consultant
Microsoft Consulting Services, Microsoft Corporation

"This is a terrific reference. Quantifying and organizing the world of the Internet is an awesome task, and Harley Hahn has done it!"

Bob Rieger
Founder and Senior Vice President, Operations
Netcom On-Line Communications Services

"*The Internet Complete Reference* is the best Internet book ever written. Superb and authoritative, a must-read for personal and business users alike. No matter how many other Internet books you have, you NEED this book. "

Sam Albert, President
Sam Albert Associates
Scarsdale, New York

"Hahn has produced by far the most comprehensive and illuminating work on the Internet. This is indeed an indispensable guide for any serious user to the world's largest computer network."

Samuel Ko
Internet Book Reviewer, Vancouver, Canada

"At last, a single volume that not only explains what the Internet is, and why it is so exciting, but also provides detailed instructions on how to use all the major Internet services, and includes exhaustive listings of useful archive sites and a wealth of 'insider' information. Hahn writes with wit, clarity and depth that will entertain and inform both the curious novice as well as the experienced net dweller. Copies of this book should be in close proximity to any window onto the Internet!"

Mark Schildhauer, Ph.D.
Technical Coordinator, Social Sciences Computing Facility
University of California, Santa Barbara

"Finally, a book that painlessly releases the many gifts of the Internet! *The Internet Complete Reference* empowers the reader with a complete understanding of the facilities, terminology, and concepts necessary to participate in the Internet. Unique in its approach, this book covers everything, even the fun stuff, representing the Internet as it should be portrayed. I strongly recommend this book to anyone who is curious about the Internet, and needs a thorough, hands-on-oriented explanation that can be easily understood without a technical degree."

Stewart I. Alpert
Member of Technical Staff, Networking Specialist
The Santa Cruz Operation (SCO)

"*The Internet Complete Reference* is the best in its class: a masterful exposition of everything you need to know to use the Internet. The writing is superb: the text easy to understand. The catalog of Internet resources is itself worth the price of the book. If you use or want to use the Internet, this book is indispensible."

Hal Topper, Director of Consulting Relations
IBM Corporation

"If you want to tap into the Internet's wealth of resources, this straight-forward guide will empower you. The Internet is vastly complex and difficult--which is why mortals need *The Internet Complete Reference*."

David Flack
Editor-In-Chief
Unixworld

"The Internet has promised to change the way people communicate on a global basis, and in the same breath threatened to do so in virtually undocumented fashion. Harley Hahn's *The Internet Complete Reference* changes all that by presenting more useful techniques and tips for using and understanding this global tangle of wires than anything we've seen in print. It's readable AND we can find stuff in it. Hint: This book will be the handbook for Internet users for many years to come."

Jack Rickard
Editor/Publisher
Boardwatch Magazine

"*The Internet Complete Reference* is an excellent tool for new Internet users and for anyone involved with training new users. Harley Hahn has succeeded in explaining Internet concepts in plain language, making it easy for the uninitiated to get started on the network. Just what every new user needs: a pro at their shoulder offering advice. The Internet resource catalog is great!"

Susan Calcari
Info Scout (Internet resource specialist)
InterNIC Information Services, General Atomics

"*The Internet Complete Reference* is an excellent introduction to the Internet for the novice. With the increasing access of K-12 schools on the Internet, this book should be a winner."

Bart Miller, Associate Professor
Computer Sciences Department
University of Wisconsin-Madison

The Internet Complete Reference

Harley Hahn
Rick Stout

Osborne **McGraw-Hill**

Berkeley   New York   St. Louis   San Francisco
Auckland   Bogotá   Hamburg   London   Madrid
Mexico City   Milan   Montreal   New Delhi   Panama City
Paris   São Paulo   Singapore   Sydney
Tokyo   Toronto

Osborne **McGraw-Hill**
2600 Tenth Street
Berkeley, California 94710
U.S.A.

For information on software, translations, or book distributors outside of the
U.S.A., please write to Osborne McGraw-Hill at the above address.

The Internet Complete Reference

890 DOC 9987654

ISBN 0-07-881980-6

Contents

Introduction

The Internet is, by far, the greatest and most significant achievement in the history of mankind. What? Am I saying that the Internet is more impressive than the pyramids? More beautiful than Michelangelo's David? More important to mankind than the wondrous inventions of the industrial revolution?

Yes, yes and yes.

Do I expect you to believe this? Of course not—not right now anyway. After all, the Internet is just a computer network and—let's face it—most of what we use computers for is pretty dull.

However, for years now, people have been connecting computers into networks. At first, the networks were a lot of trouble but, by the early 1990s, the engineers and scientists finally figured out how to make it all work most of the time. And now we have the Internet: a worldwide network connecting millions of computers and millions of people. What is amazing is that, within a few short years, the Internet has changed our civilization permanently and has introduced us to two completely unexpected ideas.

First, tens of thousands of people have been laboring to build the Internet. They have worked alone, in small groups, and within organizations, but always like so many ants in a global anthill. Most of these people are only doing their jobs but, just like the ants, they serve the common good while having no conception of the order and the compelling forces that drive their work.

Is it possible that there is a biological urge for us to create computers in the first place? I think not. But, once we did develop these strange and wonderful slaves, there most certainly was an irresistible genetic program for us to follow. Deep inside us, there is a voice that we hear only subconsciously and only as a species; a voice that commands us to take these computers, connect them into networks and . . . communicate.

This leads to the second great idea: that when we connect computers we invariably create something that is much more than the sum of its parts. As we said, the Internet consists of millions of computers and millions of people, but the computers are not important: we use them only to run our programs (that is, to follow our instructions). What is important is something that we cannot yet understand, the answer to the question: What happens when millions of people gather in a safe place to talk and to share? That is what the Internet is all about and that is what we are just beginning to find out.

I sense that we are near the beginning of a great and important change in human affairs. Personally, I don't understand this change. Indeed, I suspect that it is beyond the capabilities of any of us to completely appreciate what is happening, and that the best we can do right now is ride the wave. What I do understand is that there is a reason why we built the Internet and that, as human beings, we have an obligation to learn how to use it and to participate.

Now, about this book.

Before we start, I want to make sure that you understand an important point: the Internet is easy to use, but is difficult to learn. I have filled this book with technical details, important details that you must understand and master in order to use and enjoy the Internet. Still, don't for a moment think that you need to be a computer expert. The Internet is not for nerds, but just as surely it is not for dummies: it is for those people who are willing to think and to learn.

If you ever have one of those days where everything seems to go wrong, and the computer just won't cooperate, and you can't understand what is happening, take a moment and try to see things in a larger sense. As a human being, you share the birthright of intelligence, curiosity and, above all, the ability to learn. However, these gifts are not free. By your very nature, you can not only learn, you are compelled to learn if you are to remain happy and fulfilled. Still, as the saying goes, there is no royal road to knowledge.

To use the Internet, you will have to expend some time and some effort (actually, a great deal of time and effort). But if you do, here is what I promise you:

> I promise that I will stay with you.
> I promise that I will guide you.
> I promise that I will teach you the new words and the new
> ideas and the new skills that you need to use the Internet.

I want you to know that I have taken a great deal of time to organize this book so as to lead you from one topic to another in a way that will make sense to you (at least, in retrospect).

However, you do not need to read everything from start to finish (although I think you should). Actually, this book comes with only two simple instructions:

1. Start by reading Chapter 1.

2. When in doubt, do something enjoyable that requires effort.

I have not forgotten what it is like to be faced with a large, inconceivable mass of unknowable information. What you are holding in your hands is much more than a computer book. It is a connection—a real connection—between you and me. Your job is to put in the effort and the time. My job is to open the doors.

Buy this book. I am on your side.

— Harley Hahn

Acknowledgments

The writing and production of this book required the help of a large number of people. The most important is my co-author, Rick Stout, who worked a great many hours researching and creating the list of Internet access providers, the master list of Usenet newsgroups, and the extensive catalog of Internet resources. If you enjoy using the catalog, you may want to send Rick a small gift on his birthday (October 9th). Don't worry if you've missed it, Rick will accept presents all year round.

The second person I would like to thank is my senior technical reviewer, Michael Peirce, of the Computer Science Department at Trinity College in Dublin, Ireland. Michael read over every word of this book and tested every example, sending me back hundreds of "suggestions". Without Michael's careful reviewing and comments, this book would have been finished in half the time (and would have been about one third as good).

On the publisher's side of the aisle, the person with the most profound influence on this book was the Acquisitions Editor Scott Rogers. Throughout the course of the writing, Scott provided daily demonstrations of patience and skill that would be difficult to overpraise. In the chain of evolution, computer book editors generally lie somewhere between diamond-backed rattlesnakes and landlords, making Scott's exceptional enthusiasm and respect for writers all the more valuable.

Perhaps the most important of Scott Roger's contributions was his willingness to actually acquire an Internet account and learn how to send and receive electronic mail, proving, as he so properly observes, that "anyone can do it".

If you ever find yourself near the Osborne McGraw-Hill offices, we recommend that you pay them a visit. Walk up to the receptionist (her name is Ann) and ask nicely if you might be allowed to go in and shake Mr. Rogers hand. (However, don't be disappointed if he happens to be in a meeting and you are forced to settle for an autographed picture.)

Aside from Scott, there are a number of other people at Osborne McGraw-Hill who deserve special mention. The Project Editor, Kelly Barr, was always ready to do whatever was necessary to make this book as good as possible. As you can see, he succeeded. (Although, between us, it may not have actually been necessary for Kelly to sleep with the manuscript under his pillow each night for four months. Still, there are many tricks of the publishing trade that I still can't figure out, and one does not argue with success.)

The other Osborne people that deserve mention are Sherith Pankratz (Scott Roger's Editorial Assistant) for her excellent editorial assisting; Marcela Hancik (Production Supervisor) for all the extra work she did designing special formats for our tables and figures; Peter Hancik, also in production, who contributed significantly (and at short notice) to the design of the catalog; Ann Kameoka (Publicity Manager) who made sure that everyone in America knew about this book; and Ann Pharr, the most courteous receptionist west of the Mason-Dixon line. Finally, there is the big man in charge, Jeff Pepper (Editor-in-Chief), whose conception of this book was so immaculate as to assure his place in literary history.

By the way, as you can tell from their names, Marcela and Peter Hancik are married. If you are like me, you are probably wondering what the home life of two book design people must be like. To satisfy your curiosity, here is a typical dinner conversation:

> Marcela: Thanks for setting the table Peter, but why did you place the salt shaker so far to one side?
>
> Peter: Well, Marcela, I wanted to align both the salt and pepper shakers flush left in order to balance them with the napkins.
>
> Marcela: But what about the forks?
>
> Peter: Well, I decided to go with the regular dinner forks which, as you know, are three point sizes larger than the salad forks. But since all the cutlery is monospaced, I had to put the plates on the floor to make sure that everything else fit within the whitespace around the outside of the tablecloth.
>
> Marcela: Looks good. Let's go with your design and if there is any problem, we can change it in the next reprinting at breakfast.

On a more punctuational note, I would like to take a moment to acknowledge the excellent copy editing of Lunaea Hougland, queen among copy editors. Time after time, Lunaea has always managed, under great pressure, to ensure that each comma was in its exact place. In fact, take a few minutes now and check out some of the commas in this book. See what I mean? You will never see "hel,lo" or "ball,oon" or even "ranny,gazoo"—not in one of Lunaea's books.

Throughout the research and writing of this book, there were many Internet people from all over the world who contributed greatly. Here they are:

Andy Gruss:	Carnegie-Mellon University, Pennsylvania
Bart Miller:	University of Wisconsin, Madison
Bob Hawley:	Bell Laboratories, New Jersey
Brian K. Reid:	Digital Equipment Corp, Palo Alto, California
Brooke Jarrett:	La Mesa, California
David Gaudines:	Concordia University, Montreal, Canada
David Lemson:	University of Illinois
Desiree Madison-Biggs:	Netcom Online Communication Services
Dick Martin:	TRW Corporation, San Diego, California
Hugh Evans:	European Space Research & Tech Centre, The Netherlands
Iain Lea:	Siemens AG, Erlangen, Germany
Jason Heimbaugh:	University of Illinois
Jeff Inglis:	Middlebury College, Middlebury, Vermont
Kai Rohrbacher:	University of Karlsruhe, Germany
Kent C. Ritchie:	University of Minnesota
Larry W. Virden:	Chemical Abstracts Service, Ohio
Len Rose:	Pagesat; Netsys, Palo Alto, California
Marc Tamsky:	University of California at Santa Barbara
Mark Schildhauer:	University of California at Santa Barbara
Mary A. Axford:	Georgia Institute of Technology
Mary Leyva:	GTE California, Whittier, California
Michael Mealling:	Georgia Institute of Technology
Michael Peirce:	Trinity College, Ireland
Michael Shuster:	Technical University, Vienna, Austria
Paola Kathuria:	ArcGlade Services, London, England
Paul Vixie:	Vixie Enterprises, Redwood City, California
Peter ten Kley:	Utrecht, The Netherlands
Peter Wemm:	Demaas Proprietary Limited, Perth, Australia
Randy Bush:	PSGnet, Portland, Oregon
Rhett Jones:	University of Utah
Rick Broadhead:	York University, Toronto, Canada
Robert W. Hill:	University of Kansas, Lawrence, Kansas
Scott Yanoff:	University of Wisconsin
Stewart Alpert:	Santa Cruz Operation
Tom Gerstel:	Ithaca College, Ithaca, New York

I would like to take a moment to offer special thanks to three wonderful people who helped Rick and me tremendously by sending us a great number of insightful comments, as well as hidden treasures, from around the Internet. These people are Mary Axford (ace reference librarian) at the Georgia Institute of Technology, Peter ten Kley in The Netherlands, and Rick Broadhead at York University in Toronto, Canada.

Finally, I would like to thank my wonderful wife Kimberlyn, whose beauty and grace are surpassed only by her intelligence and talents.

CHAPTER 1

Introduction

In this chapter we introduce you to the Internet and lay bare our intentions for this book.

We start by asking the seminal question, "What is the Internet?" From there, we advance from the philosophical to the practical by offering our suggestions on how to best use and enjoy this book.

Next, we tackle an oft-asked query: "Do I need to learn Unix?" (as well as the only slightly less-asked query, "What is this Unix thing anyway?").

What is the Internet?

The Internet is the name for a group of worldwide information resources. These resources are so vast as to be well beyond the comprehension of a single human being. Not only is there no one who understands all of the Internet, there is no one who even understands most of the Internet.

The roots of the Internet lie in a collection of computer networks that were developed in the 1970s. They started with a network called the Arpanet that was sponsored by the United States Department of Defense. The original Arpanet has long since been expanded and replaced, and today its descendents form the global backbone of what we call the Internet.

FUN TIP: Technology marches on: The first experimental network using Internet-like technology involved four computers and was built in 1969. This was 56 years after the invention of the zipper, 37 years after the introduction of the first parking meter, and 13 years prior to the development of the first IBM personal computer.

It would be a mistake, though, to think of the Internet as a computer network, or even a group of computer networks connected to one another. From our point of view, the computer networks are simply the medium that carries the information. The beauty and utility of the Internet lie in the information itself.

Take a moment and skim through the catalog that comprises the last part of this book. Notice the enormous variety. As we start to work together, this is how we want you to think of the Internet: not as a computer network, but as a huge source of practical and enjoyable information.

But this is only the beginning. We would also like you to develop an appreciation of the Internet as a people-oriented society. Put simply, the Internet allows millions of people all over the world to communicate and to share. You communicate by either sending and receiving electronic mail, or by establishing a connection to someone else's computer and typing messages back and forth. You share by participating in discussion groups and by using the many programs and information sources that are available for free.

Does this mean that we are saying that the Internet resources will become as important to you as your telephone and your post office? Yes, that is exactly what we are saying.

In learning how to use the Internet, you are embarking upon a great adventure. You are about to enter a world in which well-mannered people from many different countries and cultures cooperate willingly and share generously. They share their time, their efforts, and their products. (And you will, too.)

For one more moment, take another look at the catalog. Each of those items is there because some person or some group volunteered their time. They had an idea, developed it, created something worthwhile, and then made it available to anyone in the world.

Thus, the Internet is much more than a computer network or an information service. The Internet is living proof that human beings who are able to communicate freely and conveniently will choose to be social and selfless.

The computers are important because they do the grunt work of moving all the data from place to place, and executing the programs that let us access the information. The information itself is important because it offers utility, recreation, and amusement.

But, overall, what is most important is the people. The Internet is the first global forum and the first global library. Anyone can participate, at any time: the Internet never closes. Moreover, no matter who you are, you are always welcome. You will never be excluded for wearing the wrong clothes, having the wrong colored skin, being the wrong religion, or not having enough money.

A cynic might say that the reason the Internet works so well is that there are no leaders. Actually, there is some truth to this. As unbelievable as it sounds, nobody actually "runs" the Internet. Nobody is "in charge" and no single organization pays the cost. The Internet has no laws, no police, and no army. There are no real ways to hurt another person, but there are many ways to be kind. Perhaps, under the circumstances, it is only natural for people to learn how to get along. (Although this does not stop people from arguing.)

What we choose to believe is that, for the first time in history, unlimited numbers of people are able to communicate with ease, and we are finding it is in our nature to be communicative, helpful, curious, and considerate.

That is the Internet.

Using the Internet

Using the Internet means sitting at your computer screen and accessing information. You might be at work, at school, or at home, using virtually any type of computer (including a PC or a Macintosh).

A typical session might begin with you checking your electronic mail. You can read your messages, reply to those that require a response and, perhaps, send a message of your own to a friend in another city.

You might then read a few articles in some of the worldwide discussion groups: Jokes from one of the humor groups, or perhaps recipes for a dinner that you are planning for the weekend. Maybe you are following a discussion about Star Trek or philosophy or literature or aviation.

After leaving the discussion groups, you might play a game, or read an electronic magazine, or search for some information on a computer in another country.

This is what using the Internet means and this is what we will show you how to do in this book.

How to Use This Book

The title of this book contains the word "Complete", so let's take a minute to discuss exactly what we mean by that.

We promise to teach you everything you need to know to be able to use the Internet and its basic resources. With the catalog that forms the latter part of the book, we offer a large, representative list of all that is available. Our intention is that for most people, most of the time, this book will be all that you need.

Since the purpose of this book is to show you how to understand and use the Internet, we need to start by providing the proper background and by teaching you some technical details. This is what we will do in Chapters 2, 3, and 4. These chapters explain the basics: how the Internet is organized, how you connect to the Internet and, most important, how to understand Internet addressing.

As you know, when you use a telephone you need a number to dial, and when you mail a letter you need a postal address. Similarly, the Internet has its own official "addresses". Each person and each computer has its own address, and Chapter 4 explains how the system works. Almost everything is built around these addresses, so it is crucial that you read this chapter at the beginning of your Internet career.

Once you have read Chapters 2, 3, and 4, you can go wherever your interests take you. If you want to send electronic mail, start with Chapters 5 and 6. If you want to participate in a Usenet discussion group, read Chapters 9, 10, and 11, and then whichever of chapters 12, 13, 14, or 15 is appropriate. Once you understand the basics, there is no special order for learning.

If you are not sure where to start, take a look at the catalog. Read through until you find something that interests you, and then turn to the chapter that shows you how to access that service.

If you are using this book to teach a course, you can follow the chapters in order. We have carefully designed the material to cover the most important topics first. You can start from the beginning and teach your students (or yourself) as much as you want, leaving the rest for another day.

Do You Need to Know Unix?

The answer is no, but read on.

Unix is a family of operating systems (master control programs) that are used to control computers. Virtually all types of computers can run Unix. Conversely, there are many variations of Unix that run on all sizes of computers.

When we use the word "Unix" as a general term, we mean more than an operating system. We are actually referring to an entire culture with its own language, technical terms, conventions, traditions, and a wide variety of computer-oriented facilities. To many people, the Unix culture is intimately connected to the Internet. Some people consider the Internet to be part of the Unix culture. Other people consider Unix to be part of the Internet culture.

The truth of it is that the Internet very much has a life of its own. Most of the Internet computers use Unix, but the details are hidden from you. Thus, you do not need to learn Unix *per se* in order to use the Internet.

However, you do need to understand the rudiments of using your own computer system. You must know how to start work, enter commands, use the keyboard (and mouse if you have one), and stop work when you are finished. It is also helpful—almost indispensable—to be able to manipulate data files, so you can save and retrieve information, and so you can create and edit your own information. For example, when you send someone a message using electronic mail, it is convenient to be able to use a text editing program to compose the message ahead of time.

When you use a Unix system to access the Internet, it is often difficult to say where Unix ends and the Internet begins. The programs that you will be using will all be Unix programs, perhaps not built into the operating system itself, but very much a part of the Unix culture. For example, many people read their electronic mail using the Unix mail program.

However, you are perfectly justified in keeping your knowledge of Unix to a practical minimum and spending your time, instead, exploring and using the Internet. There is no need for you to become a Unix expert.

Should You Learn Unix?

Having just told you that you do not need to become a Unix expert in order to use the Internet, we are now going to tell you that, yes, we do recommend that you learn Unix. You do not need to, but you ought to, and here is why.

As we mentioned earlier, Unix is a lot more than an operating system. It is actually a large, worldwide culture that is intimately connected to the Internet. On its own, Unix has a lot to offer and, hence, is worth learning.

Moreover, if you are using Unix to access the Internet, there are a great many advantages to having some technical knowledge of the underlying system. For example, you will probably want to send messages, either to another person or to a discussion group. To compose such messages, you should be able to use one of the Unix text editing programs.

As you work with the Internet, you may also want to collect information that you find on computers around the world. This means that you will need to know how to create and manipulate data files.

There are many such examples and they all illustrate the fact of life that, regardless of what anyone tells you, it is always a good idea to learn something about the technical details of your computer system. People have gone to a great deal of trouble to make computer systems easy to use and to make certain parts of the Internet easy to use, but—and this is an important but—you will be far better off if you learn more than the bare minimum.

Wait a minute, we hear some of you saying, that may be fine for the people who are using Unix-based systems, but I use a PC (or a Macintosh or some other fine non-Unix system) to access the Internet, and the people who set everything up promised it would be user friendly. They told me that I could select anything I wanted from a menu and that I wouldn't have to spend any time learning anything as evil and horrible as Unix. Besides, computers scare me.

Well, far be it from us to deny you the comfort and security of your own fears. Moreover, the last thing that we would do in a family-oriented book like this would be to disagree contentiously with those misguided souls who firmly believe that the future of computing lies in the populace-at-large selecting items from menus.

Suffice it to say that, as you use the Internet, you will from time to time connect to remote computers which will be running Unix. At such times, a basic knowledge of the Unix culture and its tools can provide practical benefits.

In order to help you, we recommend the book *A Student's Guide to Unix* by Harley Hahn (McGraw-Hill College Division, 1993). That book and this Internet book complement one another nicely.

Hint

Learning the fundamentals of Unix will make your work with the Internet more comfortable and will expand your possibilities.

CHAPTER 2

Understanding the Internet

In order to use the Internet well, you need to understand something of what it is and how it works. In this chapter, we will start with the idea of computer networks, introduce a few basic ideas and terms, and then move quickly to the Internet itself.

After explaining the basic services upon which the Internet depends, we will survey the large variety of resources you will be able to use. If you have been wondering, "What can the Internet do for me?", this is the chapter that will explain it all.

If you hang around the Internet at all, you will from time to time hear people talk about something mysterious called "TCP/IP". At the end of this chapter we will discuss TCP/IP and show you what it really is. (What? You have never heard of TCP/IP? Obviously, you have been hanging around the wrong sorts of people. Well, we can fix that.)

Our Friend the Network

The term *network* refers to two or more computers connected together. There are a number of reasons to connect computers into networks but the two most important (from our point of view) are:

■ to allow human beings to communicate

■ to share resources

Once you start using the Internet, you can send messages to anyone else on the Internet. You can even send messages to people who use other networks that are connected to the Internet.

As for sharing, computer managers arrange networks so resources that are expensive or difficult to maintain can be used by anyone on the network. For instance, a manager might attach a costly printer to a network so that everybody who needs it can use the same printer. On the Internet, we share information resources rather than pieces of hardware. For example, the items in the catalog (later in this book) are some of the many resources that can be used by anyone on the Internet.

A *local area network*, or *LAN*, is a network in which the computers are connected directly, usually by some type of cable. When we connect LANs together, we call it a *wide area network* or *WAN*. Most wide area networks are connected via leased telephone lines, although a variety of other technologies, such as satellite links, are used as well. The wide area connections for most of the Internet travel over some telephone system or another. Indeed, the bottleneck in establishing Internet service within developing countries is usually due to the lack of a reliable telephone system.

Here is a typical example of a network. Imagine yourself sitting in a room full of computers in the Social Sciences Computing Facility at a major university. (The weather outside happens to be cold and rainy, but since it has nothing to do with the example, we won't mention it.) Your computer is connected in a LAN to all the other computers in the room and to the computers within people's private offices throughout the building. This arrangement is shown in Figure 2-1.

There are a number of other LANs on the campus. For example, the Psychology department has its own network of computers, as does the Math

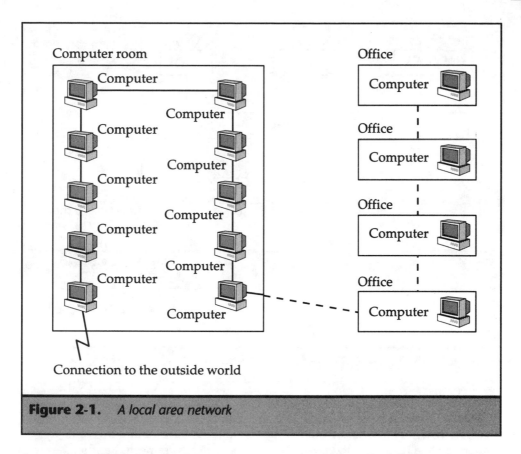

Figure 2-1. *A local area network*

department, the Computer Science department, and so on. Each of these LANs is connected to a high-speed link, called a *backbone*, to form a campus-wide WAN. This is shown in Figure 2-2.

Although we have used a university as an example, many types of organizations use similar arrangements: companies, governments, research facilities, other types of schools, and so on. If an organization is small, it may have only a single LAN. Large organizations may have multiple LANs connected into one or more complex WANs. Such organizations usually have a full-time staff to care for and feed the networks.

How are the LANs connected? By special-purpose computers called *routers*. The job of a router is to provide a link from one network to another. We use routers to connect LANs (to form WANs) and to connect WANs (to form even larger WANs). In other words, you can consider the computers within the Internet to be connected into LANs and WANs by a large number of routers. However, there is more to the picture that we need to explain. But first, let's take a moment to talk about hosts and terminals.

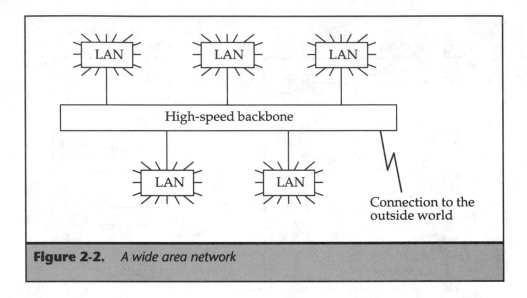

Figure 2-2. *A wide area network*

Hosts and Terminals

There are two meanings for the word *host* that you should know about.

First, within the Internet, each separate computer is called a host. For example, you might tell someone that he can find the information he wants by connecting to a host in Switzerland. If your computer is connected to the Internet, then it too is a host even though you may not share any resources with the rest of the Internet.

You may also see a computer referred to as a *node*. Here is why. If you draw a diagram of points and lines to represent the connections within a network, each computer will be a point and each connection will be a line. In the part of mathematics that deals with such diagrams, each such point is called a "node". Network specialists have borrowed this term to refer to any computer that is connected to a network. Thus, "node" is a more technical synonym for "host".

Hint

If you are trying to see if someone is a nerd, ask him "What is the technical term for a computer that is attached to the Internet?" If he says "node", chances are, he is a nerd.

The second meaning of the word "host" has to do with how certain computer systems are set up.

In general, there are two ways you might use a computer. You might have it all to yourself, or you might share it. For example, when you use a PC or a Macintosh, you are the only person using the computer. Single-user computers, especially the more powerful ones, are often called *workstations*.

Some computers, however, are made to support more than one user at the same time. These multi-user systems are often referred to as host computers.

A large mainframe computer, for example, can act as a host for hundreds of users at the same time. More commonly, a smaller computer, perhaps one that looks no larger than a PC, will act as a host for a small group of users. The Unix operating system, which we discussed in Chapter 1, is a multi-user system. Although some people use a Unix computer as a personal workstation, many Unix computers are used as hosts to support multiple users.

When you have your own computer, you interact by using the keyboard, screen, and (possibly) a mouse. These devices are part of the computer. With a multi-user computer, each person has his or her own *terminal* to use. A terminal has a keyboard, screen, perhaps a mouse, and not much more. All the terminals are connected to the host, which provides the computing power for everybody. This arrangement, called a *time-sharing system*, is shown in Figure 2-3.

Thus, there are two meanings for the word "host". Within the Internet, each computer is called a host. Within a time-sharing system, the main computer that supports each user on a separate terminal is also called a host. Of course, if such a computer were connected to the Internet, it would be both a time-sharing host and an Internet host.

Client/Server Systems

As you know, one of the principle uses of a network is to allow the sharing of resources. Much of the time, this sharing is implemented by two separate programs, each running on different computers. One program, called the *server*, provides a particular resource. The other program, called the *client*, makes use of that resource.

For instance, say that you are working with a word processing program that is running on your own PC. You tell the program that you want to edit a particular file that is stored on another computer on your network. Your program will pass a message to that computer asking it to send the file. In this case, your word processing program is the client while the program that accepts the request and sends the file is the server. More precisely, it is a file server.

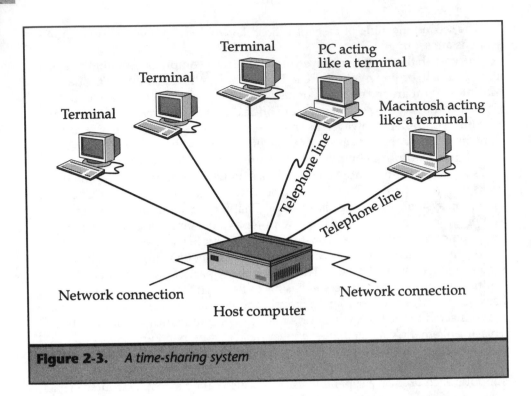

Figure 2-3. *A time-sharing system*

On local area networks, where the hardware is close by and visible, it is common for people to use the word "server" to refer to the actual computer that runs the server program. For example, one can imagine a network manager showing around the President of the United States and, while pointing to an otherwise nondescript beige machine, proudly announcing, "This is our file server." (To which proper etiquette requires that the President respond by nodding his head and saying, "How nice for you.")

On the Internet, hardware is normally not visible, and the terms "client" and "server" usually refer to the programs that ask for and provide services.

Here is an important example. Many Internet sites provide a service called a "Gopher" (which we will discuss in more detail later). Briefly, a Gopher allows you to select items from menus. Each time you select an item, the Gopher performs the required task. For instance, if the item describes a particular piece of information (such as "News of the Day"), the Gopher will retrieve this information and display it for you.

When you use a Gopher, two different programs are involved. First, there is the program that provides your interface. This is the program that interprets your keystrokes, displays the menus, and generally makes sure your requests are carried out. This program is called the Gopher client.

The other program is the one that supplies whatever it is the Gopher client has requested on your behalf. This program is called the Gopher server.

The beauty of this system is that the client and server programs do not necessarily run on the same computer. Indeed, more often than not, the client and server programs reside on different computers. For instance, you may be sitting in front of a PC in Rio Linda, California, using a Gopher to read the "News of the Day" at the National Security Agency in Virginia, three thousand miles away.

In this case, the Gopher client is a program running on your PC, while the Gopher server is a program running on a supercomputer on the other side of the country.

All of the Internet services make use of this *client/server relationship*. Learning how to use the Internet actually means learning how to use each of the client programs. Thus, in order to use an Internet service, you must understand:

1. How to start the client program for that service

2. How to tell the client program which server to use

3. Which commands you must use with that type of client

Your job is to start the client and tell it what to do. The client's job is to connect to the appropriate server and to make sure that your commands are carried out correctly.

Each type of Internet client has its own commands and conventions. For example, the commands you use with a Gopher client are different from the commands you use with, say, an Archie client (another Internet service). Fortunately, there are only a handful of basic services you need to learn.

As you learn about Internet services, you will find that the client programs—such as Gopher or Archie—have many different commands. However, you will also find that you actually need to learn only a few of the most important commands.

The Internet client programs all come with their own built-in help functions. Thus, if a situation arises in which you require an esoteric command that you have not yet learned, you can use the built-in help to find what you need.

And, of course, you always have this book.

X Window and X Clients

There is one special type of client/server system, called X Window, that you may need to know about. If your computer uses X Window, this section will explain the basic concepts you should understand. As you will see, X Window gives you certain advantages when you use the Internet.

If you are not an X Window user, feel free to move on to the next topic. Still, X Window is used widely, and you may want to skim this section to see what all the fuss is about.

X Window is used with Unix systems that support graphical user interfaces. A *graphical user interface*, or *GUI*, allows you to operate a computer by using not only your keyboard, but also a mouse or some other type of pointing device. With the help of your mouse, you select items from menus and manipulate objects on the screen. You can run more than one program at the same time, each of which can reside in its own rectangular area called a *window*.

Of course, the idea of a GUI (pronounced "goo-ee") is nothing new. If you have ever used a Macintosh, or a PC running Microsoft Windows or OS/2, then you have used a GUI.

As we mentioned in Chapter 1, there are many different types of Unix. X Window was developed in order to provide a standard set of tools for programmers who develop graphical applications and a standard interface for users to interact with those applications. For convenience, X Window is usually referred to as X. For example, a friend tells you that he has a program you might want to use. You might ask him, "Does it run under X?"

Although there is nothing in the design of X that says it must be used with Unix, in practice, this is almost always the case.

Hint

At one time, an operating system named V (the letter *V*) was developed at Stanford University. To work with V, the programmers created a windowing system which they named W. Later, the W system was sent to someone at MIT who used it for a new windowing system which he named X. The modern X Window system, first developed in 1984, grew out of this initial effort. Today, the current version of X Window is version 11 release 5, usually written as "X11 R5". Note that the official name of this system is the singular "X Window", not the plural "X Windows".

In X terminology, the three devices you use to interact with your computer—the keyboard, the screen, and the mouse—are referred to as a *display*. X lets you run more than one program at the same time on a single display. As part of the graphical user interface, each program resides in its own window on the screen. When you want to switch from one program to another, you use the mouse to move from one window to another.

When you use X, the details of maintaining the GUI for all the programs that are running are handled by a single program called a *display server* or an *X server*.

For example, say that you have four programs running at the same time, each of which resides in its own window. As you work, you can move the windows or even change their size. Now, suppose one of the programs needs to draw a circle on the screen. Rather than doing the work itself, it sends a message to the X server—the program that is controlling the screen—telling it to draw a circle of a particular size at a particular location. The X server actually does the work.

This division of labor has several important advantages. First, it means that the entire GUI is controlled by a single program that will ensure that everything works as it is supposed to. For example, the window in which a program is running may be partially obscured by another window. The program itself does not need to know this, nor does it care. The X server will handle the details.

Second, when a programmer designs a new program, he or she does not have to worry about the user interface. All that is necessary is for the program to call on X in the standard way whenever such work needs to be done. This makes for smaller, more reliable programs that are portable from one X system to the next.

Since all X servers provide the same functions, a program that is written to depend on an X server will run under any X system. You can, for instance, find an X program anywhere on the Internet, copy it to your computer, and run it under your own graphical user interface. Once you learn how to use the Internet file transfer service, you will be able to acquire many such graphical programs for free.

The third advantage of X is that the graphical user interfaces are more or less standard.

The part of the system that provides the look and feel of your interface is called the *window manager*. (Technically, the window manager is itself a program that runs on top of X.) There are two widely used window managers, named *Motif* and *Open Look*. There are differences between the two, but they are not profound.

If you saw identical programs running on two computer screens, one using Motif and one using Open Look, the main thing you would notice is that the appearance of the windows and other graphical elements would be somewhat different. In addition, you would use the mouse and the menu system slightly differently.

However, the basic concepts are the same no matter which window manager you use. Certainly, the programs inside the windows would not change. Basically, if you know how to use one X system, you know how to use them all.

Figure 2-4 shows the screen of an X display in which you are running five different programs at the same time, each one in its own window. Notice that, as these programs execute, they depend on a single X server program to maintain the user interface.

By now, this should all remind you of the client/server relationship that we discussed in the previous section. Indeed, this is the case. The programs you run are clients and, as such, are referred to as *X clients*. They request services—managing the user interface—from the X server program you run on your own computer. In other words, the term "X client" is a synonym for any program that runs under the auspices of X Window.

The most powerful feature of X Window is that the X clients do not have to run on the same computer as the X server. When you use X, the X server program will always run on your computer, the one in front of you. However, the X clients can run on any computer that is connected to your computer via the network. In Figure 2-4, you can see that of the five X clients, three are running on remote computers.

If you are an X user, the Internet is important to you in three ways. First, as we mentioned above, there are many X clients available for you to copy and use.

Second, you can use the Internet to connect to another computer and run an X program that will be displayed on your screen. (However, if the remote computer is far away, the response may be too slow for your liking.)

Finally, as an X user, you may have an advantage when you use Internet services.

As we explained in the previous section, you access Internet services by using a client program. For example, to utilize a Gopher server, you use a Gopher client.

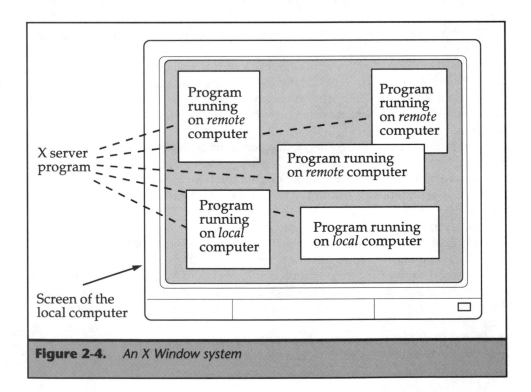

Figure 2-4.   *An X Window system*

If you are an X user, you may be able to find more powerful X-based clients to use instead of the regular character-oriented clients. For instance, many installations have a graphical version of the Gopher client that runs under X. If this is the case where you work, you will be able to use a graphical Gopher client with a spiffy user interface while the peasants down the hall (whose computers do not run X) will have to make do with the regular client program.

Hint

Some people will tell you that the X Window system is confusing because the server program runs on your computer, while the client programs may run on remote computers. Such people maintain that this is the opposite of what you might expect.

As you now know, the idea behind X is simple, so don't let silly people confuse you. When you meet such a person, be graceful and remind yourself, "There but for the grace of God, go I."

Explain to this person that you already understand X just fine, thank you. If he looks disappointed at not being able to mix you up with a lot of unnecessary technical details, you might gently suggest that he use his time to perform some socially useful function like writing a letter to the editor or learning a new Unix command.

Four Important Services Provided by the Internet

The software that supports the Internet provides a large number of technical services upon which everything else is built. Most of these services operate behind the scenes, and you do not need to understand them. (However, we should take a moment to recognize the many overworked network administrators who do take the time to understand the details.)

Nevertheless, there are four important Internet services we do need to talk about. You don't need to know the details, but you need to know they exist. In this section, we will describe these services. In the following section, we will discuss the many Internet resources—available to you—that depend on these services.

First, the mail service reliably transmits and receives messages. Each message is sent from one computer to another on its way to a final destination. Behind the scenes, the mail service ensures that the message arrives intact at the correct address.

The next service, called *Telnet*, allows you to establish a terminal session with a remote computer. For example, you can use Telnet to connect to a host on the other side of the world. Once the connection is made, you can log in to that computer in the regular manner. (Of course, you will need a valid user account and password.) Telnet also allows two programs to work cooperatively by exchanging data over the Internet.

The word "telnet" is often used as a verb. For example, you might tell someone, "If you telnet to this computer, you will be able to use a computerized reference book."

The third service is called *FTP* (file transfer protocol). FTP allows you to transfer files from one computer to another. Most of the time, you will use FTP to copy a file from a remote host to your computer. This process is called *downloading*. However, you can also transfer files from your computer to a remote host. This is called *uploading*. In addition, should you find it necessary, FTP will allow you to copy files from one remote host to another.

Hint

To remember the difference between "uploading" and "downloading", think of the remote host as floating above you in the sky. You send files up, and you ask for files to be sent down.

The last Internet service you should understand is the general client/server facility we discussed earlier in the chapter. A client program can connect to another computer and ask for the help of a server program.

For example, the Gopher system we mentioned earlier works in just this way. Your Gopher client displays a menu for you. When you make a selection from the menu, the client connects to the appropriate server—no matter where it is on the Internet—and procures the service you requested.

A Quick Tour of the Internet Resources

Having covered the basic framework of the Internet, let's take a look at all it has to offer. There are many Internet resources, and new ones are added whenever some clever person can figure out a new way to use the network.

In this section, we will examine all the important Internet resources. In later chapters, we will discuss each resource in detail, showing you how it works and the best way to use it. Indeed, many of these resources demand an entire chapter to themselves.

The purpose of this section is to give you an idea of exactly what the Internet has to offer you. If you take a look at the entries in the catalog, later in the book, you will see that each one depends on one or more of these resources. Table 2-1 at the end of this section contains a summary of each of the Internet resources and tells you what chapter to read for more details.

Electronic Mail

As an Internet user, you can send and receive messages from anyone else on the Internet. Moreover, you can do the same for other mail systems—such as CompuServe or MCI Mail—that have connections with the Internet.

However, mail does not mean simply personal messages. Anything that can be stored in a text file can be mailed: computer (source) programs, announcements, electronic magazines, and so on.

When you need to send a binary file that cannot be represented as regular text—such as compiled computer programs or graphical pictures—there are facilities for encoding the data into text. At the other end, the recipient simply decodes the textual data into its original format.

Thus, you can mail virtually any type of data to anybody. The Internet mail system is the backbone (and original motivation) for the network itself.

Remote Login

As we described in the previous section, you can telnet to a remote computer anywhere on the Internet. Once you have established a connection, you can log in (as long as you have a valid account on that computer). Since most Internet computers use Unix, we borrow the terminology for logging in from Unix itself.

The name by which an account is known is called a *userid* (pronounced "user-eye-dee"). The secret code you must enter, to prove that it is really you, is called your *password*. As long as you have a valid userid and password, you can log in to any computer on the Internet.

As a public service, many Internet systems are set up to allow anybody to log in using a special *guest account*. For example, in the United States, there is a system that will display weather reports from around the country. Anyone can log in to this system and check out the weather.

FUN TIP: If you are working in a room without a window, it is often faster to check the weather by connecting to the weather server (which is in Michigan) than by going outside.

Finger Service

Most Internet computers offer a facility that allows you to ask for information about a particular user. This service is known by the descriptive name of *finger*. As you will see, people on the Internet are often known by their userid. You can use finger to find out the name of the person behind the userid. For example, you might find out that the userid **harley** is registered to Harley Hahn.

We use the word "finger" not only as a noun, but also as a verb. For example, you might overhear someone ask, "Who is userid **tln**?", to which a second person answers, "I don't know. Why don't we finger him and find out?"

Depending on how the finger service has been set up on the computer you contact, you may be able to find out other information about the person: phone number, office address, and so on. In addition, some finger systems will tell you when the person last logged in and if they have unread mail. This can come in handy when you need to check if someone has received an important message.

There is also a way for you to customize part of what people see when they finger your userid. You can specify certain information that you want displayed. For example, a professor might specify his office hours. Somebody giving a party might give directions to her house. You can display this information whenever you want just by fingering that person's userid.

You can also finger a computer rather than a userid. In this case, the computer will respond by showing you a summary of all the userids currently logged in.

Finally, some systems use finger to support a public request for certain specific information. For example, there is a particular userid and computer at the University of Washington at Seattle that you can finger to display information about recent earthquakes.

FUN TIP: In a more esoteric vein, you can finger a computer at Carnegie-Mellon University in Pennsylvania to find out if the Coca Cola machine is empty or not. By fingering a different userid, you can check if the candy machine is empty.

Usenet

Usenet—the name is a contraction of "User's Network"—is one of the main reasons people use the Internet. Usenet itself is not an actual network. It is a system of discussion groups in which individual articles are distributed throughout the world.

Usenet has literally thousands of discussion groups, so there is definitely something for everyone. (See the list of groups in Appendix G.)

At each Internet site, the system administrator decides whether or not to carry the Usenet discussion groups. Thus, Usenet is not available everywhere. Moreover, even sites that do provide this resource will not carry every conceivable discussion group.

FUN TIP: Reading Usenet articles—especially the jokes—is a good way to spend time that would otherwise be wasted on work.

Anonymous FTP

As we mentioned in the previous section, the FTP service allows you to copy files from one computer to another. Anonymous FTP is a system in which an organization makes certain files available to the general public. You can access such a computer by using a userid of **anonymous**. No special password is required.

Anonymous FTP is one of the most important Internet services. Virtually every possible type of data is stored somewhere, on some computer, and it is all available to you for free. For example, many of the programs used on the Internet are created and maintained by individuals who then distribute the programs worldwide via Anonymous FTP. You can also find electronic magazines, archives of Usenet discussion groups, technical documentation, and much, much more. As a full-fledged Internet user, you will come to depend heavily on Anonymous FTP.

Archie Servers

There are thousands of Anonymous FTP servers around the world offering a vast number of files. The role of the Archie servers is to make the whole system manageable by helping you find what you need.

Suppose you want a particular file, for instance, a program you have heard about. You can use an Archie server to tell you what Anonymous FTP sites store that file. Once you know the names of the sites, it is a simple matter to use FTP and download the file.

If you consider the world of Anonymous FTP as an enormous, worldwide library that is constantly changing, you can think of the Archie servers as the catalog. Indeed, without the Archie servers, most of the Anonymous FTP resource would be unreachable.

The name "Archie" was coined from the idea of an "archive server". In deference to the well-known propensity of computer people to engage in whimsical personification, we usually refer to "Archie" as if it were human. For example, you might ask someone, "Do you know which Anonymous FTP sites carry the electronic magazine called *The Unplastic News*?" "No, I don't," replies your friend, "Why don't you ask Archie?"

Talk Facility

The Talk facility establishes a connection between your computer and someone else's. You can then use this connection to type messages back and forth (until you get bored).

The great thing about the Internet Talk facility is that it makes it possible to hold a conversation with someone no matter how far away they are. The other person sees what you type as you type it, and you can both type at the same time without your messages getting mixed up.

Internet Relay Chat

Internet Relay Chat (IRC) is like a Talk facility for more than one person at the same time. As you might imagine, IRC is used heavily and offers a lot more than simple conversation.

You can take part in public conversations with a large number of people. These conversations are loosely organized around various topics or ideas. Alternatively, you can use IRC to arrange a private conversation with people of your own choosing, much like a telephone conference call.

Hint

If you find yourself spending large amounts of time using a Chat facility, remind yourself that you are talking to the type of people who will spend a large amount of time using a Chat facility.

Gopher

Gophers provide a series of menus from which you can access virtually any type of textual information, including that provided by other Internet resources. There are many Gopher systems around the Internet, each one administered locally. Each Gopher contains whatever information the local Gopher people have decided to share.

While some Gophers are standalone systems, most Gophers are set up to connect to other Gophers. For example, say you are using a Gopher in California. With a simple menu selection, you can connect to another Gopher in Africa or in South America. What makes Gophers so powerful is that, no matter what Gopher you are using and no matter what information you are using, the interface is always the same simple menu system.

Veronica and Jughead

Nobody really knows how many Gophers there are in the world. Suffice it to say there are a great many, all of which have their own series of menu items offering information and services.

Veronica is a tool that keeps track of many Gopher menus from around the world. You can use Veronica to perform a search and look for all the menu items that contain certain keywords (whatever you specify). Jughead does the same thing for a specific group of Gopher menus.

The result of a Veronica or Jughead search is a custom menu, containing whatever items were found. Selecting any item from this menu automatically connects you to the appropriate Gopher, wherever it may happen to be. In fact, unless you specifically ask, you will not even know what computer you are using or what country it is in. Amazing.

Wais Servers

Wais servers provide another method of finding information that is spread around the Internet. Wais is able to access any of a large number of databases. To start, you tell Wais which databases you want to search. Next, you specify a list of one or more keywords to search for. Wais will search every word in every article in all the databases that you specified.

The result of a Wais search is a list of the articles, culled from the various databases, that are likely to be of interest to you. Wais presents them as a menu, with the most relevant items first. From this list, you can ask Wais to display whichever articles take your fancy.

The name "Wais" stands for "Wide Area Information Service" and is pronounced "Wayz".

World-Wide Web

The World-Wide Web—often called "the Web"—is a hypertext-based tool that allows you to retrieve and display data based on keyword searches. What makes the Web so powerful is the idea of hypertext: data that contains links to other data.

For instance, as you are reading some information, you will notice that certain words and phrases are marked in a special way. You can tell the Web to jump to one of those words. It will follow the link, find the relevant information, and display it. In this way, you can jump from place to place, following logical links in the data.

White Pages Directories

Within the often overwhelming world of the Internet, nothing is more important than a person's electronic address. Once you know someone's address, you can send mail, have a Talk conversation, or even find out more about the person by using Finger.

What do you do when you want to contact someone, but you don't know his or her address? You use one of the White Pages Directories. The name, of course, reminds us of the standard telephone book.

However, the electronic counterpart is actually much different, principally because there is no single Internet directory. After all, as we explained in Chapter 1, nobody actually "runs" the Internet. With no single person or organization in charge, it is not surprising that there is no central source of names and addresses.

Rather, there are a number of different White Pages Directories—special-purpose servers—that you can search for a name. When you lose that bar napkin on which you carefully wrote someone's name during your last vacation abroad, there is a chance that a White Pages Directory may help you track down your quarry's electronic address, as long as he or she is an Internet user. (And why would you be associating with anyone who is not an Internet user?)

Electronic Magazines

The Internet is host to a variety of magazines that are published electronically. That is, the articles are stored as text files that are accessible to one and all. Some electronic magazines are scholarly journals of interest mainly to specialists. Other magazines are of general interest.

FUN TIP: If you want to read a strange collection of weirdness and esoterica, subscribe to The Unplastic News.

There are two ways in which electronic magazines are distributed. Some maintain a mailing list. When a new issue comes out, it is sent to you as a mail message. Other magazines are stored in well-known Anonymous FTP sites. You can download copies, including back issues, whenever you want.

Aside from the ease of distribution, electronic magazines have two important advantages over their conventional paper and ink counterparts. First, electronic magazines do not contain advertisements showing people who are richer and better looking than you, having more fun than you will ever have.

Second, electronic magazines do not contain irritating subscription cards that fall out when you turn the page.

Mailing Lists

A mailing list is an organized system in which a group of people are sent messages pertaining to a particular topic. The messages can be articles, comments, or whatever is appropriate to that topic.

All mailing lists—and there are thousands of them—have someone in charge. You can subscribe or unsubscribe to a list by sending a message to the appropriate address. Many mailing lists are "moderated", which means that someone decides which material will be accepted. Other lists will accept and send out messages from anybody.

If you feel lonely and ignored, subscribing to just a handful of mailing lists is guaranteed to keep your electronic mailbox filled to the Plimsoll line.

Internet BBSs

A BBS, or Bulletin Board System, is a repository for messages and files, often devoted to a particular topic. To use a BBS, you connect to it and select items from a series of menus.

Typically, a BBS will be maintained by a single person or by an organization. There are countless BBS systems in the world, most of them reachable by telephone. The Internet has many BBSs that you can reach by the more refined method of Telnet.

Games

Where would a computer system be without games? Of course, there are many computer games you can download via Anonymous FTP and run on your own computer. However, there are also special network games that take advantage of the Internet facilities.

For example, you can play the games of Chess and Go with other people from around the Internet. Or you can play the game of Diplomacy by electronic mail. Perhaps you would enjoy a simulated spaceship dogfight based on Star Trek. Maybe you would like to use your skills as a programmer to play Core War, a game in which players devise assembly language programs (for a virtual computer) that attempt to dominate and control the computer's memory.

Whatever your inclination, you can find it on the Internet. Just don't tell your boss or the powers that be. Certainly, the vast amounts of money and time that are spent maintaining the Internet are not expended merely so we can amuse ourselves with games. (Or are they?)

MUD

A MUD, or Multiple User Dimension, is a computer program that provides a virtual reality. To participate in a MUD, you telnet to a MUD server, take on a role, and explore. As you do, you interact with other users who are playing their own roles. In other words, a MUD allows you to exercise your fantasies and pretend, just like playing Dungeons and Dragons or visiting a singles' bar.

Rest assured, MUDs are at once complex, fascinating, and addictive. As you walk around the virtual reality of a particular MUD, you can talk with people, solve puzzles, explore strange places (like caves), and even create some reality of your own.

FUN TIP: Join a MUD and hide from reality. Forget to graduate. Lose your job. Ignore your friends, family, and loved ones. In other words, lots of good, clean fun.

Chapter	Resource	Description
5, 6	Electronic Mail	Send and receive messages
7	Remote Login	Connect to and use a remote host
8	Finger Service	Show information about a user
9–15	Usenet	Vast system of discussion groups
16	Anonymous FTP	Public access to data archives
17	Archie Servers	Search Anonymous FTP archives
19	Talk Facility	Converse with one person
20	Internet Relay Chat	Converse with a group of people
21	Gopher	Menu-based information
21	Veronica, Jughead	Search for Gopher menu items
22	White Pages Directories	Search for a user's address
23	Wais Servers	Search indexed databases
24	World-Wide Web	Access hypertext information
25	Mailing Lists	Information distributed by mail
*	Electronic Magazines	Magazines, journals, newsletters
*	Internet BBSs	Share information and messages
*	Games	Fun and diversions
*	MUD	Multi-person virtual reality

Table 2-1.   *Summary of the Internet Resources*

*see the catalog later in this book

What is TCP/IP?

In order to complete this chapter and our overview of the Internet, we need to spend a few moments talking about *TCP/IP*. As you know, the Internet is built on a collection of networks that cover the world. These networks contain many different types of computers, and somehow, something must hold the whole thing together. That something is TCP/IP.

The details of TCP/IP are highly technical and are well beyond the interest of almost everybody, but there are a few basic ideas you will want to understand.

To ensure that different types of computers can work together, programmers write their programs using standard *protocols*. A protocol is a set of rules that describes, in technical terms, how something should be done. For example, there is a protocol that describes exactly what format should be used for a mail message. All Internet mail programs follow this protocol when they prepare a message for delivery.

TCP/IP is the common name for a collection of over 100 protocols that are used to connect computers and networks. We have already mentioned two of the TCP/IP protocols, Telnet and FTP (file transfer protocol).

The actual name "TCP/IP" comes from the two most important protocols: *TCP* (Transmission Control Protocol) and *IP* (Internet Protocol). Although you don't need to know the details, it is useful to have an appreciation for what these protocols are and how they hold the Internet together.

Within the Internet, information is not transmitted as a constant stream from host to host. Rather, data is broken into small packages called *packets*.

For example, say that you send a long mail message to a friend on the other side of the country. TCP will divide the message into packets. Each packet is marked with a sequence number and with the address of the recipient. In addition, TCP inserts some error control information.

The packets are then sent over the network, where it is the job of IP to transport them to the remote host. At the other end, TCP receives the packets and checks for errors. If an error has occurred, TCP can ask for that particular packet to be resent. Once all the packets are received correctly, TCP will use the sequence numbers to reconstruct the original message.

In other words, the job of IP is to get the raw data—the packets—from one place to another. The job of TCP is to manage the flow and ensure that the data is correct.

Breaking data into packets has several important benefits. First, it allows the Internet to use the same communication lines for many different users at the same time. Since the packets do not have to travel together, a communication line can carry all types of packets as they make their way from place to place. Think of a highway in which separate cars all travel on a common road even though they are headed for different places.

As packets travel, they are routed from host to host until they reach their ultimate destination. This means the Internet has a lot of flexibility. If a particular connection is disrupted, the computers that control the flow of data can usually find an alternate route. In fact, it is possible that, within a single data transfer, various packets might follow different routes.

This also means that, as conditions change, the network can use the best connection available at the time. For example, when a particular part of the network becomes overloaded, packets can be routed over other, less busy, lines.

Another advantage of using packets is that, when something goes wrong, only a single packet may need to be retransmitted, rather than the entire message. This greatly increases the overall speed of the Internet.

All of this flexibility makes for high reliability. One way or another, TCP/IP makes sure the data gets through. In fact, the Internet runs so well that it may take only a few seconds to send a file from one host to another, even though they are thousands of miles apart and all the packets must pass through multiple computers.

Thus, there are several answers to the question, "What is TCP/IP?" The technical answer is that TCP/IP is a large family of protocols that are used to organize computers and communication devices into a network. The two most important protocols are TCP and IP. IP (Internet protocol) transmits the data from place to place, while TCP (transmission control protocol) makes sure that it all works correctly.

The best answer, though, is that the Internet depends on thousands of networks and millions of computers, and TCP/IP is the glue that holds it all together.

Hint

If you wonder what the route looks like from your computer to another, you can use the **traceroute** command. This command is not available on all systems. If your computer does have **traceroute**, just enter the command name followed by the address of an Internet host. (We will discuss Internet addresses in Chapter 4.) For example, to see the current route from your computer to the host whose address is **rtfm.mit.edu**, enter:

```
traceroute rtfm.mit.edu
```

The output will show you each step in the path between the two computers. If you try this command at various times, you may see different routes as conditions change.

CHAPTER 3

How to Connect to the Internet

It's stimulating to read about the wonderful things you can do once you have access to the Internet. But how do you get that access? Lying in bed at night, with visions of computerized sugarplums dancing in your head, it's natural to feel like Moses, perched on Mount Pisgah, staring down wistfully at the Promised Land. Shakespeare, of course, had the same problem when he was unable to get an Internet account. "Why," he asked plaintively, "should a man whose blood is warm within, sit like his grandsire cut in alabaster?"

Exactly. We couldn't have put it better ourselves.

So, in this chapter, we will show you how to connect to the Internet.

The Internet Connection

To start, let us consider the question, what does it mean to have access to the Internet?

This means you are using a computer that is a part of a network connected to the Internet. In practical terms, this means you can use the Internet resources we described in Chapter 2.

When you are using such a computer, we say that you are *on* the Internet. In common usage, the Internet is often referred to *the Net*.

For example, say that you are walking in the park one day and you happen to meet the woman (or man) of your dreams. Well, one thing leads to another and, after an enjoyable hour together, you part in a mutual flourish of good will and telephone numbers. However, you realize that the incipient relationship might well do better if you could correspond by electronic mail, so you ask, "Are you on the Net?"

Hint

Determining up front if a potential mate has access to the Internet can help you decide quickly and easily whether or not the relationship will be worthwhile.

It is important to realize that a great many people who do not have full Internet access are able to communicate with Internet users. For example, many people use some type of non-Internet mail service that can exchange messages with the Internet. It is also common to find people who are not on the Internet who have access to the Usenet discussion groups.

What you should understand is that while mail and Usenet are important, they are not everything. We want you to have access to all the Internet resources, including Telnet and FTP. In other words, we want you to be able to use a computer that has a full TCP/IP connection with the Internet.

Types of Internet Connections

Before we discuss the various ways in which you might obtain access to the Internet, we need to talk about the two different types of Internet connections.

First, you might use a computer that is directly connected to the Internet. For example, you might be using a PC or Macintosh or workstation that is part of a network connected to the Internet. In such a case, your computer will be a full-fledged Internet host, with its own electronic address.

The other way to connect to the Internet is by using a terminal that is connected to an Internet host. In this case, the terminal itself—not being a computer—is not on the Internet. You simply use the terminal to access a computer that is on the Internet.

Here is an example. Imagine that you are being taken on a tour of a building used by many Internet users. First, you are taken to the Computer Room, in which you see 40 different PCs, connected in a network. You are told that this network is connected to the Internet, so that all the PC users can access the Internet directly.

Next, you are taken to another room, the Terminal Room. Here you are shown 40 terminals, all of which are connected to a time-sharing computer in a closet at the back of the room. This computer is also connected to the Internet.

Thus, each PC user has his or her own Internet host. Each PC has its own Internet address and is a self-sufficient, standalone system.

In the Terminal Room, life is different. Each person is actually using the same computer, the one in the closet. These users access the Internet by logging in to the time-sharing system that provides Internet access. Thus, they all share one computer that has a single Internet address.

Telephone Connections

In the last section, we explained that you can use either a computer or a terminal to access the Internet. The example we used described two rooms. One contained 40 PCs connected into a network. The other contained 40 terminals connected to a time-sharing host. What they had in common was that all the devices—computers and terminals alike—were connected directly using some type of cable. This type of arrangement is called a *hard-wired* connection.

The main advantage of a hard-wired connection is its permanence. All you have to do is turn on your PC or terminal, and your connection is ready to use. The disadvantage, of course, is the lack of flexibility. If you want to move the computer or terminal to another location, you must deal with the cables.

A more flexible system is one in which the computer or terminal uses a *dial-up* connection over a telephone line. In such cases, you can work anywhere you want, as long as you have access to a phone line.

To use a dial-up connection, you need a hardware device to convert computer signals to telephone signals and back again. In technical terms, the computer signal is "digital" and the telephone signal is "analog". A device that converts from digital to analog is called a *modulator*. A device that converts from analog to digital is called a *demodulator*. When we connect computers over a phone line, we must be able to send data in both directions. For this, we use a *modem*, a "modulator-demodulator".

There are various types of modems. A modem can be a separate box that attaches to your computer (usually via a "serial" cable). Or a modem might be an adaptor board that is installed in the computer. Figure 3-1 compares a hard-wired connection to a dial-up connection. Notice that a dial-up connection requires a modem at each end of the telephone line.

Hint

When you buy a modem for your own computer, you will encounter a bewildering number of technical terms. However, all you have to do is remember the following four guidelines:

- The speed of a modem is measured in *bits/second* or *bps*. Fast modems transmit at speeds up to 14,400 bps (often abbreviated as 14.4K bps). Slow modems transmit at 9,600 bps.

- Buy only a 14,400 bps modem. If you buy anything slower, you will be disappointed. No matter what anyone tells you, do not buy one of the very slow 2,400 bps modems (which were the standard about four or five centuries ago).

- You will see a number of strange terms that describe various functions: V.32bis, V.42, MNP 5, and so on. Don't worry about them. All modern 14.4K bps modems come with the right stuff. Just buy a 14.4K bps modem and you'll be okay.

- Modem manufacturers lie about the speeds of their modems, but they all lie in the same way, so it's okay. For example, it is commonly claimed that a 14,400 bps modem, using what is called "data compression", can achieve an overall performance of 57,600 bps (four times the nominal speed). Similarly, it is said that a 9,600 bps modem with data compression will perform at 38,400 bps. These numbers are highly inflated theoretical maximums that are completely divorced from reality. Most people do not know this, but now you do.

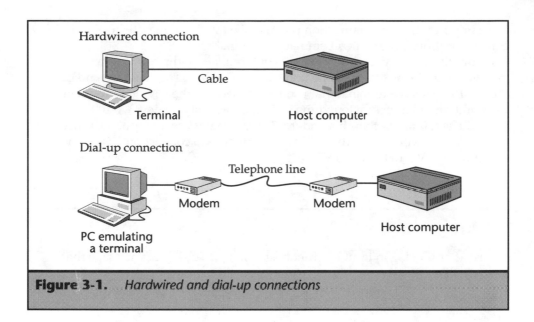

Figure 3-1. *Hardwired and dial-up connections*

Connecting a Terminal Over a Telephone Line

A common need is to be able to use a PC or Macintosh to access a remote computer over a telephone line. For instance, you may have an Internet host at work or at school that you want to use from your home.

Such hosts are usually configured to accept connections to a terminal. That is, you can connect a terminal to the telephone line (using a modem) and dial the remote host. Once you make the connection, you can work on your terminal in the usual manner.

The thing is, your PC or Macintosh is a full-fledged computer and not merely a terminal. But the remote host is set up to communicate only with terminals.

The solution is to run a program on your computer that *emulates* a terminal. In other words, you run a program that makes your computer act just like a terminal. As far as the remote host is concerned, your computer really is a terminal.

The standard way to do this is to use a *communications program* on your computer. Such a program will handle all facets of communicating over a telephone line. The program will help you keep a list of phone numbers, set up the proper communication options, dial the number, and generally run the show.

Most important, once a connection is established with the remote host, the communications program will emulate a terminal.

Now, there are many different types of terminals, and most programs will give you a choice as to which one you want to emulate. The standard choice is the *VT-100*. This is a terminal that used to be made by the Digital Equipment Corporation. It has been a long time since DEC actually made a VT-100, but nevertheless, it has become a standard. Indeed, many host computers, especially Unix systems, will assume that, by default, all dial-up connections use some variation of a VT-100, such as VT-102.

Hint

Just after you log in to an Internet host, you may see a short message that looks like this:

```
TERM = (vt100)
```

What is happening is that the system is asking you what type of terminal you are using. The name in parentheses, **vt100**, is the default. If you are using (or emulating) a VT-100 terminal, simply press the RETURN key. Otherwise, type the name of a different terminal and press RETURN. It is important to initialize your terminal type properly. Otherwise, the output on your screen may end up scrambled.

To summarize, if you want to use the Internet from your home, you need four things: a computer, a modem, a communications program, and the telephone number of a remote host to which you can connect. (We will give you guidelines on how to find such a system later in the chapter.)

Hint

There are two things that will make your work with a remote host a lot easier. First, choose a communications program capable of storing the information that is displayed on your screen. As new data is displayed, the previous data will scroll off the top of the screen. It is a great convenience to be able to ask your communications program to recall previous lines of output that otherwise would be gone for good. Sometimes such a facility is called a *scroll buffer*.

Second, use a computer system that will run more than one program at the same time. For example, with a PC, you should run your communications program under Microsoft Windows or, better yet, OS/2. This allows you to keep an Internet session in its own window while you do other work. Moreover, such systems enable you to cut and paste from one part of the screen to another, which can be a great advantage.

For example, let's say that you are reading a Usenet discussion group in which someone has mentioned the address of a computer that has a great new service. You would like to enter a command to connect to that computer to try out the service.

Normally, you would type the command, including the computer's address. However, Internet addresses can be long and must be typed exactly right. Instead, you can "cut" the address from one part of your screen and "paste" it back as part of the command. That way, you do not have to worry about making a spelling mistake. Once you become a veteran Internet user, you will find that such situations arise frequently.

Connecting a Computer over a Telephone Line

Most of the time, using a computer that emulates a terminal is the best way to connect to an Internet host over a telephone line. But remember, your computer is merely acting like a terminal. It is not on the Internet itself.

In certain situations, your computer should be an actual Internet host. For example, you might have a small company that cannot afford a network with a dedicated full-time Internet connection. However, you may still want your own Internet host. In such cases, there is a way to establish a full Internet connection over a telephone line.

To do so, you must first arrange for some other Internet host to act as your connection point. You then install a set of programs on your computer called *PPP* (Point to Point Protocol). Once a telephone line connection is made between the two computers (using modems, of course), PPP will endow your computer with TCP/IP capabilities. This enables your computer to be a real Internet host with its own official electronic address.

You will also hear about an older system called *SLIP* (Serial Line Internet Protocol). Both PPP and SLIP will work, but if you have a choice, use PPP.

When you use these systems, there are two ways to set up the telephone line. First, you might use a regular dial-up line with a standard modem, just as we described in the previous section. Although your computer will be considered an official Internet host, it will not be connected at all times. When such a system is set up, you must arrange for the computer that acts as your connecting point to save mail messages that arrive when you are not connected.

The alternative is to use a *dedicated phone line*. As the name implies, a dedicated phone line is always connected. This, of course, costs a lot more than a regular line. However, for a small company, a dedicated phone line using PPP can be a relatively economical way to establish Internet access. This connection can then provide Internet access for the rest of the computers in the company. Nevertheless, if there is to be a fair bit of Internet traffic, you are probably better off with a faster connection.

Later in the chapter, we will explain how certain companies are in the business of providing commercial access to the Internet. Many such companies offer, not only a host computer to which you can connect with a PC, but PPP or SLIP service for customers who require a full Internet connection.

Hint

If you want to access the Internet from your home, use a communications program and emulate a terminal. Establishing a PPP or SLIP connection sounds nice, but it is expensive and requires a fair bit of technical knowledge. When you use a terminal emulator to connect to a remote host, someone else is taking care of all the details for you. When you use PPP or SLIP, you will have to manage the system yourself.

Are You Already On the Internet?

Now that we have described the various ways in which a computer or terminal can connect to the Internet, we will spend the rest of the chapter explaining how you might go about getting Internet access for yourself.

To start, we will discuss the easiest solution of all. It may be that you already have access to the Internet and do not know it.

How can this be? Well, if you are using a computer that is part of a network in a university, you are probably already on the Internet. Virtually all universities have Internet facilities. Alternatively, if you work at a company or organization whose computers are networked together, there may be an Internet connection already set up. This is especially true for computer companies.

To find out if you are on the Internet, ask your system manager or a friendly computer nerd. The rule of thumb is that the type of person who would use a computer on a Saturday night is the type of person who would know if your installation is on the Internet. (Don't laugh. Once you learn how to use the Internet, you will find yourself using the computer on a Saturday night.)

Another way to check your Internet status is to enter a command that tries to connect to a remote computer and see what happens. You might try the **telnet** command (see Chapter 7). When you do, be sure to specify the name of a computer outside of your local area.

The **telnet** command is basic to the Internet. If you enter a **telnet** command and you see a message like this,

```
telnet: Command not found.
```

it means that the command does not exist on your system and you are probably not on the Internet.

If you find that you are already on the Internet, you can bask in the glow of your good fortune and move on straightaway to Chapter 4. Otherwise, read on for some ideas to help you get Internet access.

Using the Internet for Free

If you do not already have Internet access, there are several ways to get it. Let's start with those that are free.

As we mentioned, virtually all universities—and a lot of companies—have access to the Internet. To get free access, all you have to do is convince somebody to give you an account on one of the computers. If you belong to such an organization, find out who is in charge of computer accounts. Then ask around discreetly and find out what criteria one has to meet to get an account. In some universities, for example, any student who asks will be given a computer account, but you have to ask. In other schools, you need to be taking a particular class or working on a particular project. Sometimes if you work for a professor you can get an account. Similarly, some companies give Internet access to any employee. Other companies give it only to qualified employees (such as researchers).

Hint

If anyone asks you why you want Internet access, the correct response is, "I want to be able to access free information that will be of great benefit to our organization." The incorrect response is, "I want to be able to read the jokes on Usenet."

If you are not a student, but there is a university nearby, you may be able to get someone to give you free access to one of the university computers. This may involve some finesse and will certainly demand discretion. But, as one of our readers, you are certainly not lacking in either of these qualities.

Community Computing and the Freenets

An alternative to sponging off a local university is to look for a free community computing service. Such services offer computer-mediated communication facilities to anyone for no charge. All you need is a computer with a modem. Using your modem, you connect to a host computer over a telephone line. The best known community computing services are the *Freenets* (sometimes spelled "Free-Nets"). There are a number of Freenets in the United States and Canada that offer all sorts of information services as a general service to the community. The original system is the Cleveland Freenet. Look in the catalog later in this book for more details. (Note: Not every such system has the word "Freenet" in its name.)

The Freenet system has given rise to an organization called the *National Public Telecomputing Network* or *NPTN*. The NPTN is a non-profit organization dedicated to providing free public access to a large variety of information.

Community computing services usually have an electronic mail system that can communicate with the Internet. Most of them also carry the Usenet discussion groups. Moreover, they have a number of unique information offerings, including special programs for schools. Perhaps their best feature is the price: they are free.

However, there are a number of important disadvantages. First, these systems do not have full Internet access. For example, a Freenet user cannot usually connect to a remote Internet host or use a Gopher.

Second, unlike the Internet, community computing services have Someone-In-Charge, just like the commercial services such as CompuServe or Prodigy. This means that there are rules and there is censorship (although it is usually not heavy-handed).

The Freenets are typically connected to a local university computer system. This is how they exchange mail messages with the Internet and how they access Usenet. Although their users cannot connect to remote Internet systems, Internet users can use a Freenet by using Telnet. See the catalog for details.

Public Internet Access

When all else fails, you can always get Internet access by paying for it. There are many companies or organizations that will provide such access. Typically, you must pay a certain amount per month, along with a fee for *connect time*, that is, the actual time you are using the service. Here are some suggestions as how to find such a service and how to make a good choice.

First, try a local university. Some universities offer Internet access to the general public. Even if you have to pay, the service will probably cost a lot less than from a commercial provider.

If this is not available, check the list in Appendix A. You will notice that some companies offer services to a large area, for example, the entire United States. Other companies are regional in scope.

Start by looking for a company that offers a phone number that is a local call for you. You can expect to spend many hours on the Internet, so it is best not to have to pay long distance charges.

> ### Hint
>
> You will find that some companies have toll-free numbers that you can use from anywhere in the country. However, if you compare closely, you will find there is a surcharge for using such a number. This surcharge may amount to more than paying for the long distance yourself.

After you have found a few likely Internet providers, call them and ask:

- What are your rates?
- Do you offer full Internet service?

Make sure that you do not have to settle for only electronic mail and Usenet. You want the whole set of resources that we described in Chapter 2. The question to ask is, "Do you offer Telnet and FTP?"

You should plan on spending many hours a month glued to the Net. Thus, look for a company that offers as many hours of connect time as you need for a flat fee. If you are not yet an experienced Internet user, take our word for it: the Internet will consume a lot of your time, and paying an hourly rate can get expensive. You may find that paying a long distance charge to call a flat-fee service is cheaper than using a local service that charges by the hour. This is especially true if you call at night or on the weekend when the rates are lower.

> ### Hint
>
> When choosing an Internet provider, look for:
>
> - The full range of Internet services
> - A local phone number to dial
> - A flat monthly fee for unlimited connect time
>
> If you must call long distance (within the U.S. or Canada), check with the various phone companies to see which will offer you the most economical calling plan.

If you have access to Usenet and to electronic mail, there is a discussion group specifically to help people find Internet access in their area. The name of the group is **alt.internet.access.wanted**. This is the place to send a query like "Can someone please tell me where to find public Internet access in the Fuzzballville area?" If you do not have access to Usenet, perhaps you can find a friend to ask the question for you.

Finally, there is a large, up-to-date list of companies that provide public dial-up Internet access. This list is posted regularly to several Usenet groups and is also available by Anonymous FTP and by mail. See Appendix A for the details. If you are just starting, you might prevail upon an experienced Internet person to help you download this list.

One service to watch for is *Public Access Unix*. As we discussed in Chapter 1, most Internet computers run Unix. In some cities, there are companies that provide commercial access to a Unix system. These systems will typically offer some type of Internet access.

CHAPTER 4

Internet Addressing

Whatever you do, make sure you read this chapter. Here is why: Every computer that is on the Internet has its own unique address. Likewise, every person who uses the Internet has his or her own address. You must learn how to understand such addresses.

In the non-computer part of our life, we need to remember different types of information to be able to communicate with someone: a postal address, perhaps a separate street address, a home telephone number, a business telephone number, a fax number, and so on.

On the Internet, there is only one type of electronic address. Once we know someone's Internet address, we can send mail, transfer files, have a conversation, and even find out information about that person. Conversely, once you start to use the Internet, you need give other people only one simple address in order for them to be able to communicate with you.

Moreover, when someone wants to tell you where on the Internet certain information or a particular resource is to be found, he or she does so by giving you the address of a computer. For examples, look at the various entries in the catalog.

For these reasons, understanding the Internet addressing system is crucial to using the Internet. In this chapter, we will show you everything you need to know.

Standard Internet Addresses

On the Internet, the word *address* always refers to an electronic address, not a postal address. If a computer person asks for your "address", he or she wants your Internet address.

Internet addresses all follow the same form: the person's userid, followed by an @ character (the "at" sign), followed by the name of a computer. (Every computer on the Internet has a unique name.) Here is a typical example:

```
harley@fuzzball.ucsb.edu
```

In this case, the userid is **harley**, and the name of the computer is **fuzzball.ucsb.edu**. As this example shows, there are never any spaces within an address.

You will remember that in Chapter 2 we explained that each person has a user name called a userid (pronounced "user-eye-dee"). It is this userid that we use as the first part of someone's address. If you have a Unix system, your userid will be the name that you use to log in.

The part of the address after the @ character is called the *domain*. In this case, the domain is **fuzzball.ucsb.edu**.

Thus, the general form of all Internet addresses is:

```
userid@domain
```

As you might imagine, a userid by itself is not necessarily unique. For instance, within the entire Internet, there are probably a number of people lucky enough to have a spiffy userid like **harley**.

What must be unique is the combination of userid and domain. So, although there may be more than one **harley** on the Internet, there can be only one such userid on the computer named **fuzzball.ucsb.edu**.

If you read an Internet address out loud, you will see that using the @ character is appropriate. For example, let's say that you wanted to send mail to the person at the address we just mentioned. The command to do so is:

```
mail harley@fuzzball.ucsb.edu
```

As you enter this command, you can say to yourself, "I am sending mail to Harley, who is at the computer named **fuzzball.ucsb.edu**".

Sometimes an address in this form is called a *fully-qualified domain name* or *FQDN*.

Understanding a Domain Name: Sub-Domains

In the last section, we used **harley@fuzzball.ucsb.edu** as an example of an Internet address. We said that **harley** is the userid and **fuzzball.ucsb.edu** is the domain. Each part of a domain is called a *sub-domain*. As you can see, sub-domains are separated by periods. In our example, there are three sub-domains: **fuzzball**, **ucsb** and **edu**.

The way to understand a domain name is to look at the sub-domains from right to left. The name is constructed so that each sub-domain tells you something about the computer. The rightmost sub-domain, called the *top-level domain*, is the most general. As you read to the left, the sub-domains become more specific.

In our example, the top-level domain **edu** tells us that the computer is at an educational institution. (We will explain the meanings of the various top-level domains in a moment.) The next sub-domain, **ucsb**, tells us the name of this institution (the University of California at Santa Barbara). Finally, the leftmost sub-domain is the name of a specific computer named **fuzzball**. Thus, as you enter the command:

```
mail harley@fuzzball.ucsb.edu
```

you can say to yourself, "I am sending mail to Harley, at a computer named **fuzzball**, at the University of California at Santa Barbara, which is an educational institution."

When you type an address, you can mix upper- and lowercase letters. (With computers, *uppercase* refers to capital letters; *lowercase* means small letters.) For example, the following two addresses are equivalent:

```
mail harley@fuzzball.ucsb.edu
mail harley@FUZZBALL.UCSB.EDU
```

There are two common variations that you will see. First, some people type only the top-level domain in uppercase:

```
mail harley@fuzzball.ucsb.EDU
```

Other people like to emphasize the site of the computer:

```
mail harley@fuzzball.Ucsb.Edu
```

In all cases, the uppercase letters are optional.

Hint

As a general rule, use all lowercase for Internet addresses. There is really no need to mix in uppercase letters.

If you see an address in which some of the letters are uppercase, it is always safe to change them to lowercase.

If you do decide to use some uppercase letters, it is best not to change the userid. On some systems, it may make a difference (although it's not supposed to).

Variations on the Standard Internet Address Format

All Internet addresses follow the standard format:

> `userid@domain`

However, there are several variations you may encounter. The example we have been using has three sub-domains:

> `harley@fuzzball.ucsb.edu`

You will often see addresses that have more sub-domains in order to be more specific. Here is an example:

> `scott@emmenthaler.cs.wisc.edu`

In this case, the userid is **scott**. The domain refers to a computer **emmenthaler** that is part of the Computer Science department at **wisc**, the University of Wisconsin (presumably, an educational institution).

Many Internet sites use patterns to name their computers. For example, you might see computers named after cartoon characters, mythical heroes, local landmarks, or whatever. At the University of Wisconsin, many computers are named after types of cheese.

FUN TIP: As all students of world geography know, it is the custom within the United States for each state to have its own nickname. For example, California is the Golden State and New York is the Empire State. The reason that there are so many "cheese" computers at the University of Wisconsin is that Wisconsin is known as the Dairy State.

(In other words, by the time someone got around to Wisconsin, all the good nicknames were already taken.)

Some Internet addresses have only two sub-domains (the minimum). Here are two examples:

> `rick@tsi.com`
> `randy@ucsd.edu`

When you see an address with only two sub-domains, it can mean two things. First, it might mean that the organization is so small that it only has one computer on the Internet. In the first example, this is the case. The top-level domain **com** tells us that this is a commercial organization. (More about top-level domains in a moment.) The other sub-domain, **tsi**, is the name of the computer. It happens that this is a company named Technology Systems Integrators—talk about snazzy names—that has only one computer on the Internet.

The second example is from an organization that has a great many computers. In such organizations, one computer is usually used to send and receive all the outside mail. In our example, the name of this computer is **ucsd.edu**. The system administrators for this organization have simplified the mailing addresses by arranging for everyone to be able to receive mail addressed to **ucsd.edu**. Here is how it works.

In general, the term *gateway* refers to a a link between two different systems. In this case, we have a *mail gateway*. The computer **ucsd.edu** acts as the mail link between the internal network and the outside world. The mail gateway has a list of userids and local addresses. When a message arrives, the gateway can check the list and forward the message to the appropriate local computer.

For example, say that a person has a userid of **melissa** on a computer named **misty**. Normally, her address would be:

```
melissa@misty.ucsd.edu
```

However, in order to make her mail address simpler, she registers with the mail gateway. From then on, she can receive mail at **melissa@ucsd.edu**. When mail arrives, the gateway will automatically forward it to the **misty** computer.

So, as you can see, an address with only two sub-domains usually means that the organization is very small (like **tsi.com**) or very large (like **ucsd.edu**).

We explain this in detail because there may be times when you need a person's exact address, for example, to contact them using the **talk** command (which we discuss in Chapter 19). In such cases, the simplified mail address may not work and you will have to ask for a longer, more specific one.

To conclude this section, we will discuss one final form of Internet addressing that you will sometimes see in a mail address. In this form, a % (percent) character is used as part of the address. In such addresses, the % character will be to the left of the @ character. For example:

```
melissa%misty@ucsd.edu.
```

The idea is that the computer that receives the message (in this case, **ucsd.edu**) will look at everything to the left of the @ character (in this case, **melissa%misty**) and try to make sense out of it.

Usually, the % character will separate a userid from the name of a local computer. In this example, userid **melissa** uses a local computer named **misty**. Within the local network, there may be several different connections from the mail gateway to that computer. At the time the message is received, the mail gateway will choose the best path to use to deliver the mail.

Hint

Many people know their Internet address even though they don't understand it. Some organizations have more than one way to address mail, and the system managers will usually tell their users which address works best.

So don't worry too much about the variations. When you send mail to someone, just use whatever address the person gave you.

Top-Level Domains

As we mentioned earlier, the way to understand an address is to read it from right to left. The top-level domain will be the most general specification. In the example we looked at earlier:

```
mail harley@fuzzball.ucsb.edu
```

the top-level domain of **edu** told us that the computer was at an educational institution. We also looked at another address:

```
rick@tsi.com
```

in which a top-level domain of **com** indicated a commercial organization.

In general, there are two types of top-level domains: the old-style *organizational domains* (as in these two examples) and the newer *geographical domains*.

The organizational domains are based on an addressing scheme that was developed before the days of international networks. It was intended to be used mainly within the United States.

The idea was that the top-level domain would show the type of organization that was responsible for the computer. Table 4-1 shows the various categories. All of these categories have been around since the beginning of the Internet, except **int**, which is a relatively recent addition for certain organizations that span national boundaries (such as NATO).

Once the Internet expanded internationally, new, more specific top-level domains became necessary. To meet this need, a new system of geographical domains was developed in which a two-letter abbreviation represents an entire country. There are many such top-level domains—one for every country on the Internet—and they are all listed in Appendix F. For quick reference, Table 4-2 shows a representative sample.

As an example, take a look at the following address:

```
michael@music.tuwien.ac.at
```

Here we have a computer at the Technical University of Vienna in Austria (top-level domain of **at**).

Some countries use a sub-domain, just to the left of the top-level domain, to divide it into categories. For example, you might see **ac**, referring to an academic organization or **co**, referring to a commercial company. In our example, we see that the Austrian address uses **ac**.

Domain	Meaning
com	commercial organization
edu	educational institution
gov	government
int	international organization
mil	military
net	networking organization
org	non-profit organization

Table 4-1. *Organizational Top-Level Domains*

Domain	Meaning
at	Austria
au	Australia
ca	Canada
ch	Switzerland ("Confoederatio Helvetia")
de	Germany ("Deutschland")
dk	Denmark
es	Spain ("España")
fr	France
gr	Greece
ie	Republic of Ireland
jp	Japan
nz	New Zealand
uk	United Kingdom (England, Scotland, Wales, Northern Ireland)
us	United States

Table 4-2. *Examples of Geographical Top-Level Domains*

For the most part, the geographical domains are simply the standard two-letter international country abbreviations (the ones that everybody in the world knows about except the Americans). The exception to this scheme is Great Britain. Its international abbreviation is **gb**, but it also uses a domain name of **uk** for "United Kingdom".

This makes sense when you remember that Great Britain includes England, Scotland, and Wales, while the United Kingdom also includes Northern Ireland. (Evidently, the English use some alternative meaning of the word "united" that nobody else understands.)

As you can see from Table 4-2, the United States does have a geographical domain name (**us**), although it is not used much. Outside the U.S., though, geographical names are used almost exclusively. However, regardless of what type of top-level domain your organization uses, you can communicate with any address on the Internet. Both types of top-level domains are recognized everywhere.

Hint

Within Britain and New Zealand, the order of domains is often reversed. For example, you may see an address like:

```
peirce@uk.ac.oxford.compsci
```

The mailing system is supposed to reverse the domains in order to communicate with the outside world. However, occasionally, a reversed address will escape from one of these countries, like a bad table wine exported for foreign consumption. If you need to use such an address outside its native environment, be sure to reverse the sub-domains:

```
peirce@compsci.oxford.ac.uk
```

so that they form a standard address.

Pseudo-Internet Addresses

There are many organizations that would like to be on the Internet, but do not have the time or the money to maintain a permanent Internet connection. As an alternative, they make an arrangement with a nearby Internet site that agrees to act as a mail gateway for them. For example, a small company might make such an arrangement with an Internet access provider (see Chapter 3) or a university.

With such an arrangement, the organization can be given a mailing address that looks just like a standard Internet address. However, such organizations are not really on the Internet.

For example, say that the Marlinspike Consulting Company would like an Internet address. It contracts with SnowyNET, a local Internet access provider, to handle its mail. As part of the arrangement, the company is given the name **marlin.com**, which is registered with the Internet addressing system.

Any mail that is sent to **marlin.com** is automatically routed to the SnowyNET gateway computer. At certain times, the **marlin.com** computer connects over a telephone line with the SnowyNET computer and picks up its mail. At the same time, any mail from **marlin.com** that is being sent outside the company is passed to the SnowyNET computer.

A user at a company with such an arrangement will use an address that looks just like a standard Internet address. For example, someone might tell you that his address is:

```
tintin@marlin.com
```

When you see such an address, there is no way of knowing whether or not it represents a real Internet address. In other words, there is often no way to tell, from the address alone, if a computer is actually on the Internet. If you see a name like:

```
small-company-name.com
```

you might be suspicious. However, even some large companies use a gateway rather than a real Internet connection.

If it is important that you know for sure if a computer is on the Internet, you can use the **host** command, described in the next section.

IP Addresses: The host Command

So far, we have talked about Internet addresses in which each sub-domain is a name. However, underneath the socially acceptable veneer of this system lies a typical computer trick: the real Internet addresses are actually numbers, not names. For example, the computer **ucsd.edu**, which we mentioned earlier, is actually called **128.54.16.1**.

Names, of course, are easier for people to use, but every time you use a domain address, your system has to turn it into a number, although the details are hidden from you. You will recall that in Chapter 2 we explained that the part of the Internet that moves data packets from one place to another is called IP (Internet Protocol). For this reason, the numeric version of an address is called an *IP address*. For example, the computer **ucsd.edu** has an IP address of **128.54.16.1**.

As you can see, an IP address looks something like the domain addresses that we have already discussed in that there are several parts separated by periods. However, the parts of an IP address do not correspond directly to sub-domain names, so don't read too much into the pattern.

You can use an IP address anywhere you would use a regular address. For example, the following two mail commands are equivalent:

```
mail randy@ucsd.edu
mail randy@128.54.16.1
```

The part of the Internet that keeps track of addresses is called the *Domain Name System* or *DNS*. DNS is a TCP/IP service that is called upon to translate domain names to and from IP addresses. Fortunately, it is all done behind your back, and there is no reason to bother with the details.

In rare cases, it may be that your system has trouble understanding a domain address. If this happens, the IP address may work better (as long as it is correct).

If you would like to test the DNS, you can use the **host** command. There are two formats:

```
host standard-address
host IP-address
```

If you specify a standard address:

```
host ucsd.edu
```

DNS will display the IP address. If you specify an IP address:

```
host 128.54.16.1
```

DNS will display the standard address.

Hint

Only real Internet hosts have IP addresses. Thus, you can use the **host** command to check if a computer is on the Internet. If **host** displays an IP address, the computer you specified is on the Internet.

If you specify an address that is not on the Internet, you will see:

```
Host not found.
```

If this happens, be sure to check your spelling.

> ### Hint
>
> The **host** command is not available on all systems. If your system does not
> have **host**, try **nslookup**.

UUCP

If you have read the previous sections of this chapter, you have learned just
about everything you need to know about standard Internet addresses.
However, we do want to spend a little time discussing addressing schemes used
by a few other networks. As an Internet user, you can exchange mail with these
networks, and it will help to understand what type of addresses they use.

To start, we will discuss the Unix-based UUCP network. We will then
conclude the chapter by explaining what addresses to use to send mail to other
popular networks: CompuServe, MCI Mail, FidoNet, and Bitnet.

All Unix systems come with a built-in networking system called *UUCP*.
UUCP is a family of programs. The name UUCP comes from one of these
programs, **uucp**. The **uucp** program copies files from one Unix system to another.
Thus, the name means "Unix to Unix copy".

(You may see a similarity here to the name TCP/IP. As we mentioned in
Chapter 2, TCP/IP is a large family that is named after its two most important
members, TCP and IP.)

UUCP is not as powerful as TCP/IP. For example, UUCP does not provide
remote login. Moreover, as we will see in a moment, the UUCP mail facility is
slower and more awkward than the Internet system. However, UUCP does have
an important advantage. It is a standard part of Unix (free with most systems),
and it runs cheaply and reliably over dial-up or hardwired connections.

UUCP works by allowing Unix systems to connect together to form a chain.
For example, say that you are using a computer named **alpha**. Your computer is
connected to another computer named **beta**. This computer is connected to
gamma, which is, in turn, connected to **delta**.

You decide to send mail to a person with a userid of **murray**, who uses **delta**.
You send out the message from your computer, **alpha**. UUCP will pass the
message from **alpha** to **beta** to **gamma** to **delta**, where it will be delivered to
userid **murray**.

Our example involves four computers and three different connections. It
might be that these are hardwired connections, in which the computers are
joined by a cable. Typically, though, the connections are made over a telephone
line. At certain intervals, each computer calls (or is called by) its neighbor. When

they connect, they swap whatever mail they have waiting for one another. Some of the mail will be for local users. Other mail will need to be forwarded to another computer.

The system works well in that it provides an economical way to send mail from computer to computer over large distances. However, there is an important limitation: since many UUCP connections are made over a telephone line at certain predefined times, mail delivery can take hours or even several days.

Compare this to the Internet, in which connections are permanent and messages are transmitted quickly, often within seconds, usually within minutes.

UUCP Addresses and Bang Paths

Many sites which at one time depended on UUCP for mail and file transfer now use the faster, more dependable Internet. However, there are still a lot of UUCP installations, so it is a good idea to know something about their addresses.

To send mail to a UUCP address, you must specify the route that you want the message to take. For example, you must say, "I want this message to go to computer **beta**, and from there to computer **gamma**, and from there to computer **delta**. At that point, I want the message delivered to userid **murray**."

To do so, you construct an address consisting of each of these names in turn, separated by ! (exclamation mark) characters. For example, here is a **mail** command that will send a message to the userid we just described.

```
mail beta!gamma!delta!murray
```

When you create such a message, your system will store it until contact is made with computer **beta**, at which time the message will be sent on its way.

In Unix terminology, one of the slang names for the ! character is *bang*. Thus, a UUCP address that specifies multiple names is sometimes called a *bang path*. When a Unix person reads such a path out loud, he will pronounce the ! character as "bang". For example, you might hear someone say, "I tried to send you mail at beta bang gamma bang delta bang murray."

Hint

In Unix systems, the program that reads and interprets the commands you enter is called a *shell*. Some shells—especially the C-Shell—recognize a ! character as part of a facility called *history substitution*. This facility allows you to recall and edit previously entered commands. As you might imagine, history substitution can be a real time-saver.

However, this means that the ! character has a special meaning, and when you type this character as part of a UUCP address, it will cause an error. For example, if you enter:

```
mail beta!gamma!delta!murray
```

the C-Shell will try to interpret the command as a history substitution request. We won't go into the details here, except to say that you will see an error message like this:

```
gamma!delta!murray: Event not found.
```

(The term "event" refers to a previously entered command.)

Thus, if you are using a shell like the C-Shell, you need to tell it that the ! characters in a UUCP address are to be taken literally. To do so, you preface each ! with a \(backslash) character:

```
mail beta \!gamma\!delta\!murray
```

The \ characters are not really part of the address. They are there only to tell the shell not to misinterpret the ! characters.

Simplified UUCP Addressing

As we explained in the previous section, UUCP is inexpensive and accessible to anyone with a Unix system, a modem, and another computer with which to connect. Indeed, before the Internet became so popular, many people used to send mail over a large, worldwide UUCP network. Today, many of these people have migrated to the Internet, but there are still a large number of computers out there that are reachable only by UUCP.

One of the problems with UUCP addressing is that the addresses can be long. Moreover, you must specify an exact path from one computer to the next. In the last section, we looked at a sample **mail** command that specified a path via three different computers.

```
mail beta!gamma!delta!murray
```

This is okay, albeit inconvenient, as long as you know which path to use. But many UUCP paths are much longer. Moreover, how do you know how to construct the path?

Let us say, for example, that you have a friend with a userid of **albert** who uses a computer called **gendeau.com**. If he were connected to the Internet, you would send mail to him by using the command:

```
mail albert@gendeau.com
```

But suppose he is on the UUCP network, not the Internet. How do you have any idea what path to use from your computer to his?

In general, this is a big problem with UUCP because the path to a computer depends very much on where you start. Thus, if your friend wants to correspond with people at different locations, he might have to give each of them a different address. The nice thing about the Internet and the Domain Name Service is that all you have to do is specify the address of the destination. The system automatically figures out the best route to take.

To bring the same sort of convenience to UUCP, an undertaking was started called the *UUCP mapping project*. This project regularly publishes "maps" of data that are sent to many key UUCP computers. When UUCP mail reaches these computers, they can look at the maps and decide the best route to use. In essence, this allows you to use a UUCP address that is similar to an Internet address, and let the system do the work.

Thus, on occasion, you may see an address that uses a top-level domain of **uucp**. For example, a friend might say that you can send mail to him at:

```
albert@gendeau.uucp
```

When you use such an address, it is a signal to the mail routing software to find the name in the UUCP mapping data and figure out the best path to use.

Your computer may be able to do this itself, or it may send it to another computer to do the job.

In practice, all you have to do is get the address right and it should work. But if you ever have a choice between a UUCP address and an Internet address, choose the Internet one.

Sending Mail to Other Networks

The Internet has mail gateways to a large number of other networks. As long as you know the right way to address mail, you can send messages to people on these networks via their gateway. Some of these gateways serve commercial networks that charge for their services. Nevertheless, as an Internet user, you can use the gateway for free.

FUN TIP: No matter how much a user of a commercial network might pay for the privilege of sending and receiving mail from you, as an Internet user you will never have to pay anything.

To conclude this chapter, we will discuss a few of the most widely used mail gateways, and we will show you how to use an Internet-style address to send mail to users on these networks.

Let's start with two popular commercial systems, CompuServe and MCI Mail. A user from one of these networks has his or her own account number. Here is how to convert this account number into a proper Internet address.

A CompuServe account number consists of a sequence of digits, with a comma somewhere within the sequence. For example:

```
12345,678
```

To send mail to a CompuServe account, all you have to do is replace the comma with a period, and use a domain of **compuserve.com**. Thus, to send mail to the user with the account we just mentioned, use the command:

```
mail 12345.678@compuserve.com
```

MCI Mail addresses are similar. Customers can use either an identification number or a user name. For instance, you might have a friend whose MCI Mail number is **12345** and whose user name is **hhahn**. To send him mail, use either the number or the name with a domain of **mcimail.com**. For example:

```
mail 12345@mcimail.com
mail hhahn@mcimail.com
```

Another address that you might encounter is one from FidoNet. This is a worldwide network of personal computers that connect via telephone lines. (In principle, FidoNet is not unlike UUCP.) To reach FidoNet, you use a domain name that ends in **fidonet.org**. The name of the actual FidoNet computer is specified as a series of sub-domains. It works like this.

In the terminology of FidoNet, a computer name consists of three parts: a zone number, a net number, and a node number. The zone number is followed by a : (colon) character, and the net number is followed by a / (slash) character. For example, someone might tell you that his FidoNet computer is:

```
1:234/567
```

In this case, the zone number is **1**, the net number is **234**, and the node number is **567**.

To specify a FidoNet computer name from the Internet, you use these same three numbers, in reverse order, according to the following pattern:

```
fnode.nnet.zzone.fidonet.org
```

Within FidoNet, users are known by their full names. You separate each part of the name with a period. Thus, a person named Rick Shaw will have a user name of **Rick.Shaw**.

For example, to send mail to Rick Shaw at the FidoNet computer **1:234/567**, use:

```
mail Rick.Shaw@f567.n234.z1.fidonet.org
```

The last type of mail address that we will mention is for Bitnet users. Bitnet is a collection of different networks based in the United States, Canada, Mexico, and Europe. To send mail to a Bitnet user, you need to know his or her user name and host computer. From the Internet, the address is simple. Use a top-level domain of **bitnet**. To the left, put the name of the Bitnet host.

For example, say that you want to send mail to a friend whose Bitnet user name is **lunaea**. The name of her computer is **psuvm**. Use the command:

```
mail lunaea@psuvm.bitnet
```

If you hang around with Bitnet people, you will notice that a lot of the computer names end in "vm". This is because they are IBM mainframe computers using the Virtual Machine operating system.

The name **bitnet** is not an official Internet domain and is an example of what is called a *pseudo domain*. (The name **uucp** that we saw in the previous section is also a pseudo domain.) When you use an address with a pseudo domain, the mailing software on your system must recognize the domain, rewrite the address, and send the message to a computer that knows how to send mail to that particular network. In this case, your mailing program would have to send the message to a *Bitnet/Internet gateway*.

On some systems, this type of address will not work because the local mail software is not set up to recognize a **bitnet** pseudo domain. If this is the case on your system, you can send the message directly to any one of the Bitnet/Internet gateways. Here are several of them:

cornellc.cit.cornell.edu
cunyvm.cuny.edu
mitvma.mit.edu
pucc.princeton.edu
vm1.nodak.edu

There are two addressing formats that you can use. The preferable one uses the UUCP bang path notation:

```
gateway!computer.bitnet!userid
```

For example, to send a message to **lunaea** on the Bitnet computer named **psuvm**, you might use the address:

```
cornellc.cit.cornell.edu!psuvm.bitnet!lunaea
```

The second format uses the % notation that we discussed earlier:

```
userid%computer.bitnet@gateway
```

For example:

```
lunaea%psuvm.bitnet@cornellc.cit.cornell.edu
```

Technically speaking, this form of the address is not officially supported by the Internet, although such addresses will usually work.

CHAPTER 5

The Internet Mail System

By far, electronic mail is used more than any other Internet resource. Indeed, for many people, mail *is* the Internet.

In this chapter, we examine the Internet mail system and discuss everything you need to understand to get started. We begin with the basic ideas and concepts: What are the programs that provide the Internet mail service? What technical words and terms will you encounter?

From there, we complete the orientation by giving you some guidelines for using the mail system, and by explaining the technical information that is a part of each message. In Chapter 6, we will show you how to use the standard Unix mail program. At that time, we will give you a hands-on description of how to send and receive messages.

The Basis for the Internet Mail System: The Transport Agent

The Internet mail system is the most important Internet resource. Every day, countless messages are sent from one part of the Internet to another. As you would expect, many of these messages are personal notes from one user to another. However, the mail system is a general service that can transport all types of information: documents, publications, computer programs, and so on. The only requirement is that the data be stored as ASCII characters. (That is, the data uses the regular set of characters that are on our keyboard.)

In some circumstances, it is possible to mail non-textual data, such as pictures or sound recordings. We will discuss this later in the chapter.

There are millions of people who use the Internet, and the mail system transports just about every type of information you can think of. One of the most significant uses of this system is to allow people who are not near one another to work together on a project. There are many projects in which people from all over the world participate. Indeed, it is common for someone to collaborate with a person who is thousands of miles away, someone they will never actually meet in person.

For example, as this book was being written (in California), each chapter was sent to a person in Austria and a person in Ireland, both of whom would review the work and make comments. The text of the chapter was sent as a mail message. Each reviewer would insert his comments right into the text and mail it back. Because the Internet is so fast and reliable, it was not uncommon for us to be able to review an entire chapter within a day. If we had to depend on standard post office mail, the whole process would take weeks.

The same process is carried out, on a much larger scale, when decisions have to be made that affect the Internet itself. A new idea or a new standard is proposed, after which anyone can participate in a worldwide debate via the mail system. (The Usenet discussion groups are also used for this purpose.)

Thinking about all of this, it is natural to wonder what holds it all together. After all, the Internet connects thousands of different networks, each of which has its own mix of computers and software. How can all these different systems work together to exchange mail?

The answer is that the delivery of mail is standardized by a system called *SMTP*. SMTP, which stands for *Simple Mail Transfer Protocol*, is part of the TCP/IP family of protocols. It describes the format of a mail message and how messages are handled as they are delivered. Every Internet computer runs a mail program that works behind the scenes to ensure that messages are addressed and transported in an orderly fashion. This program, called a *transport agent*, follows the SMTP protocol and, in doing so, provides your mail link to the outside world.

On most systems, the transport agent runs by itself in the "background", always ready to respond to whatever requests it may receive. In Unix terminology, such a program is called a *daemon* (yes, that is the correct spelling).

Every Unix system has various daemons that are lurking in the background, quietly providing services for you. In theory, it doesn't matter what transport agent your system uses as long as it knows how to send and receive mail using SMTP. Most Unix systems use a daemon named **sendmail**.

Now, are we saying that you should understand the technical details of how all of this works? Of course not. What we do want you to appreciate is that the mail system works because everybody's network has at least one computer running a transport agent that sends and receives mail according to the SMTP standard. We also want you to know that on many systems the transport agent is a daemon named **sendmail**. You will never work with **sendmail** directly, but at least you know its name and what it does.

Your Interface to the Mail System: The User Agent

As a user, you do not interact with your system's transport agent; it operates behind the scenes sending and receiving mail via SMTP. The mail program that you use is called a *user agent*. It is the user agent that acts as your interface into the Internet mail system, allowing you to read your messages, compose new messages, delete messages that you have already read, and so on.

There are many different user agents used on the Internet. In this section, we will survey the more common ones. We will start with mail programs that you might find on a Unix system. We will then discuss the user agents that you might use on a PC or Macintosh. Which user agent you use to compose and read mail depends on what is available on your system. On some systems, there is only one mail program. On other systems, you have a choice.

The most widely available user agent is the Unix **mail** program. This program comes with all Unix systems and, hence, is available to most Internet users. Your system may have a better program—we will discuss a few of these in a moment—that you prefer to use. This is fine. Still, **mail** is a standard tool and is the only user agent that you can expect to see on every Unix system. For this reason, we will discuss **mail** in more detail, in Chapter 6.

Hint

The first version of **mail** was primitive, offering only the most rudimentary functions to compose and read messages. Some years ago, newer versions of **mail** were developed that are far more powerful. At first, there were two versions of this program, one for each of the two main types of Unix: Berkeley Unix (known as BSD, for Berkeley Software Distribution) and System V Unix (which came from AT&T's Bell Labs).

The BSD user agent was called **mailx**. The System V user agent was called **Mail**. (As you may know, Unix distinguishes between upper– and lowercase letters, so **Mail** and **mail** are two different names.)

At first, users had to remember to use the right name. If they used **mailx** (or **Mail**), they would get the new version of the program. If they used **mail**—the more logical name—they would get the old version.

These days, the old program is rarely used. When you use the **mail** command, you will automatically get either **mailx** or **Mail**, whichever is on your system. There is a way to use the old program, but you will never need it.

Common Unix–Based Mail Programs

We have mentioned that **mail** is the standard Unix user agent, the program that you use to send and receive mail. However, other systems have other user agents and, if one is available, you may choose to use it instead of **mail**. Moreover, some Unix systems have more than the standard **mail** program. There are many such programs, and we won't discuss all of them. However, let's take a quick look at the most widely used ones.

One of the most popular user agents is named *Elm*. Elm is a full–screen program that is easy to use and simple to learn. Even if you have never used electronic mail before, you may be able to start using Elm without any documentation. As you become more experienced, you will find that Elm offers a lot more than simple mail handling. There are all kinds of advanced features that you can customize.

Another popular program that is even easier to use is *Pine*. Pine uses a simple menu-driven interface that is especially suited to beginners or casual users. Pine is great for inexperienced people who have to sit down at a computer and start using mail immediately. The menu items are clear and easy to understand, and you can display help information at any time. However, if you find yourself sending and receiving a lot of mail, you will be better off using a different user agent. Although Pine is easy to learn, the other programs are more powerful and streamlined, and better suited to experienced users.

Another user agent that you may encounter is called *MH* (*Message Handler*). MH is actually a set of relatively simple, single–purpose programs. Instead of using one main program for everything, you use different programs for each task.

For example, when you use **mail** or Elm, you enter one command to start the program. Everything that you want to perform is done under the auspices of this program.

With MH, you use separate commands for each task. For instance, to check what messages are waiting for you, you use the **scan** command; to display messages, you use the **show** command; to reply to a message, you use the **repl** command; to compose a new message, you use the **comp** command; and so on.

The disadvantage of such a system is obvious: you have to memorize a lot of commands. However, the advantages of MH are important. You don't have to stop what you are doing just to do work with your mail. You can enter an MH command at any time, to perform one simple task for you.

For example, say that you have just read a message from a friend in which he asks if you have the latest version of a particular document. To check this, you need to enter a file command to examine the document.

If you are using **mail** or Elm, you would have to stop the program—or at least put it on hold—in order to enter the file command. With MH, you are not working within a mail program. You can simply enter the file command, look at the information, and then enter the MH command to send a reply.

Another user agent that you may find on your system is *Mush*. The name stands for *Mail User's Shell*. Mush is flexible in that it can be used in two different ways: either like **mail** (which uses a line–oriented interface), or like Elm (using the entire screen at once). There is also a commercial product, called Zmail, that is based on Mush.

So far, all the user agents we have mentioned will work on any Unix system. If you are an X Window user (see Chapter 2), there will probably be one or more X user agent programs that you can use to take advantage of the graphical user interface. For example, if you are using a Sun workstation, you can use a graphical mail program called Mailtool.

The last user agent that we want to mention in this section is *Rmail*. This is a mail facility that is built into some versions of Emacs. *Emacs* is a complete working environment that is based on a powerful text editing program. Within the Emacs environment, you can not only edit text, you can develop programs, read Usenet articles, and, using Rmail, send and receive mail. We won't go into the details of Emacs in this book. However, we do want you to be aware of

Rmail. If you are an Emacs user, you will find that Rmail, being integrated into the Emacs environment, is particularly convenient.

PC- and Macintosh–Based Mail Programs

In Chapter 2, we explained how you might access the Internet in two ways. First, you can use a terminal that is connected to a host computer. Second, you can use a computer of your own—such as a PC or a Macintosh—that is connected to a local area network. With such systems, there are a separate class of user agents that can be used to access the mail system. In this section, we will discuss two such user agents: Pegasus and POP-based programs.

Pegasus, also called *P-Mail*, is a user agent for local area networks. Pegasus works just like Novell's Message Handling Service (MHS). However, Pegasus, which was developed by David Harris of Dunedin, New Zealand, is distributed free over the Internet and does not require Novell's MHS software. Pegasus runs on both PCs and Macintoshes and is used widely throughout the Internet.

The other PC and Macintosh mail programs that enjoy wide use are based on a different system. As we discussed earlier, Internet hosts follow a protocol called SMTP when they send and receive mail. SMTP is part of the TCP/IP family and describes how messages are to be handled. For networks that have PCs and Macintoshes, there is another protocol that is sometimes used named *POP*, or *Post Office Protocol*. These personal computers can run a special POP-based user agent that takes advantage of this protocol.

Within each Internet site, there are one or more computers that act as mail repositories. Messages that are waiting to be read are kept on these computers. With POP-based user agents, the messages are not kept in a central computer; they are sent to your personal computer where they are stored until you read them.

To see how this works, imagine a message is being sent to you from the other side of the world. At each step along the way, the message is sent from computer to computer over the Internet. Eventually, the message arrives at the computer in your network that acts as a mail repository. At this point, your POP-based user agent can ask this computer to send the message directly to your PC or Macintosh, where you can read it.

As the message travels through the Internet, the computers use SMTP, the standard TCP/IP mail protocol. But on the last leg of its journey, the message is sent to your PC or Macintosh using the POP protocol.

POP-based user agents have certain important advantages. First, they give you complete control over your mail, because messages are stored on your personal computer. This means, for example, that you can keep a lot of messages around without having a system manager complain that your messages are taking up too much disk space. (This becomes especially important when you are paying for disk space on the central computer. Storage space on your own computer will be free.)

Second, POP–based user agents are written to make use of the special features of the computer. For example, a Macintosh mail program will take advantage of the Macintosh mouse–driven graphical user interface, with its easy-to-use pull–down menus. Similarly, a DOS program written for a PC can make use of the special features of the computer, such as the function keys. Many users find it easier to use a program that was written specifically for their type of system than to learn how to use, say, the standard Unix **mail** program.

Some of the common POP–based user agents that you may encounter are Eudora, NuPop, and Popmail/PC. Such programs offer the advantage of being native to your environment. However, there are some important disadvantages.

The first problem is one of security. In some organizations, PCs and Macintoshes must be shared. For example, there may be a computer room that anyone can use. With a regular mail program, no one can read your mail unless they are able to log in under your account. In other words, unless someone knows your password—or unless you walk away from your terminal leaving it logged in—your messages are safe from prying eyes.

With a POP–based user agent, all your messages are moved from the central repository to your computer and stored in regular files. If someone else has access to the same computer, he or she will be able to look at your messages. (Remember, PCs and Macintoshes do not use passwords.) Thus, a POP system works best when each computer is used only by a single individual.

Aside from the security problem, there are also logistical considerations. Since all your mail is moved to your personal computer, what happens if you want to access the messages from another computer? With the regular mail system, it is possible to log in to any computer on the network and access your mail. With a POP–based system, messages that have been moved to a particular computer are accessible only on that computer.

Say, for example, that you routinely use two different computers, a standard desktop PC at work and a small portable computer at home. With the home computer, you use a modem to connect to the network over a phone line. You use the portable computer to access the network and the Internet from home. However, once you move messages to a POP–based system on your office PC, you will not be able to read these messages unless you are at the office.

Some versions of POP–based programs solve these problems by allowing you to read mail from the central repository without having to first move it to your computer.

> ## Hint
>
> POP–based user agents are best for people who use only one PC or
> Macintosh that they do not have to share. In such cases, a POP–based
> program can provide a powerful, easy-to-use mail service.

Finally, there is one last consideration we would like to mention. The most
common user agent programs used on the Internet are the ones we mentioned in
the previous section, **mail** (which is actually part of Unix) and Elm. It is always a
good idea to be able to use standard tools. For instance, if you know how to use
mail, you will know how to send and receive mail on any Unix machine.

If you only know how to use a PC or Macintosh POP–based program, your
options are limited. Although this may seem fine today, both you and the
Internet will be around for a long time, and you never know what systems you
will be called upon to use. For example, if you are a student, you might use
Eudora (a Macintosh POP–based user agent) at your university. But after
graduation, you could easily find yourself working for a company that doesn't
have Macintoshes.

Basic Terminology

In this section, we will discuss the technical terms that you will encounter as you
use the Internet mail system.

To start, we will remind you that when you see the word "mail", it always
means electronic mail, and the word "address" always refers to an Internet
address. On those rare occasions when it is necessary to talk about regular post
office mail, the reference will be explicit. Thus, if someone on the Internet asks
"What is your address?", give him or her your electronic address.

Sometimes, you will see post office mail referred to facetiously as *snail mail*.
For example, it is common for people who send in articles to Usenet discussion
groups to put their addresses at the end of the articles. Sometimes you will also
see "Snail Mail:" followed by a postal address. The name, of course, refers to the
fact that post office mail is much slower than electronic mail.

After you read a message, there are several ways to dispose of it. First, you
can get rid of the message by *deleting* it. While you are still within the mail
program, it is usually possible to get back a deleted message should you change
your mind. However, once you quit the program, all the deleted messages are
gone for good.

If the message requires an answer, you can *reply* to it. Your mail program will make it easy for you to send a response back to the same userid that sent the original message. If you want, you can even include parts of the original message in your response.

Another option is to *save* a message. This means that the contents of the message are copied to a file that you can keep as long as you want. You may also choose to *forward* a message. The mail program will make a copy of the message and send it to whomever you specify.

Hint

Be careful what you write to other users. Most mail programs make it easy to forward mail, and some people love to do so. Don't be surprised if a private note to a colleague ends up in someone else's mailbox.

When you send a message, you specify the userid of the recipient. At the same time, you can specify more than one recipient and the message will be sent to each one.

You can also designate one or more userids to receive copies of the message. There are two types of copies, regular ones and *blind copies*. Normally, each recipient will see the names of all the other userids who were sent a copy of the message. A blind copy is a secret copy that no one else knows about.

Here is an example. Let's say you work with four people whose userids are **curly**, **larry**, **moe**, and **harley**. (Sounds like an interesting place to work, doesn't it?) You send a message to **curly**, with a copy to **larry** and **moe**, and a blind copy to **harley**. The result will be that all four people will get the same message. Everybody will know that a message was sent to **curly**, and that both **larry** and **moe** were sent copies. However, **curly**, **larry**, and **moe** will not know that **harley** also received a copy.

Hint

When you receive a message, there is no way for you to tell if a blind copy was sent to someone else.

You might ask, is there any difference between specifying several recipients and specifying a single recipient with several copies? For instance, does it make any difference if you send a message to **curly**, **larry**, and **moe**, or if you send a message to **curly** with copies to **larry** and **moe**?

The answer is, unless you send blind copies, there really is no difference. However, you might want to be aware of the political implications of how you choose to address mail. If **curly** were the boss, for example, it might be better to name him as the main recipient and specify the other two as receiving copies.

From time to time, you may find yourself sending mail to the same group of people repeatedly. In such cases, you can establish what is called an *alias*, a name to represent the group. Each time you send a message to the alias, it will automatically be sent to the members of the group.

For example, say that you define an alias of **executives** to represent the three names, **curly, larry, and moe**. Sending mail to **executives** is the same as sending mail to each of the three userids. Of course, as the need arises, you can change or even delete the alias. How you define and manipulate aliases depends on which mail program you use. You will have to check the documentation for your particular mail program for details.

A similar, but more formal arrangement is the *mailing list*. As the name implies, this is a list of userids who are all to receive mail. Mailing lists can be quite large and, as a rule, will have some person who acts as an administrator.

Within the Internet, there are literally thousands of different mailing lists devoted to different subject areas. For example, there is a mailing list that is used to discuss visual arts. To join a list, you send mail to the administrator. From time to time, messages—usually articles of some type—will be mailed out to everybody on the list. We will discuss mailing lists in detail in Chapter 25.

Automatically Forwarding Your Mail

If you are using the Unix system, there is a way to have all your mail forwarded automatically to another address. For example, if you have several accounts on different computers, you may find it more convenient to have all your mail go to just one computer. Similarly, if you are visiting a place where you have been given a guest computer account, you may decide to temporarily forward your mail to that account.

To forward mail automatically, create a file named **.forward** in your home directory (the **.** character is part of the name). Within this file, put a single line that contains the address to which you want to forward mail. For example, you might use:

```
harley@fuzzball.ucsb.edu
```

From now on, all mail that comes to you on the computer with the **.forward** file will be redirected to the specified address. At any time, you can stop the forwarding by removing the **.forward** file.

Mail Tips and Guidelines

Before we get down to the nitty gritty of reading and sending mail, let's cover a few important tips that might save your life.

If you have never used electronic mail, it will take awhile for you to appreciate how different it is from regular mail or from talking on the telephone.

Because it is so easy to send a quick note anywhere over the Internet, people often forget how permanent such a note can be. As a general rule, don't send messages that you would not want to see a year from now. It is easy for someone to save your messages in a file and then, when you least expect it, dredge them up again.

Hint

When someone does something to make you angry—such as send you a stupid message—resist the temptation to lash out by mailing back something abusive or sarcastic. You will be far better off to save the message and wait a day before replying. Once you mail a message, there is no way to get it back, even if it has not yet been delivered.

(Of course, we realize that nobody will follow this advice, but don't say we didn't warn you.)

Even more important, do not assume that mail is private. Although you may send what you think is a private message, it is easy for the recipient to send a copy to someone else, and you will have no way of knowing what has happened. Similarly, do not assume that any message you have received is private. Someone else may have received a blind copy. When it comes to private matters, such as love letters, you will find that it is more romantic and much safer to follow traditional customs (such as using your employer's Federal Express number).

Another variation that you may encounter is the form letter. One time, we had received a lot of mail overnight and our mailbox (the system file that holds unread mail) had grown quite large. When we read the mail, we noticed that one of the messages was from **root**, which, on Unix systems, is the userid of the system manager. The message informed us that our mail was taking up a lot of space and would we please dispose of it.

Being the perfect users that we are, we immediately sent a polite reply saying that we would take care of the problem right away. (It is always a good idea to respect the wishes of the system manager.) Later, we found out the original letter was automatically generated by a program that, each night, looked for users with large mailboxes.

Imagine what the system manager thought when he received a strange message in which we promised to "take care of the problem" right away. He had not asked us to do anything; the form letter had simply been sent off under the auspices of his userid. To this day, we are still embarrassed at our obvious naivete (so please don't tell anybody).

Another situation in which you might encounter form letters occurs when you are working with a program and the computer goes down unexpectedly. Later, when the computer is restarted, you may be sent a message telling you that some of your data was saved automatically.

A common example is the Unix **vi** text editor. If you are using **vi** to edit a file and the computer goes down, you will later receive a form letter. This message will tell you that your file was saved and will give you instructions for recovering your data.

The last point that we would like you to appreciate is that electronic mail has a lot in common with talking on the phone or in person: it is quick, simple, and usually informal. For this reason, mail messages often have a sense of immediacy missing in conventional correspondence. What is missing, of course, are the inflections and personality of your voice and your body language.

Because of this, you will find that it is easy to get carried away with the informality of the system and insult someone accidentally. What you mean to be humorous might be taken literally and cause offense. For this reason, there is a convention that whenever you write something in jest that might be misunderstood, you should include a *smiley* at the end of the remark.

A smiley is a tiny picture of a smiling face, drawn with punctuation characters, for example:

```
:-)
```

(To see the face, turn your head sideways to the left.)

For example, say that a friend has just sent you a message telling, in great detail, how he met the woman of his dreams the night before at a Unix singles club. If he were to tell you this story in person, you could make a funny remark without risking an insult (wink, wink, nudge, nudge). By mail, however, it is easy to be insulting where no offense was meant.

Thus, if you want to reply in a humorous vein, it is best to include a smiley:

```
Oh yes, I know who you mean...
She'll go home with anyone who has a real IBM computer :-)
```

This ensures that your friend will be able to appreciate the subtle irony of your comment.

We will talk about smileys more in Chapter 10.

Understanding Mail Headers

Mail messages have a standard format consisting of two parts: the *header* and the *body*. The header consists of a number of lines of information at the beginning of the message. The body is the actual text of the message.

Figure 5–1 contains a sample message that we will examine. The format of the header may vary on your system, but the general idea will be the same.

```
From rick@tsi.com Wed Mar 31 14:47:02 1993
Received: from hub.ucsb.edu by engineering.ucsb.edu
        id AA15594 to harley; Wed, 31 Mar 93 13:19:25 PST
Received: from fuzzball (fuzzball.ucsb.edu) by hub.ucsb.edu;
        id AA11868
        sendmail 4.1/UCSB-2.0-sun
        Wed, 31 Mar 93 13:23:58 PST for harley@cs.ucsb.edu
Received: by fuzzball (5.57/UCSB-v2)
        id AA07200; Wed, 31 Mar 93 13:17:40 PST
Received: from tsi.com by ucsd.edu; id AA03169
        sendmail 5.67/UCSD 2.2 sun via SMTP
        Wed, 31 Mar 93 13:23:52 -0800 for harley@fuzzball.ucsb.edu
Received: by sdcc12.UCSD.EDU (4.1/UCSDGENERIC.3)
        id AA25582 to harley@fuzzball.ucsb.edu;
        Wed, 31 Mar 93 13:23:51 PST
Date: Wed, 31 Mar 93 13:23:51 PST
From: rick@tsi.com (Rick Stout)
Message-Id: <9303311840.AA06711@tsi>
X-Mailer: Mail User's Shell (7.1.2 7/11/90)
To: harley@fuzzball.ucsb.edu
Subject: this is the subject
Cc: addie@nipper.com kim@nipper.com
Status: RO

Harley: I have set up an appointment for next Thursday afternoon.
Please let me know if you can attend.

-- Rick
```

Figure 5-1.   *A sample mail message*

The first line of a header will always start with the word **From**. This line shows the address of the userid that sent the message. In our case, the message was sent by **rick@tsi.com**.

The rest of the lines do not necessarily come in the same order. Nor will they always be present. It depends on how the sending and receiving mail systems are configured and what options are set with your particular mail program. At the bare minimum, you will always see an initial **From** line, a **Date** line, and a **Subject** line.

In our example, the initial **From** line is followed by fourteen lines of technical information containing five **Received** statements. These lines show us the path the message took, as well as the times, dates, and what programs were being used at each step of the way. You can usually ignore these lines. However, if you take a moment to look closely, you can see some interesting information. For example, along the way, the message was handled by two different versions of **sendmail** (the daemon that we mentioned earlier that acts as the transport agent).

Now take a look at the fourth **Received** statement. It contains the line:

```
Wed, 31 Mar 93 13:23:52 -0800 for harley@fuzzball.ucsb.edu
```

This is the time and date that the message was received at this particular location. Notice that the time is given as **13:23:52**. This is because the Internet uses a 24–hour time system. The time here represents 1:23 PM (and 52 seconds).

Notice also, the notation **-0800** after the time. This is an important convention that you should understand.

Since the Internet spans the globe, its users are in many different time zones. It is often important to know what time a message was sent, but local time might be confusing. Suppose, for example, that you are in Vienna, Austria, and you receive a message from someone in California. The time on the message is **13:23:52**. How do you know if this is California time or Austrian time?

This same problem arises in many situations. So, as a solution, the Internet has adopted Greenwich Mean Time as a standard. Sometimes you will see this written as GMT. You may also see it referred to as Universal Time or UT (which is the newer, more official name). Whatever you see, just remember that whenever a standard reference time is needed, GMT is used.

The line that we listed above indicates that a local time is being used but that the local time is 8 hours less than GMT. In other words, the message was sent at 1:23 PM local time, which is 9:23 PM GMT.

Aside from GMT or UT, there are other common acronyms that you will see in North America. These names refer to local North American time zones and are shown in Table 5–1.

Now, to return to the header of our sample message, we see that the next line starts with the word **Date**. This shows the time and date that the message was sent.

Acronym	Time Zone
UT	Universal Time (same as GMT)
GMT	Greenwich Mean Time
EST	Eastern Standard Time
EDT	Eastern Daylight Time
CST	Central Standard Time
CDT	Central Daylight Time
MST	Mountain Standard Time
MDT	Mountain Daylight Time
PST	Pacific Standard Time
PDT	Pacific Daylight Time

Table 5-1. *Summary of Time Zone Names*

The following line is another **From** line. This shows extra information about the userid that sent the message. In this case, we can see that this userid is registered to Rick Stout. This means that either Rick or someone using his account sent the message.

Following this is a **Message–id** line that shows the unique identification tag that was assigned to this message. You can ignore this line.

Next, there is an **X–Mailer** line. This shows us the program that the sender used as his user agent. We see here that he is using the Mail User's Shell (Mush). Again, this is information you can ignore.

The **To** line which follows shows us the address to which the message was sent. Of course, this will be your address. If the message was also sent to other people, their addresses will appear on this line as well.

Following this is the **Subject** line. As you will see in Chapter 6, you specify a subject—a short description—each time you compose a message. Whatever you specify is included in the header as the **Subject** line. When you read mail, your mail program will show a summary containing the subject of each message. Thus, you can see at a glance what messages are waiting to be read.

> ## Hint
>
> If you are sending mail to someone who is too busy to pay attention to everything he or she receives, it behooves you to make your subject interesting. Many busy people will ignore mail that looks unimportant.
>
> For example, a subject of "Your office is being demolished tomorrow" will get a faster response than "Upcoming infrastructural modification".

After the **Subject** line, we see the **Cc** line. This shows us the userids that have received copies of the message. In this case, there are two such userids, **addie@nipper.com** and **kim@nipper.com**. The **Cc** designation is the traditional abbreviation for "copy sent to". (Originally—in the days of typewriters—"cc" stood for "carbon copy".) If no copies were sent, this line will be omitted.

Although copies are indicated, blind copies are not. After all, they are secret. So remember, somebody you do not know about may have also received a copy of the message.

The last line in our example is the **Status** line. This is inserted by the mail reading program. Not all programs include such a line. There are two different statuses that you are likely to see. A status of **R** indicates that you are currently reading the message for the first time. A status of **RO** (as in this example) means that the message has already been read and is considered to be old mail.

*FUN TIP: It is common for mail programs to display a **Status** line. However, it is rare to find someone who knows what the different designations mean, because they are not documented anywhere.*

*Here, then, is a great way to make some easy money. Go into a bar where a lot of Internet people hang out. Offer to bet anyone in the place that you can name a mail header line that they don't understand. Lose the first few bets by offering easy lines, like **From**, **Subject**, and **Date**.*

*Then, ask for long odds for one last bet. You can now clean up by asking, "What does a status of **RO** mean?"*

The Difference Between Text and Binary Data

In the next section, we will discuss how you might be able to use the Internet mail system to send and receive all types of data—pictures, sound recordings, and so on—as opposed to simple messages. Before we approach this topic, we need to take a moment to talk about how computer people classify different types of data.

The term *data* refers to any type of information that might be stored or processed by a computer. It is helpful to consider data as being of one of two basic types: text or binary data.

Text consists of ordinary characters: letters, numbers, punctuation, and so on. Special characters, such as the space and tab, are also considered to be text. An example of text is a message that you might type at your keyboard and mail to a friend. Within this chapter, all the data we have looked at is text.

A file that contains such data is called a *text file*. Another name for such a file is an *ASCII file*. (The name comes from the *ASCII code*, a specification that defines how all the various characters are represented as computer data. ASCII stands for "American Standard Code for Information Interchange". We will not go into the details here.) Any data that is not simple text is referred to as *binary* data. An example of binary data is a file that contains a picture. Such a file does not contain characters. Rather, it contains information that represents the many small dots that make up the image. A file that contains binary data is called a *binary file*.

Where does the name "binary" come from? Imagine an image that you might display on your computer screen. The image consists of many tiny dots. For the sake of this example, let us say that each dot is either black or white.

You might wonder, how are such images stored in files? And, is it possible to mail such a file to a friend so he can display the same image on his computer?

A binary file that stores a picture does not contain the actual image in the way that a photograph album might hold a picture. Rather, the binary file stores the data necessary to recreate the image. Here is how it works.

Each individual dot is encoded as one of two numbers, either a 0 or a 1. In our example, a 0 might represent a white dot; a 1 might represent a black dot. In other words, a file that contains an image actually consists of a long string of 0's and 1's. The program that displays the image must be able to read and understand this type of data in order to recreate the picture on your screen.

In computer science terms, an element that can contain either one of two values—such as 0 or 1—is called a *bit*, which is an abbreviation for "binary digit". In technical terms, we can say that our file contains a large number of bits, all of which have a value of either 0 or 1. Each of these bits represents either a black dot or a white dot.

In this context, the word "binary" indicates that only two different values are used. In computer terms, any data that does not consist of characters and must be represented by sequences of bits is called "binary data".

Our explanation is, of course, simplified to the level of a normal human being. Actual computer scientists (and other related fauna) recognize many different types of bit patterns and, hence, many different types of binary data. For instance, to store color pictures, you would need to use the 0's and 1's differently. To store sound recordings, you need to use yet another method of encoding 0's and 1's.

The important idea is that it is relatively simple for a program to work with text data. With binary data, each individual bit is important, and manipulating the data becomes more complex.

All the Internet user agents (mail programs) can send and receive text messages: messages you can type at your keyboard and display on your screen in individual characters. However, many mail programs cannot handle binary data. Those that do use a special protocol that we will discuss in the next section.

Hint

When you transfer data from one computer to another, there will be times when it is necessary for you to know whether you are working with text or binary data. As a general rule, data consisting of ordinary characters is text; anything else is binary data. However, there are times when you may get fooled.

It is obvious that a file that holds a picture or a sound recording is a binary file, since such information cannot be represented by ordinary characters. What you may not know is that many common computer tools—such as word processors and spreadsheet programs—also store data as binary files. Although the information may look like characters when it is on your screen, the program puts in special non-character codes when it stores the data in a file. For example, a word processor will use such codes to indicate italics or boldface words. Thus, files such as word processing documents or spreadsheets are stored in a special format and are properly classified as binary files.

Using MIME to Mail Binary Data

In the last section, we explained that binary files contain data that does not consist of simple text. There is a way to mail such files using the Internet mail system. However, it will only work if the mail programs at each end are set up to handle binary data.

In Chapter 2, we explained that the Internet uses a large family of protocols to ensure that all the different types of computers and programs can work together. Each protocol is a set of rules and specifications that describes how something should be done. For example, SMTP (Simple Mail Transfer Protocol) describes how mail is transported. Using SMTP, Internet hosts can send and receive messages that consist of text.

To enable people to send binary data, another protocol, named *MIME*, was developed. The name stands for *Multipurpose Internet Mail Extensions*. A system using MIME can include binary data along with a regular message. The whole thing is then transported (using SMTP) to the destination computer. At the other end, another MIME system will make the binary data available to the recipient. Of course, everything is automatic, so you don't need to care about any of the details.

In order to send or receive binary data, all you need is a user agent (mail program) that supports MIME. We say this rather glibly, but the fact is, many mail programs cannot use MIME. This is the case for the standard Unix **mail** program that we will discuss in detail in Chapter 6. However, MIME is important, and you will find that modern versions of the newer mail programs do support it.

The way to send binary data is to store it in a file that is combined with a regular text message. We say that you *attach* the file to your message. If your mail program supports MIME, it will have a facility for creating such attachments when you compose a message. Typically, all you need to do is specify the name of the binary file you want to attach. In this context, the data you include is often referred to as *richtext*, the idea being that the data is "richer" than ordinary text.

As you would expect, when you receive a message that contains richtext, you cannot process it properly unless your mail program supports MIME. If so, the mail program will tell you that the message contains an attachment. When you read the message, your program will separate the attachment and store it in a file. If your mail program does not support MIME, you may see the binary data as part of the message, but it will be gibberish.

It is important to understand that the sender and recipient do not have to use the same mail program. All that is necessary is that both programs support MIME.

So, for example, say that you are using a computer system that will let you record a voice note and store it in a file. When you play back the file, you hear your voice. You have a friend elsewhere on the Internet who has the same type of computer. Both of you use mail programs that support the MIME protocol.

To send a voice note to your friend, you start by recording the message and saving it in a binary file. Next, you send a message to your friend, attaching the binary file to the message. When your friend receives the message, he saves the attachment in a file of his own. He can now use his computer to play back that file and hear your voice.

Hint

If you do not have access to MIME, there is another way to send a binary file via the mail system. You can use a program named **uuencode** to encode the binary data as text. You can then mail the text as part of a regular message. At the other end, the recipient can take the text and, using a program named **uudecode**, convert it back to its original binary format. Both **uuencode** and **uudecode** are available on many Internet hosts. These programs are discussed in Chapter 18.

CHAPTER 6

The Standard Unix Mail System Program

In Chapter 5, we looked at the Internet mail system. We saw that Internet computers send and receive messages using a system called SMTP (Simple Mail Transport Protocol).

At each Internet computer, there are two important programs that provide the mail service. The first program, the transport agent, works behind the scenes to provide your mail link to the outside world. The second program, the user agent, is your personal mail program. This is the program that you use to send and receive messages.

Of all the user agents that we mentioned in Chapter 5, the most important is the standard Unix mail program, named **mail**. This program is the only mail program that is available on all Unix systems. Even if you use a different program, you should at least skim the rest of the chapter. The hints and techniques are generally applicable to all mail programs. In particular, you should know about the **from** and **biff** commands.

Hint

No matter what mail program you use, it is a good idea to have some familiarity with the standard Unix **mail** program. It is the only program that you can count on finding on all Unix systems.

Orientation to the Unix mail Program

To start, here is a quick overview of how the **mail** program works. In the following sections, we will discuss sending and receiving mail in more detail.

The **mail** program is used in two ways. To send mail, you type **mail**, followed by one or more addresses to which the message should go. For example, if you want to send a message to **harley@fuzzball.ucsb.edu**, you would enter:

```
mail harley@fuzzball.ucsb.edu
```

Most systems are set up to ask you for the subject and possibly some other information. You can then enter your message, one line at a time.

When you are finished, start a new line and press CTRL-D. (That is, hold down the "CTRL" key and press D.) In Unix, this key is called the **eof** (end of file) key. You use CTRL-D to tell a program that there is no more data to be processed. Once you press CTRL-D, the message will be sent on its way and the **mail** program will end.

To read your mail, enter the simple command:

```
mail
```

If you have messages waiting, **mail** will show you a list and wait for you to enter a command. By using the various commands, you can read and respond to messages as you see fit. When you are finished, you enter the command **q** (for quit) and the **mail** program will end.

If there are no messages, you will see a message telling you that there is nothing for your userid. For example:

```
No mail for harley
```

Specifying Mail Addresses

To send mail, you enter the **mail** command followed by the names of one or more addresses. For example:

```
mail harley@fuzzball.ucsb.edu
mail addie@nipper.com  mschuster  mpeirce
```

The first example sends mail to a single address. The second example sends mail to three addresses.

Let's talk for a minute about the addresses in the second example. The first address (**addie@nipper.com**) is a full address, with a userid and a domain. The second and third addresses only have userids. This is because they are used on the same computer as the person sending the message, so a full address is not necessary. Here is how it works.

Say that your full address is **harley@fuzzball.ucsb.edu**. As a general rule, when you send mail to someone, you can omit the sub–domain names that you have in common. For example, say that you want to send a message to **kim@furface.ucsb.edu**. Since the last two sub–domains (**ucsb** and **edu**) are the same, you can leave them out and send mail to **kim@furface**. Another way to say this is that, because **fuzzball** and **furface** are on the same local network, you need only put in the computer name.

Now, if you are on the exact same computer as someone, the entire domain will be the same and you only have to use the userid. Thus, if you logged in to **fuzzball** and you want to send a message to someone else at **fuzzball**, you need specify their userid only. For example, if your full address is **harley@fuzzball.ucsb.edu** and you want to send a message to someone whose full address is **mschuster@fuzzball.ucsb.edu**, all of the following commands will work.

```
mail mschuster@fuzzball.ucsb.edu
mail mschuster@fuzzball
mail mschuster
```

For more information about Internet addresses, see Chapter 4.

Hint

If you send mail to an invalid address, the message will be sent back to you with an explanatory note. This may take awhile, as the message may have to make its way to the destination computer before it is discovered that the address is wrong.

In such cases, we say that the message has *bounced*. For example, you might call a friend and say, "Can you please tell me your address again. I sent mail to the address you gave me yesterday, but the message bounced."

Composing a Message

Once you enter the **mail** command, you will see the following:

```
Subject:
```

You are being asked to enter the subject of the message. Type anything you want and press RETURN. Whatever you type will become part of the **Subject** line in the header.

Hint

Keep your subject descriptions short. As a rule of thumb, do not use more than 40 characters.

Depending on how your program is set up, you may now see the line:

```
Cc:
```

You are being asked if you want to send copies of the message to anyone. If so, type the address (or addresses) and press RETURN. Otherwise, just press RETURN.

You can now enter your message, one line at a time. When you are finished, press RETURN (to move to a new line) and then press CTRL–D. As we explained earlier, CTRL–D is the **eof** (end of file) key. It indicates to the program that there is no more data.

Figure 6-1 shows a sample message being sent. The words that the user typed are shown in boldface. (The % character in the first line is the Unix shell prompt: the signal that the shell is waiting for a command.) Notice that, at the end of the message, after we press CTRL–D, **mail** responds by displaying **EOT**. This stands for "End of Transmission". It is confirmation that the message is complete and is ready to send on its way.

If you are composing a message and you change your mind, you can cancel it by pressing CTRL–C twice in a row. On Unix systems, CTRL–C is called the **intr** (interrupt) key and is used to abort a program that is running. When you want to abort a message, **mail** makes you press CTRL–C twice. The first time you press it, you will see:

```
(Interrupt--one more to kill letter)
```

This is a safeguard for your own protection. If you pressed CTRL–C by mistake, you can ignore the message and continue typing. However, if you really did want to abort the message, press CTRL–C again to confirm.

```
% mail kim@furface.ucsb.edu
Subject: An important visit
Cc:

Kim:
    The great scholar T.L.Nipper will be visiting next week
and giving a guest lecture. Do you want me to get tickets?

-- Harley
CTRL-D
EOT
```

Figure 6-1.   *Sending a message*

The Tilde Escape Commands

As you are typing a message, there are certain commands that you can send to the **mail** program. In order to make sure that these commands are not mixed up with what you are typing, they all start with a ~ (tilde) character. When a single character is used in this way—to signal that what follows is to be treated in a special manner—it is called an *escape character*. For this reason, these commands are called the *tilde escapes*. When you enter a tilde escape, it must always be at the beginning of a line.

Table 6-1 summarizes the tilde escapes. In this table, we use the standard Unix command syntax. That is, we describe the technical format of the commands by using the standard Unix conventions. The words in italics represent names you must fill in. When you see "...", it means you can fill in more than one name.

Command	Description
~?	help: display a summary of the tilde escapes
~b *address*...	add addresses to "Blind copy" line
~c *address*...	add addresses to "Copy" line
~d	read in the contents of `dead.letter` file
~e	invoke the text editor
~f *messages*	(follow–up) read in messages
~h	edit all the header lines
~m *messages*	read in messages, shift right one tab
~p	display (print) current message
~q	quit (same as pressing CTRL-C twice)
~r *file*	read in the contents of a file
~s *subject*	change the "Subject" line
~t*address*...	add new addresses to the "To" line
~v	invoke an alternative editor (usually same as ~e)
~w *file*	write current message to a file
~! *command*	execute shell command, then return to message
~| *command*	pipe current message through a filter

Table 6-1. *Summary of the* **mail** *Tilde Escapes*

We won't go over each of the tilde escapes. You should take some time and experiment for yourself. However, we do want to give some examples of the most important ones.

First, the **~?** command will display a summary of all the tilde escapes, much like the one in Table 6-1.

The most important command is **~e**. This will start a text editor to allow you to make changes before you send off the message. As we explained in the previous section, you compose a message one line at a time. While you are typing a line, you can backspace and make corrections. However, you cannot backspace to change something on a previous line.

The solution is to use **~e**. This will start your default text editor. Normally, this will be either **vi**—the standard Unix text editor—or **emacs**. Now you have full control to modify any line and add as much text as you want. (Of course, you have to know how to use the editor.) When you are finished, simply quit the editor in the usual fashion. You will now be back in **mail** where you can press CTRL–D to send your message.

Another useful tilde escape is **~h**. This tells **mail** that you want to modify the header lines of your message. You will be prompted for several items in turn. You can change them as you wish. If you don't want to make any changes to a particular item, just press RETURN.

First, you will see:

```
To:
```

You can now change the address to where the message should be sent. Next, you will see:

```
Subject:
```

You can now change the subject of the message. After this, you will see:

```
Cc:
```

You can specify the addresses of any people to whom you want to send copies of the message. Finally you will see:

```
Bcc:
```

Here you can specify if you want to send blind copies to anyone. (Remember, if you don't want to change anything, just press RETURN.)

After you have finished with each item, you will be returned to where you were and you can continue typing your message.

If you would like to change one of the header lines directly, instead of looking at each one in turn, you can use one of the following commands: **~t** (To:), **~s** (Subject:), **~c** (Cc:), or **~b** (Bcc:). Simply specify the new information as part of the command. For example, to change the subject of your message to "Important new item", use:

```
~s Important new item
```

If you decide to send a blind copy of the message to a friend whose address is **kim@furface.ucsb.edu**, enter:

```
~b kim@furface.ucsb.edu
```

The next two tilde escapes come in handy when you are responding to mail that you have just read. We will discuss the details later in the chapter, but for now we will explain that after you read a message, **mail** makes it easy for you to send a reply to the userid that sent the message. When you do, it is often helpful to include a copy of the original message as part of your reply.

To do so, use the **~f** (follow–up) tilde escape. If you use **~f** by itself, **mail** will insert the message you have just read into the text of your reply. If you want to include a different message, just specify its number. (Each message that you read will have a number.) For example, to include message #4, enter:

```
~f 4
```

The **~m** command will do the same thing, except that each line of the included message will be prefaced with a tab character. This way, when the person reads your reply, the included text will be indented.

A nice way to reply to a message is to include all or part of the message in your reply, and insert your remarks within the old message. For example, say that you receive the following message:

```
Do you want to go to a movie later today?

If so, would you like to see the film version of
the book 'A Student's Guide to Unix'?
```

You can use ~f to insert this message into your reply. Next, use ~e to start your text editor. Edit the message and add whatever comments you want. The reply might look like this:

```
> Do you want to go to a movie later today?

Yes, that would be wonderful

> If so, would you like to see the film version of
> the new book 'A Student's Guide to Unix'?

I have already seen it. It is great.
Actually, I wouldn't mind seeing it again.
```

Hint

Here is how to make your replies look nice. First, use ~f to insert the original message into your reply. Next, use ~e to bring the whole thing into your text editor. Now delete all the header lines from the message along with anything extraneous that you do not wish to quote in your reply.

Now, for whatever lines remain, place the two characters "> " (greater-than, followed by a space) at the beginning of each line. (Some mail programs will do this for you, but **mail** won't, so you have to do it yourself.) If you are using the **vi** editor, the following command will insert these characters in front of every line in the editing buffer:

```
:%s/^/> /
```

Now move to each place where you want to reply to a specific point and insert your reply. When you are finished, quit the editor and press CTRL-D to send the message on its way.

Another tilde escape that will include text as part of a new message is ~r (read). This will insert the contents of an existing file into your message. Just type ~r followed by the name of the file. For example, to read in a file named **memo**, enter:

```
~r memo
```

The last two commands we want to mention are ~! and ~ |. These are used to access Unix commands while you are composing a message. The ~! tilde escape will pause the **mail** program, execute the Unix command that you specify, and then return to **mail**. For example, to pause briefly in order to display the time and date, enter:

```
~! date
```

With ~ |, you also specify the name of a Unix command. In this case, **mail** will send the contents of your message to this command to be processed. Your message will be replaced by the output of the command. For example, say that you have typed in a message that consists of a long list of items. Now that the list is typed, you want to use the Unix **sort** command to sort the list before you send the message. Enter:

```
~| sort
```

Your message will be replaced by the output of the **sort** command, that is, by all the lines in sorted order. If, at this point, you want to make a few small changes, you can use ~**e** to start the text editor.

Hint

Unix has a command named **fmt** that was developed specifically to format mail messages. The **fmt** command will read data and, line by line, format it to be as close to 72 characters per line as possible. The command will not change spacing at the beginning of a line or in between words. It will also leave blank lines untouched. In other words, **fmt** will format a message nicely as long as you use a blank line to separate paragraphs.

To format an entire message before you send it, enter the tilde escape:

```
~| fmt
```

If you want to check the results, you can use ~**p** (print) to display the message, or ~**e** to start the text editor.

A Shortcut When Sending Mail

We have described how to send mail: you enter the **mail** command, along with one or more addresses. You then specify a subject, after which you compose your message. When you are finished, you press CTRL-D to send the message.

To speed things up, there are several shortcuts you can take by using variations of the **mail** command. First, you can specify the subject as part of the command. To do so, type **mail**, followed by **–s**, followed by the subject. For example:

```
mail -s Meeting harley@fuzzball.ucsb.edu
```

This command specifies that the subject line in the header should be "Meeting". If you want a more complex subject line, with spaces or punctuation, enclose it in single quotes. For example:

```
mail -s 'Meeting next week' harley@fuzzball.ucsb.edu
```

When you specify a message in this way, **mail** will not prompt you for a subject. If you want to change the subject, you can use the ~s or ~h tilde escape.

Hint

In Unix terminology, the **–s** is called an *option* or a *switch*. According to Unix rules, all options must come at the beginning of the command. Thus, make sure you specify the addresses after the **–s**. For example, it will cause a mistake if you enter:

```
mail harley@fuzzball.ucsb.edu -s 'Meeting for next week'
```

The **mail** program will think that you are trying to send a message to three different addresses—**harley@fuzzball.ucsb.edu**, **–s**, and **'Meeting for next week'**—and everything will go awry.

Another shortcut you can take is to prepare the message ahead of time and store it in a file. You can then tell **mail** to take the message directly from the file.

To do so, type a < (less-than) character at the end of the command, followed by the name of the file.

For example, say that you have a message stored in a file named **memo**. You want to send the message to **harley@fuzzball.ucsb.edu** with a subject of "Meeting next week". Enter the following command:

```
mail -s 'Meeting next week' harley@fuzzball.ucsb.edu < memo
```

What we are doing is telling Unix that the input for the **mail** program should come from a file rather than from the keyboard. In Unix terminology, we are "redirecting the standard input" from a file named **memo**. In this case, the entire message will be read from a file, so you do not need to press CTRL–D to indicate the end of the message.

How Can You Tell If a Message Has Been Read?

In general, there is no way to tell if a message has been read by the recipient. However, there is a command that will work some of the time.

This command is called **finger**, and we will discuss it in detail in Chapter 8. Basically, **finger** will display public information about any userid on the Internet. (There are sites, however, that prohibit **finger** access to their systems. This may be for security reasons, or simply to lower the number of network requests that come to that system.)

Some versions of **finger** will tell you if there is mail waiting for that userid that has not been read. For example, let's say that you send mail to a friend whose address is **harley@fuzzball.ucsb.edu**. You can display information about him by using:

```
finger harley@fuzzball.ucsb.edu
```

If the **finger** system on his computer displays mail information, the output will contain something like this:

```
New mail received Thu Apr  1 12:39:22 1993;
unread since Thu Apr  1 08:16:11 1993
```

(We will look at the full output of the **finger** command in Chapter 8.)

Hint

Most systems have a command that will check to see what mail is waiting for you. For example, Unix has the **from** command (which we will meet in the next section). Such commands display a list of messages along with their subjects.

Once a person has used the **from** command, it looks to **finger** as if he or she has already read the mail, even though this is not the case. More precisely, **finger** looks at the last time the file that holds the mail was accessed. Since **from** must access this file, it may erroneously look as if the mail has been read, when all that has happened is that the person has displayed the list of unread messages.

Keep this in mind before you fire off an angry message saying, "I know that you have already read my last message. Why haven't you replied?"

Checking to See If Mail Has Arrived: from, biff

Aside from using **mail**, Unix systems have two other commands that will tell you if mail is waiting. The first command is **from**. Simply enter:

```
from
```

You will see a summary of all the messages that are waiting. Here is an example:

```
From addie@nipper.com Wed Mar 31  23:49:50 1993
From kim@fuzzball Thu Apr 1 06:42:25 1993
From mschuster@music.tuwien.ac.at Thu Apr 1 06:50:48 1993
From addie@nipper.com Thu Apr 1 08:22:28 1993
From mpeirce@shamrock.tcd.ie Thu Apr 1 08:23:45 1993
```

As you can see, **from** displays the first line of the header for each message. If there are no messages waiting, **from** will not display anything.

The second command, named **biff**, works in the background, monitoring the file that holds your mail. Whenever a new message arrives while you are logged

Hint

When messages arrive for you, they are stored in a file until you can read them. Each user has his own such file. The **from** command works by looking in this file to see if there are any messages.

On some systems, the mail file is actually removed when there are no more messages in it. When a new message arrives, the file is re-created. Thus, when you have no mail, the **from** command will find not an empty file, but a missing file. You will see a warning similar to:

```
Can't open /usr/spool/mail/harley
```

Literally, this means that **from** cannot find your mail file. What it really means is that you have no mail.

in, **biff** will let you know automatically. That way, you do not have to keep typing **from** to check for new mail.

The **biff** command sets an on/off switch. When the switch is on, **biff** will keep track of your mail. When the switch is off, **biff** will not check for new mail. Most of the time, it is handy to keep **biff** on. However, if you are working on something special, and you do not want to be interrupted whenever mail arrives, you may want **biff** off.

To turn **biff** on, enter the command followed by **y** (for "yes"):

```
biff y
```

To turn off **biff**, use **n** (for "no"):

```
biff n
```

If you want to check the current status, enter the command name by itself:

```
biff
```

You will see either:

```
is y
```

or:

  is n

If you are using an X Window system (see Chapter 2), there is an X client

---

Hint

When you log in to a Unix system, you may see a message telling you that you have mail. This message is coming from your shell (the command processor), not from **biff**.

---

named **xbiff** that you may want to use. The **xbiff** program draws a picture of a mailbox for you. When mail arrives, the flag on the mailbox will go up. To put the flag back down, move the mouse pointer to the box and click.

*FUN TIP: The command **biff** was named after a dog named Biff who belongs to a programmer named Heidi Stettner. In the summer of 1980, Heidi was working for a professor of Computer Science at the University of California at Berkeley as she waited to enter graduate school. At the time, Heidi used to bring Biff to her office, not far from where the Unix programmers were working on the new version of BSD (Berkeley Unix).*

*There is an apocryphal story that the **biff** command was named after Biff the Dog because he used to bark whenever the mailman came. That is not true. The real story is that Biff was a universal favorite among the professors and students. Biff had his picture on the bulletin board that had photos of all the Computer Science graduate students. (In fact, Biff even got a B in a compiler class, but that is another story.)*

*One day, one of the graduate students who was working on Unix—John K. Foderero—decided to create a command and name it after Biff. He came up with the idea to write a program to check for mail. Ever since, Unix people around the world have been using **biff** to check their mail.*

(By the way, as we write this book, Biff the Dog is retired and still living in Berkeley, California, with Heidi. Like many of the elderly Unix pioneers, Biff spends most of his time sleeping.)

Reading Your Mail: An Overview

To read your messages, enter the **mail** command by itself:

```
mail
```

If there is mail waiting for you, **mail** will display information about each message, in the form of short one–line descriptions called header summaries. Here is an example:

```
*Mail version 2.18 5/19/83.  Type ? for help.
"/usr/spool/mail/harley": 5 messages 5 unread
>U 1 addie@nipper.com Wed Mar 31   23:49   35/1204 "Re: seeing a movie"
 U 2 kim@fuzzball Thu Apr ? 06:42   138/5518 "Re: having dinner"
 U 3 mschuster@music.tuwien.ac.at Thu Apr ? 06:50   46/1592 "Students Guide To Unix"
 U 4 addie@nipper.com Thu Apr ? 08:22   48/1595 "Something interesting about TLN"
 U 5 mpeirce@shamrock.tcd.ie Thu Apr ? 08:23   343/16810 "Schedule for Chapter 5"
```

Each line in the summary represents one message that is waiting for you. If a header summary is very long, it may be broken onto more than one line on your screen. As you can see, the messages are numbered and, in this example, we have five messages.

After the number, you see the name of the userid that sent the message. For example, message #1 was sent by **addie@nipper.com**. Next is the date and time. After this are two numbers, separated by a / (slash) character. This tells you the size of the message: the number of lines and the number of characters. In our example, message #1 consists of 35 lines and has a total of 1204 characters. Finally, you see the subject of the message.

Once the header summaries are displayed, you will see an **&** (ampersand) character displayed on a line by itself:

```
&
```

This is a *prompt*, a signal from **mail** that it is waiting for you to enter a command. You can now enter one command after another to read and process your mail. When you are finished, enter the **q** (quit) command to stop the program.

When you see the **&** prompt, there are many commands that you can enter. These are summarized in Tables 6-2 and 6-3. Table 6-2 shows all the important commands, listed in alphabetical order. Table 6-3 shows the same commands organized by function.

Abbreviation	Full Name	Description
?	—	display command summary
!	—	execute a single shell command
	—	display the next message
-	—	display the previous message
RETURN	—	display the next message
number	—	display message #*number*
d	delete	delete message
dp	—	delete current msg, display next msg
e	edit	use text editor on message
h	headers	display header summaries
l	list	list names of all available commands
m	mail	send new message to specified userid
n	next	display the next message
p	print	display (print) messages
pre	preserve	keep messages in system mailbox
q	quit	quit `mail`
r	reply	reply to sender & all other recipients
R	Reply	reply to sender only
s	save	save messages to specified file
sh	shell	pause `mail`, start a new shell
to	top	display top few lines of message
u	undelete	undelete previously–deleted messages
w	write	same as s, only do not save header
x	exit	quit `mail`, neglect any changes
z	—	show next set of header summaries
z-	—	show previous set of header summaries

Note: On some systems, the meaning of **r** *and* **R** *is reversed.*

Table 6-2.   *Summary of Important* **mail** *Commands: Alphabetical Order*

Abbreviation	Full Name	Description
Stopping mail		
q	quit	quit mail
x	exit	quit mail, neglect changes
Help		
?	—	display summary of commands
l	list	list names of commands
Header		
sh	headers	display header summaries
z	—	show next header summaries
z-	—	show previous summaries
Displaying Messages		
+	—	display the next message
-	—	display the previous message
RETURN	—	display the next message
number	—	display message #number
n	next	display the next message
p	print	display (print) messages
to	top	display top lines of messages
Replying and Mailing		
m	mail	compose a new message
r	reply	reply to sender & recipients
R	Reply	reply to sender only
Processing a Message		
d	delete	delete messages
dp	—	delete current msg, display next msg
e	edit	use text editor on messages
pre	preserve	keep messages in system mailbox
s	save	save messages to specified file
u	undelete	undelete previously–deleted msg
w	write	same as s, do not save header

Table 6-3.   *Summary of Important* **mail** *Commands: Organized by Function*

Abbreviation	Full Name	Description
Shell Commands		
!	—	execute a single shell command
sh	shell	pause `mail`, start a new shell

*Note: On some systems, the meaning of **r** and **R** is reversed.*

Table 6-3.   *Summary of Important **mail** Commands: Organized by Function (continued)*

The Current Message and Message Lists

In a moment, we will cover the most important **mail** commands. These are the commands that you will use to read and process your messages. Before we talk about these commands, we must discuss the idea of "message lists".

Take another look at the header summary that we showed you in the last section.

```
Mail version 2.18 5/19/83.  Type ? for help.
"/usr/spool/mail/harley": 5 messages 5 unread
>U  1 addie@nipper.com Wed Mar 31  23:49   35/1204 "Re: seeing a movie"
 U  2 kim@fuzzball Thu Apr ? 06:42  138/5518 "Re: having dinner"
 U  3 mschuster@music.tuwien.ac.at Thu Apr ? 06:50  46/1592 "Students Guide To Unix"
 U  4 addie@nipper.com Thu Apr ? 08:22  48/1595 "Something interesting about TLN"
 U  5 mpeirce@shamrock.tcd.ie Thu Apr ? 08:23  343/16810 "Schedule for Chapter 5"
```

Notice that message #1 is marked with a > (greater-than) character at the far left. The > character indicates the **current message**. The current message is the default when you enter any command that does not specify a message explicitly.

For example, as we will see in a moment, the **d** command deletes messages. The command:

```
d 3-5
```

will delete messages #3 through #5. If you do not specify a message number, the command will act on the current message. For instance, if you enter:

d

it will delete the current message (in our case, #1).

When you start reading your mail, the current message starts at message #1. As you read one message after another, the current message will change to the last one that was read. Thus, at any time, any command that does not specify a message explicitly will act on the last message that you read.

Sometimes, though, you will want to specify a message. For most commands, you can use one or more message numbers. For instance, in the second last example, we specified that we wanted to delete messages #3 through #5. The way that you specify messages is by using a *message list*.

A message list can include a combination of one or more numbers, a range of numbers, or characters that have special meanings. Table 6-4 shows the various ways in which you can specify a message list.

Most of the time, you will use commands that specify only one number or no message list at all. However, being able to specify a message list can come in handy, and you should be familiar with the idea and how it works. Table 6-5 shows a number of message list examples using the **d** (delete) command.

Specification	Meaning
. [a period]	the current message
n	message number n
n-m	all messages from n to m inclusive
^ [a circumflex]	the first message
$ [a dollar sign]	the last message
* [an asterisk]	all messages
userid	all messages from specified userid
/pattern	all messages containing pattern in subject
:n	all new messages
:o	all old messages
:r	all messages that have been read
:u	all messages that are still unread

Table 6-4.   *Summary of the Various Ways to Specify a Message List*

Command	Meaning
d	delete the current message
d .	delete the current message
d 3	delete message
d 3-5	delete messages 3 though 5 inclusive
d ^	delete the first message
d $	delete the last message
d *	delete all messages
d harley	delete all the messages from userid harley
d /hello	delete all the messages with "hello" in subject
d :n	delete all the new messages
d :o	delete all the old messages
d :r	delete all the messages that have been read
d :u	delete all the messages that are still unread

Table 6-5. *Examples of Specifying a Message List*

Hint

Within a message list, **:r** refers to the messages that you have already read. The **:u** specification refers to unread messages. How do you keep track of which are which?

Take another look at the header summary at the beginning of this section. Notice that there is a **U** character at the left–hand side of each header line. This indicates that the message is as yet unread. Once you read a message, the U will disappear.

Displaying a Message

The **mail** program is designed to make it easy to display one message after another. Here is how it works. When you start reading, the current message is set to #1. To read the first message, just press RETURN. To read the next message, press RETURN again. In other words, you can read all your messages, one after another, just by pressing RETURN.

If you want to jump right to a particular message, you can use the **p** (print) command. For example, to display message #4, enter:

```
p 4
```

For convenience, if you enter a number by itself, **mail** assumes that you want to display that message. So an easier way to display message #4 is to enter:

```
4
```

In addition, you can enter + (the plus character) to display the next message and – (the minus character) to display the previous message.

When a message is too long to fit on one screen, **mail** will call upon a *paging program* to display the message for you. A paging program displays data one screenful at a time. At the bottom of each screen, the program displays a note showing you there is more to come. Which paging program is used depends on how your system is set up. The three most common paging programs are called **more**, **pg**, and **less**.

Hint

When you are using **more** or **less**, you advance to the next screenful of data by pressing the SPACEBAR. With **pg**, you must press the RETURN key.

Saving and Deleting Messages

Once you have read a message, you must dispose of it in some way. You have two choices. You can delete it, or you can save it to a file. To delete a message, use the **d** command. To save a message, use the **s** command. If you enter the command without a message list, it will act upon the current message, the one you have just read.

For example, say that you have just read a message. To delete it, simply enter:

```
d
```

Of course, you can use the **d** command with a message list. For example, to delete message #2 and messages #5 through #7, enter:

```
d 2 5-7
```

If you change your mind about deleting a message, you can undelete it by using the **u** command. However, you must do so before you quit the program. For example, to undelete message #6, enter:

```
u 6
```

When you save a message, you must specify the name of the file. For example, to save the message you have just read to a file named **important**, enter:

```
s important
```

If you save a message to a file that already exists, **mail** will append the new data to the end of the file. If the file does not exist, **mail** will create it for you automatically.

Hint

If you want to save messages from a number of people, use files named after the people. For example, every time you read a message from someone named Harley, you can save it by entering:

```
s harley
```

This keeps all your correspondence nicely cataloged.

Each time you send a message to Harley, send yourself a copy. You can then save the copy to the **harley** file. In this way, you will be able to save all of your messages—sent and received—pertaining to a particular person.

When you quit the **mail** program (by entering the **q** command), **mail** will remove all the deleted and saved messages from your mailbox. In addition, **mail** will save any messages that you have read, but not deleted or saved, to a file named **mbox**. Messages that you have not yet read are left alone.

Why is **mail** so anxious to dispose of your messages? When messages arrive, they are kept in a file, called your *mailbox*, that is part of the system area of the disk. On many systems, the disk storage used by your mailbox is not counted against your personal quota. However, when **mail** dumps messages into a file, it is stored in one of your directories. (In Unix, files are kept in directories.) The storage space is now yours.

Hint

On some systems, especially those for which you pay real money, the storage space used by unread messages in your mailbox is billed to you.

If you have accidentally made too many deletions, or have mucked up your messages in some other way, you can quit **mail** by using the **x** (exit) command instead of **q**. This tells **mail** to ignore any changes that were made during the current session.

If you do not have time to dispose of some of the messages that you have read, you can use the **pre** (preserve) command before you quit. This tells **mail** not to move certain messages to the **mbox** file. For example, to keep messages #2 through #6 in your mailbox, enter:

```
pre 2-6
```

To keep all of the messages in your mailbox, use:

```
pre *
```

Hint

Normally, you read the messages that are being kept in your system mailbox. However, there will be times when you may want to read messages that are stored in one of your own personal files. For example, you may have saved a number of messages from a friend named Harley in a file named **harley**. Or the **mail** program may have saved some of your messages to a file named **mbox** (as we described above).

In such cases, use the **–f** (file) option when you enter the mail command. After the **–f**, type the name of the file. The **mail** program will read messages from this file. For example, to read messages that are stored in a file named **harley**, enter:

```
mail -f harley
```

To read messages stored in a file named **mbox**, use:

```
mail -f mbox
```

Replying to a Message

To reply to a message, use the **r** command. The **mail** program will automatically use the address of the person who sent the message. You can now type your message, one line at a time, just as we described earlier. And, since you are composing a message, you can use all the tilde escapes. In particular, you can use the **~m** command to include the original message within your response. (See the discussion earlier in the chapter.)

Here is an example. You have just read a message that says:

```
Do you want to go to a movie tonight?--Kim
```

To reply, you enter:

```
r
```

You can now start typing your message:

```
Yes, I would love to go to a movie,
as soon as I finish writing Chapter 6.
--Harley
```

After typing the last line of the message, press CTRL-D to send it on its way.

If copies of the original message were sent to other userids, using the **r** command will automatically send your reply to everyone. In some cases, this may be what you want. However, there are times when you may want to reply only to the person who sent the message. For example, if you are replying to a memo that was sent to thirty different people, you will probably not want your reply to be sent to all thirty people. In such cases, use the **R** command, instead of **r**. This tells **mail** to reply only to the sender of the message.

Hint

Once you start receiving electronic mail, it is easy to end up with a lot of messages in your mailbox (or in your **mbox** file). In order to keep things from getting out of hand, dispose of a message as soon as you have read it. Follow these three simple guidelines:

- As soon as you have read a message, decide if you want to reply and do so right away.
- Before you move on to the next message, either delete or save the one you just read.
- When it doubt, throw it out.

Composing New Messages and Forwarding Mail

As you are reading messages, you may decide to send a new message to someone. You do not have to stop the **mail** program and enter a new command. Rather, you can use the **m** command to start a new message.

For example, say that you are reading your messages and you realize that you need to send a quick note to a friend whose address is **harley@fuzzball.ucsb.edu**. Simply enter:

```
m harley@fuzzball.ucsb.edu
```

You will be asked to enter the subject of the message. After doing so, you can type your message in the regular manner. Once you press CTRL–D to send the new message, you will be back where you left off, reading your messages.

Some mail programs have a command to forward a message to another person. The **mail** program does not, but it is not too hard to do so yourself.

First, use the **m** command to start a new message to the person. Second, use the **~f** tilde escape to include the message you have just read in the new message.

For example, say that you have just read message #3 and you want to forward it to a friend whose address is **kim@nipper.com**. Start by entering the command to send a new message:

```
m kim@nipper.com
```

You will see:

```
Subject:
```

Type the subject and press RETURN. Now enter the command to copy the current message into the new one:

```
~f
```

You will see:

```
Interpolating: 3(continue)
```

This tells you that **mail** has copied message #3 to your new message. Now press CTRL–D. You will see:

```
EOT
```

The new message is on its way, and you can go back to reading your mail.

Hint

When you forward a message in this way, you may want to use the ~e tilde escape to start the text editor after you use ~f. This will allow you to remove the old header lines before you send the message.

CHAPTER 7

Using Telnet to Connect to Remote Hosts

One of the wonderful things about the Internet is that it is as easy to use a computer on the other side of the world as it is to use a computer across the hall. In this chapter, we will explain how to connect to a remote computer by using the Telnet service.

We will discuss how there are two principal ways to use a remote computer. First, you can log in to any Internet host on which you have an account. For a Unix computer, for instance, this means that you must have a userid and password. Once you log in, you can use the computer in the regular manner.

Second, there are many Internet computers that offer some type of public Telnet access. Such systems are available to anyone and do not usually require a password. Many of the resources in our catalog are accessed in just this way.

As you will see, Telnet is so transparent that it is easy to forget that you are separated from the remote computer. The main limitation you will find is that, when Internet network traffic is high, the response from a faraway computer may slow down slightly. Still, it is common for experienced users to work with several different Internet computers, moving from one to another smoothly and easily.

An Overview of Telnet Connections

In Chapter 2, we explained that the Internet offers a large number of services that are based on standard protocols. The service that allows you to connect to a remote Internet host is called *Telnet*.

To use Telnet, you run a special program, called **telnet**, on your computer. This program uses the Internet to connect to the computer you specify. Once the connection is made, **telnet** acts as an intermediary between you and the other computer. Everything you type on your keyboard is passed on to the other computer. Everything that the other computer displays is sent to your computer, where it appears on your screen. The end result is that your keyboard and screen seem to be connected directly to the other computer.

In Telnet terminology, your computer is called the *local* computer. The other computer, the one to which the **telnet** program connects, is called the *remote* computer. We use these terms no matter how far away the other computer actually is: whether it is across the world or in the same room.

As we mentioned in Chapter 2, we often refer to Internet computers as hosts. So, using Telnet terminology, we can say that the job of the **telnet** program is to connect your local computer to a remote Internet host.

One last comment on terminology: We often use the word *telnet* as a verb. For example, say that you are visiting a friend in a distant city. You might ask him, "Can I use your computer for a minute? I want to telnet to my computer to read my mail."

Starting the telnet Program

To make a remote connection, you use the **telnet** program. There are two ways to start this program. In this section, we will show you how it is done most of the time. In the next section, we will discuss an alternate method.

To start **telnet**, enter the name of the command followed by the address of the remote host to which you want to connect. For example, say that you want to connect to a computer named **fuzzball**, whose full address is **fuzzball.ucsb.edu**. Enter:

```
telnet fuzzball.ucsb.edu
```

If you are connecting to a computer on your local network, you can usually get by with just the name of the computer, instead of the full address. For example:

```
telnet fuzzball
```

Hint

As we explained in Chapter 4, all Internet hosts have an official address known as an IP address. This address consists of several numbers separated by periods. For example, **128.54.16.1** is the official IP address of the computer whose standard address is **ucsd.edu**.

Some systems have trouble dealing with certain standard addresses. If you encounter such a problem with **telnet**, try using the IP address. For example, either of the following commands will connect to the same host:

```
telnet ucsd.edu
telnet 128.54.16.1
```

For more information about IP addresses and Internet addresses in general, see Chapter 4.

When the **telnet** program starts, it will initiate a connection to the remote host that you specified. As **telnet** is waiting for a response, you will see:

```
Trying...
```

or a similar message. Once the connection is made—which might take a few moments if the host is far away—you will see a message like the following:

```
Connected to fuzzball.ucsb.edu.
Escape character is '^]'.
```

(We will explain the reference to an "escape character" later.)

If, for some reason, **telnet** is unable to make the connection, you will see a message telling you that the host is unknown. For example, say that you want to connect to the remote host **nipper.com**. However, you mistakenly enter:

```
telnet nippet.com
```

You will see:

```
nippet.com: unknown host
telnet>
```

At this point you can either specify another host name or quit the program. We will explain how to do this in the next section.

Hint

The **telnet** message "unknown host" can be misleading. There are many reasons why **telnet** might not be able to make a remote connection. The three most common are:

- You spelled the address of the computer wrong.

- The remote computer is temporarily unavailable.

- You specified the name of a computer that is not on the Internet.

Another problem that may arise is that your local network may, for some reason, not be able to make connections to certain parts of the Internet. One reason is that a particular host may be off-limits for security reasons. Another reason is that some locations just do not have a way to connect to other locations.

For example, one of our friends in Ireland complains that he is unable to connect to computers in Australia. In such cases, **telnet** will display a message like:

```
Host is unreachable
```

If this happens to you, double-check that you are entering your **telnet** command correctly. You can also ask your system manager if there is some trick that you don't understand about making such a connection. However, if the remote host is really unreachable from where you are, there is not much you can do about it. (The best advice we could give to our Irish friend is that many people never connect to Australian computers and still live full, rich lives.)

Once **telnet** makes a connection, you will be interacting with the remote host. At this point, most hosts display some type of informative message, usually identifying the computer. If you are expected to log in, you will see the standard prompt. For example, if you have connected to a remote Unix computer, you will see:

```
login:
```

You can now log in in the regular manner. Type your userid and press RETURN. You will then see:

```
Password:
```

Now enter your password and press RETURN again. (Note: When you type your password, it will not be displayed. This prevents someone else from finding out your password by watching you log in.)

We mentioned earlier that some remote hosts are set up to offer a public service. In such cases, you will not have to use a password when you log in. For example, in Chapter 17, we will discuss an important Internet resource called Archie servers. When you log in to a public Archie server, you use a userid of **archie**. Once you enter this userid, the Archie program will start automatically.

Some public hosts are even more automatic. As soon as you use **telnet** to connect, the remote program starts by itself; you do not even need to log in.

When you are finished working with the remote computer, all you need to do is log out in the regular manner. The connection will break and **telnet** will stop automatically.

A Second Way to Start telnet

In the previous section, we mentioned that there are two ways to start the **telnet** program. The first way is to enter the **telnet** command along with the address of the remote host. For example:

```
telnet fuzzball.ucsb.edu
```

The second way is to start **telnet** without specifying a host. Simply enter:

```
telnet
```

The program will start, but will not make a connection. You will see:

```
telnet>
```

This is the **telnet** prompt. It means that the program has started and is waiting for you to enter a command. To make a connection to a remote host, type **open**, followed by the address of the host. For example:

```
open fuzzball.ucsb.edu
```

The connection will be made just as if you had specified it when you entered the **telnet** command.

In the previous section, we gave an example in which a **telnet** command had a bad address. In the example, the remote host was named **nipper.com**, but we mistakenly entered:

```
telnet nippet.com
```

What happens in such a case is that **telnet** tries to make the connection. When it can't, it gives up and displays its prompt, waiting for you to enter a command. In this case, you would see:

```
nippet.com: unknown host
telnet>
```

You can now enter an **open** command with the correct address:

```
open nipper.com
```

If this address doesn't work, you can try another one. If you decide to give up, enter:

```
quit
```

This will stop the **telnet** program.

Summary of Starting and Stopping telnet

There are two ways to start **telnet**. Either enter the command with the address of a remote host:

```
telnet fuzzball.ucsb.edu
```

or enter the command by itself:

```
telnet
```

and then, at the **telnet>** prompt, enter an **open** command:

```
open fuzzball.ucsb.edu
```

There are two ways to stop **telnet**. If you are connected to a remote host, log out in the regular manner and **telnet** will stop automatically. Otherwise, at the **telnet>** prompt, enter the quit command:

```
quit
```

Connecting to a Specific Port Number

Within the Internet, there are numerous hosts that offer Telnet access to public services. Indeed, if you look in the catalog, you will see that a large number of the items are accessed via Telnet.

Many of these hosts require that you specify a particular *port number* when you make the connection. The port number identifies the type of service that you are requesting. Here is an example.

There is a host at the University of Michigan that provides American and Canadian weather reports. The name of this computer is **downwind.sprl.umich.edu**. When you connect to this computer, you must specify a port number of 3000. This tells the system that you want to use the weather service.

All you have to do is enter the port number at the end of the **telnet** command:

```
telnet downwind.sprl.umich.edu 3000
```

(Be sure to leave a space between the address and the port number.) If you connect by using an **open** command at a **telnet>** prompt, type the port number in the same way:

```
open downwind.sprl.umich.edu 3000
```

When you connect to a remote host that uses a port number, the program that you want will usually start automatically, and you will not have to log in. When you end the program, the connection will be broken and **telnet** will stop by itself.

Hint

In this chapter, we describe the **telnet** command as it is used on most Unix systems. On certain other types of systems, the format is slightly different. For example, on some VAX computers running the VMS operating system, you need to put **/port=** in front of the port number:

```
telnet downwind.sprl.umich.edu /port=3000
```

If you have any problems on your computer, the best idea is to check the local documentation for **telnet**.

More About Port Numbers

In computer terminology, the term *port* refers to a connection between two devices or systems. For example, you might attach a printer to a port on the back of your computer. In a Unix system, we say that each terminal is connected to its own port on the host.

This same idea is used in Internet terminology. In Chapter 2, we explained that the Internet uses a protocol called TCP (Transmission Control Protocol) to transfer data from one host to another. Whenever TCP connects one Internet host to another, a port number is used to identify the type of connection. In fact, there is an Internet organization—called the Internet Assigned Number Authority—that maintains the official list of port numbers (of which there are many), and makes sure that unique numbers are allotted when necessary.

By default, ordinary Telnet connections are made using port number 23. In other words, when you do not specify a port number, the **telnet** program automatically connects using port 23. Thus, the following two commands will make the same type of connection:

```
telnet fuzzball.ucsb.edu
telnet fuzzball.ucsb.edu 23
```

In order for you to connect to a remote host, it must be running a program that is prepared to communicate over the port that you use. When such a program is waiting for a connection, we say that it is *listening* on that port.

Thus, we can say that for a host to support regular Telnet connections, it must have a program that is listening on port 23. Another way to put it is that when you use **telnet** to make a regular connection to a remote host, it contacts that host and checks to see if there is a program listening on port 23.

It is only when you want to use a Telnet connection over a different port that you need to specify an actual port number. Many Internet systems use different port numbers to offer some type of special service.

In the previous section, we connected to a remote host named **downwind.sprl.umich.edu** using port number 3000. We did this in order to use a special-purpose program that displays weather reports. In order for this connection to work, the weather report program must be running on this computer and it must be listening on port number 3000. In most cases, a host can support more than one connection to a specific port at the same time. This particular weather server, for example, will support up to 100 simultaneous users.

The telnet Escape Character: CTRL-]

As you interact with a remote host, there is a way to put your work on hold and enter commands directly to **telnet**. For example, if you are having problems with the remote host, you can pause your work session, return to **telnet**, and enter the **quit** command.

The way to do this is to press a special key combination CTRL-]. That is, hold down the CTRL key and press] (the right square bracket). When you press this key, it sends a signal to **telnet** to pause the remote connection and display the prompt:

```
telnet>
```

You can now enter any **telnet** command that you want. (We will discuss the most important ones in a moment.) With some commands, **telnet** will automatically resume the connection after carrying out the command. Otherwise, you can resume the connection at any time by pressing RETURN at the **telnet>** prompt.

The technical term for a key like CTRL-] is an *escape character*. Many programs allow you to use an escape character to request a special service or to indicate that what follows is to be interpreted differently.

In the world of Unix, there is a convention that CTRL keys are often indicated by using the ^ (circumflex) character. For example, CTRL-C would be written as **^C**.

Now we can make sense out of the message that appears whenever **telnet** makes a remote connection:

```
Escape character is '^]'.
```

You are being reminded that the Telnet escape character is CTRL-]. It is possible to change this to another character, but there is usually no reason to do so.

Using telnet Commands

Any time you are at the **telnet>** prompt, there are a number of different commands that you can use. In this section, we will go over the most important ones. Before we do, we will remind you that if instead of entering a command you just press RETURN, **telnet** will resume the remote connection.

To display a summary of the various **telnet** commands, you can enter the **?** character:

```
?
```

Here is a typical summary:

```
Commands may be abbreviated. Commands are:

close     close current connection
display   display operating parameters
mode      try to enter line-by-line or character-at-a-time mode
open      connect to a site
quit      exit telnet
send      transmit special characters ('send ?' for more)
set       set operating parameters ('set ?' for more)
status    print status information
toggle    toggle operating parameters ('toggle ?' for more)
z         suspend telnet
?         print help information
```

Out of all these commands, the most important ones are **?**, **open**, **close**, **quit**, and **z**.

The **open** command tells **telnet** to make a connection to a remote computer. Enter **open** followed by the address of the computer. For example:

```
open fuzzball.ucsb.edu
```

The **close** command terminates a remote connection without stopping the **telnet** program. Here is how this might come in handy.

Let's say that you are working with a remote host and something goes wrong. For some reason, the host seems to be ignoring your commands. No matter what you type, nothing happens, and you can't even log out. One solution is to press CTRL-], wait for the **telnet>** prompt, and then enter the **close** command. You can now reestablish a connection to the same host. You can, of course, also connect to a different host.

The **quit** command stops **telnet**. If a remote connection is active, **telnet** will terminate it.

> ### Hint
>
> Before you close a connection or quit **telnet**, remember to log out from the remote host. Most hosts will log you out automatically when the connection drops, but it is better to do so yourself. This way, you can ensure that whatever program you were using is terminated properly and all your data was saved.

All the other commands (except **z**, which we will discuss in the next section) are less important, and you will probably never need to use them. For the most part, they control various technical aspects of the communication session that you almost always ignore. If you want to display a summary of the **send**, **set**, or **toggle** command, enter the command followed by a **?** character. For example:

```
send ?
```

> ### Hint
>
> If you want more technical information about **telnet**, see the documentation for your particular system. If you are using a Unix computer, you can display the **telnet** entry in the online manual by using the command:
>
>
>
> ```
> man telnet
> ```

Job Control

Unix systems support a facility called *job control* that allows you to pause a program, work with another program, and then return to the first one. We won't go into all the Unix details here, but we will explain how, if your system has job control, **telnet** will cooperate.

At the **telnet** prompt, you can enter the **z** command. (We will explain the name in a moment.) This tells **telnet** to pause itself and return you to the shell (the program that reads and processes your commands). This allows you to enter regular commands in the middle of a remote session.

The **z** command will only work if your shell supports job control. If you are using a modern shell, such as the C-Shell or the Korn shell, this will be the case. If you are using the older Bourne shell, there will be no job control and the **z** command will not work. In fact, **telnet** itself may freeze.

The program with which you are currently working is said to be in the *foreground*. When you put a program on hold, we say that it is in the *background*.

When you enter the **z** command, **telnet** will put itself in the background and return you to your local shell. You can now enter as many regular Unix commands as you want. For example, you can check your mail, display the time and date, or whatever. When you want to resume your remote connection, enter the command:

```
fg
```

This tells Unix to reactivate (move to the foreground) the last program that was put on hold. You can now resume your remote connection.

You might wonder, why is the **telnet** job control command named **z**? The answer is that, on Unix systems that support job control, you can move your current program into the background by pressing the **susp** (suspend) key. Usually, this key is CTRL-Z. Thus, the **telnet** command is named after the Unix **susp** key.

Hint

Many hosts will automatically log you out if nothing happens for a specified period of time. For example, a system might log you out if you have not typed anything for 15 minutes. Remember this when you use the **z** command to put a **telnet** session on hold. If you do not resume the session before it times out, you may be disconnected automatically.

Figure 7-1 shows an example of how this all works. The commands that we entered are printed in boldface.

At the beginning of the example, we are logged in to a computer named **nipper**. You can see the shell prompt:

```
nipper%
```

At this prompt, we enter a **telnet** command to connect to the remote computer whose address is **fuzzball.ucsb.edu**:

```
nipper% telnet fuzzball.ucsb.edu

Trying...
Connected to fuzzball.ucsb.edu
Escape character is '^]'.

ULTRIX V4.2A (Rev. 47) (fuzzball)
Welcome to the Fuzzball System.

login: harley
Password:

Last login: Sun Apr 18 00:09:58 from nipper.com
fuzzball% date
Sun Apr 18 00:24:54 CDT 1993
fuzzball% CTRL-]
telnet> z
Stopped

nipper% mail
No mail for harley
nipper% fg
telnet fuzzball.ucsb.edu

fuzzball% logout
Connection closed by foreign host.

nipper%
```

Figure 7-1.   *Using Job Control with **telnet***

```
telnet fuzzball.ucsb.edu
```

Once the connection is made, **fuzzball** displays the standard Unix login prompt and we log in with a userid of **harley**. Notice that, for security reasons, the password is not displayed as we type it. After the login is complete, the remote host displays some information followed by a shell prompt:

```
fuzzball%
```

We now enter the **date** command to display the time and date.

At this point, we decide to return to **nipper** temporarily and check our mail. First, we press CTRL-]. This puts the remote connection on hold and returns us to **telnet**. You can see the prompt:

```
telnet>
```

Next, we enter the **z** command. This places the **telnet** program in the background. We now see the **nipper** shell prompt. We use the **mail** command (see Chapter 6) and find out that there are no messages.

We now enter the **fg** command. The shell responds by displaying the last command that was placed in the background (in this case, it was the **telnet** command). The shell then brings this program back into the foreground, which automatically resumes the remote connection. Once again, we can see the **fuzzball** shell prompt.

Finally, we enter the **logout** command to log out from **fuzzball**. The remote connection is closed automatically, and **telnet** stops. We are left where we started, at the **nipper** shell prompt.

Hint

On Unix systems, the shell is the program that reads and processes your commands. There are several Unix shells and you may be able to choose the one you want.

When a shell is ready to accept a command, it displays a prompt. If you have an account on more than one computer, it is a good idea to customize your prompts so that they contain the name of the computer. (We won't go into the details here.) In this way, your shell prompt is always a handy reminder of which system you are using.

Traditionally, the last character of a prompt is used to show the type of shell you are using. The **%** character (as in our example) indicates the C-Shell. The **$** character indicates the Korn shell or Bourne shell.

CHAPTER 8

Fingering the World

With so many millions of people using the Internet, it is important to be able to find out something about a particular person. The *Finger* service allows us to do just that. With a simple command, we can find out certain public information about another Internet user.

In this chapter, we will explain how the Finger service works. We will show how to ask for information and how to control the public information about yourself. We will also explain how this same service is used to offer various other types of information.

135

What Is the Finger Service?

In Chapter 2, we explained that a client/server system consists of two programs: a client that requests a particular resource and a server that provides that resource. In many cases, the client and server programs run on different computers, allowing you to access a remote host. The Finger service is a client/server system that provides three main types of information.

First, you can display certain public information about any user on an Internet host. What you see will vary from host to host. Indeed, some hosts will not display any user information at all, for security reasons. In general, though, most hosts will share basic information about their users.

All you need to know is which host a person uses, and either their userid, first name, or last name. With this information, you can use the Finger service to display some or all of the following:

- the person's userid

- the person's full name

- if the userid is currently logged in

- the last time someone logged in using that userid

- whether or not mail has been read

- a phone number

- an office location

- information that the person has specifically prepared for the general public (for example, a list of office hours)

Hint

If you know the name of a person, but you do not know which host he or she uses, you may be able to find them by using the techniques described in Chapter 22.

A second way to use the Finger service is to check to see who is currently using an Internet host. You can display a summary that shows some or all of the following information for each userid that is logged in:

- userid

- full name

- when the userid logged in

- how long it's been since there was activity on that terminal

- phone number and office information

- name of computer or terminal server from which the person logged in

The third use for Finger is to communicate with certain Internet hosts that have been set up to offer other, more esoteric types of information. For example, one host will return current information about earthquake activity.

Most of the time, the basic reason to use Finger is to display public information about a particular person. To use this service, you run the **finger** program. For example, to display information about the person whose userid is **harley** and who uses a computer whose address is **fuzzball.ucsb.edu**, you can enter:

```
finger harley@fuzzball.ucsb.edu
```

Notice that we are using a standard Internet address. As we explained in Chapter 4, the Internet has one main addressing system and we use it for everything. Thus, you use the same address to finger someone as you would to send mail.

Here is a typical example of output from such a request. For now, don't worry about the details. Just take a quick look to get an idea of how the service works.

```
[fuzzball.ucsb.edu]
Login name: harley              In real life: Harley Hahn
Phone: 202-456-1414
Directory: /usr/harley          Shell: /bin/csh
Last login Wed Apr 21 21:20 on ttyp4
No unread mail
Project: Writing books about the Internet and Unix.
Plan:
You can find me in my office most days.
If you have any questions about the Internet catalog,
please send them to Rick Stout.
```

The **finger** program is the client. It acts on your behalf and sends a request for information to the appropriate host, in this case, **fuzzball.ucsb.edu**. On the remote host, a Finger server is waiting for such requests. In our example, the Finger server on **fuzzball.ucsb.edu** honored the request and sent back the information displayed above.

On most Internet systems, some type of Finger server is always running, ready to handle any request for information that may arrive. On Unix systems, the server is usually named **fingerd**, "finger daemon" (although you don't really need to know the name). As we explained in Chapter 5, a daemon is a program that runs in the background, usually to provide a service of general interest.

When we talk about the Finger service, it is common to use the word *finger* as a verb. For example, suppose you are at a party and you meet the woman (or man) of your dreams. As you leave, you might say, "By the way, if you need my phone number, just finger me." (Yes, Internet people really do talk this way.)

Displaying Information About a Person

To display information about a person, you need to know the address of his or her computer. You also need either a userid, a first name, or a last name. Enter the **finger** command using the following format:

```
finger name@address
```

For example, say that you want to display information about a friend of yours named Ben Dover who works at the Hong Fatt Noodle Company. His userid is **bdover**, and the address of his computer is **noodle.com**. You can use any of the following commands:

```
finger bdover@noodle.com
finger ben@noodle.com
finger dover@noodle.com
```

The first command specifies the userid. This is the best thing to use if you know it. If not, you can try the first name or last name as we did in the second and third commands. However, if you do so, for example, by using **ben**, there will often be more than one person with the same name. In such cases, the Finger server will display information about all the users with that name.

When you specify a first or last name, it does not matter if you mix upper- and lowercase letters. For example, the last two commands are the same as:

```
finger Ben@noodle.com
finger DOVER@noodle.com
```

Hint

When specifying a name in a **finger** command, the rule is that first and last names can mix upper- and lowercase letters, but userids must be exact. Since virtually all userids are lowercase, the best idea is to always use lowercase for everything.

If you are fingering someone who uses a computer on your local network, you usually need to specify only the name of the computer. For example, say that you are logged into a computer named **fuzzball.ucsb.edu**. You want to finger a friend named Michael Schuster, who has a userid of **mschuster** on a local computer named **furface.ucsb.edu**. You can use any of the following commands:

```
finger mschuster@furface
finger michael@furface
finger schuster@furface
```

If you want to finger someone on the same computer that you are using, you may omit the address entirely. For example, say that you want to finger someone named Artie Choke who has a userid of **achoke** on your computer. You can enter any of the following commands:

```
finger achoke
finger artie
finger choke
```

If a Finger server cannot find the person you want, it will not say so directly. Rather, it will respond with a message that contains no personal information. For example, say that on the **noodle.com** host, there is no person that can be identified by the name of "addie". However, you enter the command:

```
finger addie@noodle.com
```

The remote Finger server will return a message that looks like this:

```
[noodle.com]
Login name: addie            In real life: ???
```

Don't be fooled. This message does not mean that there is an "addie", but that no information is available. The message really means that, as far as this Finger server is concerned, "addie" does not exist.

Understanding Finger Information

In this section, we will take a look at the type of information Finger will show you about a person. From system to system, there is some variation as to what you will see. What we have here is a typical example.

To start, you enter the following command:

```
finger harley@fuzzball.ucsb.edu
```

As soon as it confirms that the address is valid, **finger** will display the name of the remote host:

```
[fuzzball.ucsb.edu]
```

A moment later, you will see the output from the remote Finger server:

```
Login name: harley              In real life: Harley Hahn
Phone: 202-456-1414
Directory: /usr/harley          Shell: /bin/csh
Last login Wed Apr 21 21:20 on ttyp4
No unread mail
Project: Writing books about the Internet and Unix.
Plan:
You can find me in my office most days.
If you have any questions about the Internet catalog,
please send them to Rick Stout.
```

Let's go over this output line by line.

The first line shows basic information about the user. In this case, we see the userid **harley** (euphemistically referred to as **Login name**) and the full name **Harley Hahn**.

There is an important variation of this line that you might encounter. It looks like this:

```
Login name: harley  (messages off)  In real life: Harley Hahn
```

The notation **(messages off)** indicates that the person has chosen to not allow other people to send messages to his screen. Unix users can set this preference by using the **mesg** command.

In Chapter 19, we will discuss the **talk** program, which enables you to have a conversation with another Internet user. When you enter a **talk** command, it sends a message to the other person telling him that you want to start a conversation. If you have trouble getting through, try fingering the person. It may be that he or she has set "messages off" for privacy. If that is the case, the person will not get your **talk** message.

The second line of the Finger output is straightforward: it shows the person's telephone number.

The next line shows technical Unix information about the person's account. The directory is the name of the person's home directory, the place that has been allotted to that user to store his files. The next piece of information is the name of the user's shell. In this case, we see that this person uses the C-Shell (if we recognize that **csh** is the name of the C-Shell program).

Hint

You can ignore the technical Unix information that is part of a Finger report. There is probably some set of circumstances in which it is important to know the name of someone's home directory and shell. However, in over 250 years of using the Internet, we have yet to come across such a situation.

Following the technical Unix information, we see the last time that the person logged in. In this case:

```
Last login Wed Apr 21 21:20 on ttyp4
```

This tells us that the last login was on Wednesday, April 21 at 9:20 PM. (Remember, Unix and the Internet use a 24-hour clock.) The last part of the line tells us that the person's terminal was connected to the port named **ttyp4** on his computer. Most of the time, this is another piece of information you can ignore.

If the person logged in from another computer (using Telnet, for instance) or from a terminal server, you may see the name of that machine. For example:

```
Last login Tue Apr 20 20:33 on ttyp0 from unix1.tcd.ie
```

If you finger someone who is currently logged in, you will see a message similar to the following:

```
On since Apr 26 09:21:12 on ttyp1 from engrserv
```

In this example, the person logged in on April 26 at 9:21 AM and is still logged in. He or she is using a terminal that is connected to port **ttyp1**. The connection was made via **engrserv** (which happens to be a terminal server).

Hint

Before you use the **talk** command to have a conversation, finger the person. This way, you check if he or she is logged in and is accepting messages.

After the time of the last login, we see some information about mail. In this case, we see:

```
No unread mail
```

This tells us that the person has read and disposed of all his mail. Such information can come in handy when you send a message to someone and you want to know if he or she has read it yet. However, there are some limitations you should know about. We will discuss this whole topic in more detail later.

After the mail status, we see the final two items, called **Project** and **Plan**. These two pieces of information are important because you can change them whenever you want to display public information. We will discuss this in a moment.

Is There a Way to Find Out If Someone Has Fingered You?

No.

The .plan and .project Files

In the previous section, we looked at an example of output from a **finger** command. At the end of the output, we saw the following:

```
Project: Writing books about the Internet and Unix.
Plan:
You can find me in my office most days.
If you have any questions about the Internet catalog,
please send them to Rick Stout.
```

Here is where this information comes from and why it is important.

The original **finger** command was developed at the University of California at Berkeley as part of their Unix project. (The author, by the way, was Earl T. Cohen.) At the time, the command was designed to display information that was appropriate to Berkeley's academic environment.

One of the built-in facilities allowed professors and students to tell everyone what projects they were working on. For example, a professor might describe his research interests. A graduate student might describe her dissertation topic.

Another facility allowed people to show their current plans. For instance, a professor might show his office hours and when he planned to be out of town at conferences.

The way it works is that the Finger server looks for two special files named **.project** and **.plan** in your home directory. (In Unix, each user has a home directory for storing his or her files.) If the **.project** file is present, Finger will display the first line. If the **.plan** file is present, Finger will display the entire file.

In Unix, a file that begins with a . character is called a *dotfile*. When we talk about such files, we pronounce the . character as "dot". For example, we might ask someone, "Have you changed your dot-plan file to reflect your new office hours?"

The significance of dotfiles has to do with the Unix command that lists the names of your files. You use this command—named **ls**—to keep track of your files. Now, certain files, like the ones that hold your project and plan information, are needed by most people. It is a bother to see these same names every time you list your files.

To solve this problem, the **ls** command will not display file names that begin with a **.** character unless you use a special option. Thus, you can have files with names such as **.project** and **.plan** and not have to look at them every time you list your files. Unix uses many different dotfiles, although we won't go into the details here.

All you have to do is create these files in your home directory and put in whatever information you want. Remember, Finger will display the entire **.plan** file, but only the first line of the **.project** file. If Finger cannot find a **.plan** file, it will display the message:

```
No Plan.
```

(However, there is no such message for **.project**.)

This message can be puzzling if you don't know exactly what it means. Many people do not create a **.plan** file. When you finger such people, you will see the standard public information followed by **No Plan.**, as if to imply that this person is wandering aimlessly through life (although this may well be the case).

How do you create your own **.project** and **.plan** files? You need to be able to use a text editor program. The standard Unix text editor is named **vi**. Another common editor is Emacs. In addition, there are a number of other such programs that you will find. Unfortunately, the details of using a text editor are well beyond the scope of this book.

We strongly recommend that you do learn to use a text editor. For example, it is helpful to be able to compose mail or to create articles to send to a Usenet discussion group.

Hint

When you start learning Unix, you will encounter people who will tell you that the **vi** editor (or Emacs) is difficult to learn. Some people will even tell you to stay away from these programs.

Don't listen to such people; they are just being ignorant. They are also being shortsighted. It is impossible to use the Internet fully without being able to edit your own files. Take it from us, you do need to learn to use a text editor. Yes, it may take a little practice, but you will be glad you put in the effort.

In the meantime, ask someone who does know how to use a text editor to help you prepare your **.project** and **.plan** files.

*FUN TIP: You can put whatever information you want in your **.project** and **.plan** files. For example, if you are having a party, you can put directions for getting to your home in your **.plan** file. You can then mail an invitation to your guests, telling them, "If you need directions, just finger me." You will find that people use their **.plan** files to hold all types of imaginative items: jokes, poems, comments on life, drawings made up of computer characters, and so on.*

Interpreting Information About Mail

In a previous section, we mentioned that a Finger server may tell you something about a person's mail. In this section, we will discuss the type of information that Finger might show you and how to interpret what you see.

First off, you should understand that the output of a **finger** command depends on the Finger server at the remote host. Thus, you will get different types of output depending on which host you finger. Some Finger servers are happy to display mail information. Other systems won't even mention it.

As we explained in Chapter 6, each userid's mail is kept in its own file called a mailbox. When you finger a user, the Finger server can examine the user's mailbox. The server can check the last time the mailbox was accessed and compare it to the last time that new mail arrived. Here are some of the messages that you might see from a Finger server.

If the mailbox is empty, you will see:

```
No unread mail
```

If the mailbox is not empty, there are three different messages you might see. If you see something like:

```
Unread mail since Wed Apr 21 20:46:13 1993
```

it means that the mailbox has not been accessed since the mail arrived. If you see:

```
Mail last read Wed Apr 21 20:46:13 1993
```

it means that the mailbox has been accessed since the mail arrived. Finally, if you see:

```
New mail received Wed Apr 21 20:46:13 1993;
unread since Wed Apr 21 20:32:32 1993
```

it means that the mailbox has been accessed, but new mail has arrived since then.

The problem is that it is possible to access your mailbox without reading any mail, which leads to misleading Finger messages.

How can this be? In Chapter 6, we explained that you can use the **from** program to check if mail is waiting for you. The **from** program does not show you your mail: it merely displays a one-line summary of each message. However, to do so, it must access your mailbox. Thus, whenever you check your mail with **from** (or a similar program), the access time on your mailbox is updated, and the Finger server will jump to the conclusion that you have read your mail.

On some systems, it is possible to specify a list of commands that are executed automatically whenever you log in. Some people use this facility to have a program like **from** display a mail summary each time they log in. So be careful: what a Finger server tells you about someone's mail may be misleading.

Here is a real-life example of how you might get into trouble.

You are conducting delicate diplomatic negotiations between the Syldavians and the Bordurians regarding joint military exercises to be held on St. Vladimir's Day (which is tomorrow). These two countries have a long history of conflict, so the utmost tact is necessary.

At 18:00 (6:00 PM), you send a confidential message to the Bordurian ambassador. You wait impatiently for an answer. Finally, at 22:00 (10:00 PM), you decide to finger the ambassador to see what is happening. You enter:

```
finger musstler@kurvi-tasch.bo
```

As part of the output, you see:

```
Mail last read Wed Apr 21 20:46:13 1993
```

Aha, you say, the ambassador read his mail at 20:46 (8:46 PM), which was over an hour ago. Obviously, he has decided not to respond to my message.

You inform the Syldavians who quickly conclude that the Bordurians must be planning a preemptive strike. Hoping to beat the Bordurians at their own game, they immediately declare war.

Later, you find out that the Bordurian ambassador had not actually seen your message. When he logged in, a **from** command had automatically accessed his mailbox. However, at the time you fingered him, he was busy looking at the jokes in a Usenet discussion group and had not yet got around to reading his mail.

Imagine your embarrassment.

Fingering a Remote Host

So far, all our examples have specified the name and address of a particular user. Finger responds by displaying information about that user. Another way to use Finger is to specify only the address. The Finger server at that remote host will display a summary of all the users who are logged in.

The format of such a command is:

```
finger @address
```

For example:

```
finger @fuzzball.ucsb.edu
```

Hint

When you finger a remote host, remember to put the @ character before the address. If you leave out this character, Finger will think that you are referring to a user on your own computer. For example, if you enter:

```
finger fuzzball.ucsb.edu
```

it looks as if you are fingering a user named **fuzzball.ucsb.edu** on your own computer. You will see a message like this:

```
Login name: fuzzball.ucsb.edu       In real life: ???
```

If you want to finger a computer that is on the same local network as your computer, you can usually abbreviate the address. For example, say that you are logged in to **fuzzball.ucsb.edu** and you want to finger the computer named **furface.ucsb.edu**. You can probably use:

```
finger @furface
```

If you want to finger your own computer, just omit the address:

```
finger
```

When you finger a remote host, the format of the output may vary, depending on how the host's Finger service is implemented. Let's take a look at two typical examples of what you might see. Here is the first one:

```
Login       Name            TTY  Idle    When        Office
pmason      Perry Mason     *p0          Fri 08:32   Law Building 260
dstreet     Della Street     p1      2d  Wed 08:55   Law Building 261
hburger     Hamilton Burger  p2      10  Fri 09:16   City Hall 3rd Fl
atragg      Arthur Tragg     p5    3:05  Fri 09:05   City Hall 5th Fl
```

The first two columns show each person's userid (**Login**) and name. The third column shows the port to which the user's terminal is connected. (In Unix, "TTY" is often used as an abbreviation for "terminal".) The important thing about this column is that if the person has chosen to disallow messages, you will see an * (asterisk) character. This is the case with the userid **pmason**.

The next two columns give you vague information about the person's activity. The **When** column shows the day and time that the person logged in. The **Idle** column shows how long it has been since there was some activity on that terminal. If the entry is blank, it means that the person is actively using his or her terminal. Otherwise, you will see the amount of time that the terminal has been idle. You may see a number of minutes (such as **10**), a number of hours and minutes (**3:05**), or even a number of days (**2d**).

Finally, the last column shows general information under the heading of **Office**. Such information varies because, on most systems, people can change this whenever they want.

Here is a second example of output from fingering a remote host. This example shows another common format that you may see:

```
Login       Name            TTY  Idle    When        Where
freddy      Frederick Bean   p3          Mon 16:09   pigpen.bean.com
jinx        Jinx T. Cat     *p1    2:08  Mon 12:47
wiggins     Mrs. Wiggins     p4       5  Mon 10:47   cowbarn.bean.com
```

The main difference between the two examples is in the rightmost column. In this case, the **Where** column shows how the person has logged in. If the entry is empty, the person is using a terminal that is connected directly to the host. Otherwise, the person connected via the computer or terminal server whose name is given. For a discussion of such connections, see Chapter 3.

Finally, we will mention that if you finger a host that no one is using, you will see a message like:

```
No one logged in.
```

Changing Your Personal Finger Information

We have already explained how you can use your **.project** and **.plan** files to display whatever you want. On some systems, you are also allowed to change some of the other information that Finger gives out on your behalf.

If you have a Unix system, try using the **chfn** (change finger) command:

```
chfn
```

If this command is enabled on your system, it will prompt you for information.

Here is an example. Say that your userid is **harley**. You enter the **chfn** command, and you see the following:

```
Changing finger information for harley.
Default values are printed inside of '[]'.
To accept the default, type <return>.
To have a blank entry, type the word 'none'.

Name [Harley Hahn]:
```

The name in square brackets is the current value. If you want to change it, simply type a new name and press RETURN. Otherwise, just press RETURN.

On some systems, all you can change is your name. On other systems, you will be prompted to enter further information, such as a room number, an office phone number, and a home phone number. If you do not want to specify one of these items, simply enter the word **none**. Out of courtesy, you should at least allow people to know your name.

> ## Hint
>
> Some systems are set up so that you cannot modify your own Finger information. If this is the case on your computer, you will have to ask your system manager to make changes.

A Guide to Safe Fingering

So far, we have described how the Finger service will provide you with all kinds of information about Internet users. However, there are a few limitations and warnings you should understand.

First, not all Internet hosts provide a Finger service. For the most part, this service is offered only by Unix hosts, and there are many non-Unix computers on the Internet. Although there are a few non-Unix hosts that have a Finger service, it is unusual. If you finger a host and nothing seems to happen, it is likely that you have reached a non-Unix computer, such as a PC or Macintosh, or a VAX running VMS.

Second, many Unix hosts will not respond to Finger requests from outside their local network. This may be for security reasons or simply to cut down on network traffic. For example, let's say you try to finger someone who uses the host named **microsoft.com**:

```
finger billg@microsoft.com
```

You will see:

```
[microsoft.com]
connect: Connection refused
```

This means that the Finger server on the remote host is refusing to connect to an outside host. Alternatively, some Finger servers will accept such connections, but will not return any information.

The second limitation is that not all remote hosts that support Finger will display all possible information. For example, some hosts will not show the contents of **.project** and **.plan** files to outsiders.

Perhaps most important is that you should not always believe what you see. As we explained earlier, it is possible for users on some systems to change their public Finger information, even their full names.

Hint

If you are unsuccessful fingering someone by using a real name, it may be because he or she has changed this information. This is a fairly common phenomenon among students whose sense of humor has not yet fully matured.

For example, if you finger a university computer to ask about a userid named **ugrad-56**, and you see that the real name is "James T. Kirk", you might be a tad suspicious.

The final limitation we want to mention is one that we discussed earlier in the chapter. It is possible that Finger will tell you that someone has read his or her mail, when all he or she has done is display a summary. Don't jump to conclusions.

Special-Purpose Finger Services

So far, we have looked at the ways in which you can use Finger to display information about people. A number of Internet hosts use this same service to offer other kinds of specialized information.

If you look through the catalog, you will see that some of the entries tell you to finger a special userid and address to get certain information.

In case you are interested in experimenting, here are a few of the special-purpose Finger services that were available at the time we wrote this chapter.

To display recent baseball scores:

```
finger jtchern@sandstorm.berkeley.edu
```

To display a summary of recent auroral activity (the Northern Lights):

```
finger aurora@xi.uleth.ca
```

To display chart listings of the top songs as ranked by *Billboard* magazine:

```
finger buckmr@rpi.edu
```

To display information about recent earthquake activity around the world:

```
finger quake@geophys.washington.edu
```

To display daily press releases from NASA (National Aeronautics and Space Administration):

```
finger nasanews@space.mit.edu
```

To display a seasonal hurricane forecast for the Atlantic region:

```
finger forecast@typhoon.atmos.colostate.edu
```

Coke Servers

Perhaps the most unexpected use of Finger is called a *coke server*. This is a Finger service that is used to display information about a non-traditional device, such as a vending machine. The first coke server was set up in the mid-1970s for the Computer Science department at Carnegie-Mellon University.

At the time, the main terminal room housed a Coca-Cola machine that was frequently used by programmers to replenish their natural stores of caffeine. A recent expansion of the department had placed some of the programmers' offices on a different floor from this room. To their consternation, they would often make the long trip to the machine only to find that there was no Coke. Or, almost as bad, they would find Coke that had been in the machine only a short time and was still warm.

Their solution was to install switches in the machine to sense how many bottles of soda were available and to keep track of how long each bottle had been cooling.

To allow people to check the machine from a distance, a special Finger server was set up. When you fingered a particular userid (**coke**), it would display the status of the contents of the machine.

In the early 1980s, this system was discontinued. However, in recent years, it has been resurrected and the term "coke server" is now used to denote any similar system, whether or not it provides soft drink information. Carnegie-Mellon inhabitants can now use a coke server to check not only a soft drink machine, but a candy machine as well.

If you would like to try this for yourself, there are three ways to finger the CMU coke server. First, to display basic information about the state of the soft drink machine, enter:

```
finger @coke.elab.cs.cmu.edu
```

The output will look like this:

```
Coke Server Ver 0.99 2-26-93
Coke:           Cold:    7   Warm:    0          Buttons
Diet coke:      Cold:   14   Warm:    0         C: COLD
Sprite:         Cold:    1   Warm:    0   C: EMPTY    D: COLD
                                          C: COLD     D: COLD
                                          C: COLD     D: COLD
                                                      C: COLD
                                                      S: COLD
```

For basic information about the candy (M&M) machine, you can finger the **mnm** userid:

```
finger mnm@coke.elab.cs.cmu.edu
```

The output looks like:

```
Coke Server Ver 0.99 2-26-93
Information may not be correct, use at your own risk.
The M&M machine is 46% full.
```

Finally, to see a pictorial representation of both machines, you can enter:

```
finger bargraph@coke.elab.cs.cmu.edu
```

Here is some sample output:

```
Coke Server Ver 0.99 2-26-93
M&M information may not be correct, use at your own risk.
    M & M                     Buttons
    /-----\           C: CC......................
    |     |        C: ............  D: CCCCC........
    |     |        C: C...........  D: CCCC.........
    |*****|        C: CC..........  D: CCCCC........
    |*****|                         C: CC...........
    \-----/                      S: C............
       |          Key:
       |          0 = warm; 9 = 90% cold; C = cold; . = empty
       |          Leftmost soda/pop will be dispensed next
    ---^---
```

If you want to look at a couple of other Coke servers, you can try the following commands to access Finger servers at the Rochester Institute of Technology.

```
finger drink@csh.rit.edu
finger graph@drink.csh.rit.edu
```

For more on Coke servers, look in the catalog.

The University of Wisconsin has a computerized Coke machine. To buy a drink, you log in to the terminal next to the machine and select the appropriate command. Of course, you need to pay money in advance to have credit in your account.

If you want to see the instructions for this system, finger the following server:

```
finger coke@cs.wisc.edu
```

If you are a researcher looking for some interesting work, apply for a grant to implement an Internet coke server at your location.

CHAPTER 9

Introduction to Usenet

Like most people, you will probably soon find that Usenet is your favorite part of the Internet. You will be able to partake in discussions with people from all over the world on virtually every conceivable topic.

In this chapter, we will discuss the basics and lay the foundation for using Usenet. We will start by answering the question: What is Usenet? We will then cover the basic terminology that you will encounter.

After this orientation, we will discuss how Usenet data is transported from place to place and how it is organized. We will show you how Usenet works and what you can expect to find.

In the following chapters, we will build on this foundation and show you the details of accessing Usenet.

What Is Usenet?

Usenet is a large collection of discussion groups involving millions of people from all over the world. Each discussion group is centered around a particular topic. Jokes, recipes, mathematics, philosophy, computers, biology, science fiction—just about any subject you can think of has its own group.

In total, Usenet has over 5,000 different discussion groups. Many of these are of regional or local interest. For example, there is a discussion group for discussing restaurants in the San Francisco Bay Area. Nevertheless, over 2,500 groups are of general interest and are read by people throughout the world.

One of the first questions that people ask is: What does it cost to use Usenet?

The answer is that Usenet is free. You may have to pay something for Internet access (as we discussed in Chapter 3), but there is no charge for Usenet per se. In fact, if you have access to an Internet computer at no cost, then everything, including Usenet, is free.

Just about every topic of interest to human beings has a place in some Usenet discussion group or another. When the need arises for a new group, there are well established procedures to form one. Unlike the commercial services (such as Compuserve), there is no central authority that controls Usenet. Thus, when the users decide that there should be a new discussion group, one is formed.

This system has two important results: First, new groups can be created in a timely manner whenever the need arises. Second, there are a great many groups devoted to esoteric subjects. To sample this variety, take a quick look at the list of groups in Appendix G.

FUN TIP: No matter what your interests, there will be Usenet discussion groups for you.

Basic Terminology

The original Usenet network was conceived in order to display notices and news items. The idea was to create a computerized version of a bulletin board. Usenet soon outgrew its original blueprint, but the legacy of an electronic news network remains. Although Usenet is used primarily for discussion groups, we still talk about it using news-oriented terms.

For example, Usenet itself is often referred to as *the news*, or *netnews*, even though there is little real news in the sense of a newspaper. For example, you might hear someone say, "I picked up a recipe for groat cakes while reading the news yesterday." What he means is that he found the recipe in one of the Usenet discussion groups.

Similarly, the Usenet discussion groups are usually referred to as *newsgroups* or, more simply, *groups*. Within each newsgroup, the individual contributions are called *articles* or *postings*. When you submit an article to a newsgroup, we say that you *post* the article. Thus, you can imagine the following conversation between two Usenet people in Fargo, North Dakota, both of whom happen to be named Mike:

Mike: These are great groat cakes. New recipe?
Mike: Yes, I saw an article in the vegetarian cooking
     group about how nutritious groats are,
     so I posted a request for recipes. The next day, someone
     from France had posted a reply with this wonderful recipe
     for groat cakes with truffles.
Mike: Well, they certainly taste good, Mike.
Mike: Thank you, Mike.

What Is It Like When You Use Usenet?

To read Usenet articles, you use a program called a *newsreader*. The newsreader acts as your interface: you tell it which newsgroups you want to read, and it presents the articles for you, one at a time. (Remember, although we call them newsgroups, they are really discussion groups.) You can read any newsgroup that you want, as long as it is carried by your local site. Most sites do not offer all the newsgroups, as it would take up too much disk space.

When you start to learn how to use your system, you may find that you have a choice of newsreader programs. If so, you can pick the one that suits you best. Newsreaders are usually complex programs with many commands, so it is a good idea to settle on one that you like and take some time to learn it well.

There are a number of different newsreaders, and each one works in its own way. On Unix systems, four of the most popular newsreaders are named **nn**, **rn**, **tin**, and **trn**. If you are using a Unix system, at least one of these newsreaders should be available on your system. In Chapter 11, we will talk about reading the news. In chapters 12 through 15 we will explain how to use the newsreader programs.

One of the functions of a newsreader is to keep track of which newsgroups you want to read. At any time, you can add or delete newsgroups from your list. When you add a group to your personal list, we say that you *subscribe* to that group. Similarly, if you indicate that you do not want to read a newsgroup any longer, we say that you *unsubscribe*.

Don't misinterpret these terms. There is no formal subscription process nor is there any fee. Subscribing simply means that you tell your newsreader that you wish to follow a newsgroup. Moreover, the system is private: no one keeps track of which groups you read.

As you subscribe and unsubscribe to groups, and as you read articles, your newsreader maintains a file on your behalf. This allows the newsreader to keep track of which newsgroups are subscribed to and which articles you have read. On Unix systems, this file is named **.newsrc** and is stored in your home directory.

As you read an article, the newsreader allows you to perform many different actions. For example, you can move from one article to another, save an article to a file, mail a reply to a person who posted an article, compose an article of your own, and so on. If you were to watch someone reading the news, you would see him reading text on the screen and, from time to time, typing a command.

In Chapter 2, we explained that many services are built upon a client/server relationship. The client is a program that requests a service; the server is a program that supplies this service.

When you read the news, you are using a client/server system. The actual articles are stored and managed by a program called a *news server*. Your newsreader acts as a client. Each time you enter a command to read an article, your newsreader requests that article from the news server.

A typical setup on a local area network is to have one computer run the news server program. This computer acts as the central news repository for everybody on the network.

Imagine many people, each at his or her own computer or terminal, reading the news. Each person is interacting with his or her own newsreader program. Each of these programs acts as a client, requesting services from the central news server.

Normally, the whole thing works automatically. You tell your newsreader that you want to read an article and it just appears. However, if the news server should go down (stop working) for some reason, no one will be able to read the news until the person in charge fixes the problem.

In some large networks, such as a university, many people may want to read the news at the same time. If the news server is busy, you may experience a delay at times, when your newsreader is waiting to receive an article.

Hint

On some networks, you may not be able to read the news at peak periods. When you start your newsreader, it will display a message telling you that the work load on the news server is too high and suggesting that you try again later. For example:

```
Server hub.ucsb.edu responded with code 400, which
probably means the load on the server is too high.
Please try again later.
```

What Is the Difference Between Usenet and the Internet?

The names "Usenet" and "Internet" sound familiar and—at least at first—you may find it easy to confuse the two, so let's take a moment and get the differences straight in our minds.

The name "Usenet" was chosen as an abbreviation for "users network". However, the name is a misnomer. Usenet is not really a network in the sense that a network is a group of computers connected together. Usenet is a collection of discussion groups.

The Internet, conversely, really is a network. More precisely, the Internet is a collection of tens of thousands of networks worldwide.

In other words, the Internet is a general-purpose carrier of information, while Usenet is simply one type of service that makes use of this capability. (As you know, the Internet supports a number of other services, such as electronic mail, remote Telnet connection, file transfer, and so on.)

Now, you might ask, is every Internet computer part of Usenet? The answer is no. Many Internet computers are owned by people who, for some reason, choose to not participate in Usenet. To be a part of Usenet requires someone to act as an administrator. It also requires that a computer with a large amount of disk space be available to hold news articles and programs. Some people do not want to spend the time and money that it would take to maintain such a system.

You might also ask, are all Usenet sites part of the Internet? Again, the answer is no. The only requirements to being part of Usenet are:

1. You must furnish a computer to act as a local repository for articles

2. You must have someone to administer the system

3. You must find another Usenet site to which your computer can connect in order to swap articles back and forth.

There are many computers that are part of Usenet, but are not on the Internet; they use some other networking system. One such system, used by many Unix computers, is called UUCP. We won't go into the details, except to mention that the Internet is faster, better, and more expensive. Suffice it to say that many people participate in Usenet by using UUCP—or another networking arrangement—instead of the Internet.

Who Runs Usenet?

The fascinating thing about Usenet is that there is no central authority. Usenet is run by the people who use it.

Usenet was started in 1979 in North Carolina (USA) as an experiment. The idea was to create an electronic bulletin board to facilitate the posting and reading of news messages and notices. At first, there were only two sites: the University of North Carolina and Duke University.

Before long, other places joined and, with the explosion of networking in the 1980s and 1990s, Usenet expanded enormously. Today, there are tens of thousands of Usenet sites with more than two and a half million participants.

Each Usenet site is run by a person called the *news administrator* or *news admin*. In some places, the news administrator is the same person who manages the system. However, this need not be the case. In large installations, such as a university, the news administrator may be a staff member, or even an ambitious volunteer, who reports to the system manager.

Each news administrator is responsible for his or her own site alone and nothing else. The arrangement works because the news administrators stay in contact with one another and cooperate. Indeed, there are Usenet newsgroups specifically set up for news administrators to have discussions.

This means that, in the global scheme of things, nobody can tell anybody else what to do. This lack of central authority is what gives Usenet its charm, and is what distinguishes it from other discussion group systems (such as CompuServe or Prodigy) where there are Rules and People-In-Charge.

This is not to say that Usenet is total anarchy. There are a good number of conventions that have been developed over the years. As a responsible member of Usenet, you are expected to learn and follow these conventions. Still, when people misbehave—for example, by repeatedly sending silly or offensive messages to a newsgroup—more right-thinking people usually have to confine themselves to public criticism or to sending letters of complaint to the person's electronic mailbox.

Actually, as you will see, when there are no rules, most people choose to cooperate.

Hint

When everyone else is cooperating, acting like a jerk gets boring real fast.

How Is the News Transported?

If there is no central authority to coordinate everything, how are news articles transported all over the world? The answer is that new articles are passed from one computer to another. Eventually, a copy of each article spreads throughout the entire Usenet system.

Let's discuss a typical example in which you read articles in one of the Usenet newsgroups devoted to mathematics. As you read, you decide to post an article of your own. We will take a look at some of the details and follow your article as it travels around the world. The details may vary from one system to another, but the general idea is the same.

To begin, you log in to an Internet host and enter the command to run your newsreader program. You start by telling the newsreader which newsgroup you want to read. As we mentioned earlier, your newsreader keeps a file in which it records the newsgroups you subscribe to and the articles you have read.

In this case, you specify a particular mathematics group, so the newsreader checks its records and then connects to the news server to request the next article from this particular newsgroup. The news server responds by sending this article, which your newsreader then displays on your screen. When you finish reading an article, your newsreader requests and displays the next article.

As it happens, someone on the other side of the world has posted an article in which he asks if anybody has a shorter solution to Fermat's Last Theorem. You remember that you came up with just such a solution the other day. You enter a newsreader command to tell the newsreader that you want to compose an article in response to the one you are currently reading. (This is called a "followup" article.)

Your newsreader starts a text editor program so you can compose your article. You say that you have a wonderfully short solution to the problem, but, unfortunately, it is a bit too long to write down at this time.

After you finish writing, you enter the command to quit the text editor. The newsreader program regains control and sends a copy of your new article to the central news server. At this point, your article can be read by anyone in your local network. However, the article still needs to be transported around the world.

Now, when one news server supplies Usenet articles to another server, we say that it offers a *news feed* or, more simply, a *feed*. In order to participate in Usenet, your news server gets a news feed from another Usenet site. From time to time, your news server will connect to the server at this site in order to get its feed.

Each time this happens, your server passes on any new articles that have not yet been received at the other location. In particular, the new article that you just composed will be sent out. Later, when that news server connects to its news feed, your article will be sent to another site. This is how Usenet works: your article is passed automatically from news server to news server, one connection at a time.

What makes the system fast is that some news servers act as way stations, providing news feeds for many other servers. Once your article hits one of these way stations, it will be sent to many news servers within a short amount of time.

By the next day, your article is available on virtually every news server in the country. Within two or three days, your article is available around the globe. By the end of the week, your great mathematical accomplishment is recognized worldwide and you are famous.

You might ask, if new articles are constantly being transported from one news server to another, don't they just pile up indefinitely? This is the same question that you might ask as you ride down an escalator: Why doesn't the basement fill up with steps?

With an escalator, the answer is, of course, that the steps are recycled. With Usenet, the solution is to keep each article for a certain amount of time and then throw it away. Each news administrator decides how long articles should be kept. The news server regularly checks the articles and deletes those that are older than the specified time interval. When this happens, we say that the articles have *expired*.

The news administrator at your location can specify different expiration dates for the various newsgroups. For example, he or she may decide that the newsgroup in which people discuss general Unix questions should have a longer expiration period than the newsgroup in which people swap tasteless jokes. (Or, perhaps, the other way around.) If you want to know the exact expiration policy at your location, you will have to ask your local news administrator.

Hint

As a general rule, most newsgroup articles are kept from two days to two weeks.

How Big Is Usenet?

Since Usenet has no central administration, it is difficult to know exactly how big the system is. However, various people have implemented programs to estimate Usenet statistics.

At DEC Network Systems Laboratory, Brian Reid runs the Measurement Project, which regularly publishes such numbers. Reid uses a program called **arbitron** that many Usenet news administrators run on their systems. This program compiles data on how many people read Usenet, which newsgroups they read, and so on. (It is all anonymous, so privacy is protected.) Reid then analyzes the results mathematically and estimates statistics for the entire Usenet system.

Here are the statistics that were current at the time we wrote this chapter. (Remember, these are estimates. The actual numbers are not important. All we want is for you to have an appreciation for the size of Usenet.)

There were 76,000 different sites that carried Usenet. (Remember, not all of these are on the Internet.) At these sites, there were 8,445,000 users of all types, of whom about 29 percent (2,417,000) participated in Usenet.

Another question to ask is: How much data is actually transported on Usenet? To answer this question, the Measurement Project looks at how much Usenet data is received at its own news server. This computer, named **decwrl**, receives most of the newsgroups and provides a news feed to other servers.

On the average, **decwrl** received 26,400 new messages a day, containing a total of 56.2 megabytes (56.2 million characters) of data.

Within Usenet, the most important news feed service is provided by a company called Uunet Technologies. They also compile statistics on Usenet data. At the time we wrote this chapter, the numbers for the previous two weeks were as follows.

During that time, Uunet received 400,805 articles totaling about 744 megabytes. As we will see in Chapter 10, each Usenet article has standard information called a "header". The header contains technical data, the address of the person who posted the article, the time and date, the subject of the article, and so on. Including headers, the 400,805 articles totaled about 951 megabytes.

This all works out to about 28,700 articles a day, comprising 53 megabytes per day (68 megabytes counting headers).

You might wonder, why are the Uunet numbers higher than the Measurement Project numbers? The answer is in the way articles are counted. When you post an article, you can send it to more than one newsgroup at a time. (This is called "cross-posting".) Uunet counts an article once for every newsgroup in which the article appears. The Measurement Project counts each article only once.

To continue, Uunet reported that during this two week period, the articles came from 31,123 different Usenet sites and were submitted by 91,726 different users.

To put this all in perspective, in January 1989, Usenet had about 450 newsgroups. (There are now over 5,000.) At that time, there was an average of 1,400 articles a day, consisting of 3 megabytes of data.

FUN TIP: Usenet consists of about 2.5 million people, using 76,000 different computers, who post about 27,000 new articles every day. If all the people who participate in Usenet were laid end to end, they would stretch from Lausanne, Switzerland, to the confluence of the Tigris and Euphrates rivers.

The Mainstream and Alternative Hierarchies

As we have explained, there are a vast number of Usenet newsgroups, with new ones being added all the time. In order to make everything manageable, we use a system in which the newsgroups are collected into categories called *hierarchies*. Each hierarchy has its own name and is devoted to a particular area of interest. Table 9-1 shows the most important hierarchies.

The hierarchies in Table 9-1 are distributed all over the world (although, as we will see in a moment, not all hierarchies are carried at every Usenet site). Each newsgroup is given a name that consists of two or more parts, separated by . characters. The first part of the name is the hierarchy to which the newsgroup belongs.

For example, in the **news** hierarchy, there is a newsgroup for people who are learning how to use Usenet. In this newsgroup, you can ask any question you want about Usenet and some kind soul will answer you. The name of this newsgroup is **news.newusers.questions**.

Here is another example. In the **rec** hierarchy, there are several newsgroups devoted to various aspects of Star Trek. For discussions about Star Trek conventions and memorabilia, there is **rec.arts.startrek.fandom**. For reviews of Star Trek episodes, movies, and books, you can read **rec.arts.startrek.reviews**.

Name	Topic
alt	alternative newsgroups, many different topics
bionet	biology
bit	many topics: from Bitnet mailing lists
biz	business, marketing, advertisements
comp	computers
ddn	Defense Data Network
gnu	Free Software Foundation and its GNU project
ieee	Institute of Electrical and Electronics Engineers
info	many topics: from University of Illinois mailing lists
k12	kindergarten through high school
misc	anything that doesn't fit into another category
news	about Usenet itself
rec	recreation, hobbies, the arts
sci	science of all types
soc	social issues
talk	debate on controversial topics
u3b	AT&T 3B computers
vmsnet	DEC VAX/VMS and DECNET computer systems

Table 9-1. *The Most Important Usenet Newsgroup Hierarchies*

The two Star Trek examples each have a four-part name, with the first three parts the same. This is the system that is used to name newsgroups. The first part is the hierarchy; the other parts show categories and sub-categories. For example, the newsgroup devoted to science fiction movies is **rec.arts.sf.movies**. The group for discussing boating is **rec.boats**.

The Usenet hierarchies are divided into two categories, *mainstream* and *alternative*. The mainstream hierarchies are carried on all Usenet news servers. The alternative hierarchies are considered optional. Many Usenet sites carry them, but many choose not to.

At each Usenet site, the news administrator decides which hierarchies and newsgroups to carry. At most places, you can expect to see all the mainstream hierarchies and at least some of the alternative hierarchies. However, bear in mind that even if your site carries a particular hierarchy, it may not have all the newsgroups in that hierarchy.

The principal difference between the two types of hierarchies has to do with the way new groups are formed. As we will explain later, people must follow well-defined procedures to start a mainstream newsgroup. There must be a discussion and a vote, and enough people must express interest. The alternative hierarchy is more lax: any person who knows how to do so can start a newsgroup. Historically, the mainstream hierarchies were the original Usenet categories and were strictly controlled. The alternative hierarchies came later and were created as a less-restrictive facility.

As a whole, mainstream newsgroups tend to be more stable and are more readily accepted by news administrators. For example, an administrator who is short of disk space might decide that **news.newusers.questions** is more important than **alt.sex.bondage**. Even so, many of the alternative newsgroups are popular and are widely circulated. However, as you might imagine, you will often find ridiculous or spurious alternative newsgroups, especially within the **alt** hierarchy.

Hint

The distinction between mainstream and alternative newsgroups is made only for purposes of organization. When you read the articles, there is no real difference. The main consideration is that your news server will probably not carry all of the alternative newsgroups.

There are seven mainstream hierarchies and eleven alternative hierarchies. They are shown in Tables 9-2 and 9-3.

The Cultural, Organizational, and Regional Hierarchies

In Tables 9-2 and 9-3, we listed the mainstream and alternative hierarchies along with the number of newsgroups. In total, we had 972 mainstream groups and

Hierarchy	Number of Newsgroups
comp	459
misc	40
news	22
rec	273
sci	71
soc	87
talk	20
TOTAL	972

Table 9-2. *The Mainstream Usenet Newsgroup Hierarchies*

Hierarchy	Number of Newsgroups
alt	586
bionet	41
bit	192
biz	32
ddn	2
gnu	28
ieee	12
info	39
k12	36
u3b	5
vmsnet	32
TOTAL	1005

Table 9-3. *The Alternative Usenet Newsgroup Hierarchies*

1,005 alternative groups. We said earlier that there are over 5,000 different newsgroups. What about the others?

As you remember, the mainstream and alternative hierarchies are distributed worldwide (although most Usenet sites do not carry every single newsgroup). In addition, there are many other hierarchies that are more cultural, organizational, or local in nature. These hierarchies are usually circulated only in their areas of interest.

Cultural hierarchies contain newsgroups devoted to a particular people. For example, four of the more well-known cultural hierarchies are **de** (German [Deutsch] newsgroups), **fj** (Japanese), **aus** (Australian), and **relcom** (Russian).

Although most Usenet articles are in English, the cultural hierarchies often have articles in their own language. For such articles, you may need special software to display the non-English characters. For example, some of the Japanese articles in the **fj** hierarchy use the Kanji alphabet.

The organizational hierarchies contain newsgroups devoted to a university, company, or other organization. There are many such hierarchies. For instance, the University of California at Santa Barbara has its own hierarchy, named **ucsb**. Trinity College in Dublin, Ireland, has a **tcd** hierarchy. If you are part of a large organization, chances are it has its own hierarchy and its own newsgroups.

Within the **ucsb** hierarchy, for example, are newsgroups of interest to the university community. For instance, there is a group named **ucsb.rides** for people who need local transportation. There is also a newsgroup, **ucsb.english.eng109c**, just for one particular English class. As you might imagine, such newsgroups are created and dissolved as the need arises.

Aside from the cultural and organizational hierarchies, there are a large number of regional hierarchies. These contain newsgroups of interest to people who live in a particular area. For example, the **ba** hierarchy contains many newsgroups pertaining to the San Francisco Bay area, such as **ba.market.housing**.

Many of the cultural, organizational, and regional hierarchies are distributed throughout the world. For example, many Usenet sites carry the **fj** hierarchy. After all, there are Japanese people everywhere and not just in Japan. Similarly, someone who used to live in San Francisco may want to read a regional newsgroup to keep in touch. Moreover, if you are planning to move to a new area, you may want to post an article of your own to a regional newsgroup to ask for information, say, about housing.

Clarinet: Real News

As you know, we refer to Usenet as the "news" and to discussion groups as "newsgroups". However, Usenet itself does not carry real news in any organized fashion. There is a real news service, however. It is named Clarinet, and it has its own hierarchy, **clari**.

The Clarinet service is provided by a private company. Unlike Usenet, if your organization wants to carry Clarinet, it must pay for it. However, the costs are reasonable, and many organizations choose to offer this service.

The **clari** hierarchy contains many different newsgroups. (At the time we wrote this chapter, there were 246 such groups.) Some are of global interest, others are regional or local. Figure 9-1 shows a few of these groups. If your news server carries Clarinet, you can read any of these groups using your newsreader program. You will notice no difference from reading a Usenet newsgroup, except that virtually all the groups are for reading only; you cannot post articles to them.

FUN TIP: Although your organization must pay a subscription fee to offer the Clarinet newsgroups, there is no cost to you to read them.

Clarinet news comes from a number of different sources, including a live UPI newswire. The various newsgroups contain a wide variety of topics. They cover just about anything that is news, and the articles are continually being updated. Many of the newsgroups are devoted to specialized areas or to features, such as well-known columnists.

If your news server does not carry the **clari** hierarchy, there is still a Clarinet newsgroup that you can read. It is named **biz.clarinet.sample**. Clarinet regularly sends interesting articles to this newsgroup as free samples.

```
clari.biz.economy.world
clari.biz.market.report
clari.canada.politics
clari.feature.dave_barry
clari.feature.movies
clari.local.new_york
clari.nb.ibmclari.net.newusers
clari.news.books
clari.news.sex
clari.sports.tennis
clari.tw.science
```

Figure 9-1.   *Examples of Clarinet newsgroups*

How Many Different Newsgroups Are There?

Now that we have covered all the hierarchies, we are in a good position to answer the question: How many Usenet newsgroups are there?

The total number of newsgroups changes frequently, especially in the alternative hierarchies. At the time we wrote this chapter, there were 972 mainstream groups and 1,005 alternative groups. This makes a total of 1,977 newsgroups that are widely distributed throughout the world. (For reference, there is a list of these groups in Appendix G.)

There were also an estimated 3,325 cultural, organizational, and local groups. (We say "estimated" because many such groups never make it outside their local news server.) To this, we can add the 246 Clarinet groups.

This gives us a grand total of 5,548 different newsgroups.

FUN TIP: *Even before we could finish writing this chapter, the number 5,548 was out of date. To give you a feeling for how Usenet grows, we looked at figures from six months ago. At that time, there were 4,410 newsgroups. Thus, in the last six months, the number of newsgroups had grown by 1,139 (about 26 percent).*

Newsgroup Naming Conventions

Newsgroup names are easy to understand. The first part of the name shows the hierarchy. For example, a group whose name begins with **comp**, such as **comp.unix.questions**, is in the computer hierarchy; the groups whose names begin with **talk**, such as **talk.environment**, are debate oriented; and so on. You can expect to find the weirdest newsgroups in the **alt** hierarchy. For example, that is where you will find **alt.sex.bestiality**.

Thus, when you are checking out a list of newsgroups (such as the one in Appendix G), trying to decide what to read, start by looking at the first part of the name. To show you some examples, Figure 9-2 contains a list of the

```
alt.tasteless.jokes
rec.humor
rec.humor.funny (moderated)
```

Figure 9-2.   *Newsgroups devoted to jokes*

newsgroups devoted to jokes, and Figure 9-3 shows all the newsgroups that have something to do with sex. (As you might imagine, these newsgroups have some of the largest audiences on Usenet.)

```
alt.binaries.pictures.erotica
alt.binaries.pictures.erotica.blondes
alt.binaries.pictures.erotica.d
alt.binaries.pictures.erotica.female
alt.binaries.pictures.erotica.male
alt.homosexual
alt.politics.homosexuality
alt.politics.sex
alt.sex
alt.sex.bestiality
alt.sex.bondage
alt.sex.fetish.feet
alt.sex.masturbation
alt.sex.motss
alt.sex.movies
alt.sex.pictures
alt.sex.pictures.d
alt.sex.pictures.female
alt.sex.pictures.male
alt.sex.sounds
alt.sex.stories
alt.sex.stories.d
alt.sex.wanted
alt.sex.wizards
alt.sexual.abuse.recovery
clari.news.group.gays (moderated)
clari.news.law.crime.sex (moderated)
clari.news.sex (moderated)
rec.arts.erotica (moderated)
soc.bi
soc.motss
```

Figure 9-3.   *Newsgroups related to some aspect of sex*

After the hierarchy name, you will see categories and (possibly) sub-categories. When two newsgroups are related, they will have similar names, differing only in the last part of the name. For example:

```
alt.binaries.pictures.erotica.blondes
alt.binaries.pictures.erotica.female
alt.binaries.pictures.erotica.male
```

Another point to notice is that some of the newsgroup names end in **.d**. This means that the group is for discussion of the contents of another group. For example, we have:

```
alt.sex.stories
alt.sex.stories.d
```

The first group contains stories about sex (check it out sometime) and only stories. The second group is for people who want to discuss the stories. If you post an article to **alt.sex.stories** that is not a story, someone may remind you that all discussion should take place in the **.d** group. This is so people who want to read stories only do not have to bother with anything else.

There is another important example of this convention that you should know about. The newsgroup **rec.humor** is for people who want to post and read jokes. If you want to discuss jokes, you should post to **rec.humor.d**. This is also the place to send a request like, "Does anybody have the list of all the light bulb jokes?" (That is why we did not list **rec.humor.d** in Table 9-2. It does not contain jokes. Rather, it is a group in which people talk *about* jokes.)

Hint

In the non-moderated joke newsgroups (**rec.humor** and **alt.tasteless.jokes**), the unofficial rule is that all postings must contain at least one joke. You will often see postings in which a person has given in to temptation and made some sort of non-joke comment. However, to be polite and to adhere to the convention, he or she has also included a joke. This is called an *obligatory joke* or an *objoke*.

You will sometimes see the same prefix, "ob", used in other groups to indicate the same convention: that a person is respecting the purpose of the group by making sure to include a relevant item within an otherwise questionable posting.

Moderated Newsgroups

You will notice that some of the newsgroups listed in Figures 9-2 and 9-3 (jokes and sex) are *moderated*. This means that you cannot post articles directly to the group. You do submit articles in the regular way, but they are automatically rerouted and sent to one person, called the *moderator*, who decides what will go in the group. A moderator—who is an unpaid volunteer—will not only pass judgment on what goes into a group, he or she will often edit and organize the articles.

The reason for moderators is to minimize the number of low-quality articles in a newsgroup. Perhaps the best examples are the humor groups. The groups **rec.humor** and **alt.tasteless.jokes** are not moderated. This means that anyone can post to these groups, with the result that there is usually a lot of silliness and repetition (not to mention old jokes).

The group **rec.humor.funny**—started by Brad Templeton, the creator of Clarinet—is moderated. Only those jokes that are deemed funny by the moderator, Maddi Hausmann, are posted. People all over the world submit jokes to Maddi, and she picks the ones she thinks are best. As a result, **rec.humor.funny** is read by more people than **rec.humor** (although they are both popular).

Some moderated newsgroups offer a special type of posting called a *digest*. The moderator creates a digest by collecting submissions, questions, answers, and bits of information. He or she will edit the information into a series of interesting items and post it as one large article, the digest. Two examples of such newsgroups are **comp.sys.ibm.pc.digest** and **sci.psychology.digest**.

A digest is much like an issue of an electronic magazine, with a volume and issue number, and a table of contents. Most newsreader programs have a special command to let you jump from one item to another as you read a digest.

Overall, the idea of moderated newsgroups provides a nice balance to Usenet. Many newsgroups carry a lot of low-quality articles, and it can be nice to read a group where everything is interesting (at least by one person's standards).

Of course, moderated newsgroups are a form of censorship because one person controls everything that is posted to the group. However, Usenet has many groups, almost all un-moderated, so you need never feel deprived.

What Are the Most Popular Newsgroups?

One of the most interesting questions about Usenet is: What are the most popular newsgroups? There is more than one way to answer this question depending on what you mean by "popular".

At regular intervals, the Measurement Project (which we mentioned earlier in the chapter) sends estimates of readership statistics to several newsgroups. (If you want to read such postings, check out the group **news.lists**.)

One of the postings identifies the 40 most popular newsgroups based on the estimated number of total readers. The thing is, not all groups are carried at all Usenet sites. For example, many news administrators will not carry groups like **alt.sex** even though, as you can imagine, they are extremely popular.

Let's take a look at some of these statistics. One of the perennial favorites is **rec.humor.funny**. At the time we wrote this chapter, this newsgroup had an estimated audience of 160,000 people. The group **alt.sex.stories** had only 130,000 readers. By this standard of measurement, the humor group is a lot more popular.

However, **rec.humor.funny** is carried at 82% of all Usenet sites. The group **alt.sex.stories** is carried at only 53% of the sites. The real question is: How well would a particular newsgroup do if it were carried at all sites? In other words, instead of asking how many people read the group, we ask, how much do people like the group?

To come up with such numbers, we start with two of the statistics estimated by the Measurement Project: the total number of readers and the percentage of sites that carry each newsgroup. To compare readership equally, we divide the total readers by the percentage at that site. This gives us an estimate of how many total readers a newsgroup would have if it were carried at all sites.

For example, **rec.humor.funny** would have 160,000 divided by 0.82, or 195,122 readers. The **alt.sex.stories** group would have 130,000 divided by 0.53, or 245,283 readers. Thus, the sex group is more popular than the humor group.

To make these hypothetical readership numbers easier to understand, we normalize them, setting the most popular one to the value 100. To do so, we divide each such number by the hypothetical readership of the most popular group and multiply by 100. The most popular newsgroup is **news.announce.newusers**, which contains valuable information for new Usenet users. It has a hypothetical readership of 280,000 divided by 0.91, or 307,692.

The result is a number between 0 and 100, which provides a picture of true popularity. We call this value the *Hahn Popularity Index* or *HPI*. For example, **rec.humor.funny** has an HPI of 195,122 divided by 307,692 multiplied by 100, or 63.

Table 9-4 contains the relevant numbers with all the arithmetic done for the 25 most popular Usenet newsgroups at the time we wrote this chapter. Draw your own conclusions.

FUN TIP: *According to our research, the most popular newsgroup that could exist would be* **alt.sex.jobs.offered.newusers**.

Ranking	Newsgroup	HPI	Readers	Sites
1	news.announce.newusers	100	280,000	91%
2	misc.forsale	98	250,000	83%
3	misc.jobs.offered	92	240,000	85%
4	alt.sex	87	180,000	67%
5	news.answers	82	220,000	87%
6	alt.sex.stories	80	130,000	53%
7	alt.binaries.pictures.erotica	71	120,000	55%
8	rec.arts.erotica	67	150,000	73%
9	rec.humor.funny	63	160,000	82%
10	alt.sex.bondage	57	110,000	63%
11	alt.activism	53	110,000	68%
12	rec.humor	52	130,000	81%
13	alt.binaries.pictures.misc	49	88,000	58%
14	news.groups	47	130,000	90%
15	news.announce.newgroups	47	130,000	90%
16	soc.culture.indian	45	100,000	73%
17	news.newusers.questions	43	120,000	90%
18	comp.graphics	42	110,000	85%
19	comp.lang.c	41	110,000	88%
20	misc.jobs.misc	39	100,000	83%
21	alt.bbs	37	83,000	72%
22	misc.wanted	37	93,000	81%
23	comp.binaries.ibm.pc	37	94,000	82%
24	alt.sources	37	89,000	79%
25	talk.bizarre	36	80,000	72%

Table 9-4. *The 25 Most Popular Usenet Newsgroups*

How Are New Newsgroups Started?

A new newsgroup is created by sending a special message, called a *control message*, throughout Usenet. Various types of control messages are used by news administrators to govern the operation of Usenet. There is one particular type of control message that is used to start a new group. (There is also another type of control message to remove an obsolete or spurious group.)

When a news administrator sends out such a message, it propagates from news server to news server, just like a regular article. Eventually, the message reaches all the news administrators. Each administrator decides whether or not to create the new group on his or her system.

News administrators almost always honor an authorized request to start a new mainstream group. This is because such a message will not be issued until certain well-defined criteria are met (described below).

Conversely, a control message to start an alternative newsgroup can by issued be anyone who knows how to do it. As you might imagine, there are all sorts of requests to start bizarre alternative groups. For this reason, many news administrators will pick and choose which new alternative newsgroups they are willing to create on their systems.

Here is how a new mainstream group gets formed.

To start, somebody must have the idea for a new group. Perhaps an existing group should be split into two. Perhaps a whole new area of discussion has arisen. Most ideas are discussed in existing newsgroups or by mail for some time, as ideas are proposed, examined, and modified.

Once the idea for the new group is set, someone sends a message to the group **news.announce.newgroups** (which is moderated). At the same time, the message is cross-posted to any other relevant groups.

The moderator of **news.announce.newgroups** will post an article explaining the name and purpose of the proposed group. There is now a 30-day discussion. This discussion takes place in the newsgroup, in other, related newsgroups, and by private mail correspondence.

At the end of 30 days, if there is clear agreement as to the name and purpose of the group, the moderator of **news.announce.newgroups** will post a general request for people to vote. Anyone who is interested can vote (once) by sending an appropriate mail message to a specified address. The period of voting is set in advance and is usually between 21 and 31 days.

At the end of the voting period, the totals are posted along with a full list of everyone who voted and how they voted. (No secrets on Usenet.) There is now a five-day waiting period in which anyone can request a correction to a particular vote or to the voting procedure.

At the end of five days, the vote is finalized. The vote is successful if the new group was approved by at least two thirds of the voters, and if the yes votes were at least 100 more than the no votes. After a successful vote, the moderator of **news.announce.newgroups** sends out a control message to start the new group.

If the vote was unsuccessful, the same newsgroup cannot be brought up for discussion for at least six months.

Hint

If you want to start a new alternative group, do the following:

1. Make sure that you have at least a few months of Usenet experience, so you understand how the system works.

2. Suggest your idea by posting an article in one or more existing groups (choose appropriately), and see what other people think. The newsgroup **alt.config** is used for discussing proposals for new alternate groups. Remember, alternative groups can only be created successfully if other people cooperate.

3. Read **news.announce.newgroups** for a while to see what types of issues arise during the formation of new groups.

4. After a reasonable amount of time—and as a reader of this book we know you are reasonable—decide if you still want to start the group. If so, ask your local news administrator for help in sending out the control message.

Frequently Asked Question (FAQ) Lists

As you start to use Usenet, you will find that you have many questions. Moreover, as you read the various newsgroups, you will have questions about those topics. Many of these questions will be the same ones that everybody else asks when they first start to read a particular newsgroup.

For example, the group **misc.consumers** is often used to discuss consumer credit. A common question is: "What is the difference between Visa and Mastercard?" Here is another example. The newsgroup **rec.arts.disney** discusses anything related to the world of Disney. A question often asked by newcomers is: "Is it true that Walt Disney was frozen in cryogenic suspension after he died?"

You will find that the participants of many newsgroups have asked and answered questions like these many times. Although they are interesting to a beginner, you can imagine that experienced readers grow tired of seeing the same questions posted repeatedly.

The solution is the *frequently asked question list* or *FAQ*. An FAQ list is a document, maintained by a volunteer, that identifies and answers all the frequently asked questions for a particular newsgroup. Many newsgroups have FAQ lists which the keeper of the FAQs posts to the group regularly. It is considered good manners to refrain from asking questions in a newsgroup until you have read the FAQ list. (Although not all newsgroups have such lists.)

There are four ways to get a FAQ list. First, you can read a newsgroup regularly and, if there is a FAQ list, it will be posted eventually. (Typical intervals are every week, every two weeks, or every month, depending on the group.)

Second, you can read the newsgroup named **news.answers**, which consists of nothing but FAQ lists and related material. This is an interesting group to follow as you will see lists of the best questions and answers on topics you would normally not read about.

The third way to get a FAQ list is to use Anonymous FTP to download the FAQ list from a Usenet archive. (We will explain how to do this in Chapter 16.)

Finally, after trying the other methods, you can post a short article to the group asking if a FAQ list exists and, if so, can someone mail you a copy or tell you how to get it.

FUN TIP: Learn how to use Anonymous FTP and take a look at the FAQ lists in the Usenet archives. There are all kinds of interesting questions and answers.

CHAPTER 10

I n Chapter 9, we introduced Usenet, the worldwide collection of over 5,000 different discussion groups. We saw that people participate by posting articles to the various groups and by using a program, called a newsreader, to read the articles.

In this chapter, you will find out what you can expect to see when you participate in Usenet. We will show you what a typical article looks like and how to make sense out of all the technical information. We will then discuss the conventions of Usenet: what guidelines you will want to follow and what new terminology you can expect to encounter.

Under- standing Usenet Articles

After reading this chapter, you will be prepared to start reading the news and posting your own articles. In Chapter 11, we will discuss the general considerations. In chapters 12 through 16, we will explain how to use the various newsreader programs.

The Format of a News Article

A news article consists of three parts: a *header*, followed by the *body*, followed by an optional *signature*.

The header contains technical information about the article. There are twenty different types of lines that you might commonly see in the header, each of which contains a different type of information. We will discuss headers in detail in the next section.

The body of the article is actual text: the main part of the article.

Finally, the signature consists of a few lines that come at the end of the article. These lines are composed by the person who sent the article and are automatically appended to the end of every article that he or she posts. We will talk more about signatures later in the chapter.

Figure 10-1 contains a typical Usenet article. In this article, the header consists of the first 13 lines. The body of the article consists of the next 7 lines. In this case it happens to be a joke. Most articles are longer than our example, and, in most cases, the body is larger than the header. Finally, the last 4 lines of our sample contain a signature. In our example, the signature identifies the person who posted the article, along with an address, phone number, and a short quotation.

The Header

All Usenet articles must have a header. A header consists of special lines that are at the beginning of the article. As we mentioned in the last section, there are twenty different types of lines that you might see in a header. These are listed in Table 10-1.

Not all articles will contain all twenty lines. For example, the article in Figure 10-1 contains only 13 header lines. These are the ones that you will see most often. For reference, though, we will briefly describe all twenty types of header lines, so you will be able to understand them when you see them.

```
Path: ucsbcs1!mustang.mst6.lanl.gov!nntp-server.caltech.edu!
news.claremont.edu!uunet!news.univie.ac.at!email!mich
From: mich@music.tuwien.ac.at (Michael Schuster)
Newsgroups: rec.humor
Subject: The Secret of Life
Summary: Advice for understanding life.
Keywords: life, philosophy
Message-ID: <1993May14.073130.15261@email.tuwien.ac.at>
Date: 14 May 93 07:31:30 GMT
Distribution: world
Sender: news@email.tuwien.ac.at
Organization: Tech Univ Vienna, Dept of Realtime Systems, AUSTRIA
Lines: 12
Nntp-Posting-Host: idefix.music.tuwien.ac.at

Here is some great advice I just found:

"When you get serious about foolishness,
you are getting into serious foolishness."

By the way, does anyone know who first said it?


—
Michael Schuster            |"I love you for your beauty; love me
TU Vienna, Austria          | though I am ugly"
mich@music.tuwien.ac.at     | - Miguel Cervantes, Don Quixote
+43/1/12345
```

Figure 10-1. *A typical Usenet article*

Before we do, you should understand that whether or not you even see the header at all depends on which newsreader program you are using and how it is configured. Some newsreaders show you all of the header lines by default. Other newsreaders will not show you any header lines unless you ask for them; you will see only the body and the signature. However, all newsreaders allow you to customize how articles should be displayed. Moreover, as you are reading an article, it is always possible to ask your newsreader to redisplay the current article along with the full header.

We will now discuss the various types of header lines that you might see. Few articles contain all twenty of the standard header lines, but it is a good idea to know what they mean. The header lines in Figure 10-1 are the most common ones.

Please do not think that you have to memorize what all these header lines mean. Just read through this section once. Then, whenever you see an article with a header line that puzzles you, you can look it up.

(From time to time, you will encounter other types of header lines that we do not mention here. Almost always, these will be non-standard lines that you can safely ignore.)

Header Line	Description
`Approved:`	identifies moderator who posted article
`Control:`	contains special administrative commands
`Date:`	time and date that article was posted
`Distribution:`	recommendation for where to send article
`Expires:`	recommendation for when to remove article
`Followup-To:`	shows where followup articles will be sent
`From:`	userid and address that posted the article
`Keywords:`	one or more words to categorize the article
`Lines:`	size of the body + signature
`Message-ID:`	unique identifier for the article
`Newsgroups:`	newsgroups to which the article was posted
`NNTP-Posting Host:`	name of Internet host that posted article
`Organization:`	describes the person's organization
`Path:`	shows transit route of the article
`References:`	identifies article to which followup refers
`Reply-To:`	address to send personal replies
`Sender:`	address of computer that sent out article
`Subject:`	short description of contents of article
`Summary:`	one-line summary of article
`Xref:`	local cross-posting information

Table 10-1.   *Header Lines for Usenet Articles*

Approved: In Chapter 9, we explained that some newsgroups are moderated. This means that when you post an article to such a group it is not sent to the group directly. Rather, it is sent to a person, called the moderator, who decides which articles should be sent to the group. Within moderated groups, the **Approved:** header contains the mail address of the moderator. This header line is also used with some types of **Control:** messages.

Control: This type of header line contains special commands that are used by news administrators to control the Usenet system. For example, there is a special type of control line that is used to start a new newsgroup. It is possible that you will never see a **Control:** line, as it is not used in regular articles.

Date: This line shows the time and date that the article was posted. The time will often be in Greenwich Mean Time [GMT], which is the standard Internet time. (Sometimes GMT is called Universal Time.) We discuss the Internet time conventions in Chapter 5.

Distribution: When you post an article, you will be given a chance to specify where you would like the article to be sent. Some news posting programs will show you several choices and ask you to pick one. Typical choices include: your organization, your region, your country, and worldwide. On some systems, you are not shown the choices, you just have to know what is available in your area. As an example, here are the choices you will see when you post an article at the University of California at San Diego.

`local:`	local to the site
`ucsd:`	local to the UCSD campus
`uc:`	all the University of California campuses
`sdnet:`	local to San Diego County
`ca:`	everywhere in California
`usa:`	everywhere in the USA
`na:`	everywhere in North America
`world:`	everywhere in the world

Here is another example, from one of the computers at the Technical University in Vienna, Austria.

`inst182:`	local to department #182 (Institut #182)
`tuwien:`	local to university (Technische Universitaet Wien)
`at:`	everywhere in Austria
`europe:`	everywhere in Europe
`world:`	everywhere in the world

(Note: The international domain name for Austria is **at**.) An important point to understand is that a distribution does not guarantee where the article will be propagated. It is only a suggestion. Each news administrator decides which distributions he or she will accept. Some administrators will accept every newsgroup they can get. Thus, for example, we have had the experience of logging in to a computer in Palo Alto, California, and reading local news from Edmonton, Canada.

Expires: As you know, each news administrator sets up the time intervals that articles will be retained on his or her news server. The **Expires:** header is used when you want to recommend a different expiration date.

For example, if an article is announcing an upcoming seminar, it makes sense to have the article expire the day after the seminar. Most articles do not use this header line. They expire according to whatever the local policy happens to be.

Remember though, it is the news administrator that has the last word about when articles expire. If you see an article with a particular expiration date, it's completely possible that the article may disappear before then.

Followup-To: From time to time, you may wish to submit an article in which you respond to a previous article. In such cases, your posting is called a *followup* article. All newsreader programs make it easy to compose and post followup articles.

Normally, a followup article will be sent to the newsgroup (or newsgroups) in which the original article appeared. However, in certain cases, a person may want to control where the followup articles should go. In such cases, a **Followup-To:** header line specifies to which newsgroup such articles should be sent.

For example, say that you post a joke to the **rec.humor** newsgroup. It happens to be the type of joke that you know people will want to comment on. However, the **rec.humor** newsgroup is supposed to be for jokes only. Discussion of jokes should be in the group **rec.humor.d**. So, when you post the joke to **rec.humor**, you can use a **Followup-To:** header line that specifies **rec.humor.d**. This technique is also handy when you post an article to more than one newsgroup and you want to direct all subsequent discussion to a single group.

There is one special followup designation that you should know about. If this header line specifies **poster**, it means that followup articles cannot be sent to the newsgroup. Rather, you should continue the discussion with the person who posted the article by sending a message to his or her personal mail address. (Your newsreader program will help you do this easily.)

From: This header line is an important one, as it tells you who posted the article. You will always see the person's mail address. In most cases, you will also see their real name.

Although knowing who posted an article is mildly interesting, the **From:** header line comes in handy in another way. There is a computer named **rtfm** at MIT that provides a Usenet archive service. Many important Usenet articles—such as the FAQ (frequently asked question) lists—are stored on this computer, and you can access them by using Anonymous FTP (see Chapter 16). This computer also offers another service called the Usenet address server. A program on **rtfm** scans all the Usenet articles and stores the **From:** header lines in a database. You can send a mail request to this server, asking it to search for a particular person. You will get a response, also by mail. If that person has posted to Usenet, there is a good chance that the address server can find him. The details are explained in Chapter 22.

(By the way, the old name of the **rtfm** computer was **pit-manager**, in case you ever see the name somewhere. The significance of the new name, **rtfm**, will become clear later in the chapter.)

Hint

It is possible for people to change the real name that is shown on the **From:** header line. So, if something looks suspicious, be suspicious.

Keywords: This line contains one or more words or terms that categorize the content of the article. Some people look at this line to decide if they want to read the article (although not as often as the **Subject:** line).

Lines: This header line is straightforward. It shows the total number of lines in the article. The number of lines includes the body and the signature, but not the header.

Message-ID: On this line is a unique identifier that is generated automatically by the program that sent out the article. The last part of this name is the address of the computer the article was posted from. This information is used only by news programs. You can ignore it.

Newsgroups: This is an important line. It shows to which newsgroups the article has been posted. When you are using a newsreader program to look at an article, it has obviously been posted to the newsgroup you are reading. However, if you are looking at an old article—that you have previously saved, that someone has mailed to you, or that you have found in the Usenet archive mentioned above—it is handy to know what newsgroup the article originally appeared in.

When you post an article, you will be asked to specify what group you want to send the article to. If you want, you can specify more than one group. This is called *cross-posting*. When an article is cross-posted, you will see more than one group name in the **Newsgroups:** header line.

It is considered good etiquette to post to only one or, at most, a small number of newsgroups. It is not appreciated when people send an article to many newsgroups, most of which have only a tenuous relationship to the topic under discussion. For example, if you have a Unix question, send it to the most appropriate of the many different Unix-oriented newsgroups. Do not send it to every Unix group.

Hint

When you cross-post an article, make sure you put a **Followup-To:** line in the header to direct subsequent discussion to one specific group. If appropriate, consider using **Followup-To: poster** to direct all subsequent discussion to your personal mailbox.

NNTP-Posting Host: In Chapter 2, we explained that TCP/IP is a large family of protocols. (A protocol is a set of technical rules.) We have already discussed TCP (Transmission Control Protocol) and IP (Internet Protocol), which are used to transport data, and SMTP (Simple Mail Transfer Protocol), which is used for sending mail messages. In Chapter 16, we will talk about FTP (File Transfer Protocol), which is used to copy files from one Internet host to another.

The TCP/IP protocol that is used to transport Usenet articles is called NNTP (Network News Transfer Protocol). The **NNTP-Posting Host:** header line shows the name of the Internet computer from which the article was posted.

In other words, this line will show you what computer the person was using when he or she posted the article. The **Sender:** line, if present, shows the name of the computer—often the news server—that actually sent out the article over Usenet.

*FUN TIP: If you want to hunt for forgeries (which do turn up from time to time), look for articles in which the name of the computer in the **NNTP-Posting Host:** header line does not agree with the name of the computer in the **From:** line. It is easy to change the **From:** line, but not so easy to modify the **NNTP-Posting Host:** line.*

Organization: This header line contains a short phrase describing either the organization to which the person who posted the article belongs, or the organization that owns the computer. The purpose of this header line is to help identify the person who posted the article. The **From:** line does contain an address, but sometimes such addresses can be difficult to understand.

Path: The information on this line consists of a number of computer names, separated by ! (exclamation mark) characters. The names show the path that the article took, from one computer to another, to reach your news server. (To understand the path, read it from right to left.)

Although we consider this information to be a single header line, it is often so long as to be broken onto more than one line on your screen. This is the case in Figure 10-1 where you can see the names of the computers that the article passed though on its way from Vienna, Austria, to Santa Barbara, California.

You can safely ignore the information in this header line.

References: You will see this header line only in followup articles. It will contain the identifier from the **Message-ID:** of the original article. The example in Figure 10-2 is a followup article and, as such, contains a **References:** header line.

This is another header line that you can ignore. It is used by newsreader programs to group related articles together. In this way, after you read an article, your newsreader can show one followup article after another. A connected series of articles is called a *thread*. As you read the news, you will often want to *follow* a thread, reading one followup article after another.

Reply-To: This header line has the same format as the **From:** line. If this line is present, replies that are mailed to the person who posted the article will be sent to this address. This header line is handy when you want private mail replies to go to a different address than the one from which the article was posted.

Sender: The purpose of this header line is to show the name of the computer from which the article was posted. A **Sender:** line will be generated automatically whenever the information in the **From:** line might be misleading: for example, if someone manually enters his own **From:** line rather than having the news posting program do it automatically.

For example, let's say that you are visiting a friend and using her account to post an article. Normally, the news posting program will put in a **From:** header line with her name and address. Instead, you delete this line and put in your own **From:** line, showing your name and address.

The news program will automatically generate a **Sender:** header line containing the name of the computer that posted your article. Sometimes you will also see the name of the system userid that sends out news articles.

```
Path: ucsbcs1!nipper.ucsb.edu!harley
From: harley@nipper.ucsb.edu (Harley Hahn)
Newsgroups: rec.humor
Subject: Re: The Secret of Life
Summary: Advice for understanding life.
Keywords: life, philosophy
Message-ID: <1993May16.060436.13827@nipper.ucsb.edu>
Date: 16 May 93 10:13:12 GMT
References: <1993May14.073130.15261@email.tuwien.ac.at>
Distribution: world
Lines: 22


In article <1993May14.073130.15261@email.tuwien.ac.at>
mich@music.tuwien.ac.at (Michael Schuster) writes:

> Here is some great advice I just found:
>
> "When you get serious about foolishness,
> you are getting into serious foolishness."
>
> By the way, does anyone know who first said it?

Yes, this advice was first offered by the
Canadian philosopher, Tim Rutledge.

Obligatory Joke...

Hahn's Maxim: If something is worth doing,
  it's worth doing to excess.


—

Harley Hahn
writer of Unix and Internet books
```

Figure 10-2. *A typical Usenet followup article*

Yes, as you suspect, this feature makes it a little more difficult for bored undergraduates to get away with forging news articles.

Subject: This line is, by far, the most important header line. It contains a short description of what the article is about. You create this description when you compose the article.

Why is this header line so important? Of course, it helps you understand what the article is about. But there is another reason. Some newsreader programs ask users to choose which articles they want to read based on the contents of **Subject:** header lines. This means that the contents of this header line are often the only thing a person sees before choosing whether or not to read the article.

Hint

When you post an article, it behooves you to take a moment to make your **Subject:** line interesting and accurate. Most people will not even look at an article unless the subject description takes their fancy.

Summary: This heading provides a brief, one-line summary of the article. This information is useful in followup articles. The **Summary:** header line is not used all that often, but you can put one in if you want.

Xref: When an article is cross-posted to more than one newsgroup, the **Xref:** header line (which stands for "cross-reference") will show which newsgroups contain the article. It will also show the local article numbers that identify the article in each group. You can ignore this line. It is only there to be read by your local newsreader program.

The Signature

We said earlier that a Usenet article has three parts: the header, the body, and the signature. We have already talked about the header (the technical information) and the body (the actual text of the article). Let's take a moment now to discuss the signature.

The signature is an optional addendum that shows information about the person who sent the article. For example, here is the signature from our sample article:

```
Michael Schuster         |"I love you for your beauty; love me
TU Vienna, Austria       | though I am ugly"
mich@music.tuwien.ac.at  | - Miguel Cervantes, Don Quixote
+43/1/12345
```

If you want to use a signature, you have to create it for yourself and store it in a file. Your news posting program will automatically append the signature to the end of each article you post before it is sent out. Here is how it works on Unix systems. (If you are using a non-Unix computer, you will have to read the documentation for your system.)

You use a text editor to create a file called **.signature** in your home directory. (The **.** character is part of the name.) Whenever you post an article, the news program looks to see if such a file exists. If so, the program appends the contents of the file to the end of your article.

You might ask, What if I don't know how to use a text editor program? Well, this is why, in Chapter 1, we suggested that you learn some Unix. At the very least, you should learn how to use a text editor so you can create things like a **.signature** file. You could, we suppose, ask someone else to help you create your **.signature** file. However, it wouldn't do you much good because if you can't use a text editor, you won't be able to compose an article in the first place.

(However, we don't want to be too dogmatic. You do not need to use a text editor if you just want to read articles, only if you want to post articles of your own, or if you want to reply to someone via a mail message.)

The standard information to put in a signature is the name of the person who posted the article and a mail address. You may also see the name of the organization to which the person belongs, a postal address, and perhaps even a telephone number. The signature in our example shows that the article was posted by Michael Schuster, at the Technical University of Vienna. We see his mail address and his phone number. To the right, we see an interesting quotation.

The wonderful thing about signatures is that people use them in highly imaginative ways. Michael Schuster, as you can see, includes a quotation. You will see many quotations, as well as witty sayings, drawings, jokes, and so on. Indeed, there are people who collect interesting signatures and post them, from time to time, as a unique bit of Usenet art (usually in the **rec.humor** newsgroup).

FUN TIP: *To see some bizarre signatures—as well as comments about bizarre signatures—try the* **alt.fan.warlord** *newsgroup.*

Hint

It is not considered imaginative to use your signature to publicize your religion.

Once you start reading a lot of Usenet articles, you will find that it is annoying to have to read long signatures. For this reason, it is a well-established convention that signatures should be no more than four lines long. Some news posting programs will actually enforce this rule by removing any extra lines. Although this may seem restrictive, you will be amazed at how much a creative person can do with only four lines.

FUN TIP: When it comes to creating a signature, anything goes. Aside from the four-line limit, there are no rules, so use your imagination.

Followup Articles

As we mentioned earlier, a followup article is one that is posted in response to a previous article. Creating a followup article is easy. Say that, as you are reading the news, you come across an article to which you would like to follow up. All you have to do is enter a simple command to your newsreader program.

The newsreader will make everything as easy as possible by creating a skeleton for the new article (a header and part of the body), and will start your text editor so you can compose your response. Once you finish, you quit the text editor. The newsreader will take over again and post the article for you automatically.

Take another look at the sample followup article in Figure 10-2. This article is a response to the one in Figure 10-1. Notice that the body of the new article starts with two lines that identify the previous article and the person who posted it. These lines are followed by the original article with each line marked with a > character.

Most of this was generated automatically by the newsreader program. All we had to do to create the followup was compose our response.

When you read an article, there are three ways to tell if it is a followup. First, the subject will be the same as the original article, but will have the characters **Re:** in front of it. For example, in Figure 10-2, we see:

```
Subject: Re: The Secret of Life
```

This is automatically put in by the newsreader when it creates the header of a followup article. However, no more than one **Re:** will be put in. Thus, a followup to a followup has only a single **Re:**.

The second thing to look for (if your newsreader is displaying the header for you) is a **References:** line. This contains the identifier of the original article. Although you can ignore this header line, it does serve as official notice that you are reading a followup article. If you are reading a followup to a followup, you may see more than one identifier in the **References:** line.

Finally, a followup article will often have all or part of the previous article embedded. As we mentioned, the lines from the previous article will be prefaced with a special character. Most of the time, this will be > (a greater-than sign) as in our example. However, you will sometimes see other characters.

In such cases, we say that the followup article is *quoting* those parts of the original article. In our example, the followup article quoted six lines from the original article.

Finally, if you see more than one > character, it means that an article is quoting something that has already been quoted. For example, you might see:

```
In article <1993May17.002440.8495@nipper.ucsb.edu>
harley@nipper.ucsb.edu (Harley Hahn) writes:

> In article <1993May14.222614.9344@unix1.tcd.ie>
> mepeirce@unix1.tcd.ie (Michael Peirce) writes:

>> Does anyone know the name of the computer that
>> contains the Usenet archives?

> Yes, it is pit-manager.mit.edu.

Actually, the name has been changed to rtfm.mit.edu.

--
Jonathan Kamens     Geer Zolot Associates        jik@GZA.COM
```

In this example, a reader named Michael Peirce had originally posted an article asking for the name of a particular computer. Another reader, Harley Hahn, sent a followup article answering the question. A third reader, Jonathan Kamens, then created a second followup article in which he corrected Hahn's answer.

> ## Hint
>
> It can be tiresome to read followup articles in which someone has quoted a long article and added only a small comment to the end. This is especially true when you are following a thread in which person after person has quoted the same long passages.
>
> For this reason, when you create a followup article, it is considered good manners to delete as much superfluous material as possible. For example, if the original article was 100 lines long, but you are really responding to only 3 lines, you should delete all but those 3 lines.
>
> Actually, some newsreader programs will refuse to post a followup article if the quoted material is longer than the reply.

FUN TIP: From time to time, you will see people propagate a long series of silly followups, called a cascade. *Sometime when you have nothing to do and you are in a capricious mood, check out the* **alt.cascade** *newsgroup (which is devoted to such creative endeavors).*

Usenet Acronyms

What would a computer system be without acronyms? Those marvelous little abbreviations that make you feel so important when you know what they mean and like a piece of cheese when you don't.

Not to fear. Table 10-2 contains all of the common Usenet acronyms. Not every possible acronym that you will ever encounter, but enough to get you through the night.

Study this table well. These acronyms are used in all manner of Internet discourse, not only within Usenet articles, but in mail messages and when using the **talk** command (see Chapter 19) to have a conversation with somebody.

(BTW, the MUD that we mention is a multi-user game. For more information, look under the "Games" category in the catalog.)

Usenet Slang

Aside from the acronyms and abbreviations in Table 10-2, you will also encounter a great number of slang terms and expressions. So, to finish this chapter, let's take a few minutes and look at the world of Usenet slang.

Acronym	Meaning
BRB	be right back
BTW	by the way
CU	see you (good-bye)
FAQ	frequently asked question
FAQL	frequently asked question list
FOAF	friend of a friend
FYI	for your information
IMHO	in my humble opinion
IMO	in my opinion
MOTAS	member of the appropriate sex
MOTOS	member of the opposite sex
MOTSS	member of the same sex
MUD	multiple user dimension
Ob-	[as a prefix] obligatory
Objoke	obligatory joke
OS	operating system
PD	public domain
SO	significant other (spouse, boy/girlfriend...)
ROTFL	rolling on the floor laughing
RTFM	read the [expletive]* manual
WRT	with respect to

Table 10-2.   *Common Communication Acronyms*

*Expletive deleted by Editor as a public service.

To begin, let us recall the new terms that we have already mentioned in this chapter and in Chapter 9.

Usenet is a vast system of discussion groups. We refer to Usenet as the *news*, or *netnews*, although there is little real news (in the sense of a newspaper). Thus, the discussion groups are often called *newsgroups*.

A computer that acts as the Usenet repository for an organization is called a *news server*. The person who manages the news server is the *news administrator*.

Each news server gets its information from another news server on a regular basis. This arrangement is called a *news feed*.

The items that you read in a Usenet discussion group are called *articles* or *postings*. A Usenet article is divided into three parts: the *header*, the *body*, and an optional *signature*. When you send an article to Usenet, we say that you *post* it. An article that is sent to more than one newsgroup is *cross-posted*. Usenet articles are kept for a predetermined time. When the time is up, the articles *expire* and are deleted from the news server.

An article that responds to another one is called a *followup article*. When a followup article contains parts of the original, we say that it *quotes* the original article. A series of followup articles is called a *thread*. When you read such articles, we say that you are *following* the thread.

There are over 5,000 different newsgroups, organized into *hierarchies*. *Mainstream hierarchies* are carried by all news servers. *Alternative hierarchies* are carried according to the wishes of the news administrator.

Moderated newsgroups are managed by a person called a *moderator*. The moderator decides which articles will be posted to the group. Such groups sometimes offer a collection of edited articles called a *digest*.

To read Usenet articles, you use a program called a *newsreader*. When you tell your newsreader that you want to start reading a particular group, you are *subscribing* to that group. Similarly, when you tell your newsreader that you do not want to read a group any longer, you are *unsubscribing*.

(We hope you have all of that, because there will be a short quiz at the end of the class.)

Now here are some new terms. We mentioned in Chapter 9 that anyone who knows how to do so can try to start a newsgroup in one of the alternative hierarchies. As you might imagine, such latitude often leads to newsgroups that don't really exist. Someone might send out a control message to start a new group, but for one reason or another, the group never gets started.

Nevertheless, the name of such a newsgroup may find its way into the master list on your news server. When your newsreader tries to find this group, it won't be there. Your newsreader will proudly inform you that it has found a *bogus* or empty newsgroup.

Another term you will encounter frequently is *flame*. This refers to a followup article (or a personal mail message) in which someone says something critical about someone else. The word "flame" is also used as a verb, as in "Scott posted an article without a joke to **rec.humor**, and he was flamed from all over the world."

All too frequently, a *flame war* will start, in which people send a large number of argumentative articles and mail messages excoriating one another. In such situations, it doesn't take long for other people to jump in and criticize the critics which, as you might imagine, only fans the flames. Although flame wars can have their interesting moments, most such postings are usually the written equivalent of each person sticking his or her tongue out at someone else. Eventually, the thread will die out, more from boredom than anything else.

As you are reading an article, you can tell your newsreader that you want to skip all the rest of the articles in the same thread. We say that you *kill* or *junk* the thread. Some newsreaders will let you specify that all articles with certain subject descriptions should be killed automatically. Your newsreader will save such requests in a special *kill file*.

Some newsgroups discuss works of art such as movies or novels in which it would be reasonable to discuss the plot. Of course, if you have not yet seen the movie or read the book, you may not want to find out the plot ahead of time. For this reason, it is customary that, when an article gives away a plot, the **Subject:** header line should say that the article contains a *spoiler*.

Well, that's it: about all the basic terminology that you need to know to understand Usenet. However, to complete your basic education, there are five more terms that you need to know. They are so important, in fact, that we will give them their own sections.

Foo, Bar, and Foobar

There are three marvelous words that you will see a lot: *foo*, *bar*, and *foobar*. These words are used as generic identifiers throughout Usenet as well as within the world of Unix.

The idea is that whenever you want to refer to something without a name, you can use "foo" (or less often, "foobar"). When you want to refer to two things without a name, you use "foo" and "bar". Nobody knows how this tradition got started, but it is used a lot.

For example, say you are reading an article in the **comp.unix.questions** newsgroup. (This is the group to which you can send Unix questions for experienced people to answer.) Someone is asking about editing files. You read:

```
Can anyone tell me how to move more than one file
at a time?  For example, say that I want to move two
files named foo and bar.  I tried using the command
'mv foo bar' but I got an error message...
```

Or you might read the following in the **rec.arts.movies**:

```
...Can anyone remember the musical in which Frank Sinatra
played an old-time Chicago gangster?  The name is something
like "Foobar and the Seven Hoods"...
```

So, where do these strange words come from? The word "foobar" derives from the World War II acronym FUBAR, which meant "fouled up beyond all recognition".

The word "foo", however, seems to have a more robust history. There is no doubt that foo owes much of its popularity to foobar. Nevertheless, it seems to have been used on its own even earlier. For example, in a 1938 cartoon, Daffy Duck holds up a sign that reads, "Silence is Foo" (which is absolutely correct). Some authorities speculate that foo might have roots in the Yiddish "feh" and the English "phoo".

RTFM

The term *RTFM* embodies the single, most important idea in the world of the Internet. RTFM means that before you ask anyone else for help, you should try to solve your problem by checking with a book or manual.

This idea is not mean-spirited; after all, the Internet in general (and Usenet in particular) is filled with people who will be glad to help you with anything. However, it is usually faster, and a whole lot more satisfying, if you can find an answer for yourself. Moreover, all new users tend to ask the same questions, and it is understandable that many people feel that, before you post an article asking for help, you should at least check all the standard references.

This raises the question: What are the standard references? Books, of course, like this one, as well as technical manuals. However, there is one standard reference for all Unix systems that is so important that you *must* know how to use it.

Every Unix system has an *online manual*. This is a computerized facility that will display help information about any Unix command. Using the online manual takes a while to learn, and we won't go into all the details here—for that you will have to read a Unix book. However, here is a brief summary.

To display information about a Unix command, you use the **man** (manual) command. Enter **man** followed by the name of the command you want to learn about. For example, to learn about the Unix **cp** (copy files) command, enter:

```
man cp
```

Unix will display a technical description of the specified command, one screenful at a time. As you are reading, you can move to the next screenful by pressing SPACE (on some systems) or the RETURN key (on other systems). (Try both and see which one works on your computer.) If you want to quit reading, press the **q** key.

Hint

To display help about the **man** command itself, enter:

```
man man
```

To display the official manual for a newsreader program, use **man** with the name of your newsreader. For example, to display the manual for the **nn** newsreader, enter:

```
man nn
```

Please don't forget: Before you post a question that has something to do with a Unix command, you should at least use **man** to check the online manual. Then, if you still don't have the answer you need, feel free to ask the Net. If you do ask a question that is answered in the manual, you may be gently (or not so gently) reminded to RTFM.

Now, where does this funny term come from?

Originally, RTFM was an acronym that stood for:

Read the [expletive]* manual

However, through the years, RTFM has taken on the more refined meaning of "try to find the answer yourself before you ask someone". (Actually, you will be surprised how often you *can* find the answer yourself.)

The term RTFM is also used as a verb, as in:

```
...Does anyone know how to save previously read articles to a file
using the foo newsreader?  I RTFM'ed, but I couldn't find the
answer anywhere.
```

Expletive deleted by Editor as a public service.

By now, it should make sense to you why the computer that contains the Usenet archives is named **rtfm.mit.edu**. This computer holds copies of all the frequently asked question (FAQ) lists. (And in Chapter 16, we will show you how to use Anonymous FTP to access such information.)

Smileys

The last Usenet term we will discuss in this chapter is both cute and useful. It is called a *smiley*, and it is used to indicate irony. Here is how it works:

When you talk with someone in person, you use your body language and voice inflections to project all kinds of non-verbal messages. For example, you can jokingly insult someone and get away with it (at least sometimes) as long as you get across the idea that you are just kidding.

With a Usenet article, this is not possible. Moreover, Usenet articles are read all over the world and not everybody will be able to appreciate the subtle nuances of another culture's humor. For example, it is altogether possible that someone outside the U.S. would not understand that the word "mother" is not always a term of endearment.

Enter the smiley.

A smiley is a small drawing, using only regular characters, that looks like a face. Here is the basic smiley:

```
:-)
```

To see the bright, smiling face, all you need to do is turn your head to the left.

We use smileys to make sure that someone does not accidentally misinterpret a potentially ambiguous remark. Putting a smiley at the end of a sentence is sort of like saying "just kidding".

For example, say that you are participating in a flame war in **rec.food.cooking**. You might post a followup article with the sentence:

```
How do you expect people to use your recipe for
Consomme aux Pommes d'Amour when you don't even
know how to cook groat cakes :-)
```

Hint

Smileys are used all over the place, not only within Usenet articles. For example, you can use them in private mail messages and when you are having a conversation by using the **talk** command (see Chapter 19).

Throughout the years, people have developed many different smileys. Indeed, there are vast collections of all kinds of strange smileys that are occasionally posted to **rec.humor**. Figure 10-3 shows a few such creations.

Smiley as Typed	Smiley Sideways	Meaning
: -)		smiling
: - D		laughing
; -)		winking
: - (		frowning
: - I		indifferent
: - #		smiley with braces
: - {)		smiley with a mustache
{ : -)		smiley with a toupee
: - X		my lips are sealed
= : -)		punk rocker
= : - (		real punk rockers don't smile

Figure 10-3.   *A selection of smileys*

Now, to conclude this chapter, here is one last smiley:

 %-)

Turn your head to the left. This is the face of a writer of Internet books who has stayed up all night to finish a chapter.

CHAPTER

11

Reading and Posting Usenet Articles

In Chapter 9, we talked about Usenet. We saw that there are over 5,000 different discussion groups with more than 2,500,000 participants. In Chapter 10, we took a look at what Usenet articles look like and what technical terms and conventions you can expect to see.

In this short chapter, we will explain what you will need to understand in order to read and post Usenet articles. We will start by showing you how your newsreader program is part of a client/server system. We will then talk about the various newsreaders. We will show you how to decide which newsreader will suit you best and how to go about learning it.

To conclude the chapter, we discuss a couple of technical topics that pertain to reading the news.

In the following chapters, we discuss the four most popular Unix newsreaders: **rn** (Chapter 12), **trn** (Chapter 13), **nn** (Chapter 14), and **tin** (Chapter 15). After you finish this chapter, turn to whichever newsreader chapter pertains to you, at which time you will (finally) start to read the news.

Understanding Your Newsreader as a Client

Like most Internet services, news reading makes use of a client/server arrangement (see Chapter 2). In this case, the news server resides on one computer in your network. Its job is to maintain a news feed to the outside world and to manage the large number of articles and newsgroups that are received every day.

To read the news, you use a client program, called a *newsreader*. The newsreader acts as your interface to Usenet, enabling you to choose the newsgroups to which you want to subscribe, select articles from these newsgroups, and display the articles one page at a time. As you read, you can preserve an article by saving it to a file, by mailing a copy to someone, or by printing it. You can also respond to the article, either by sending a private mail message to the author or by posting a followup article of your own.

As we explained in Chapter 10, news articles are transported throughout the Internet using a protocol called NNTP (Network News Transfer Protocol). As articles arrive at a news server, they are stored in a standard format. To read the news, all you need is a newsreader program that knows how to understand this format. There are many different newsreader programs, written to operate on a large variety of computers.

However, as we discussed in Chapter 1, many of the people who use the Internet use some type of Unix computer. In this chapter, we will discuss the general ideas involved in reading the news and posting your own articles. In the following chapters, we will discuss the details of the four most popular Unix-based newsreaders.

How Does a News Server Keep Track of All the Articles?

In order to keep track of the articles in each newsgroup, your news server assigns an identification number to each article. The numbers are assigned in the order in which the articles arrive.

The numbering starts with **1**. Whenever a new article comes in, it is assigned the next number in the sequence. For instance, say that a new article arrives for the **rec.humor** newsgroup and it happens that the last such article was number **1055**. So, the new article will be assigned the number **1056**. The next article in this newsgroup will be called **1057**, and so on. Some newsreaders show you this number each time you read an article. Other newsreaders don't bother.

Once an article expires, its number is removed from the system. Thus, on a particular day, the **rec.humor** newsgroup might have articles **1055** through **2110** available. Eventually, the numbers will reach some specific maximum value and the numbering will start again with **1**.

Each newsgroup has its own, separate set of numbers, so it is possible that more than one group may have articles with the same numbers. For example, both **rec.humor** and **comp.unix.questions** newsgroups may have articles numbered **1056**.

Your newsreader uses these numbers to keep track of which articles you have read for each newsgroup. To do so, it maintains a file named **.newsrc** in your home directory. This file contains the names of each newsgroup. For each name, there is information showing if you are subscribed to that newsgroup and which articles you have already read.

You don't have to worry about what's in your **.newsrc** file unless you want to. The first time you read the news, your newsreader will create a **.newsrc** file for you. From then on, your newsreader will maintain the file on your behalf, making changes whenever you read an article and whenever you subscribe or unsubscribe to a newsgroup. In addition, each time you start the newsreader, it will check if new newsgroups are available and update your **.newsrc** file appropriately.

Some people like to use a text editor to make changes to their **.newsrc** file directly. For this reason, there is a section at the end of this chapter that will show you the format of the **.newsrc** file along with some suggestions for making your own changes should you decide to do so.

Hint

In Unix, it is customary for files that contain initialization commands or special information to be given names that start with a . (period) character and end in **rc**. For example, we not only have **.newsrc** for newsreaders, but **.mailrc** for the mail program, **.cshrc** for the C-Shell, **.exrc** for the **ex** and **vi** text editors, **.bashrc** for the Bash shell, and so on.

The initial **.** means that these are hidden files: when you use the **ls** command to list your files, you do not see such names unless you specifically ask for them. When you pronounce names like this, the **.** character is called "dot". For example, you might hear someone ask, "What newsgroups are in your dot-news-r-c file?" For this reason, such files are sometimes called *dotfiles*.

The **rc** is an old convention that originally stood for "run commands", that is, commands to be run automatically each time a program started.

How to Choose a Newsreader

In this section, we will introduce you to the most popular Unix newsreaders—**rn**, **trn**, **nn**, and **tin**—and show you how to choose among them (if you have a choice).

Although these are the most widely used newsreaders, you may have other choices if you work within either X Window or Gnu Emacs. As we explained in Chapter 2, X Window provides a graphical user interface in which you can use more than one program at a time, each within its own window. There are a few newsreaders specifically designed to take advantage of X Window, the most notable being **xrn** and **xvnews**. Both of these newsreaders are modelled on **rn** (described below), but have extra features.

Another special working environment is Gnu Emacs. We won't go into the details here; suffice it to say that this system is built around a popular text editor named Emacs. There are two newsreader facilities designed specifically for Gnu Emacs, **gnus** and **gnews**.

For the most part, however, the most widely used newsreaders are the four that we mentioned above and those are the only ones that we will talk about in detail.

rn is the oldest of these newsreaders. It was designed by Larry Wall of the System Development Corporation in Santa Monica, California, and first released on April 8, 1983. (Wall is also the creator of Perl, a popular programming language.) Today, **rn** is maintained by Stan Barber, at the Baylor College of Medicine in Houston, Texas.

Wall developed **rn** to replace an older newsreader named **readnews**. He carefully designed **rn** to minimize the interaction between the program and the user in order to display information as quickly as possible. The result is a newsreader that displays as many articles as possible in the shortest amount of time.

trn was created by Wayne Davison of the Borland Corporation, in Scotts Valley, California. **trn** is a modern variation of **rn** that was first released on July 21, 1990. It has all the features and commands of **rn** along with some extra capabilities. Unlike **rn**, **trn** groups articles into threads and makes it easy to choose which ones you want to read. (As you may remember from Chapter 10, a thread is a connected series of articles: that is, the original article and all of the followups.)

nn was created by Kim Storm of Texas Instruments A/S in Denmark. It was first released in Denmark in 1984, in Europe in 1988, and worldwide in July 1989. Today, **nn** is maintained by Peter Wemm of Demaas Proprietary Limited in Australia.

Storm designed **nn** at a time when Usenet had grown large and there were far more articles than people could read. Even confining yourself to your favorite newsgroup often meant that you still had too many articles to wade through. Thus, Storm designed **nn** to make it easy to scan a large number of articles quickly. The trade-off is that this newsreader requires more user interaction than **rn** or **trn**.

tin is the newest of the four newsreaders. It was created by Iain Lea and first released on August 23, 1991. By that time, Usenet had become huge. Not only were there more articles than a single person could deal with, there were too many newsgroups. People who liked to follow a variety of discussions quickly found that the number of newsgroups was unmanageable. Lea designed **tin** to make it easy to scan a large list of newsgroups and choose which ones you want to investigate. Once you choose a group, **tin** presents you with a list of articles—somewhat like **nn**—and lets you choose which ones you want to read.

In other words, **rn** presents articles quickly and with a minimum of user interaction. **trn** works like **rn** except that it makes it easy to work with entire threads. **nn** requires more interaction, but lets you scan a large number of articles quickly. **tin** requires the most interaction, but it allows you to handle a large number of newsgroups comfortably and, within each group, to scan a large number of articles.

So, if you have a choice of newsreaders, here is how to decide: If you want to read most of the articles in a few newsgroups, use **rn** or better yet, **trn**, if it is available. If you want to pick and choose articles from a small number of newsgroups, use **nn**. And if you want to read a large number of newsgroups, use **tin**.

> **Hint**
>
> To choose which newsreader is best for you, consider how you like to read a newspaper.
>
> If you read a newspaper thoroughly, one page after another, you are an **rn** person.
>
> If you read each article as a whole, turning pages back and forth when necessary, you should use **trn**.
>
> If you scan the newspaper quickly, reading only what strikes your fancy, choose **nn**.
>
> And finally, if you like to read a whole pile of newspapers, zipping through one after the other, you will love **tin**.

As you can imagine, one of the eternal arguments among Usenet people is: Which newsreader is the best? As you can see, the best newsreader for you depends a great deal on how your eye and your mind assimilate information. When it comes to newsreaders, one size does not fit all.

If you would like to use a newsreader that is not on your system, show this discussion to your system manager, and ask him or her (nicely) to install the program. All of the newsreader programs that we mentioned in this section are available for free via Anonymous FTP (which is explained in Chapter 16).

FUN TIP: You might be wondering, which is our favorite newsreader? Well, we use all four newsreaders at various times. However, if we had to pick a single newsreader to take with us to a desert island, we would choose **tin**.

(Don't laugh about reading the Usenet news on a desert island. The Pagesat company of Palo Alto, California, offers an economical Usenet system that receives data via a small, personal satellite dish. For details, you can send a message to: **pagesat@pagesat.com**.*)*

What Do the Newsreader Names Mean?

Are you wondering what the names of the newsreaders mean?

Before **rn**, there was an older newsreader named **readnews** which was slow and displayed information line by line. Larry Wall designed **rn** to be a fast, screen-oriented replacement for **readnews**. The name **rn**, then, indicates that this "read news" program is sleeker than the older one it replaced.

The name **trn** is straightforward: it stands for "threaded **rn**". (You will remember that **trn** is like **rn** except with the added capability of working with entire threads of articles.)

If you use **rn**, you will see that it shows you at least the first page of each article that you scan. Kim Storm designed **nn** so that the only articles you need to look at are the ones you choose to read. He felt that the less news you read, the better. Thus, the name **nn** stands for "No News (is good news)".

Finally, we have **tin**. This newsreader was developed by Iain Lea and based on an older newsreader named **tass**. Lea chose the name **tin** to stand for "Tass + Iain's Newsreader".

Learning How to Use a Newsreader

In Chapters 12, 13, 14, and 15, we will discuss the details of the four newsreaders that we mentioned earlier: **rn**, **trn**, **nn**, and **tin**. At that time, you will actually start reading the news. Of course, you only need to learn about the newsreader that you are using, so you can skip the other chapters.

Before you start learning about your particular newsreader, we want to spend a little time going over some general information that will help you no matter which newsreader you use.

You will find that all the newsreaders have many, many commands. Don't bother trying to learn them all straight off. And don't for a minute worry because there are so many commands that you don't understand right away. Begin by learning the basics: how to start the newsreader, how to select a newsgroup to read, and how to read articles. All the newsreaders are designed so that, most of the time, you need only a small handful of commands.

You will find that the better you are at controlling your newsreader, the more enjoyment you will get out of Usenet. Eventually, you will want to learn most of the commands, but the best way to do it is one command at a time.

At various times, you will have different commands available to you. For example, when you are selecting which newsgroup you want to read, you will have different commands at your disposal than when you are reading an article. In the newsreader chapters, we will provide you with summaries of the various commands. You can use these summaries for reference as you read the news.

In addition, each newsreader has a special help command that you can use at any time. With **rn**, **trn**, and **tin**, the command is **h**. With **nn**, the command is **?**. Whenever you want, all you have to do is press this key to display a summary of whatever commands are available.

> **Hint**
>
> The way to learn to use a newsreader well is the same way that you get to Carnegie Hall: practice, practice, practice. Fortunately, reading the news is a lot more fun (and a lot quieter) than learning to play the violin.

In the following chapters, we will introduce you to each newsreader in turn. However, we don't have the room (or the stamina) to discuss every detail. Once you start reading the news, you may want to check the official documentation for your particular newsreader. To do so, you can use the online Unix manual. Here is how it works.

All Unix systems come with a built-in reference manual. This manual contains a file of information about each command. To display this information, you use the **man** command. Simply enter **man**, followed by the name of the command you want to learn about.

For example, to learn about the **ls** command (which you can use to list the names of your files), enter:

```
man ls
```

To learn about using the **man** command itself, enter:

```
man man
```

Thus, to read the official reference manual for your newsreader, all you have to do is enter **man** followed by the name of the newsreader program. The choices are:

```
man rn
man trn
man nn
man tin
```

Bear in mind that what you will see is reference information, not introductory tutorials.

When you enter such a command, **man** uses a *paging program* to display the output. The job of a paging program is to display information, one screenful at a time. The custom is to speak of each screenful of information as a *page*. Thus, we can say that the job of a paging program is to display information, one page at a time. After you read what is on the screen, you press a key to display the next page.

Like much of Unix, paging programs come with many options and a large number of commands. We won't go into the details here. Suffice it to say that the **man** command on your system will be using one of three common paging programs: **more**, **pg**, or **less**.

If you are using **more** or **less**, you can display the next page by pressing SPACE. If you are using **pg**, you press the RETURN key instead. In all three programs, you can press **h** for help and **q** to quit. If you are not sure which pager you are using, try pressing both SPACE and RETURN and see what happens.

Posting Your Own Articles

As a Usenet reader, you will, from time to time, desire to post your own articles. You may want to request some information, respond to a previous article, offer a new idea for discussion, or whatever. Each newsreader has its own commands to make posting articles simple and easy. We will go over the details in the appropriate chapters.

In this section, we would like to take a moment to discuss a few of the general considerations that apply to posting articles.

First, you must understand that it is not easy to post an article of any type unless you know how to use a text editor program. After all, your article does not even exist until you create it. Moreover, if you are responding to someone else's article, you will probably want to edit what they have said and intersperse your own comments. All of this requires a text editor.

The two most common Unix text editors are **vi** and **emacs**. Life being what it is, both of these editors are complex and take a while to learn. If you are using a PC or a Macintosh that attaches to a remote host by emulating a terminal (see Chapter 3), you may be able to compose an article on your own computer and send it to the host. Still, it is not the best way to compose articles.

The real truth is that, if you are using a Unix system, you must learn how to use a Unix text editor. You can get by without one, but your life on the Internet will not be as satisfying as it could be. (For example, it will not be easy to post your own articles to Usenet.)

If you are not sure which text editor to learn, use **vi**. It is a universal standard, available on every Unix system. Don't be intimidated by people who may tell you that **vi** is difficult to learn. All it takes is a little practice with a good book.

> ## Hint
>
> If you are using a Unix system, it is well worth your time to learn how to use the **vi** text editor.

FUN TIP: The best book to use to teach yourself **vi**—*and Unix in general—is Harley Hahn's* A Student's Guide to Unix *(McGraw-Hill 1993).*

The next point to appreciate is that the best way to respond to an article may not be to send a followup. All the newsreaders make it easy to send a personal mail message to the author of an article. Ask yourself if your response will be interesting to everyone or just to the author. Unless what you plan to say is of general interest, send a response by mail.

> ## Hint
>
> In many cases, you are better off responding to an article by sending personal mail to the author than by posting a followup that will go all over the world.

When you do send a followup, your newsreader will make it easy for you to quote (include) all or part of the previous article within your followup. It is considered good manners to edit out all the parts of the original article that are not relevant to your followup. (See Chapter 10 for the details.) It is irritating to read a followup in which the writer has quoted a long article and merely added a few new lines to the end.

> ## Hint
>
> When you create a followup, make it easy for other people to read by deleting all the superfluous parts of the original article.
>   As a rule of thumb, the quoted part should not exceed your own comments. (Some newsreader programs enforce this rule.)

When you first start to post articles, you will probably want to send out a practice article or two. You should know that there are special newsgroups just for such tests. In Table 11-1, we have listed the practice newsgroups in the mainstream and alternate hierarchies. Notice that all the names end in **.test**. At any time, you can send a posting to one of these groups just for practice.

If your location has a local hierarchy (see Chapter 9), you may find a local test newsgroup. If so, this is the best place to send a practice posting. Otherwise, you should use **misc.test** (to practice posting to a mainstream newsgroup) or **alt.test** (to send to an alternate group).

*REMEMBER: When you want to practice posting an article, use only the **.test** newsgroups. It is considered bad manners to send a test posting to a non-test newsgroup.*

We mentioned in Chapter 10 that you can append a signature to the end of your postings by creating a file named **.signature** in your home directory. When you are testing out a new signature, you can send a practice posting to one of the newsgroups in Table 11-1 to see if the new signature looks okay.

*FUN TIP: There are a number of computers around the Internet that run a special program that looks for and responds to postings in the **.test** newsgroups. When you post such an article, you may get an automated response, by mail, telling you that your practice article has arrived at such-and-such computer.*

Newsgroup Name
`alt.test`
`bit.listserv.test`
`biz.test`
`gnu.gnusenet.test`
`u3b.test`
`vmsnet.test`
`misc.test`

Table 11-1. *Newsgroups for Practice Postings*

The last point that we would like to mention is that Usenet has a number of conventions and traditions that you should understand before you post an article. We discussed these ideas in Chapter 10. In particular, before you send a question to a newsgroup, make sure that you have read the FAQ (frequently asked question) list if that group has one.

If you are a new user, there is a group set up especially for you called **news.newusers.questions**. There are a number of regular postings sent to this newsgroup that explain all about Usenet. This is also the place for you to send a question that pertains to using Usenet. If you have a question about using a newsreader, you can send the question to the **news.software.readers** group.

The Format of the .newsrc File

As we explained earlier in the chapter, your newsreader maintains a file named **.newsrc** to keep track of which newsgroups you subscribe to and which articles you have read.

In this section, we will describe the format of this file in case you know how to use a text editor and you want to modify your own **.newsrc** file. If you don't care about such matters, you can skip this section with impunity.

There are two reasons why you would want to edit your own **.newsrc** file. First, you will probably want to change the order of the newsgroups. When you start your newsreader, it will look at the newsgroups in the order it finds them. By editing your **.newsrc** file, you can put the more interesting groups first. You can use your newsreader to make such changes, but it is usually a lot easier to edit the **.newsrc** file and do it yourself.

The second reason to edit your **.newsrc** file is to make changes to your newsgroup subscriptions—that is, to specify exactly which newsgroups you want to read. With some newsreaders, you will automatically start off by being subscribed to every newsgroup. Normally, you would want to unsubscribe to all but a few of these groups. Again, this is something that you can do with your newsreader, but if you want to make a lot of changes, editing the file by hand is faster.

Hint

Before you edit your **.newsrc** file, make a backup copy. That way, if you ruin the original, you can restore it from the copy. This is good advice for editing any important file.

The **.newsrc** file is kept in your home directory and maintained automatically by your newsreader. All of the newsreader programs use this same file and there is only one format.

There is one line for each newsgroup. This line contains:

- the name of the newsgroup

- a colon (:) or an exclamation mark (!)

- a space

- a list of numbers

If there is a colon after the newsgroup name, it means that you are currently subscribed to that group. An exclamation mark means that you are unsubscribed. The list of numbers indicates which articles you have already read. (You will remember that each article is given an identification number as it arrives at the news server.) The list of numbers contains single numbers or ranges of numbers, separated by commas.

Here are a few lines from a typical **.newsrc** file:

```
alt.fan.wodehouse: 1-819
rec.humor.funny: 1-8192
rec.humor: 1-41234,41236,41239
comp.unix.questions! 1-6571
misc.books.technical!
```

You can see that the first three newsgroups are subscribed and the last two are unsubscribed. Within each newsgroup, you can also see which articles have been read. For example, in the **rec.humor** group, articles 1 through 41,234, plus articles 41,236 and 41,239, have been read.

If there are no numbers on a line, it means that no articles have ever been read in that newsgroup. This is the case with **misc.books.technical**.

Finally, notice that although **comp.unix.questions** is unsubscribed, it was subscribed at one time. You can tell because articles 1 through 6,571 have been read.

Hint

Edit your **.newsrc** file to put all your favorite newsgroups first.

Rot-13

As we explained in Chapter 9, Usenet has no central authority and no enforceable rules. The content of the newsgroups is decided by the participants, and decorum is maintained by voluntary cooperation.

One of the problems that arises, from time to time, is what to do about offensive articles. This occurs most often in the humor newsgroup **rec.humor** (and to a lesser extent in **rec.humor.funny**), where you will see jokes that some people would consider to be offensive.

One side of the argument is that Usenet should be free and unfettered, and that anyone has the right to post anything he or she wants. The other side is that it is fundamentally wrong to post articles that are patently offensive. Like all such arguments, this one has never been resolved and, no doubt, never will be.

The solution is to encode a potentially offensive article so that, unless you use a special command, the text looks like gibberish. The person who posts the article describes in the subject line who the article may offend. For example, you might see an article that looks like gibberish with a subject line of **Offensive to Americans**, or **Offensive to Italians**, or whatever.

You can decide whether or not you want to read the article. If you do, you type the special key—the exact command depends on which newsreader you are using—and the article will be decoded for you. Since someone would have to use a special command to decode the article, no one can complain about being offended accidentally.

The scheme that is used to encode an article is called *rot-13*. Rot-13 uses a simple replacement cipher in which each letter is rotated 13 positions within the alphabet. The actual substitutions are shown in Figure 11-1. For reference, Table 11-2 shows the commands to use in order to read a rot-13 encoded article. Note that rot-13 only encodes letters of the alphabet, not numbers or punctuation.

The rot-13 convention is used to encode more than offensive jokes. It is used for any article that you would not want someone to read by accident, for example, articles that give away the ending to a book or movie, answers to puzzles, answers to riddles, and so on.

On a Unix system, you can encode or decode text by using the **tr** (translate) command. We won't go into the details, but if you want to experiment, here is the actual command that will encode and decode text using the rot-13 scheme. (In this command, *infile* is the name of your input file, and *outfile* is the name of your output file.)

```
tr '[a-m][n-z][A-M][N-Z]' '[n-z][a-m][N-Z][A-M]' <infile >outfile
```

a → n	n → a	A → N	N → A
b → o	o → b	B → O	O → B
c → p	p → c	C → P	P → C
d → q	q → d	D → Q	Q → D
e → r	r → e	E → R	R → E
f → s	s → f	F → S	S → F
g → t	t → g	G → T	T → G
h → u	u → h	H → U	U → H
i → v	v → i	I → V	V → I
j → w	w → j	J → W	W → J
k → x	x → k	K → X	X → K
l → y	y → l	L → Y	Y → L
m → z	z → m	M → Z	Z → M

Figure 11-1. *The substitutions used to implement the rot-13 code*

If you want to practice decoding, use this **tr** command on the following rot-13 encoded text:

```
"Jura lbh trg frevbhf nobhg ohyyfuvg,
lbh ner trggvat vagb frevbhf ohyyfuvg."
-- Pnanqvna cuvybfbcure, Gvz Ehgyrqtr
```

If you want to post a rot-13 encoded article, there is no built-in way to do it with most newsreader programs. You will have to create the article with a text editor, and then use a command like **tr** to encode the article before you post it.

Newsreader	Command
rn	CTRL-X
trn	CTRL-X
nn	D
tin	d

Table 11-2. *Commands to Read a Rot-13 Encoded Article*

FUN TIP: *Unlike* **rec.humor** *and* **rec.humor.funny**, *the jokes in* **alt.tasteless.jokes** *are never encoded using rot-13. This is because, by definition, all the jokes in this newsgroup are supposed to be tasteless. Indeed, if you send a joke in good taste to* **alt.tasteless.jokes**, *you will be told that such jokes should be sent to* **rec.humor**.

If you complain about a joke in **alt.tasteless.jokes**, *no one will pay any attention. There is no rot-13 protection, so, if you are easily offended, don't even look at this newsgroup. (By now, you probably can't wait to read it, so move on to the next chapter.)*

## CHAPTER 12

The rn Newsreader

We have talked a lot so far about Usenet, the worldwide system of over 5,000 discussion groups. In this chapter, we will explain how to read the news and post your own articles using the **rn** newsreader. Before you read this chapter, you should at least have skimmed the previous Usenet chapters. In particular, you should understand what Usenet is and how it works (Chapter 9), the conventions and terminology that we use (Chapter 10), and what is involved in using a newsreader (Chapter 11).

Overview of the rn Newsreader

The **rn** program was designed to allow you to read articles quickly and easily with a minimum amount of intervention on your part. The program has many options and commands, and offers far more functionality than you will ever need. The best strategy is to learn the basics and practice for a while. Once you have some experience, you can look at the command summaries at the end of this chapter and experiment with whatever looks interesting.

To introduce you to **rn**, we will start with an overview of how it works. In the following sections, we will go into the details.

You start the program by entering the **rn** command. **rn** checks with your **.newsrc** file (see Chapter 11) and sees if there are any new newsgroups that are not in your file. If so, **rn** will present them to you, one at a time, and ask if you want to subscribe.

Once this is done, you are now ready to start reading articles. **rn** shows you the name of a newsgroup. You can accept this newsgroup or choose a different one. Once you choose a newsgroup, **rn** displays the first page of the next unread article in the group. You can now decide whether or not you want to read the entire article. If so, **rn** will display the article, one page at a time. If not, **rn** will show you the first page of the next unread article.

When you finish an article, **rn** displays the next one automatically. When the last article in a newsgroup is read, **rn** moves to another group. At any time, you can skip to a new article or to a different group.

In other words, at all times you are in one of three situations: selecting a newsgroup, examining an article, or paging through an article. (If this seems a bit confusing, don't worry. It will all make sense when we explain the details and when you try **rn** for yourself.)

At each point, there are a large number of commands available. As a reference, we have summarized all of the important **rn** commands in three lists at the end of the chapter.

Figure 12-1 shows the commands you can use when selecting a newsgroup. Figure 12-2 shows the commands for when you are looking at the first page of an article and deciding whether or not you want to read the whole thing. And Figure 12-3 shows the commands that are available as you are paging through an article. You may want to take a minute now to glance at these summaries.

Within the summaries, we have followed the common Unix convention of indicating CTRL keys by using the ^ character. For example, when you see ^**N**, it means CTRL-N. (That is, hold down the CTRL key and press **N**.)

In the following sections, we will discuss the details of using **rn**. As you read, and as you practice, there are four things that we would like you to remember.

First, at any time, you can press **h** (help) to display a summary of whatever commands are available.

Second, to stop a command that is in progress, press CTRL-C.

Third, for all single-letter commands, you do not have to press the RETURN or ENTER key after you type the command. All you have to do is press a single key and **rn** will react immediately. For example, to display help information, you press the **h** key. (When a Unix program reads input in this way, we say that it is operating in *cbreak mode*.)

Finally, whenever you have to choose a command, **rn** will present you with a short list of possibilities. For example, you might see:

    [ynq]

In this example, you can press **y** (for yes), **n** (for no), or **q** (for quit). If you are not sure what the choices mean, or if you would like to see what other choices you have that are not listed (and there are always some), press **h** (for help).

Whenever you are called upon to type a command, you can press SPACE to select the first choice by default. In the example above, pressing SPACE would be the same as pressing **y**. The defaults are set up so that the first choice is usually the one you want.

Hint

You can read news all day just by pressing SPACE.

Starting and Stopping

If you have never read the news before, you will not have a **.newsrc** file (see Chapter 11). Before you use **rn** for the first time, you can enter the command:

    newsetup

This will create a brand new **.newsrc** file for you containing every newsgroup that is carried by your news server. If you start **rn** for the first time without first using **newsetup**, **rn** will run **newsetup** for you automatically.

When your **.newsrc** file is created, every newsgroup will be marked as being subscribed. You will probably want to unsubscribe to most of the groups. There are two ways to do this. First, you can do it within **rn** by using the **u** command, although this will probably take a long time. The more practical method is to use a text editor, such as **vi** or Emacs, and edit the **.newsrc** file for yourself (see Chapter 11).

To start **rn**, you enter the **rn** command. There are several forms of this command and a large number of options that you can use. In this section, we will cover the most important of these choices.

First, you can enter the command by itself:

    rn

This starts **rn** in a straightforward manner.

To stop **rn**, press the **q** (quit) key. If you are reading or selecting an article, you may have to press **q** more than once to work your way back out to the top level. (This will make sense when you do it.)

Alternatively, if you are at the top level (selecting a newsgroup) you can quit **rn** by pressing the **x** (exit) key. This not only quits **rn**, but restores your **.newsrc** file to what it was when you started. This abandons all the changes you may have made, such as subscribing or unsubscribing to newsgroups, reading articles, and so on.

Most of the time, there is a better way to start **rn**. You will remember we said that the first thing **rn** does is check to see if there are any new newsgroups that are not listed in your **.newsrc** file. Although this is a good idea, it can be time consuming, perhaps taking several minutes each time.

Most people prefer to have **rn** skip this check. You can do so by using the command:

    rn -q

The **-q** is called an **option** or a **switch**. In this case, the **-q** option tells **rn** to perform a "quick" start. Using **-q** will start **rn** faster, but will not, of course, pick up any new newsgroups. For this reason, we recommend starting **rn**, say, once every week or two, without the **-q** option, just to see what is new.

There may be times when you are interested in reading only a single newsgroup. In that case, you can specify that group directly. For example, say that you only want to read the **rec.humor** newsgroup. You can enter:

    rn -q rec.humor

There is one special way to use the **rn** command that we want you to know about and that is by using the **-c** (check) option:

```
rn -c
```

This will not allow you to read news. Rather, **rn** will simply report to you if there is any unread news in the groups to which you are subscribed and then stop. You can use this command whenever you want to check quickly if there is any news.

Hint

You can place a **rn -c** command in your initialization file so that each time you log in you will be told if there is news that you have not yet read. The name of your initialization file depends on which shell you are using. For the C-Shell, the file is **.login**; for the Bourne or Korn shell, it is **.profile**; and for the Bash shell, it is **.bash_profile**.

Selecting a Newsgroup

When you start **rn**, it begins by showing you the first five subscribed newsgroups that have unread articles. The groups will be in the order that **rn** finds them in your **.newsrc** file. Here is a typical example:

```
Unread news in rec.humor.funny           11 articles
Unread news in rec.humor                 83 articles
Unread news in alt.tasteless.jokes       67 articles
Unread news in comp.unix.questions      121 articles
Unread news in biz.clarinet.sample        3 articles
etc.
```

Once **rn** has issued this initial report, it looks at the master list of active newsgroups on the news server to see if there are any new groups that are not in your **.newsrc** file (unless you have started **rn** with the **-q** option). If **rn** finds any such groups, it lists them one by one and asks you if you want to subscribe. For example:

```
Newsgroup alt.death-of-superman not in .newsrc--subscribe? [ynYN]
```

You must now make a choice. Press **y** (or SPACE) to subscribe or **n** to unsubscribe. If you want to subscribe or unsubscribe to all new groups without looking at all the names, press **Y** or **N** respectively. If you do press **y**, you will be asked to specify where you want the new group to be placed in your **.newsrc** file. Press **h** (help) to see your choices and pick the one you want.

Once **rn** has finished these preliminaries, you are ready to select the newsgroup that you want to read. **rn** will present the first newsgroup with unread articles and ask you if you want to read it. For example:

```
********   11 unread articles in rec.humor.funny--read now? [ynq]
```

At this point, you have a large number of commands that you can use. We have summarized the most important commands for you in Figure 12-1.

The simplest thing to do is to press SPACE (or **y**) and read the newsgroup. Otherwise, you can press **n** to go on to the next group with unread articles, or **q** to quit **rn**.

Selecting and Reading Articles

As soon as you have selected a newsgroup to read, **rn** will look for the first unread article and display the first page.

As you are reading an article, there are many different commands that you can use. These are summarized in Figures 12-2 and 12-3. You can use the commands in Figure 12-2 whenever you are reading an article. Figure 12-3 shows extra commands that you can use as you page through a long article.

In this section, we will discuss the most important commands. If you want to learn how to learn **rn** well, take a look at the summaries from time to time and make a point of trying out new commands.

Whenever an article is finished, the bottom line of the screen will have a message like the following:

```
End of article 8278 (of 8284)--what next? [npq]
```

If you want to read the next article, simply press SPACE or the **n** key. If you want to jump to the next article in the thread—in other words, if you want to follow the discussion—press **^N** (CTRL-N). If you want to kill (junk) all the articles in this thread, press **k**. And, if you want to quit reading this newsgroup, press **q**.

If you press ^N, this command will become the default. The next time you finish an article you will see a message like:

```
End of article 8279 (of 8284)--what next? [^Nnpq]
```

The beauty of this is that it allows you to read through an entire thread, article by article, just by pressing SPACE.

In fact, once you have read all the articles in a newsgroup, pressing SPACE will go to the next group. If you ever get to the point where you have read the last article in the last newsgroup, pressing SPACE a couple more times will quit **rn**. Thus, as we said earlier, it is possible to read the news all day long just by pressing this one key.

When an article is long, **rn** will display the first page. On the bottom line, you will see a message like this:

```
--MORE--(45%)
```

This tells you that there is more to come and shows you how far you are into the article. (In this case, you are 45% of the way through.) At this point, you can press SPACE to read the next page, or use any of the commands that we discussed above.

Before we leave this section, there are three more points that we would like to mention. As you are reading, you can display a list of all the articles remaining in the newsgroup by pressing the = (equal sign) key. **rn** will show you a list of article numbers and subjects. For example:

```
8280 New light bulb joke
8281 Riddle about Unix
8282 Silly computer names
8283 Joke about acquisition editors
8284 Horrifyingly Funny...
```

If you would like to jump directly to one of these articles, just type the number and press RETURN (or ENTER). For example, to read article number **8283**, enter: **8283**.

If you want to save a copy of an article in a file, use either the **s** (save) or **w** (write) command followed by the name of a file. The only difference is that the **w** command will not save the header.

For example, to save the current article, without the header, to a file named **harley**, enter:

```
w harley
```

If this is a new file, you will be told that it does not already exist (in which case **rn** will be glad to create it for you) and you will be asked the following:

```
use mailbox format? [ynq]
```

In other words, would you like to save the file in a format that you can later read with your mail program? If you are not sure what this means, don't worry about it: just press **n** (no).

The final point we want to mention is that, from time to time, you may see an article that is encoded in a special format known as rot-13 (see Chapter 11). This format is used to encode articles that are potentially offensive. When you see a rot-13 article, look at the subject and decide if you want to read the article. If so, press **^X** (CTRL-X) and **rn** will do the decoding for you.

Replying to an Article

When you find an article to which you want to reply, ask yourself if it is better to send a personal mail message to the author of the article or a followup article to the newsgroup.

To mail a message to the author, you can use the **r** or **R** command. The difference is that the **R** command places the original article in your message so that you can include parts of it in your reply.

Once you type one of these commands, **rn** will ask you if you have a prepared file that you want to include in your message:

```
Prepared file to include [none]:
```

If not, simply press RETURN. If you have prepared a file in advance, type its name and press RETURN.

rn will now set up the message for you and start the text editor. You can now edit and compose whatever you want. When you are finished composing the message, quit the text editor in the usual manner. **rn** will regain control and ask you how to proceed:

```
Send, abort, edit, or list?
```

If you want to send the message, enter **s**. If you want to forget about the message completely, enter **a**. If you want to re-edit the message, enter **e**. Or if you want to display the message, enter **l**.

Posting an Article

If you would like to reply to an article by posting a followup, you can use the **f** or **F** command. The difference is that the **F** command places the original article in your followup article so that you can include parts of the original article in your reply.

If you use the **F** command, **rn** assumes that you want to send a followup article on the same topic. However, if you use **f**, **rn** will ask you if you want to specify a different topic:

```
Are you starting an unrelated topic? [yn]
```

If you press **y**, you will be given a chance to specify a new topic. Otherwise, press **n**.

You will now see a standard warning, designed to remind one and all that posting articles costs somebody money:

```
This program posts news to thousands of machines throughout
the
entire civilized world.  Your message will cost the net
hundreds
if not thousands of dollars to send everywhere. Please be sure
you know what you are doing.

Are you absolutely sure that you want to do this? [ny]
```

Enter **y**. (As a reader of this book, you *always* know what you are doing.)

From here on, the procedure is just as we explained it in the previous section. You will be asked if you have a prepared file to include. **rn** will then start the text editor for you. When you are finished, **rn** will ask you to confirm that you want to send the message.

That's all there is to it.

There may be times when you are not reading the news and you would like to post an article. You do not have to start **rn** at all. You can invoke the news posting program directly by using the **Pnews** command. (Yes, the first letter is an uppercase P. This is to make it difficult for someone to post an article by accident.)

You can enter this command by itself or with the name of a newsgroup. For example:

```
Pnews
Pnews misc.test
```

If you do not specify a newsgroup, **Pnews** will ask you for one.

The **Pnews** program works just like we described above. (In fact, when you use an **f** or **F** command, **rn** calls on **Pnews** to do the work.) The only difference is that, in this case, **Pnews** begins by asking you what distribution you want. The distribution (described in Chapter 10) tells the news servers that handle your article where you want the article sent. If you are not sure what distribution to use, choose the smallest one that looks adequate.

Commands to Use While Selecting a Newsgroup

Basic Commands

h	display help information
q	quit rn
x	quit rn, abandon all changes to your .newsrc file
v	show what version of rn you are using

Start Reading Articles

SPACE	perform the default command
y	read the current newsgroup
=	same as y, but start by showing list of subjects

Controlling Newsgroups

n	go to next newsgroup with unread articles
p	go to previous newsgroup with unread articles
^	go to first newsgroup with unread articles
$	go to the end of the list of newsgroups
g *newsgroup*	go to specified newsgroup
/ *pattern*	scan forward for newsgroup name containing pattern
? *pattern*	scan backward for newsgroup name containing pattern
/	scan forward for previous pattern
?	scan backward for previous pattern
u	unsubscribe to the current newsgroup
l *pattern*	list unsubscribed newsgroups containing pattern
L	list current state of newsgroups in .newsrc file
c	mark all articles in newsgroup as read [catch up]

Entering a Unix Command

! *command*	execute the specified Unix command
!	pause rn and start a shell

Figure 12-1. *Commands to use while selecting a newsgroup*

Commands to Use While Selecting an Article

Basic Commands

h	display help information
q	quit this newsgroup
SPACE	perform the default command
n	go to next unread article
^N	go to next unread article on same subject
k	mark as read, all articles with same subject [kill]
u	unsubscribe to the newsgroup
c	mark all articles in newsgroup as read [catch up]

Re-displaying the Current Article

^R	re-display the current article
v	re-display the current article with header [verbose]
^X	decode the current article using rot-13
X	decode the current page using rot-13
b	go back one page
^L	re-display the current page

Moving to Another Article

-	re-display the last article that was displayed
N	forward: to next article, whether read or unread
$	forward: to end of the last article
p	backward: to previous article
^P	backward: to previous article with same subject
P	backward: to next article, read or unread
^	backward: to first unread article

Figure 12-2.   *Commands to use while selecting an article* (continued next page)

Using Article Numbers

=	display list of all the unread articles
number	go to article with the specified number
#	display the number of the last article

Searching for an Article

/ *pattern*	forward: look for subject containing pattern
/	forward: repeat search using previous pattern
? *pattern*	backward: look for subject containing pattern
?	backward: repeat search using previous pattern

Responding to an Article

r	send a mail message to author of article [reply]
R	same as r, include original message
f	start Pnews program to create a followup article
F	same as f, include original message

Saving an Article

s *file*	save entire article to specified file
w *file*	same as s, but do not save the header [write]

Entering a Shell Command

! *command*	execute the specified shell command
!	pause rn and start a shell

Figure 12-2. *Commands to use while selecting an article (continued)*

Commands to Use While Paging Through an Article

As you are paging through an article, you can use all the commands in Figure 12-2 as well as the extra commands in this list.

Paging Through an Article	
h	display help information
q	go to the end of the article
SPACE	display the next page
RETURN	display one more line
d	display half a page more
b	go back one page
x	decode the next page using rot-13
j	mark article as read and go to end [junk]
g *pattern*	forward: search for specified pattern
G	forward: repeat search using previous pattern
^G	while reading a digest: go to next subject
TAB	with followups: skip text from previous articles

Figure 12-3.   *Commands to use while paging through an article*

CHAPTER 13

I n this chapter, we will explain how to use the **trn** newsreader to read Usenet articles. As we explained in Chapter 11, **trn** is actually an extension of **rn**. In this chapter, we discuss only those features of **trn** that are different from **rn**. Before you read this chapter, you should already know how to use **rn**. Even if you are an experienced **rn** user, you might want to take a moment to skim through Chapter 12.

The trn Newsreader

Overview of the trn Newsreader

trn does everything that **rn** does, plus more. More precisely, while you are reading the news with **trn**, you can use all of the regular **rn** options and commands as well as the extra **trn** features.

The best way to learn how to use **trn** is to first learn how to use all the **rn** features. So, if you are a beginner, read Chapter 12 and practice with **rn** until you become comfortable. If you are already an experienced **rn** user, read on.

What **trn** adds to **rn** is the ability to work with threads as separate entities. When you start **trn**, it looks pretty much like **rn**. The only difference is that, when you choose a newsgroup, you can go into thread selection mode. **trn** will organize all the articles in threads and you can choose which threads you wish to read. After you make your selections, **trn** will show you one thread at a time. Within each thread, you will see each article in order of arrival time.

For the rest of this chapter, we will discuss those features of **trn** that are different than **rn**. We assume that you are acquainted with the basic **rn** commands. If not, you can refer to Chapter 12. As we mentioned, most of **trn** is identical to **rn**. In particular, paging through an article, saving an article, and posting your own articles is all the same in both newsreaders.

For reference, Figures 13-3, 13-4, 13-5, and 13-6 at the end of the chapter contain summaries of all the basic **trn** commands. Within these summaries, we have followed the common Unix convention of indicating CTRL keys by using the ^ character. For example, when you see **^N**, it means CTRL-N. (That is, hold down the CTRL key and press N.)

Starting and Stopping

To start **trn**, you use the **trn** command. You can use all the **rn** options as well as some extra ones. In particular, you can use the **-q** (quick start) option to suppress the initial check for new newsgroups. Thus, the two most common ways to start **trn** are:

```
trn -q
trn
```

As with **rn**, you can specify a specific newsgroup to read:

```
trn -q rec.humor
trn rec.humor
```

Again, like **rn**, you can also use the **-c** option to check for unread news:

```
trn -c
```

This is a good command to put in your **.login**, **.profile**, or **.bash_profile** initialization file.

Like all newsreaders, **trn** has a large number of options. The only ones we will talk about here are **-x** and **-X**. For more information, see the **trn** description in the online manual. To read this description, use the Unix command:

```
man trn
```

When a system manager installs **trn**, he or she must choose whether to have **trn** offer all its special features by default or to behave the same as **rn**. Most system managers choose to enable **trn** completely. However, your system manager might have chosen to have **trn** act exactly like **rn**. The reasoning is that novice users will not be bothered by any of the non-**rn** features.

If this is the case for your system, you will have to explicitly turn on the **trn**-specific features by using the **-x** option:

```
trn -x
trn -q -x
```

In such cases, you should also use the **-X** option, This tells **trn** to make the **+** command the default when you are selecting a newsgroup. (This is the command that invokes the thread selector. We will discuss it in the next section.)

Thus, two common ways to start **trn** are:

```
trn -x -X
trn -q -x -X
```

As with most such options, you can combine them using a single **-** character:

```
trn -xX
trn -qxX
```

These two commands are equivalent to their counterparts above.

If the **trn** features are turned on for your system and you want to force **trn** to act like **rn** (with none of the extended features), you can use the **+x** option:

```
trn +x
trn -q +x
```

You stop **trn** the same way you stop **rn**: press the **q** (quit) key. If you are reading or selecting an article or a thread, you may have to press **q** more than once to work your way back out to the top level.

Alternatively, if you are at the top level (selecting a newsgroup), you can quit **trn** by pressing the **x** (exit) key. This not only quits **rn**, but restores your **.newsrc** file to what it was when you started. This abandons all the changes you may have made, such as subscribing and unsubscribing to newsgroups, reading articles, and so on.

Selecting a Newsgroup

Selecting a newsgroup with **trn** works much the same as with **rn**. You will be presented with the name of a newsgroup and asked to make a decision:

```
********   11 unread articles in rec.humor.funny--read now? [+ynq]
```

At this point, you press SPACE to invoke the default command. You can press **y** (yes) to read the current newsgroup, and use **n** or **p** to go to the next or previous group. You can also use the **g** command to go to a specific newsgroup, and the **/** and **?** commands to search for a newsgroup whose name contains a specific pattern. All of these commands are summarized in Figure 13-3.

However, as a **trn** user, you have two extra commands. First, you can press **+** to start the thread selector. This will be the default command, so, normally, you would just press SPACE.

The second extra command is **A**. This tells **trn** to abandon all the changes that you might have made to the current newsgroup. Use this command when you have made a mess of marking articles as read or unread, and you would rather just forget the whole thing.

Selecting a Thread

The biggest difference between **rn** and **trn** is that **trn** has a selection capability which allows you to choose which threads you want to read. You invoke this feature by using the + command at the newsgroup selection level. When you do, **trn** will display all the threads in the newsgroup, one page at a time. Figure 13-1 shows a typical screen.

The top line of the screen shows the name of the newsgroup and the number of unread articles. In this case, we are looking at the **rec.humor** newsgroup, which has 83 unread articles.

The bottom line shows us that we are 16% of the way through the newsgroup. To the right, we are reminded that the two most likely commands we would want to use are **>** and **Z** (which we will get to in a moment). Since **>** is listed first, it is the default and we can use this command by pressing SPACE.

The main part of the screen displays as many threads as will fit. Each thread is given a *thread id*, a letter or number that identifies the thread. In this case, we see four threads, labelled **a**, **b**, **c**, and **d**. For each thread, we see the author and subject of the individual articles. For example, thread **b** has eight articles, all with the same subject. You may see threads that have more than one subject because people do modify it when they compose followup articles.

trn has three different display formats that it can use to show you the list of threads. You can cycle from one format to another by pressing the **L** key. Press **L** a few times and see which one you like best.

```
rec.humor        83 articles

a Bertie Wooster      1   >Joke about Wembley
b+Reggie Jeeves       8   >Riddle about Unix
   Bingo Little
   Rosie M. Banks
   Agatha Spenser
   Dahlia Travers
   Catsmeat Pirbright
   Roderick Glossop
   Florence Craye
c Oofy Prosser        2   >Money and computers
   Gwendolyn Moon
d+Tuppy Glossop       2   >Irish water-spaniel
   Miss Dalgleish

— Select threads — 16% [>Z] —
```

Figure 13-1. *A typical* **trn** *thread selection screen*

As you read the thread summary, the idea is to look at subjects and authors and select only those threads that you want to read. After you have made your selections, you can tell **trn** to go ahead and display the articles. From that point on, everything works pretty much the same as with **rn**.

There are a large number of commands that you can use to select threads. These commands are summarized in Figure 13-4. However, we will take a moment to discuss the most basic ones.

The easiest way to select a thread is to type the thread id. For example, to select thread **b**, press the **b** key. Once a thread is selected, you will see a + character to the right of the thread id. In our example, this is the case for threads **b** and **d**.

If you want to unselect a thread, just press the thread id key again. For example, to unselect thread **b**, press the **b** key again. The + character will disappear.

Once you have selected threads from the current page, you have two choices. You can move to the next page of threads and select some more. Or you can read what you have already selected.

In other words, there are two basic strategies for selecting and reading. Some people like to skim through all the thread selection pages, make all their selections, and then read the threads. Other people like to select and read, one page at a time. The choice is yours.

To move to the next page of threads, press >. Since this is the default command, you can simply press SPACE. When you are ready to read what you have selected, press **Z**. (If you find it more convenient, you can press TAB instead.)

As we mentioned, there are a lot more commands that you can use while selecting threads. We suggest that, from time to time, you take a look at the summary in Figure 13-4 and teach yourself a new command.

Selecting and Reading Articles

While you are selecting and reading articles, everything works almost exactly the same as with **rn**. In particular, you can use the usual commands to page through an article, respond to an article, and so on.

Nonetheless, there is one major difference: **trn** is thread-oriented. First, **trn** will, by default, display articles in the order that they appear in the thread. Second, there are special thread-oriented commands that you can use to move within the thread, from one article to another.

To help you see where you are, **trn** will draw a little diagram in the upper right-hand corner of the screen. Figure 13-2 shows a typical page of an article as it is displayed by **trn**.

The top line tells us that we are reading the **rec.humor** newsgroup. We are looking at article number 8281. There are 7 more unread articles in this thread and 74 other unread articles in this group. Below, we see the header, body, and signature as we described in Chapter 10. The bottom line shows us what article

```
rec.humor #8281 (7 + 74 more)                   (1)+─[1]─[1]
From: jeeves@ganymede.com (Reggie Jeeves)       |─[1]
Newsgroups: rec.humor                           |─[2]+─[2]
[1] Riddle about Unix                           |      \─[2]
Date: Thu May 20 00:28:07 PDT 1993              \─[1]
Organization: The Junior Ganymede Club
Lines: 20
Distribution: world

Somebody at the club here was asking the following riddle, but
nobody could come up with the answer. Does anyone on the Net
know?

─> How many Unix does it take to screw in a light bulb?

Obligatory Joke:
   "I'm a Rotationist, you know."
   "Ah, yes. Elks, Shriners and all that. I've seen pictures
of them, in funny hats."
   "No, no. You're thinking of Rotarians. I'm a Rotationist,
which is quite different. We believe that we are reborn as
one of our ancestors every ninth generation."


─
R. Jeeves                               Rem acu tetigisti
jeeves@ganymede.com
                    Junior Ganymede Club
              Curzon Street, W, London, England

End of article 8281 (of 8362)—what next? [npq]
```

Figure 13-2.   *A typical* **trn** *article screen*

we are reading and suggests that we might want to use the **n** (go to next article in the thread), **p** (previous article), or **q** (quit) command.

There are, of course, many commands that we can use at this point, and they are summarized in Figure 13-5. Since the default is **n**, we can move to the next article by pressing SPACE.

The most intriguing item on this screen is the small diagram in the upper right-hand corner. This is a tree-representation of the current thread. The shape of the tree shows the form of the thread. Each number indicates a different subject. In this example, we see the relationship of the eight articles in the thread.

The original article has four followups. The first followup has one followup of its own. The third followup has two followups of its own. Moreover, when the third followup was composed, the author changed the subject. Articles with the original subject are marked as **1**, while articles with the new subject are marked as **2**, although they are all considered to be part of the same thread.

Articles that you have already read are marked with ( ). Unread articles, including the current article, are marked with [ ]. The current article will be indicated by having the brackets and the number highlighted (for example, **[1]**). The last article you read will be indicated by having only the number highlighted (for example, **[1]**). This will all make sense when you see it on your screen.

Hint

If you have trouble understanding the tree structure, you are not alone. However, there is an easy way to learn how to read these diagrams.

As you move within the tree, the diagram will change. Choose a newsgroup that has a lot of followups (**rec.humor** will do just fine) and use some of the commands described below. As you display one article after another—that is, as you move from one part of the tree to another—watch how the diagram changes. It won't take long before you understand how it all works.

With large threads, the full tree will not fit into the corner of the page, so only part of the tree will be displayed. With such threads, you can display a diagram of the entire thread, including a list of all the different subjects, by pressing **t**.

*FUN TIP: Select an extra-large thread and then use the **t** command. You will see an awesome tree diagram.*

As we mentioned above, there are special **trn** commands that allow you to move throughout the thread. Press] to go to the article that is one position to the right in the tree. (This is called the *child*.) If you are already as far to the right as you can go in this branch,] will move down and to the right.

Press [to go the article that is one position to the left in the tree. (This is called the *parent*.) If you are already as far to the left as you can go in this branch, [will move up and to the left.

The { command moves to the top left of the tree (the *root*). The } command moves all the way to the right (the *leaf*).

Finally, if you get tired of what you are reading, it is easy to jump to a completely different thread. Press < to go to the previous thread; press > to go to the next thread.

Commands to Use While Selecting a Newsgroup

Basic Commands

h	display help information
q	quit `trn`
x	quit `trn`, abandon all changes to your `.newsrc` file
v	show what version of `trn` you are using

Start Reading Articles

SPACE	perform the default command
+	invoke thread selector to read the current newsgroup
y	read the current newsgroup
=	same as y, but start by showing list of subjects

Controlling Newsgroups

n	go to next newsgroup with unread articles
p	go to previous newsgroup with unread articles
^	go to first newsgroup with unread articles
$	go to the end of the list of newsgroups
g *newsgroup*	go to specified newsgroup
/ *pattern*	scan forward for newsgroup name containing pattern
? *pattern*	scan backward for newsgroup name containing pattern
/	scan forward for previous pattern
?	scan backward for previous pattern
u	unsubscribe to the current newsgroup
l *pattern*	list unsubscribed newsgroups containing pattern
L	list current state of newsgroups in `.newsrc` file
c	mark all articles in newsgroup as read [catch up]
A	abandon all read/unread changes to this newsgroup

Entering a Unix Command

! *command*	execute the specified Unix command
!	pause `trn` and start a shell

Figure 13-3.   *Commands to use while selecting a newsgroup*

Commands to Use While Selecting a Thread

In this list, *id* refers to an article id: **a...z, 0...9**. By default, the letters **h, k, m, n, p, q**, and **y** are omitted, as they are used as commands.

General Commands

h	display help information
SPACE	perform the default command

Changing Newsgroups

q	quit current newsgroup, move to next newsgroup
N	go to the next newsgroup with unread news
P	go to the previous newsgroup with unread news

Displaying the Selection List

>	go to next page
<	go to previous page
^	go to first page
$	go to last page
L	switch among: terse/short/long form of list

Selecting/Unselecting Which Threads to Read

id	select/unselect specified thread
id-id	select/unselect specified range of threads
y	select/unselect current thread
.	same as **y**
k	mark the current thread as completely read [kill]
,	same as **k**
m	unmark the current thread
\	same as **m**
/ pattern	select all threads whose subject contains pattern

Figure 13-4. *Commands to use while selecting a thread* (continued next page)

Changing Position Within the Thread List

n	move down to the next thread
]	same as n
p	move up to the previous thread
[	same as p

Start Reading Articles

RETURN	start reading (if nothing selected, current thread)
Z	start reading (if nothing selected, all threads)
TAB	same as Z
D	mark unselected articles on page read; start reading
X	mark all unselected articles as read, start reading

Figure 13-4.   *Commands to use while selecting a thread* (continued)

Commands to Use While Selecting an Article

Basic Commands

h	display help information
q	quit this newsgroup
SPACE	perform the default command
n	go to next unread article
^N	go to next unread article on same subject
+	enter thread selection mode
k	mark as read all articles with same subject [kill]
u	unsubscribe to the newsgroup
c	mark all articles in newsgroup as read [catch up]

Re-displaying the Current Article

^R	re-display the current article
v	re-display the current article with header [verbose]
^X	decode the current article using rot-13
X	decode the current page using rot-13
b	go back one page
^L	re-display the current page

Moving to Another Article

-	re-display the last article that was displayed
N	forward: to next article, whether read or unread
$	forward: to end of the last article
p	backward: to previous article
^P	backward: to previous article with same subject
P	backward: to next article, read or unread
^	backward: to first unread article

Figure 13-5. *Commands to use while selecting an article* (continued next page)

Moving Within the Thread

>	go to the next selected thread
<	go to the previous selected thread
]	go to article's child within the thread
[	go to article's parent within the thread
}	go to the leaf of the tree
{	go to the root of the tree
t	display the entire article tree and all its subjects

Using Article Numbers

=	display list of all the unread articles
number	go to article with the specified number
#	display the number of the last article

Searching for an Article

/ pattern	forward: look for subject containing pattern
/	forward: repeat search using previous pattern
? pattern	backward: look for subject containing pattern
?	backward: repeat search using previous pattern

Responding to an Article

r	send a mail message to author of article [reply]
R	same as r, include original message
f	start Pnews program to create a followup article
F	same as f, include original message

Saving an Article

s file	save entire article to specified file
w file	same as s, but do not save the header [write]

Entering a Shell Command

! command	execute the specified shell command
!	pause trn and start a shell

Figure 13-5. *Commands to use while selecting an article (continued)*

Commands to Use While Paging Through an Article

As you are paging through an article, you can use all the commands in Figure13-5 as well as the extra commands in this list.

Paging Through an Article

h	display help information
q	go to the end of the article
SPACE	display the next page
RETURN	display one more line
d	display half a page more
b	go back one page
x	decode the next page using rot-13
j	mark article as read and go to end [junk]
g *pattern*	forward: search for specified pattern
G	forward: repeat search using previous pattern
^G	while reading a digest: go to next subject
TAB	with followups: skip text from previous articles

Figure 13-6.   *Commands to use while paging through an article*

CHAPTER

14

The nn Newsreader

We have talked a lot so far about Usenet, the worldwide system of over 5,000 discussion groups. In this chapter, we will explain how to read the news and post your own articles using the **nn** newsreader. Before you read this chapter, you should at least have skimmed the previous Usenet chapters. In particular, you should understand what Usenet is and how it works (Chapter 9), the conventions and terminology that we use (Chapter 10), and what is involved in using a newsreader (Chapter 11).

Overview of the nn Newsreader

The **nn** program was designed to solve the problem of too much news and too little time. **nn** makes it easy to choose which newsgroups you want to look at and, within those groups, which articles you want to read. **nn** is fast, easy to learn, and powerful. However, it is a complex program with many options and commands, and offers far more functionality than you will ever need.

The best strategy is to learn the basics and practice for a while. Once you have some experience, you can look at the command summaries at the end of this chapter and experiment with whatever looks interesting. Moreover, as you will see, **nn** makes it easy to display portions of the **nn** online manual from within the program itself. This means that, at all times, the official **nn** reference material is only a few keystrokes away.

To introduce you to **nn**, we will start with an overview of how it works. In the following sections, we will go into the details.

You start the program by entering the **nn** command. **nn** checks with your **.newsrc** file (see Chapter 11) and then displays the name of the first subscribed newsgroup that has unread news. You can choose to start with this group or you can select another one. Once you select a newsgroup, **nn** displays a list of unread articles. Each entry on the list is a short, one-line description of an article.

You mark only those articles that you want to read. After selecting all such articles, you type a command that tells **nn** to start displaying the articles.

As you read, there are a number of actions that you can take. You can save an article to a file, mail a copy of an article to somebody, kill all the articles with the same subject, decode the article using rot-13, and so on. You can also respond to an article, either by sending personal mail to the author, or by posting a followup article of your own.

In other words, at all times, you are either in one of two situations: selecting articles or reading articles. As you move from one newsgroup to another, you can change back and forth between selecting and reading.

At each point, there are a large number of commands available. As a reference, we have summarized all of the important **nn** commands in three lists at the end of the chapter.

Figure 14-2 shows the commands you can use at any time. Figure 14-3 shows extra commands that you can use while you are selecting articles. And Figure 14-4 shows extra commands that you can use as you are reading. You may want to take a minute now to glance at these summaries.

In the following sections, we will discuss the details of using **nn**. As you read, and as you practice, there are three things that we would like you to remember.

First, if you are typing a command that is waiting for input—and there are quite a few of them—you can cancel the command by pressing CTRL-G.

Second, for all single-letter commands, you do not have to press the RETURN or ENTER key after you type the command. All you have to do is press a single key and **nn** will react immediately. For example, to display help information, you press the **?** key. (When a Unix program reads input in this way, we say that it is operating in *cbreak mode*.)

Finally, whenever you are called upon to type a command, you can press SPACE to select the default choice. This choice will be the most reasonable thing to do in your current situation.

For example, when you are selecting articles, pressing SPACE will display the next page of the list. If you are at the end of the list, pressing SPACE tells **nn** to start reading. As you are reading, pressing SPACE repeatedly will page through an article, go to the next article, and ultimately (when there are no more articles to read) return you to the selection screen.

Moreover, when you enter a command that requires some sort of input, pressing SPACE will select the default choice, which is always a safe bet.

Hint

After selecting articles, you can read all the news you want just by using two keys. Use SPACE to page through your selected articles. If you get bored, press **k** (described later) to kill the rest of the articles in the thread.

Starting and Stopping

To start **nn**, you enter the **nn** command. There are several forms of this command and a large number of options that you can use. In this section, we will cover the most important of these choices.

First, you can enter the command by itself:

```
nn
```

This starts **nn** in a straightforward manner.

If you have never read the news before, **nn** will create a **.newsrc** file for you (see Chapter 11). Initially, your **.newsrc** file will contain all of the active newsgroups with each newsgroup subscribed. You will probably want to unsubscribe to most of the groups. There are two ways to do this. First, you can do it within **nn** by using the **U** command, although this will probably take a long time. The more practical method is to use a text editor, such as **vi** or Emacs, and edit the **.newsrc** file for yourself (see Chapter 11).

Stopping **nn** is easy. When you are finished reading the last newsgroup to which you are subscribed, **nn** will quit automatically. Of course, since there is so much news, this may never actually happen. So, at any time, you can press **Q** to quit the program immediately.

There may be times when you are interested in reading only a single newsgroup. In that case, you can specify that group directly when you start **nn**. For example, say that you only want to read the **rec.humor** newsgroup. You can enter:

```
nn rec.humor
```

You can also specify more than one newsgroup:

```
nn rec.humor alt.tasteless.jokes
```

When you are finished with those newsgroups, **nn** will quit automatically.

If you would like **nn** to ask you what newsgroup to read, you can specify **-g** as part of the command:

```
nn -g
```

The **-g** is called an *option* or a *switch*. In this case, the **-g** option tells **nn** that you want it to ask you which newsgroup to "go to". You will see:

```
Enter Group or Folder (+./~)
```

Until you are an experienced user, don't worry about all the choices this message is offering. Simply type the name of the newsgroup you want and press RETURN. You will now see:

```
Number of articles (juasne) (j)
```

Again, don't worry about the details. Just press SPACE and **nn** will start reading the newsgroup. After you are finished with that newsgroup, **nn** will quit automatically.

If you would like to have **nn** ask you for one newsgroup after another, you can use **-g** along with the **-r** (repeat) option:

```
nn -g -r
```

In general, when you enter a Unix command, you can combine options using a single - character, so in this case you can also use:

```
nn -gr
```

Using **-gr** is similar to **-g** alone except that **nn** will not quit after the first newsgroup. Rather, it will keep on asking you for one newsgroup name after another. When you are finished, press RETURN by itself (without specifying a newsgroup name) and **nn** will stop.

As we mentioned above, there are many options that you can use with the **nn** command. The only other ones we will discuss are a set of four options that can be combined in a particularly useful manner. These options are:

-m	merge all articles into an artificial "meta" group
-x	scan all articles: both read and unread
-X	look at all the newsgroups: subscribed and unsubscribed
-s	pattern show articles whose subject contains specified pattern

You can use these options together to tell **nn** to search through every article in every newsgroup looking for all the articles whose subject contains a specified pattern. The results will then be presented together as one large artificial newsgroup, called a *meta group*.

For example, say that you want to look at all the articles that contain the word "Unix" or "UNIX" in the subject. You would enter:

```
nn -mxX -sunix
```

(When you perform such a search, **nn** does not distinguish between upper- and lowercase letters. Thus "unix" will match both "Unix" and "UNIX".)

As an experiment, we used the previous command on a system that carries a large number of newsgroups. We found that **nn** took two minutes and 38 seconds to search 197,524 articles in 4,913 newsgroups. It found 849 articles that fit our specification.

To cut down on the number of newsgroups to search, you can specify newsgroup names. The special name **all** refers to all the newsgroups in a family. For example:

```
nn -mxX -sunix comp.all
```

This command searches all the newsgroups in the **comp** (computer) hierarchy.

Finally, there may be times when you do not want to actually read the news, but you just want to see if there are any new articles waiting for you. In such cases, you can use the **nncheck** command:

```
nncheck
```

If there are unread articles in the newsgroups to which you subscribe, **nncheck** will display a message like the following:

```
There are 1155 unread articles in 18 groups
```

Otherwise, you will see:

```
No News (is good news)
```

Be careful, though: this message may also mean that the news server is not working.

Hint

You can place an **nncheck** command in your initialization file so that each time you log in you will be told if there is news that you have not yet read. The name of your initialization file depends on which shell you are using. For the C-Shell, the file is **.login**; for the Bourne or Korn shell, it is **.profile**; and for the Bash shell, it is **.bash_profile**.

Displaying Help Information

One of the nice things about **nn** is that no matter what you are doing, there is plenty of help information available. First of all, when you are selecting or reading articles, you can press the **?** (question mark) key to display a summary of whatever commands are available.

We have summarized the most important of these commands in Figures 14-2, 14-3, and 14-4. Still, you may not always have this book with you—say, if you are using **nn** on your portable computer while skydiving—and remembering the **?** command is a good idea.

nn also makes it easy to read the online manual without having to stop the program. You will remember that in Chapter 11 we explained that Unix comes with an online manual you can access by using the **man** command. For example, to display the manual entry for **nn**, you would normally use the Unix command:

```
man nn
```

Well, **nn** has a unique feature. While you are using **nn**, you can enter the command:

```
:man
```

nn will stop what it is doing and display a selection list for the entire **nn** manual. The sections of the manual will be presented using the same format that **nn** uses to present news articles. You can then use the exact same commands to select and read portions of the manual as you would with news articles.

Selecting Articles

In order to allow you to select the articles that you want to read, **nn** will show you a list. This list will display the authors and subjects of all the unread articles in the newsgroup, one page at a time.

Figure 14-1 shows a typical article selection list. **nn** can present this list in five different formats. If you would like to see the different formats, press the **"** (double quote) key repeatedly, and **nn** will change from one format to another.

The top line gives us some miscellaneous information. On the left, we see the name of the newsgroup. On the right, we see statistics about the articles: the number of unread articles in this newsgroup, the total number of unread articles in all the subscribed newsgroups, and the number of subscribed groups. In our example, we are reading the **misc.forsale** newsgroup. There are 35 unread articles in this group. Altogether, there are 817 unread articles in 14 newsgroups.

The main part of the list contains one line for each article. At the far left of each line is an *article id*, a single character that identifies the article. **nn** uses the letters of the alphabet, **a** through **z**, as article ids. If your screen holds more than 26 summary lines, **nn** will also use the digits **0** through **9**. In our example, there are 19 articles, marked **a** through **s**.

```
Newsgroup: misc.forsale              Articles: 35 of 817/14 NEW

a*Reggie Jeeves      5   Works of Spinoza and Dostoevsky
b.Bertie Wooster    39   Banjolele, almost new
c.Catsmeat Potter-  33   >> reports signed by Rev. Aubrey Upjohn
d.Tuppy Glossop     29   Football boots
e Harold Pinker     11   >
f Dahlia Travers     4   Terra-cotta figure: Infant Samuel at Prayer
g.Bingo Little       8   Theatre poster of 'What Ho, Twing!!'
h Stilton Cheesewr  10   Rowing trophy from a bump supper
i Tom Travers       16   Full set of back copies: Milady's Boudoir
j Oliver Sipperly   15   Policeman's helmet from Coventry Street
k*Roderick Glossop  30   Dead fish and 23 cats
l*Dog-Face Rainsby  12   >
m*Claude Wooster    28   >
n*Eustace Wooster   13   >
o R.E.R. Psmith     19   Umbrella
p Watkyn Bassett    14   -
q Mrs. Bingo        53   Full set of Rosie M. Banks novels
r Gussie Fink-Nott  31   Picture of Madeline Bassett; Newt food
s T. L. Nipper      17   Dish with separate water compartment


-- 15:49--SELECT--help:? -----Top 5%-----
```

Figure 14-1.   *A typical article selection list*

To the right of the article id are the name of the author, the number of lines in the article, and the subject. In the subject column, a > character indicates a followup. This allows you to identify threads. In our example, articles **d** and **e** form a short two-article thread. Articles **k** though **n** form a longer four-article thread. If an article is a followup to a followup, you will see two > characters (**>>**). This is the case for article **c**.

On occasion, more than one unrelated article may have the same subject. In such cases, **nn** will display the subject for the first article and a - (hyphen) character for the others. In our example, we see this with articles **o** and **p**.

On the bottom line, **nn** displays a few items of information. At the far left is the time. In our example, it is 3:49 PM (15:49). Next is a reminder that we are looking at the article selection list, followed by another reminder, telling us that we can display a help summary by pressing the **?** key. To the right, **nn** tells us where we are in the full list. In our example, we are at the top of the list, and the last article shown is 5% of the way through the list.

Selecting an article to read is easy. All you have to do is type the article id. For example, to select article **a**, just press the **a** key. If you want to select a range of articles, type the two article ids, separated by a - (hyphen). For example, to select articles **k** through **n**, type **k-n**.

To select an entire thread, type the article id that marks the beginning of the thread, followed by an * (asterisk). For example, to select all the articles that have the same subject as article **k**, type **k***.

When you select an article, **nn** will highlight it on your screen. If your terminal or computer cannot support highlighting, **nn** will mark selected articles with an * character after the article id. In our example, you can see that we have selected articles **a**, **k**, **l**, **m**, series and **n**.

To unselect an article, simply type the article id of the article once again. For example, once article **a** is selected, you can unselect it by pressing the **a** key a second time. You can also unselect by typing a range of article ids (**k-n**) or a thread specification (**k***).

At any time, your cursor will be on one of the articles in the list. This is the current article. You can move up and down the list by using the UP and DOWN keys (the keys with the arrows). However, you can also use the **,** (comma) and **/** (slash) keys, which is a lot more handy. The **,** moves down and the **/** moves up. The **.** (period) key selects the current article. Making your selections in this way is convenient because, on many keyboards, these three keys are next to one another.

Hint

Put the middle fingers of one hand on the **,** (comma), **.** (period), and **/** (slash) keys. Next, place your thumb on SPACE. You can now zip through a list of articles with the speed of a greased jackrabbit.

Once you have selected all the articles you want, there are several ways to tell **nn** that you want to start reading. Remember that pressing SPACE moves to the next page of articles. If you are already on the last page, pressing SPACE will start reading.

Alternatively, you can press **Z** or **X** at any time. Both of these commands tell **nn** to start reading immediately. The only difference is that, after you are finished reading, **Z** returns you to the same newsgroup, while **X** goes on to the next newsgroup.

There are two basic strategies for selecting and reading. Some people like to skim through the selection pages, make all their selections, and then read the articles. Other people like to select and read, one page at a time.

Moving to Another Newsgroup

As you are selecting articles, there are several commands that you can use to move to another newsgroup. To move to the next subscribed newsgroup in the sequence, press **N** (next).

There is also a **P** (previous) command, but it does not move backwards in the sequence of newsgroups. Rather, **P** moves to the last newsgroup that you looked at. This means that if you press **P** twice, you get back to where you started.

There is also a convenient way to move back and forward: press **A** (ahead) or **B** (back). These commands will move throughout the entire newsgroup list, to both subscribed and unsubscribed newsgroups.

When you use either of these commands, you will see the name of a newsgroup followed by a set of possible commands. For example:

```
Enter alt.fan.wodehouse (UNSUB) ?  (ABGNPy)
```

In this case, we are being asked if we want to enter the newsgroup named **alt.fan.wodehouse**, which is currently unsubscribed.

At this point, you can press SPACE or **y** (yes) to move to this group. You can also press **A** to move ahead, **B** to move back, **G** to go to a specific group (see below), **N** to move to the next subscribed group with unread news, or **P** to move to the previous subscribed group with unread news.

Hint

You can quickly skim through the full list of newsgroups by pressing **A** or **B** repeatedly.

Once you find a newsgroup you like, press SPACE. You will see:

```
Number of articles (juasne)  (j)
```

Until you become an experienced **nn** user, don't worry about what this means. Just press SPACE once again and you will jump right to the newsgroup you want.

The final way to change newsgroups is to jump directly to a specific newsgroup. To do so, press **G** (go to). After you do, you will see:

```
Group or Folder (+./~ %=sneN)
```

Once again, until you become experienced, ignore the options. Just type the name of the newsgroup you want and press RETURN. You will then see:

```
Number of articles (juasne) (j)
```

Press SPACE to jump to the group.

Hint

The Fundamental Rule of **nn**:

> Whenever you are unsure what to do, press SPACE. (If only life were that simple.)

Reading Articles

While you are reading an article, there are many commands that you can use. The basic commands are summarized in Figures 14-2 and 14-4. For the most part, these commands are straightforward and, with a little practice, you will have no trouble at all. To get you started, we will introduce you to the most important ones.

As you no doubt expect, the single most important command is SPACE. Pressing SPACE repeatedly will allow you to read an entire article, from one page to the next. When the article is finished, pressing SPACE once again will move to the next article. At the end of the last article, pressing SPACE will move you back to the selection screen.

As you read, you may decide that a particular thread is boring. If so, press **k** to kill all the remaining articles in that thread. To go directly to the next article, press **n**; to go to the previous article, press **p**.

If you are not sure that you want to read an entire thread, you can mark only the first article while you are selecting. Later, when you actually read the article, you can choose to read the rest of the thread by pressing * (asterisk).

If you happen upon a potentially offensive article that was encoded using rot-13 (see Chapter 11), you can display it in decoded form by pressing **D**.

When you encounter an article that is especially interesting or useful, there are two ways that you can preserve it. First, you can save the article in a file by using the **s**, **o**, or **w** command. The **s** command saves the full header along with the article. The **o** command saves an abbreviated header along with the article. And the **w** (write) command saves the article only, with no header.

The second way to preserve an article is to use the **m** command to mail a copy of the article to someone (including yourself).

Finally, there are two ways to respond to an article. First, you can send a private mail message to the author by using the **r** command. You will be asked:

```
Include original article?
```

Press either **y** (yes) or **n** (no). In either case, **nn** will start your text editor for you, to allow you to compose your reply. If you chose yes, **nn** will include the text of the article when it starts the editor, so you can include parts of it in your reply. Once you quit the text editor, **nn** will resume control and you will see the following:

```
a)bort e)dit h)old m)ail r)eedit s)end v)iew w)rite
Action: (send letter)
```

You now have several straightforward choices. To make a choice, you must type the specific letter and press RETURN. For example, if you decide not to send your reply after all, type **a** and then press RETURN. If you press RETURN by itself, **nn** will send the letter by default, so you don't really have to use the **s** command.

The second way to respond to an article is to post a followup article of your own. We will discuss this in the next section.

Posting an Article

There are three ways to post an article with **nn**. No matter how you do it, posting an article is easy. **nn** will lead you through the process; all you have to do is answer the questions. When **nn** is ready for you to compose the article, it will automatically start your text editor. When you quit the editor, **nn** will regain control.

The first way to post an article is as a followup while you are reading the news. If you are at the article selection screen, just move to the article to which you want to respond and press **F**. If you are currently reading an article, you can press either **f** or **F**.

In either case, **nn** will ask if you want to include a copy of the original article in your followup. The default is yes, so if you want to include the original, you can press either **y** or SPACE; otherwise, press **n** (no). If you answer **y** (yes), the original article will be inserted with a > character at the beginning of each line.

Hint

It is considered polite to delete all the unnecessary lines from a followup.

The second way to post an article while you are using **nn** is by entering the **:post** command at any time. **nn** will ask you for all the relevant information: newsgroup, subject, and so on. You can use this command to post an article to any newsgroup using any subject you want.

Finally, when you are not using **nn**, you can post an article by entering the **nnpost** command. The easiest way to use this command is to enter it without any options:

```
nnpost
```

The program will ask you everything it needs to know, just as if you had entered **:post** within **nn**.

If you would like to learn more about the **nnpost** command, you can read about it in the online manual. You will see that you can use options to specify some of the information that the program needs. In addition, you will find out how to set up automated postings.

The nn Family of Commands

The **nn** program is actually one of a family of commands. The entire family is shown in Table 14-1.

We have already discussed the **nn**, **nncheck**, and **nnpost** commands. We won't cover the other commands here. However, once you get some experience, we recommend that you take a few moments and check out these commands in

Command	Purpose
nn	read the news
nncheck	check for unread articles in subscribed newsgroups
nngoback	mark specified days' articles as unread
nngrab	find all articles whose subject contains a pattern
nngrep	display all newsgroup names that contain a pattern
nnpost	post news articles
nntidy	clean and adjust your .newsrc file
nnusage	display statistics showing who has been using nn

Table 14-1.   *The* **nn** *Family of Commands*

the online manual. (You can use either the Unix **man** command, or the **:man** command from within **nn**.)

In particular, **nngrab** is a fast alternative to the cryptic combination of options that we used earlier in the chapter (**nn -mxX -s**) to search all the newsgroups for articles whose subject contains a specific pattern.

Commands That Are Always Available

Basic Commands

SPACE	do the most reasonable thing/accept the default
Q	quit nn
CTRL-G	cancel a command that is waiting for input
:post	post a new article

Displaying Help

?	display summary of available commands
:help	show extra sources of help
:man	read the nn entries in the online manual

Controlling Newsgroups

G	go to a newsgroup out of sequence
U	unsubscribe/subscribe to the current newsgroup
Y	show overview of newsgroups with unread articles

Entering Unix Commands

! *command*	execute the specified Unix command
!	pause nn and start a new shell

Figure 14-2. *Commands that are always available*

Commands to Use While Selecting Articles

In this list, *id* refers to an article id: **a...z, 0...9**

Selecting/Unselecting Which Articles to Read

id	select/unselect specified article
id-id	select/unselect specified range of articles
id*	select/unselect articles with same subject as *id*
,	move down one line
.	select/unselect current article, then move down
/	move up one line
*	select/unselect same subject as current article
@	reverse selections on current page
=.RETURN	select all articles in the newsgroup
~~	unselect all articles in the newsgroup

Displaying the Selection List

SPACE	go to next page, unselected articles become read
>	go to next page
<	go to previous page
^	go to first page
$	go to last page
"	change the layout

Start Reading Articles

SPACE	start reading (when on last page)
Z	start reading; return to newsgroup when done
X	start reading; go to next newsgroup when done

Moving to Another Newsgroup

N	move to next subscribed newsgroup in the sequence
P	move to last newsgroup that you looked at [previous]
B	move back one newsgroup in the sequence
A	move ahead one newsgroup in the sequence
G	go to a specified newsgroup

Figure 14-3.   *Commands to use while selecting articles* (continued next page)

Saving an Article

S	save full header plus article to a file
O	save abbreviated header plus article to a file
W	save article only (no header) to a file [write]

Responding to an Article

R	send mail message to author of article [reply]
F	create a followup article
M	mail a copy of the article to someone

Figure 14-3. *Commands to use while selecting articles* (continued)

Commands to Use While Reading Articles

Moving Within an Article

SPACE	forward: one page
DOWN	Same as SPACE
d	forward: one half page [down]
RETURN	forward: one line
$	forward: to last page
BACKSPACE	backward: one page
UP	same as BACKSPACE
u	backward: one half page [up]
^	backward: to first page
g*line*	go to specified line
/*pattern*	search forward for specified pattern
.	repeat previous search
TAB	followup articles: skip text from older article

Displaying an Article

h	re-display the current article with full header
D	decode the current article using rot-13
c	omit multiple spaces and tabs [compress]

Move to Another Article

SPACE	go to next article (if on last page)
n	go to the next selected article
p	go to the previous selected article
k	mark as read articles with same subject [kill]
*	go to next article with same subject

Return to Selecting Articles

SPACE	return to selecting (if last page of last article)
=	return to selecting current newsgroup
N	go to next newsgroup
X	go to next newsgroup, mark all articles as read

Figure 14-4. *Commands to use while reading articles* (continued next page)

Saving the Current Article

S	save full header plus article to a file
s	same as S
O	save abbreviated header plus article to a file
o	same as O
W	save article only (no header) to a file [write]
w	same as W

Responding to the Current Article

R	send mail message to author of article [reply]
r	same as R
F	create a followup article
f	same as F
M	mail a copy of the article to someone
m	same as M

Figure 14-4. *Commands to use while reading articles* (continued)

CHAPTER 15

The tin Newsreader

We have talked a lot so far about Usenet, the worldwide system of over 5,000 discussion groups. In this chapter, we will explain how to read the news and post your own articles using the **tin** newsreader. Before you read this chapter, you should at least have skimmed the previous Usenet chapters. In particular, you should understand what Usenet is and how it works (Chapter 9), the conventions and terminology that we use (Chapter 10), and what is involved in using a newsreader (Chapter 11).

Overview of the tin Newsreader

The **tin** program was designed to let you skim through a large number of newsgroups and a large number of articles. **tin** makes it easy to choose which newsgroups you want to look at and, within those groups, which threads of articles you want to read. **tin** is fast, easy to learn, and powerful. However, it is a complex program with many options and commands, and offers far more functionality than you will ever need.

The best strategy is to learn the basics and practice for a while. Once you have some experience, you can look at the command summaries at the end of this chapter and experiment with whatever looks interesting. When you get really good, you can read the reference material on **tin** in the online manual and find all kinds of cool stuff.

To introduce you to **tin**, we will start with an overview of how it works. In the following sections, we will go into the details.

You start the program by entering the **tin** command. The first thing **tin** does is read your **.newsrc** file (see Chapter 11) and compare it against the master list of active newsgroups. If there are new newsgroups, **tin** asks you whether or not you want to subscribe.

Once the check for new groups is finished, **tin** presents you with a newsgroup selection list. This is a list of newsgroups from which you can choose what you want to read.

After you select a newsgroup, **tin** shows you a thread selection list in which all the different threads for this newsgroup are shown. You can now choose a thread to read.

As you read, there are a number of actions that you can take. You can save an article to a file, mail a copy of an article to somebody, kill all the articles with the same subject, decode the article using rot-13, and so on. You can also respond to an article, either by sending personal mail to the author, or by posting a followup article of your own.

In other words, at all times, you are in one of three situations: selecting a newsgroup, selecting a thread, or reading articles. (Actually, **tin** has two other environments: spool directories and newsgroup indexing. However, these topics are not important to most people and we won't discuss them in this chapter.)

At each point, there are a large number of commands available. As a reference, we have summarized all of the important **tin** commands in four lists at the end of the chapter.

Figure 15-4 shows the commands you can use at any time. Figure 15-5 shows commands that you can use while you are selecting a newsgroup. Figure 15-6 shows commands you can use while you are selecting a thread to read. And Figure 15-7 shows commands that you can use as you are reading. You may want to take a minute now to glance at these summaries.

Within the summaries, we have followed the common Unix convention of indicating CTRL keys by using the ^ character. For example, when you see **^D**, it means CTRL-D. (That is, hold down the CTRL key and press **D**.)

In the following sections, we will discuss the details of using **tin**. As you read, and as you practice, there are three things that we would like you to remember.

First, in almost all circumstances, you can press **h** to display a help summary of all the commands that are currently available.

Second, for all single-letter commands, you do not have to press the RETURN or ENTER key. All you have to do is press a single key and **tin** will react immediately. For example, to display help information, you press the **h** key. (When a Unix program reads input in this way, we say that it is operating in *cbreak mode*.)

Finally, there will be times when **tin** will ask you a question and present you with a list of possible answers. In such cases, **tin** will suggest one of the answers as being the most likely response. This is the default. If you would like to accept this suggestion, you can simply press RETURN.

For example, as you are reading an article, you can save a copy of the article to a file by using the **s** command. When you do, you will see the following:

```
Save a)rticle, t)hread, h)ot, p)attern, T)agged articles, q)uit: a
```

tin is offering you five choices. All you have to do is press the key that you want. Notice that after the colon, **tin** has put a default of **a**. Since this is the default, pressing RETURN is the same as pressing **a**.

Hint

If your terminal or computer has cursor control keys—the keys with the arrows: UP, DOWN, LEFT, and RIGHT—**tin** makes great use of them.

In general, the UP and DOWN keys move up or down one line (or, where appropriate, one page).

The LEFT and RIGHT keys move up or down a logical level. For example, when you are selecting a newsgroup, RIGHT moves you to the thread selection list. When you are selecting a thread, LEFT moves you back to the newsgroup selection list.

Once you get a little practice, this will all make perfect sense. In the words of Iain Lea (Mr. **tin**), "You can drive **tin** all day long just by using these four keys."

Customizing Your Working Environment

tin has more ways to change settings and to customize your environment than you would believe. Unfortunately, modifying most of these settings requires a degree of knowledge that is well beyond that of the beginning or casual user.

However, **tin** does make it easy for you to change some of the more basic settings. At any time, you can press the **M** (menu) key. (Try it.) This will bring up a list of a limited number of settings. Each of the settings will have a number. To change a setting, type the number, press RETURN, and follow the instructions.

After you select a setting to change, you may see the following instruction:

```
<SPACE> toggles & <CR> sets.
```

This means that the setting you are modifying has a fixed number of choices. Pressing SPACE will change from one choice to the next. When you are finished, and you want to set your final choice, press RETURN ("Carriage Return").

Starting and Stopping

To start **tin**, you enter the **tin** command:

```
tin
```

If you have never used **tin** before, you will see a screen of introductory information. This screen only appears the first time you use **tin**. Read this over, but don't worry if you don't understand it all. Everything you need to know to get started is in this chapter.

When **tin** starts, it goes through a number of initialization procedures. You will see messages like:

```
Reading news active file...
Reading attributes file...
Reading newsgroups file...
```

When you see the last line, it means that **tin** is reading your **.newsrc** file and comparing it against the news server's master list of active newsgroups. If there are any new newsgroups, **tin** will add them to your **.newsrc** and ask you if you want to subscribe.

If you have never read the news before, **tin** will create a **.newsrc** file for you (see Chapter 11). Initially, your **.newsrc** file will contain all of the active newsgroups with each newsgroup subscribed. You will probably want to unsubscribe to most of the groups. There are two ways to do this. First, you can do it within **tin** by using the **U** or **u** command, although this will probably take a long time. The more practical method is to use a text editor, such as **vi** or Emacs, and edit the **.newsrc** file for yourself (see Chapter 11).

To stop **tin**, you can use the **Q** (quit) command at any time. Simply press the **Q** key. Notice that this is an uppercase (capital) letter. The **q** command is used to leave whatever part of the program you happen to be using and return to the previous level. The **Q** command will quit **tin** outright.

Most of the time, there is a better way to start **tin** than the simple command we mentioned above. You will remember that we said that the first thing that **tin** does is to check to see if there are new newsgroups that are not listed in your **.newsrc** file. Although this is a good idea, it can be time consuming, perhaps taking even several minutes each time.

Most people prefer to have **tin** skip this check. You can do so by using the command:

```
tin -q
```

The **-q** is called an *option* or a *switch*. In this case, the **-q** option tells **tin** to perform a "quick" start. Using **-q** will start **tin** faster, but will not, of course, pick up any new newsgroups. For this reason, we recommend starting **tin**, say, once every week or two, without the **-q** option, just to see what is new.

There may be times when you are interested in reading only a single newsgroup. In that case, you can specify that group directly. For example, say that you only want to read the **rec.humor** newsgroup. You can enter:

```
tin rec.humor
```

You can also specify more than one newsgroup:

```
tin rec.humor alt.tasteless.jokes
```

When you start **tin** in this way, it will show you only the newsgroups you specify.

tin has quite a few options that you can use. Most of the time, you will not need any of them except for **-q**. However, we will discuss a few of the more useful ones in case you do want to try them. For more information about options, you can read the entry for **tin** in the online manual by entering the Unix command:

```
man tin
```

If you would like to display a quick list of all the possible options, you can use the **-h** (help) option:

```
tin -h
```

Remember that we said that the first time you use **tin** it will display a screen of introductory information. If you would like to display this screen again sometime, you can do so by using the **-H** (help summary) option:

```
tin -H
```

Another option that you may want to use is **-w** (write). You use this when you want to start **tin** only for the purpose of posting a news article. We will discuss this later in the chapter.

The last option we will mention is **-z**:

```
tin -z
```

This tells **tin** to start up only if there are articles that you have not yet read. You can combine this option with the names of one or more newsgroups. This is handy when you want to check quickly if there is anything of interest to you.

For example, say that you follow the **rec.humor** newsgroup religiously. (That is, you start each day by covering your head and reading jokes.) You can use the following command to start **tin** only if there are new articles in that newsgroup:

```
tin -z rec.humor
```

> ## Hint
>
> You can place a **tin -z** command in your initialization file so that each time you log in you will automatically start **tin** if there is news that you have not yet read. The name of your initialization file depends on which shell you are using. For the C-Shell, the file is **.login**; for the Bourne or Korn shell, it is **.profile**; and for the Bash shell, it is **.bash_profile**.

The Newsgroup Selection List

To read the news, the first thing you must do is choose which newsgroup you want to look at. To help you to make a choice, **tin** will display a list of newsgroups, one page at a time. Figure 15-1 shows a typical newsgroup selection list.

The top line of Figure 15-1 tells us that we are selecting newsgroups. The number in parentheses is the number of subscribed newsgroups, in this case, 157. In the top right-hand corner, we see a reminder that we can display a help summary by pressing the **h** key.

Next, follow as many newsgroups as will fit on one page. In our example, we see 16 newsgroups at a time. The entry for each newsgroup shows us an identification number, the number of unread articles, the name of the newsgroup, and a short description. In our example, three of the newsgroups (4, 5, and 14) have no unread articles.

The descriptions to the right are taken from a master file that is distributed regularly to news servers. You will find that not all the descriptions are informative and that many of them are missing. If you decide that you would rather not see the descriptions, you can tell **tin** to omit them by pressing the **d** (description) key. To bring back the descriptions, press **d** again. If you want to change this setting permanently, you can use the **M** (menu) command to modify the **Show description** setting.

The bottom part of the group selection list is a short summary to help you with the most important commands. After we discuss the commands (in the next section), this summary will make sense. If you decide that you would like to get rid of this summary—and make more room for newsgroup names—press **H**. To bring back the summary, press **H** again.

You will notice that our example shows a few newsgroups that have no unread articles. Quite possibly, you will not want to look at such newsgroups as there is nothing to read. If so, you can press the **r** key. This tells **tin** to display only those newsgroups that have unread articles. To change back, press **r** once

```
                    Group Selection (157)

1   3   alt.1d                   One-dimensional imaging, and the think i
2   7   alt.3d                   Three-dimensional imaging
3   27  alt.abortion.inequity    The inequity of abortion
4       alt.abuse.offender.recovery Helping offenders recover
5       alt.abuse.recovery       Helping victims of abuse recover (Moder
6   10  alt.activism             Activities for activists
7   3   alt.activism.d           A place to discuss issues in alt.activi
8   27  alt.adoption             For those involved with or contemplatin
9   5   alt.aeffle.und.pferdle   German TV cartoon characters
10  17  alt.agriculture.fruit    Fruit farming and agriculture
11  143 alt.agriculture.misc     Agriculture and farming
12  27  alt.aldus.freehand       Aldus Freehand software
13  54  alt.aldus.misc           Other Aldus software products
14      alt.aldus.pagemaker      All about Aldus PageMaker
15  36  alt.alien.visitors       Space aliens on Earth! Abduction! Gov't
16  131 alt.amateur-comp         The amateur computerist


   <n>=set current to n, TAB=next unread, /=search pattern, c)atchup,
g)oto, j=line down, k=line up, h)elp, m)ove, q)uit, r=toggle all/unread,
   s)ubscribe, S)ub pattern, u)nsubscribe, U)nsub pattern, y)ank in/out
```

Figure 15-1. *A typical group selection list*

again. If you want to change this setting permanently, you can use the **M** (menu) command to modify the **Show only unread** setting.

Selecting a Newsgroup

To start reading a newsgroup, all you have to do is move to the group you want and press RETURN. While you are working with the newsgroup selection list, there are many commands that you can use. As well as the general commands, there are commands for moving within the newsgroup list, and for telling **tin** to start reading when you have selected a newsgroup.

The general commands are summarized in Figure 15-4. The newsgroup selection commands are summarized in Figure 15-5. In this section, we will discuss the most basic of these commands. When you get some time, check out Figure 15-5 and try some of the other commands.

tin displays only as many newsgroup names as can fit on a single page (that is, one screenful). To display the next page, press PAGE DOWN. To move back one page, press PAGE UP. If your terminal or computer does not have these keys (or if you prefer not to use them), you can use SPACE (PAGE DOWN) and **b** (PAGE UP) instead.

At all times, one of the newsgroup names will be highlighted. To move down or up one line, use the DOWN and UP keys. Again, if you do not have these keys (or you do not want to use them), there are alternatives. You can use **j** (down) and **k** (up) instead.

To jump directly to a particular newsgroup, you have two choices. First, you can enter its number. For example, to jump to newsgroup #4, press **4** and then press RETURN.

Second, you can press **g** (go to), type the name of the newsgroup, and press RETURN. When you use the **g** command, **tin** will suggest the last newsgroup you jumped to as a default choice. If you want to jump to this newsgroup again, just press RETURN. (It will make sense when you see it.)

Once you have chosen the newsgroup you want, you can enter it by pressing RIGHT or RETURN. Alternatively, you can press TAB to automatically enter the next group that has unread articles.

Finally, when you are finished reading the news, press **q, Q,** or LEFT to quit **tin**.

The Thread Selection List

Once you have selected a newsgroup, **tin** shows you a list of threads. You examine this list and select the first thread that you want to read. Figure 15-2 shows a typical thread selection list.

The top line shown in Figure 15-2 tells us the name of the newsgroup, in this case **soc.culture.british**. Directly after the name, **tin** shows us the number of threads, the number of articles, the number of killed (junked) articles, and the number of hot articles. (Hot articles are those that are selected automatically according to a predefined criterion. For more details, see the reference manual.) In our example, there are 69 threads of 245 articles. None of these articles are killed or hot.

Next, follow as many thread summaries as will fit on one page. In our example, we see 15 threads. The summary of each thread shows us an

```
              soc.culture.british (69T 245A OK OH)          h=help

   1  + 2   Empress of Blandings' silver medal   Clarence Emsworth
   2         A world full of loonies             Alaric Dunstable
   3  + 3   Making a proper marriage             Constance Keeble
   4  + 13  Five ways to call a pig              George Wellbeloved
   5  + 6   The best dog biscuits to use         Freddie Threepwood
   6    3   Choosing port                        Sebastian Beach
   7         Efficiency                           Rupert Baxter
   8  +      Things I don't understand            Veronica Wedge
   9  +      The Queen of Matchingham             Gregory Parsloe
  10  + 1   Poetry for Polly Pott                Ricky Gilpin
  11  +      My life with Sue Brown               Ronald Fish
  12  + 63  Does the Ickenham System work?       Galahad Threepwood
  13  +      On The Care Of The Pig               Augustus Whipple
  14  +      Housekeeping and etiquette           Mrs. Twemlow
  15    16  Canadian poetry                      Ralston McTodd
```

Figure 15-2.   *A typical thread selection list*

identification number, the number of followup articles, and the subject and author of the first article. For example, thread #12 contains the original article plus 63 followups. The thread was started by Galahad Threepwood, who has initiated a discussion as to whether or not the Ickenham System actually works. Thread #2 contains only a single article with no followups.

You will notice that, after the thread number, some of the summaries have a + (plus) character. This indicates that the thread has not yet been read. In our example, threads 2, 6, 7, and 15 have been read.

As you can see, the thread selector list shows both the subject and the author of the first article in the thread. If the subject descriptions are long, they will be truncated. To make more room, you can press **d** to tell **tin** to display only the subject. To change back and display both the subject and author, press **d** again. If you want to change this setting permanently, you can use the **M** (menu) command to modify the **Show author** setting.

Similarly, you can tell **tin** to display only the unread threads by pressing **r**. To change back and display both read and unread threads, press **r** again. If you want to change this setting permanently, you can use the **M** command to modify the **Show only unread** setting.

Selecting a Thread to Read

Once you have selected a newsgroup, **tin** shows you a list of threads. To read the articles, all you have to do is move to a thread and press RETURN.

While you are working with the thread selection list, there are many commands that you can use. As well as the general commands, there are commands for moving within the thread list, and for telling **tin** to start reading when you have selected a thread.

The general commands are summarized in Figure 15-4. The thread selection commands are summarized in Figure 15-6. In this section, we will discuss the most basic of these commands. When you get some time, check out Figure 15-6 and try some of the other commands.

tin displays only as many thread descriptions as can fit on a single page (that is, one screenful). To move throughout the list, you can use the same commands as with the newsgroup list.

To display the next page, press PAGE DOWN or SPACE. To move back one page, press PAGE UP or **b**. To move down one line, press DOWN or **j**. To move up one line, press UP or **k**.

To jump directly to a particular line, type the number followed by RETURN. For example, to jump to thread #4, press **4** and then press RETURN.

To go to the next unread thread, press **N**; to go the previous unread thread, press **P**. If you press **K** (kill), it will mark a thread as read and move to the next line.

Once you have chosen the thread you want, you can start reading by pressing RIGHT or RETURN. Alternatively, you can press TAB to automatically start reading the next unread thread.

If you want to preserve a thread, you have three choices. You can press **m** to mail a copy to someone (including yourself), **o** to print it, or **s** to save it to a file.

Finally, when you are finished reading, you can return to the newsgroup selection list by pressing **q** or LEFT. To quit **tin** completely, press **Q**.

Reading Articles

Figure 15-3 shows a typical article as displayed by **tin**. Across the top three lines, we see descriptive information about the article. **tin** automatically extracts this information from the header.

The top line in Figure 15-3 shows the time and date that the article was posted, the name of the newsgroup, and information about the thread. The second line shows the article number, the subject, and the number of responses (followups). The third line shows the author's mail address and description.

Next comes the body of the article, including, in this case, a signature.

Finally, the last three lines contain a short summary to help you with the most important commands. If you decide that you would like to get rid of this summary—and make more room for the article—press **H**. To bring back the summary, press **H** again.

As you are reading an article, there are many commands that you can use. As well as the general commands, there are commands for paging through the article, moving to a different article, preserving the article, and responding to the article.

The general commands are summarized in Figure 15-4. The other commands are summarized in Figure 15-7. In this section, we will discuss the most basic of these commands. When you get some time, take a look at Figure 15-7 and experiment for yourself.

The most important commands are those that allow you to page through the article. To display the next page, press PAGE DOWN, DOWN, or SPACE. To move back one page, press PAGE UP, UP, or **b**. To go to the first page, press **g**. To go to the last page, press **G**.

```
Sun, 23 May 1993 08:22:42        soc.culture.british        Thread 12 of 69
Article 342            Does the Ickenham System work?        63 Responses
gally@pelican.org                        Galahad Threepwood at the Pelican Club

Fred Ickenham tells me that his system is 100% successful —
in certain cases.

I believe that he advises that "The preliminary waggle is of
the essence."

Maybe so, but in my many years as a member of the Pelican Club,
I developed my own methods which, as Sue Brown and Ronnie Fish will
tell you, have been known to meet with some measure of success.

Does anyone else have any thoughts on this matter?

—
Hon. Galahad Threepwood                              Pelican Club, London
          The world will have to wait a hundred years before it
          hears the story of young Gregory Parsloe and the prawns.

<n>=set current to n, TAB=next unread, /=search pattern, ^K)ill/select,
    a)uthor search, B)ody search, c)atchup, f)ollowup, K=mark read,
    |=pipe, m)ail, o=print, q)uit, r)eply mail, s)ave, t)ag, w=post
```

Figure 15-3.   *A typical article as displayed by* **tin**

To mark this article as read and go to the next article, press **k** (kill). To mark the entire thread as being read and go to the next thread, press **K**.

If you would like to re-display the article showing the full header, press **^H**. To decode an article using rot-13 (see Chapter 11), press **d**.

To jump to the next unread article, press RIGHT or TAB. To jump to the next thread, press RETURN.

If you want to preserve the article, you have three choices. You can press **m** to mail a copy to someone (including yourself), **o** to print it, or **s** to save it to a file.

To respond to an article, you can press **r** or **R** to send a private reply by mail to the author. The difference between these two commands is that **r** will include the text of the article in your reply.

When you press **r** or **R**, **tin** will set up the header of your message and then start your text editor for you. When you are finished, quit the editor in the regular manner and **tin** will regain control. **tin** will now ask you what you want to do with your reply:

```
q)uit, e)dit, s)end: s
```

You can press whichever letter indicates your choice. As you can see, the **s** (send) choice is the default. So, if you want to send the message—which is usually the case—you need only press RETURN.

A second way to respond to an article is to post a followup article of your own. We will discuss this in the next section.

Finally, when you are finished reading, you can return to the thread selection list by pressing **q** or LEFT. To quit **tin** completely, press **Q**.

Posting an Article

There are three ways to use **tin** to post an article. First, as you are reading an article, you can press **f** or **F** to post a followup. The difference between these two commands is that **f** will include the text of the article in your followup.

When you press **f** or **F**, **tin** will set up the header of your followup and then start your text editor for you. When you are finished, quit the editor in the regular manner and **tin** will regain control. **tin** will now ask you what you want to do with your article:

```
q)uit, e)dit, i)spell, p)ost: p
```

You can press whichever letter indicates your choice. As you can see, the **p** (post) choice is the default. So, if you want to post the article—which is usually the case—you need only press RETURN.

The **i** choice will invoke an interactive spelling checker. However, it will only work if this feature is installed in your system.

The second way to post an article is to press **w** (write) at any time. This tells **tin** that you want to post a regular article (not a followup) to the current newsgroup. Thus, if you are reading the newsgroup selection list, you can post an article to any newsgroup you want by moving to it and then pressing **w**.

After you press **w**, **tin** will ask you for the subject of your article. Type the subject and press RETURN. Everything will now proceed in the regular manner.

From time to time, you may wish to post an article when you are not reading the news (that is, when you are not using **tin**). To do so, you can invoke **tin** with the **-w** (write) option. **tin** will help you post an article and then quit automatically. You will probably also want to use the **-q** (quick start) option to bypass the automatic check for new newsgroups:

```
tin -q -w
```

In general, when you enter a Unix command, you can combine options using a single **-** character, so in this case you can also use:

```
tin -qw
```

When you start **tin** in this manner, it will ask you to enter the name of the newsgroup to which you want to post. After you type the name and press RETURN, everything will proceed in the regular manner. When you are finished posting your article, **tin** will quit automatically.

Commands That Are Always Available

Controlling the Program

q	return to previous level
Q	quit `tin`

Getting Help

h	display summary of commands [help]
H	off/on: show help menu at the bottom of the screen
v	show what version of `tin` you are using
M	display a menu of configurable options

Displaying Information

PAGE DOWN	display the next page
SPACE	same as PAGE DOWN
^D	same as PAGE DOWN [down]
^F	same as PAGE DOWN [forward]
PAGE UP	display the previous page
b	same as PAGE UP [back]
^U	same as PAGE UP [up]
^B	same as PAGE UP [back]

Posting an Article

w	post an article to the current newsgroup [write]
W	display a list of all the articles you have posted

Entering Unix Commands

`!command`	execute the specified Unix command
`!`	pause `tin` and start a new shell

Figure 15-4.   *Commands that are always available*

Commands to Use While Selecting a Newsgroup

Stopping	
q	quit `tin`
LEFT	same as q
Displaying the Selection List	
DOWN	move down one line
j	same as DOWN
UP	move up one line
k	same as UP
num RETURN	go to newsgroup number *num* (for example: 4 RETURN)
1 RETURN	go to the first newsgroup in the list
$	go to the last newsgroup in the list
N	go to the next group with unread news
g	go to specified newsgroup
/	search forward for specified newsgroup
?	search backward for specified newsgroup
Start Reading a Newsgroup	
RIGHT	start reading the current group
RETURN	same as RIGHT
TAB	go to next group with unread news and start reading
n	same as TAB [next]
Controlling the Screen Display	
d	toggle: show newsgroup names/names+descriptions
r	toggle: show all newsgroups/with unread articles only
Controlling Newsgroups	
m	move newsgroup position within the list
s	subscribe to the current newsgroup
u	unsubscribe to the current newsgroup
S	subscribe to all groups that match specified pattern
U	unsubscribe to all groups that match specified pattern

Figure 15-5.   *Commands to use while selecting a newsgroup*

Commands to Use While Selecting a Thread to Read

Stopping

q	return to newsgroup selection list
LEFT	same as q

Moving From One Article to Another

DOWN	move down one line
j	same as DOWN
UP	move up one line
k	same as UP
num RETURN	go to thread number num (for example: 4 RETURN)
1 RETURN	go to the first thread in the list
$	go to the last thread in the list
/	search forward for subject containing a pattern
?	search backward for subject containing a pattern
N	go to next unread thread
P	go to previous unread thread
K	mark thread as read, then go to next unread thread

Start Reading an Article

RIGHT	start reading the current thread
RETURN	same as RIGHT
TAB	start reading next unread thread
-	return to the last thread that you read

Controlling the Screen Display

d	toggle: show subject/subject+author
r	toggle: show all threads/unread threads only

Changing Newsgroups

n	go to next newsgroup
p	go to previous newsgroup
g	go to specified newsgroup

Preserving a Thread

m	mail the thread to someone
o	print the thread
s	save the thread to a file

Figure 15-6. *Commands to use while selecting a thread to read*

Commands to Use While Reading an Article

Stopping

q	return to thread selection list
LEFT	same as q

Displaying the Article

DOWN	display the next page
UP	display the previous page
g	go to the first page of the article
^R	same as g [re-display]
G	go to the last page of the article
$	same as G
^H	re-display article showing header
d	decode the current article using rot-13
/	search forward for specified pattern
?	search backward for specified pattern

Moving to a Different Article

RIGHT	go to the next unread article
TAB	same as RIGHT
N	same as RIGHT
RETURN	go to the next thread
n	go to the next article
p	go to the previous article
P	go to the previous unread article
K	mark entire thread as read, go to next unread thread
k	mark article as read, go to next unread article
-	return to the last article that you read

Preserving the Article

m	mail the article to someone
o	print the article
s	save the article to a file

Responding to the Current Article

F	post a followup
f	same as F, include text of article
R	reply by mail to author
r	same as R, include text of article

Figure 15-7.   *Commands to use while reading an article*

## CHAPTER 16

Anonymous FTP

One of the most important and most widely used Internet services is Anonymous FTP. This remarkable service lets you copy files from thousands of different computers in all parts of the Internet. These files contain virtually every kind of information that can be stored on a computer.

Would you like a program for your PC or Macintosh? Would you like an issue of an electronic magazine? How about a frequently asked question list from a Usenet discussion group?

Just about any type of information or computer software is waiting for you on the Internet. To access it all—for free—all you have to do is read this chapter and learn how to use Anonymous FTP. And, if for some reason, you do not have access to FTP, we will show you how to request files by mail.

Once you finish this chapter, be sure to read the related material in Chapters 17 and 18. In Chapter 17, we show you how to use an Archie server to search the vast collection of Anonymous FTP hosts to find specific files. In Chapter 18, we discuss the various types of files that are used to store data for Anonymous FTP retrieval.

Understanding FTP

The term "FTP" comes from File Transfer Protocol, the underlying set of specifications that support Internet file transfer. But what we mean when we refer to FTP is more than a dry set of specifications. FTP is a service that allows us to copy a file from any Internet host to any other Internet host. As such, FTP provides a large part of the spiritual glue that holds the Internet together.

Like most Internet services, FTP uses a client/server system (see Chapter 2). You use a client program, named **ftp**, to connect to a server program on the remote computer. Conceptually, the idea is simple. Using your client program, you issue commands that are sent to the server. The server responds by carrying out whatever commands you sent. For example, you might send a command that requests the server to send you a copy of a particular file. The server responds by sending the file. Your client program receives the file on your behalf and stores it in your directory. (We discuss files and directories later in the chapter.)

As you may remember, when you copy files from a remote computer to your own, we say that you are downloading files. When you copy files from your computer to the remote one, you are uploading files. If you have trouble remembering which is which (and we always do), just imagine the remote computer floating above you in the sky. Up is away from you, down is towards you.

When we talk about FTP, we use the same terminology as for Telnet (see Chapter 7). Your computer is called the local host, and the other computer is called a remote host. Thus, in the language of the Internet, we can say that the **ftp** client program allows you to upload and download files to and from a remote host.

To use the **ftp** program, you enter the **ftp** command and specify the address of the remote host to which you want to connect. Once the program starts, you enter one command after another, to copy files back and forth as you wish. As you use **ftp**, there are a number of commands that you can enter. For example, you can explore the directories on the remote computer and change from one directory to another.

It is common to use the word "FTP" as a verb. For example, you might hear someone say, "You can get a copy of the frequently asked question lists by FTP-ing to **rtfm.mit.edu**. If that computer is too busy, there are a number of other hosts from which you can FTP the same files."

Understanding Anonymous FTP

We have already explained that FTP allows you to transfer files from one Internet computer to another. However, there is one basic restriction: you cannot access a computer unless you have proper authorization. In practice, this means that you must be able to log in to the computer. In other words, you cannot copy files to or from a computer unless you have a userid (account name) and a password.

Anonymous FTP is a facility that lets you connect to a remote host and download files without having to be registered as a user. The system manager sets up a special userid named **anonymous** that anyone, anywhere on the Internet, is allowed to use.

Hint

You cannot use Anonymous FTP with every Internet host, only with those that have been set up to offer this service.

When you use the **ftp** program to connect to an Anonymous FTP host, it works the same way as regular FTP, except that when you are asked for a userid you enter **anonymous**. When you are asked for a password, you enter your mail address or name. This is so the keepers of the system can keep track of who is accessing their files. (For a discussion of Internet addresses, see Chapter 4.)

Hint

The **anonymous** userid can only be used with FTP. You cannot connect to an Anonymous FTP host using Telnet (see Chapter 7) and use the **anonymous** userid to log in as a regular user.

When a system manager sets up a computer to be an Anonymous FTP host, he or she designates certain directories as being open to public access. The rest of the directories are off limits. Thus, it is perfectly safe for an organization to offer public access to outside users. As an extra security measure, most Anonymous FTP hosts allow you to download files, but not upload. That is, you can copy all the files you want from the remote host, but you cannot copy files to the host.

If an Anonymous FTP host does allow uploads, you may be required to copy all new files to a single designated upload directory. Later, after the system manager has had a chance to check out the new files, he or she will move them to one of the public download directories. In this way, remote users are protected against people who might upload troublesome files, such as programs with viruses.

Hint

Most system managers do not have the time to be eternally vigilant. When you download programs, please remember to take normal precautions. For example, before you run a program on your PC, check it with a virus scanning program.

As an Internet user, you can use FTP to copy files between any two Internet hosts. Realistically, though, most people only have a single Internet account and FTP is mostly used to download public files. Thus, in this chapter, we will concentrate on what you need to know to download using Anonymous FTP.

As you will see, the **ftp** client program has a large number of commands that you can use, and some of those commands have variations. We will include a list summarizing all the important commands and what they do, and we will show you how to display help information as you are working. But we will not discuss every command in detail. Rather, we will spend our time on what you need to know to use Anonymous FTP.

If you are using a Unix system and you would like more detailed information about the **ftp** program and its commands, you can display the **ftp** entry in the online manual. Use the following command:

```
man ftp
```

We discuss the **man** command and how to use it in Chapter 11.

The Importance of Anonymous FTP

At first, Anonymous FTP sounds useful, but not really all that special. Okay, so anyone on the Internet can access files on any Anonymous FTP host. What's the big deal?

The big deal is that the world of Anonymous FTP is huge. There are thousands of Anonymous FTP hosts and countless files that you can download for free. Virtually every type of information and every type of computer program is available somewhere on the Internet. Many people and organizations have generously donated disk space and computing facilities as well as their own time to make these files available. Why? Because, as we explained in Chapter 1, the tradition on the Internet is one of sharing.

Until you start using Anonymous FTP, this may not sound like much, but it is. Indeed, we can solemnly assure you that Anonymous FTP is one of the most significant inventions in the history of mankind.

If you are new to the Internet, we don't blame you one bit for thinking that such a statement is the rankest hyperbole. Still, what we say is absolutely true, for three important reasons.

First, we have already said that Anonymous FTP allows you to download virtually any type of information. Until you are an experienced Internet user, it is difficult to appreciate how important this is. Anonymous FTP provides access to the largest library of information ever accumulated. Moreover, it is a library that is always growing, never closes, covers every conceivable topic and, best of all, is free. To get a feeling for the variety of information, take a moment now to skim through the catalog in the last part of this book. Notice how many of the items are available via Anonymous FTP.

Hint

With so many different Anonymous FTP hosts and so many files, how do you know where to find a particular file? You use a service called Archie. Archie will search a special FTP database and find all the hosts that contain the file you need. For an explanation of how to use Archie, see Chapter 17.

Second, Anonymous FTP is the principal way in which software is distributed on the Internet. The reason that the Internet can exist at all is that people use programs that provide standardized services using standardized protocols. Many of these programs are distributed via Anonymous FTP and, hence, are available to anyone who wants to set up an Internet host.

For example, in order for you to access the Usenet discussion groups, your system manager must have installed the Usenet software on your computer. Where does your system manager acquire this software? Anonymous FTP.

The final reason why Anonymous FTP is so important is that it is used to archive and disseminate the technical information that defines the Internet itself. As we explained in Chapter 2, the Internet is based on a large number of protocols and conventions. Each such protocol is explained in a technical publication called a *request for comment* or *RFC*. Don't take the name too literally. Think of an RFC as a technical explanation of how something is supposed to work. RFCs are also used to offer generally useful information on a specific topic.

Each RFC is given a number and is made freely available to anyone who wants to read it. For example, RFC #1325 is a long list of answers to questions that are commonly asked by new Internet users. (This RFC is similar to the frequently asked question lists that we discussed in Chapter 9.)

Hint

If you would like to read RFC #1325, use Archie (see Chapter 17) to search for the file named **rfc1325.txt**. Once you have found an Anonymous FTP host that has this file, use the techniques described in this chapter to FTP to the host and download the file.

If you would like to download an index of all the RFCs, use Archie to search for a file named **rfc-index.txt**.

Starting the ftp Program

To download or upload files from a remote host, you use the **ftp** program. This program acts as a client and connects to the FTP server on a remote host. Once the connection is made, you will be asked to specify a userid and password. You can then enter whatever **ftp** commands you want.

There are two ways to start the **ftp** program. In this section, we will show you how it is done most of the time. In the next section, we will show you an alternate method.

To start **ftp**, enter the name of the command followed by the address of the remote host to which you want to connect. For example, say that you want to download files from the computer named **rtfm.mit.edu**. Enter the command:

```
ftp rtfm.mit.edu
```

Hint

As we explained in Chapter 4, all Internet hosts have an official address known as an IP address. This address consists of several numbers separated by periods. For example, the official IP address of **rtfm.mit.edu** is **18.70.0.224**.

Some systems have trouble dealing with certain standard addresses. If you encounter such a problem with **ftp**, try using the IP address. For example, either of the following commands will connect to the same host:

```
ftp rtfm.mit.edu
ftp 18.70.0.224
```

For more information about IP addresses and Internet addresses in general, see Chapter 4.

When the **ftp** program starts, it will initiate a connection to the remote host that you specified. Once the connection is made—which might take a few moments if the host is far away—you will see a message like the following:

```
Connected to CHARON.MIT.EDU.
220 charon FTP server (Version 6.6 Wed Apr 14 21:00:27 EDT 1993) ready.
Name (rtfm.mit.edu:harley):
```

The first line of this message shows us that we have made the connection. Notice that the name of the computer we have connected to is really **charon.mit.edu**. This is because the name **rtfm.mit.edu** is actually an alias for **charon.mit.edu**.

Such aliasing is fairly common in the world of Anonymous FTP because it allows us to use easily remembered names. It also gives system managers the flexibility to change computers whenever they want without confusing people. For example, if it should become necessary for the **rtfm** system manager to use a different computer for Anonymous FTP, he does not have to inform everybody of the change. All he has to do is ensure that the address **rtfm.mit.edu** is aliased to the new computer.

It is especially common to see names that begin with **ftp**. For example, the Electronic Frontier Foundation has an Anonymous FTP host named **ftp.eff.org**. (The EFF is a public-service organization dedicated to "the pursuit of policies and activities that will advance freedom and openness in computer-based communications". As part of its work, the EFF maintains an Anonymous FTP host that contains a wealth of interesting information, including electronic magazines.)

The public name of the EFF's Anonymous FTP host is **ftp.eff.org**. However, when you FTP to this address, you will see that it is actually aliased to another computer. At the time we wrote this chapter, the real computer happened to be **krager.eff.org**.

Now, to return to our example, the second line we saw was:

```
220 charon FTP server (Version 6.6 Wed Apr 14 21:00:27 EDT 1993) ready.
```

This tells us the name of the FTP server and the version of the FTP software that this server is using.

Hint

Notice that the message starts with the number **220**. All messages from the FTP server start with such numbers, and there is no way to get rid of them. However, the numbers are not important, and you can ignore them.

One nice thing about the numbers is that they show you exactly which messages are coming from the remote server. Lines that do not begin with a number are from the **ftp** client program.

Finally, let us look at the last line of the message:

```
Name (rtfm.mit.edu:harley):
```

This is a request from our **ftp** client program, asking us what userid we want to use to log in to the computer named **rtfm.mit.edu**. As it happens, we are currently logged in to the local host as **harley**. The **ftp** program knows this and suggests that we may want to use the same userid on the remote system. That is why you see the name **harley** in parentheses.

If you press RETURN, the **ftp** program will use this userid as a default and send it to the remote host. However, in this case (and most of the time), you will want to log in as **anonymous**. Simply type this name and press RETURN:

```
anonymous
```

You will now see:

```
331 Guest login ok, send e-mail address as password.
Password:
```

The FTP server has approved the userid **anonymous**. You are now being asked to enter your mail address as a password. Out of courtesy, it is a good idea to honor this request. Indeed, some FTP servers will not let you log in if your password does not look like a valid address.

Hint

Some FTP servers will examine your password and decide if it looks like your real mail address. If not, you may be denied access to certain public directories (and you may not even know that this is happening).

Once you have entered an approved userid and password, you will see a message like the following:

```
230 Guest login ok, access restrictions apply.
ftp>
```

This means that you are officially logged in and can use the Anonymous FTP facility.

The second line, **ftp>**, is a prompt from your **ftp** client program. Whenever you see this prompt, you can enter one of the **ftp** commands. We will discuss the various commands later in the chapter, and at that time we will show you a full Anonymous FTP session in which we download a file.

For now, we just want to mention two commands. To display a list of all the **ftp** commands, enter the **?** command. To end the FTP session, use the **quit** command.

The last point we want to cover is what to expect if your FTP client is unable to connect to the remote host. There are three ways in which this might happen. First, the FTP service may be temporarily unavailable. In such cases, you will see:

```
ftp: connect: Connection refused
```

Second, the network connection to the remote host may be inoperative. This might be a problem with the network to which the host is connected, or it might be that the host computer itself is not working. In such cases, you will see a message like:

```
ftp: connect: Host is unreachable
```

The best thing to do is try again later.

Finally, there may be a problem with the address that you specified. For example, say that you want to download files from **rtfm.mit.edu**, but you accidentally enter the wrong address:

```
ftp rtff.mit.edu
```

You will see a message like the following:

```
rtff.mit.edu: unknown host
ftp>
```

At this point, you can enter the name of another host. (We will explain how to do this in the next section.) Otherwise, you can use the quit command to stop the program.

Hint

The **ftp** message "unknown host" can be misleading. There are many reasons why your FTP client might not be able to make a remote connection. The two most common are:

- You spelled the address of the computer wrong.
- You specified the name of a computer that is not on the Internet.

A Second Way to Start the ftp Program

In the previous section, we mentioned that there are two ways to start the FTP client program. The first way is to enter the **ftp** command along with the address of the remote host. For example:

```
ftp rtfm.mit.edu
```

The second way is to start **ftp** without specifying a host. Simply enter:

```
ftp
```

The program will start, but will not make a connection. Instead, you will see:

```
ftp>
```

This is the **ftp** prompt. It means that the program is waiting for you to enter a command. To connect to a remote host, type **open**, followed by the address of the host. For example:

```
open rtfm.mit.edu
```

The connection will be made just as if you had specified the address when you entered the **ftp** command.

At the end of the previous section, we gave an example in which an **ftp** command had a bad address. In the example, the remote host was named **rtfm.mit.edu**, but we mistakenly entered:

```
ftp rtff.mit.edu
```

What happens in such a case is that the **ftp** program tries to make the connection. When it can't, it gives up and displays its prompt, waiting for you to enter a command. In this case, you would see:

```
rtff.mit.edu: unknown host
ftp>
```

You can now enter:

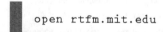

```
open rtfm.mit.edu
```

If this address doesn't work, you can try another one. If you decide to give up, enter:

```
quit
```

This will stop the **ftp** program.

Summary of Starting and Stopping ftp

There are two ways to start the **ftp** program. Either enter the command with the address of a remote host:

```
ftp rtfm.mit.edu
```

or enter the command by itself:

```
ftp
```

and then, at the **ftp>** prompt, enter an **open** command:

```
open rtfm.mit.edu
```

To stop **ftp**, wait for the **ftp>** prompt and enter the **quit** command:

```
quit
```

Directories and Files

Once you start messing around with Anonymous FTP, you will find that you need to know something about files and directories. Since most of the Anonymous FTP hosts are Unix computers, you will need to know about the Unix file system.

You may remember that, in Chapter 1, we told you it is a good idea to learn some Unix. We pointed out that even if you do not use Unix yourself, many of the computers that you will access are Unix computers. This is especially true when you are using Anonymous FTP.

In this section, we will give you the short, 75-cent tour of the Unix file system. If you want more information, the best idea is to consult a good Unix book. The one we recommend is *A Student's Guide to Unix*, by Harley Hahn (McGraw-Hill , 1993).

Within Unix, the definition of a *file* is a highly generalized one. A file is considered to be any source of input or any target of output. For our purposes, this definition is too technical, so let us be a bit more informal: a file is a collection of data.

Files can hold any type of information that can be stored on a computer: programs, documents, images, and so on. Most files are stored on disks (hard drives), but you will also hear of files stored on CDs, tapes, and floppy diskettes.

At the time a file is created, it is given a name. Whenever you want to do something with a file, you must refer to it by its name. For example, when you want to download a file using Anonymous FTP, you must be able to tell your FTP client the name of the file you want.

Unix file names can consist of letters, numbers, and certain punctuation characters. For example, earlier in the chapter we referred to two files named **rfc1325.txt** and **rfc-index.txt**.

Unix distinguishes between uppercase (capital) letters and lowercase (small) letters. This means that when you use a file name, you must use the exact letters. For example, it would be wrong to refer to the file above as **Rfc1325.txt**. To Unix, the letters **R** and **r** are completely different, so **rfc1325.txt2** and **Rfc1325.txt** are two completely different names.

For the most part, Unix file names consist of all lowercase letters. However, there will be times when you encounter a file name that has uppercase letters. When you do, be sure to type the name exactly.

Since most computers can store vast numbers of files, Unix allows us to organize them into collections called *directories*. A directory is itself a special kind of file whose only purpose is to act as a repository for other files.

When you connect to a remote Anonymous FTP host, you will usually find a large number of directories. In order to download a file, you must be able to specify the name of the directory in which the file lies, as well as the name of the file itself. Here is an example.

In Chapter 9, we explained that many Usenet newsgroups have lists of questions and answers that cover the basic topics discussed in that newsgroup. These lists are called FAQ (frequently asked question) lists. The computer **rtfm.mit.edu** acts as a Usenet archive by storing copies of the FAQ lists for people to access via Anonymous FTP.

One of the more interesting lists is from the newsgroup that discusses urban legends (**alt.folklore.urban**). The name of this file is **folklore-faq**, and it is found in the directory named **/pub/usenet/news.answers**. To make sense of this directory name, all you need to understand is how Unix organizes directories.

Being able to collect files into directories is important, as it allows us to organize large numbers of files. However, many Unix systems have thousands of files and merely collecting them into directories is not enough: we need more levels of organization.

The solution is to collect the directories themselves into other directories. This allows us to form a hierarchy in which some directories contain other directories.

For example, on the **rtfm.mit.edu** host, all the public directories are kept in a directory named **pub**. Within **pub**, all the directories that contain Usenet information are kept in a directory named **usenet**. Within the **usenet** directory, there are files and directories that contain information relating to the various newsgroups.

Now, it happens that there is a special newsgroup named **news.answers** whose purpose is to act as a repository for all the FAQ lists that are posted to the various other groups (see Chapter 9). Thus, on the **rtfm.mit.edu** host, the **usenet** directory contains a directory named **news.answers** that holds all the FAQ lists. That is why the file named **folklore-faq** is found within the directory named **/pub/usenet/news.answers**.

When a directory contains another directory, we call the second one a *sub-directory*. Sometimes, the first directory is called the *parent directory*, while the sub-directory is called the *child directory*.

Each Unix computer has one main directory that contains all the sub-directories and files on the system (at least, indirectly). This main directory is called the *root directory*. Within the root directory are a number of sub-directories; within these sub-directories are other sub-directories; and so on.

In our example, **pub** is a sub-directory of the root directory; **usenet** is a sub-directory of **pub**; **news.answers** is a sub-directory of **usenet**; and within **news.answers** lies the file **folklore-faq**.

If you were to draw a diagram of such a system, it would look like a tree with branches and sub-branches growing from one main trunk, hence the name "root directory" for the main trunk.

When we write the name of a sub-directory, we start at the root directory and show each directory that we must pass through in order to reach the ultimate branch. To understand such names, you only need to remember two things.

First, the root directory is indicated by a single **/** (slash) character. We do not actually spell the name "root". Second, we use the same **/** character to divide one directory name from another. Hence, the name **/pub/usenet/news.answers**. Literally, it means "Start from the root directory, then go to the **pub** sub-directory, then go to the **usenet** sub-directory, and finally, go to the **news.answers** sub-directory."

As you use a Unix system, you can think of yourself as sitting somewhere in the directory tree. Whichever directory you are in is called your *working directory* or *current directory*. By using the appropriate commands, you can move from one directory to another. (This is like moving from one branch of the tree to another.) As you will see, it is often convenient to move to the directory that contains the file you want before you download it.

For convenience, we often describe the location of a file by tacking its name onto the end of a sequence of directories. For example, say that you wanted to tell a friend where to find the FAQ list for the Usenet newsgroup that discusses urban folklore. All you have to do is tell him to FTP to **rtfm.mit.edu** and copy the file **/pub/usenet/news.answers/folklore-faq**.

A description like this—in which names are separated by / characters—is called a *path* or a *pathname*. Thus, in order to FTP a file anywhere on the Internet, all you need to know is (1) the name of an Anonymous FTP host that has the file, and (2) the pathname of that file. If you take a look at the catalog in the last part of this book, you will see we describe many Internet resources in just this way.

An Overview of the ftp Commands

Once you have entered the **ftp** command and established a connection with a remote host, you will see the prompt:

```
ftp>
```

At this point, you can enter an **ftp** command (of which there are many). The FTP client program will send whatever command you enter to the FTP server, which will carry out your request. The idea is to enter one command after another until you have achieved your goal (say, to download a file). Then enter the **quit** command to terminate the FTP session.

At any time, you can display a list of all the **ftp** commands by entering **?** or **help** (either will do). Figure 16-1 contains a typical response from such a command. (Do not worry if your **ftp** client does not have all these commands. It will have the most important ones.)

If you want to see a one-line summary of a particular command, enter **?** (or **help**) followed by the name of the command. For example, if you enter:

```
? quit
```

you will see:

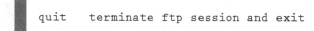

```
quit    terminate ftp session and exit
```

Notice how many commands there are in Figure 16-1. The number of commands that you see will vary, depending on what version of the FTP software you are using.

You might wonder, do I really have to learn all these commands? The answer is no. For normal Anonymous FTP sessions, all you need to know are the

At the **ftp>** prompt, you can enter **?** or **help** to display a summary of all the possible commands that are recognized by the FTP server. Here is such a summary:

```
Commands may be abbreviated.  Commands are:

!            cr           macdef       proxy        send
$            delete       mdelete      sendport     status
account      debug        mdir         put          struct
append       dir          mget         pwd          sunique
ascii        disconnect   mkdir        quit         tenex
bell 8       form         mls          quote        trace
binary       get          mode         recv         type
bye          glob         mput         remotehelp   user
case         hash         nmap         rename       verbose
cd           help         ntrans       reset        ?
cdup         lcd          open         rmdir
close        ls           prompt       runique
```

Figure 16-1.   *A list of all the* **ftp** *commands*

commands summarized in Figure 16-2. We will discuss these commands in the following section.

Hint

To stop an **ftp** command as it is executing, press CTRL-C.

The Basic ftp Commands

We can divide the **ftp** commands into several groups. First, there are the basic commands. We have already discussed **quit**, **?**, and **help** (which is the same as **?**).

The other basic command is **!** (exclamation mark). This command is used on Unix systems to enter a regular command to your local computer. Simply type the command after the **!** character and then press RETURN. The **ftp** program will put itself on hold and send the command to your Unix shell to be executed. Once the command is finished, the **ftp** program will regain control and redisplay its prompt.

Here is an example. We want to use the Unix **date** command to display the time and date on your local system. At the **ftp>** prompt, we enter **!date**. After the **date** command displays its output, we are returned to a **ftp>** prompt:

```
ftp> !date
Thu Jun 17 23:11:19 PDT 1993
ftp>
```

Basic Commands

quit	close connection to remote host, stop ftp program
?	display a list of all the ftp commands
? *command*	display one-line summary of the specified command
help	display a list of all the ftp commands
help *command*	display one-line summary of the specified command
!	local: pause ftp and start a shell
! *command*	local: execute specified shell command

Connecting

open [*host*]	establish connection to specified computer
close	close the connection to remote host, stay in ftp
user [*name* [*password*]]	set user name

Directories

cd [*directory*]	remote: change to specified directory
cdup	remote: change to parent directory
dir [*directory* [*local-file*]]	remote: display a long directory listing
lcd [*directory*]	local: change directory
ls [*directory* [*local-file*]]	remote: display a short directory listing
pwd	remote: display name of current directory

Transferring Files

get [*remote-file*[*local-file*]]	download one file
mget [*remote-file*...]	download multiple files

Setting Options

ascii	(default) set file type to ASCII text file
binary	set file type to binary file
hash	yes/no: show # for each data block transferred
prompt	yes/no: prompt for multiple file transfers
status	display current status of options

Figure 16-2. *Summary of the most useful* **ftp** *commands*

If you would like to enter more than one Unix command, you can use the **!** character by itself:

```
ftp> !
```

The **ftp** program will put itself on hold and start a new shell. You can now enter as many Unix commands as you want. When you are finished, terminate the shell and the **ftp** program will regain control. With most shells, you would press CTRL-D to terminate the shell. If this does not work, try the **exit** command.

The Connection ftp Commands

The next category of **ftp** commands are those that control the connection to the remote host: **open**, **close**, and **user**. We have already discussed the **open** command which you can use to establish an FTP connection.

The **close** command will terminate an FTP connection without quitting the **ftp** program. You can use **close** when you want to close one connection and then open another one.

If you connect to a remote host successfully, but there is something wrong with your userid or password, you may or may not lose your connection. (This depends on the FTP server.) However, if you do not lose the connection, you will not be able to do anything until you specify a valid userid and password. To do so, you can use the **user** command. Simply enter:

```
user
```

The **ftp** program will ask you to enter a userid and then a password. If you want, you can specify the information directly:

```
user anonymous harley@fuzzball.ucsb.edu
```

In this case, we are specifying a userid of **anonymous** and a password of **fuzzball.ucsb.edu**.

The Directory ftp Commands

The third group of **ftp** commands are those that you can use to move from one directory to another and to display the contents of a directory. As we explained, whatever directory you are in is called your working directory (or current directory). Once you establish an FTP connection, you have two working directories to keep track of: one on the remote host and one on your local computer.

With most Anonymous FTP hosts, you are automatically placed in the root (top-level) directory to start. To move to another directory, use the **cd** (change directory) command. Type the command name, followed by the name of the directory you want to move to, and then press RETURN. For example, if you want to move to the directory named **/pub/usenet/news.answers**, enter:

```
cd /pub/usenet/news.answers
```

If you understand the Unix file system, you can move one directory at a time by using separate **cd** commands. For example:

```
cd pub
cd usenet
cd news.answers
```

Unfortunately, the techniques that you need to use the **cd** command in this way are beyond the scope of the book. Please refer to a Unix book for more details.

At any time, you can display the name of your remote working directory by using the **pwd** (print working directory) command. For example, if you enter:

```
pwd
```

you will see a message similar to:

```
257 "/pub/usenet/news.answers" is current directory.
```

(Remember, messages from the FTP server always start with a number.)

At times, you may see a directory name that is different from what you expect. For example, on the **rtfm.mit.edu** system, if you move to the **/pub/usenet/news.answers** directory and then enter a **pwd** command, you will actually see:

```
257 "/pub/usenet-by-group/news.answers" is current directory.
```

Do not be confused. All this means is that the system manager has given the second level directory two different names: **usenet** and **usenet-by-group**. In such cases, you can use whichever name you want. The name **usenet-by-group** is more informative, but the name **usenet** is a lot easier to type.

> ## Hint
>
> In Unix, the word "print" is often used to mean "display". Thus, the job of the **pwd** (print working directory) command is to display the name of your working directory. This tradition dates back to the earliest days of Unix when terminals actually printed their output on paper.

The working directory on your local computer will be whatever directory you happened to be in when you entered the **ftp** command. When you download files, this is the directory into which they will be placed. If you want your files to be placed into a different directory, you can use the **lcd** (local change directory) command to move to a different directory before you start the downloading. For example:

```
lcd faq-files
```

This command changes your local working directory to **faq-files**. Again, you will find it helpful to read a Unix book for more information. In particular, you should learn how to make your own directories and how to move from one directory to another.

There is no **ftp** command to display the name of your local working directory. However, it is possible to do this by using the **!** command to send a **pwd** command to your local computer:

```
!pwd
```

> ## Hint
>
> Before you start an Anonymous FTP session, decide which directory on your
> local computer you want to use to store the downloaded files. Use the Unix
> **cd** command to move to that directory *before* you enter the **ftp** command.
> That way, you will not have to use a **lcd** command to ensure that your
> downloaded files are placed in the right directory.

To display the contents of a directory on the remote host, you can use two
different commands. The **ls** (list) command will display the names of all the files
in the directory. Here is some typical output. It shows the names of some of the
files in the **news.answers** directory that we discussed above:

```
esperanto-faq
feminism
finding-addresses
finding-sources
fleas-ticks
folklore-faq
fonts-faq
```

The **dir** command will display a longer listing. Along with the file name, you will
also see extra information. Here is some typical output:

```
-rw-rw-r--   8 root    3    27120 Jun  2 01:24 esperanto-faq
drwxrwxr-x   2 root    3      512 Jun 12 00:07 feminism
-rw-rw-r-- 14 root    3    28880 Jun 12 03:37 finding-addresses
-rw-rw-r-- 12 root    3    41939 Jun 16 04:04 finding-sources
-rw-rw-r-- 10 root    3    41533 Jun 16 03:30 fleas-ticks
-rw-rw-r--   8 root    3    84701 Jun 15 03:33 folklore-faq
drwxrwxr-x   2 root    3      512 Jun 18 01:46 fonts-faq
```

If you have some experience with the Unix file commands, you will recognize
this output as being from the **ls -l** command. Otherwise, don't worry about it.
You only need to understand four things.

First, the character at the far left tells whether the line describes a directory or
a file. A **d** character indicates a directory while a - (hyphen) character indicates a
file. In this example, we have two directories and five files.

At the far right of each line, we see the name of the file or directory. To the left
of the name, the time and date show you when the file or directory was updated.

Finally, the number to the left of the time and date shows the size of the file in bytes (characters). This number is not meaningful for directories.

Thus, we can see that **folklore-faq** is a file, it was last updated on June 15 at 3:33 AM, and is 84,701 characters long. You can ignore the rest of the information. You will find that it is often convenient to download a copy of a long directory listing to a file on your local computer. To do so, simply specify the name of the directory followed by the name of a file. For example:

```
ls /pub/usenet/news.answers ls.list
dir /pub/usenet/news.answers dir.list
```

Each of these commands generates a directory listing of the specified directory. The first command downloads its output to a file on your local computer named **ls.list**. (Of course, you can choose whatever names you want.)

When you use this form of the **ls** and **dir** commands, you must always specify a directory name and a local file name. If you want to download a listing of the working directory, you can use a directory name of **.** (a single period character). In Unix, a single **.** character stands for your working directory. For example:

```
ls . ls.list
dir . dir.list
```

The File Transfer ftp Commands

There are two commands that you can use to download files (that is, to copy files from the remote host to your computer). These commands are **get** and **mget**.

The **get** command allows you to download one file at a time. The **mget** (multiple get) command allows you to download more than one file at a time.

To use **get**, simply specify the name of the remote file followed by the name that you want to give the file on your local computer. For example, say that you have established an Anonymous FTP session with the **rtfm.mit.edu** computer that we mentioned earlier. You would like to download the **folklore-faq** file. You want the file to be named **urban-legends** on your computer.

First, move to the directory that contains the file:

```
cd /pub/usenet/news.answers
```

Now, enter the **get** command to download the file:

```
get folklore-faq urban-legends
```

You will see the following messages:

```
200 PORT command successful.
150 Opening ASCII mode data connection for folklore-faq (84701 bytes).
```

At this point, the file is being copied to your computer. Once the copy is complete, you will see:

```
226 Transfer complete.
local: urban-legends remote: folklore-faq
86113 bytes received in 17 seconds (4.9 Kbytes/s)
ftp>
```

You can now enter another command. (By the way, don't worry about the fact that the file is 84,701 bytes long, but we received 86,113 bytes. The extra bytes have to do with how certain characters are encoded when you copy a text file. Rest assured that the actual downloaded file will be exactly the right size.)

If you use the **get** command with only a single file name, the **ftp** program will use this name for the new file on your local computer. For example, if you enter:

```
get folklore-faq
```

the downloaded file will automatically be named **folklore-faq**.

Hint

If you already have a file with the same name on your local computer, the existing file will be replaced. Once a file is replaced, there is no way to get it back, so be careful.

If you are using a Unix system, there is a way to send the remote file as input to a Unix command on your local computer, rather than to a local file. Instead of a second file name, you type a | (vertical bar, called the *pipe* symbol), followed by the name of the command.

For example, suppose that you want to read a remote file named **README**. One way is to download it to a local file, stop or pause the FTP session, read the file, delete the file, and then resume the FTP session. Alternatively, you can read the remote file by sending it directly to a paging program (such as **more**) on your local computer:

```
get README |more
```

Notice that you cannot put a space after the | character (like you can with a regular Unix command). This is because **get** expects only two words: the remote file name (in this case, **README**) and the name of the local target (in this case, |**more**).

The **mget** (multiple get) command is used when you want to download more than one file at the same time. Type the command, followed by the names of the files you want to download. For example:

```
mget finding-addresses finding-sources folklore-faq
```

When you use **mget**, you cannot specify alternate names, so the files will be given the same names when they are copied to your local computer.

The **mget** command will transfer one file at a time. Before transferring a file, **mget** will display the file name and ask you for confirmation. For example:

```
mget finding-addresses?
```

At this point you can type either **y** (yes) or **n** (no) and press RETURN.

Hint

Remember, you can interrupt any **ftp** command, including **get** and **mget**, by pressing CTRL-C.

The nice thing about **mget** is that you can specify the name of a directory and **mget** will process each file in the directory. Similarly, if you know how to use the Unix wildcard characters, you can download all the files whose names match a particular pattern.

If you do not know about wildcard characters, we will mention briefly that an * (asterisk) character matches zero or more characters, and a **?** (question mark) character matches any single character. For example, to download all the files in the remote working directory whose names begin with the letters **fi**, you can use:

```
mget fi*
```

For more information on using such patterns, consult a Unix book. (Or experiment using the **ls** command; for example, try **ls fi*** to list the names of all the files that begin with **fi**.)

> ## Hint
>
> In this chapter, we have concentrated on downloading because that is what you usually do with Anonymous FTP. However, the **ftp** program also has commands to upload. They are **put** (to upload a single file) and **mput** (to upload more than one file at a time). These commands use the same format as **get** and **mget**.
>
> For more information, see the documentation for your local **ftp** program. If you are using a Unix computer, you can display the **ftp** entry in the online manual by entering the command **man ftp**.

Setting ftp Options

Within the **ftp** program, there are several commands that you can set to control the downloading operation. The ones we will talk about are **binary**, **ascii**, **hash**, **prompt**, and **status**.

In Chapter 5, we explained the difference between text files and binary files. Briefly, a text file (also called an ASCII file) holds ordinary characters: letters, numbers, punctuation, and so on. The files that we used as examples in this chapter are all text files.

A binary file contains information that is not textual. For example, if you want to download files that contain pictures, you will be dealing with binary files.

Be default, the **ftp** program assumes that it is working with text files. If you want to download binary files, you should tell the program before you enter the **get** or **mget** command.

The **binary** command tells the **ftp** program that you will be downloading binary files. If you want to switch back, the **ascii** command indicates that you will be downloading text files. When you use one of these commands, we say that you are setting the *representation type*.

For example, say that you are about to download some files that contain pictures. Before you do, you enter the command:

```
binary
```

When you enter this command, you will see the following message:

```
200 Type set to I.
```

The I stands for "image". (Don't worry about it.)

If you use the **binary** command and then download a text file, it will work just fine (although it may be a bit slower). It's only when you are copying binary files that you should be exact. With some remote hosts, a binary file will not be copied properly unless you have set the representation type to **binary**.

Hint

You will often encounter files whose names end in **.Z** or **.tar**. These files use special formats that we will discuss in Chapter 18. (A **.Z** file is compressed; a **.tar** file contains a collection of files called an archive.) Such files are always binary files.

Files whose names end in **.txt** are always text files.

If you are not sure what type of file you are downloading, set **binary** just to be safe.

The next option we want to mention is **hash**. This tells the **ftp** program to display a **#** character (sometimes called a hash mark) after each data block is transferred. This allows you to watch the progress when you are downloading a large file. The size of a data block depends on the nature of your FTP connection, but you will be told what it is before the downloading starts.

To turn on the **hash** option, just enter:

```
hash
```

To turn it off, enter the same command again.

The last option we want to mention is **prompt**. As we explained in the previous section, the FTP client will query you about each file whenever you use **mget**. If you turn the **prompt** option off, **mget** will automatically transfer each file without asking your permission.

To turn off the **prompt** option, just enter:

```
prompt
```

To turn it back on, enter the same command again.

Finally, if you want to display the current setting of all the options, you can use the **status** command:

```
status
```

When you do, you will see many options, most of which you can safely ignore.

A Typical Anonymous FTP Session

Figure 16-3 contains the full listing of a typical Anonymous FTP session. The commands that we entered are in boldface.

In this example, we connect to **rtfm.mit.edu**. We use a userid of **anonymous** and a password of our mail address. (As with all such systems, the password is not echoed as we type it.)

```
% ftp rtfm.mit.edu
Connected to CHARON.MIT.EDU.220 charon FTP server
  (Version 6.6 Wed Apr 14 21:00:27 EDT 1993) ready.

Name (rtfm.mit.edu:harley): anonymous
331 Guest login ok, send e-mail address as password.
Password:230 Guest login ok, access restrictions apply.

ftp> cd /pub/usenet/news.answers
250 CWD command successful.

ftp> hash
Hash mark printing on (8192 bytes/hash mark).

ftp> get folklore-faq
200 PORT command successful.
150 Opening ASCII mode data connection for folklore-faq (84701 bytes).
##########
226 Transfer complete.local: folklore-faq remote: folklore-faq
86113 bytes received in 17 seconds (4.9 Kbytes/s)

ftp> quit
221 Goodbye.
```

Figure 16-3. *A sample Anonymous FTP session*

We then change to the **/pub/usenet/news.answers** directory and turn the **hash** option on. Next, we use **get** to download the file named **folklore-faq**.

Once the downloading is complete, we enter the **quit** command to close the connection to **rtfm.mit.edu** and end the session.

Finding Your Way Around a New Computer

When you FTP to a new computer, it can sometimes take a while for you to find your way around. To orient yourself, start by looking for certain files and directories.

First, look for a directory called **pub**. Most Anonymous FTP hosts contain such a directory to hold all the public files and sub-directories.

Next, in each directory you use, look for one or more of the following files:

```
README
index
ls-1R.Z
ls-1tR.Z
```

If the **README** or **index** files exist, download and read them before you download other files. The **README** file will contain general information, while the **index** file will have a description of what is available.

Hint

The first time you use a new Anonymous FTP host, look for a file named **README** in the root directory.

The **ls-lR.Z** and **ls-ltR.Z** files will contain a detailed listing of the public files and directories. The name comes from the Unix **ls** (list) command that is used to generate such a listing. We won't go into the details except to say that the **ls-ltR.Z** file is sorted by time, showing the most recently modified files first. The **ls-lR.Z** file is sorted by regular, alphabetical order. If you want to look at such a listing, you only need to download one of these files.

The **.Z** characters at the end of these file names indicate that the files are stored in a special compressed format. When you download such files, you must process them with the **uncompress** command before you can read them. This is explained in Chapter 18.

Hint

If you FTP to a remote host that has a large number of files, the **ls-lR.Z** and **ls-ltR.Z** files may be so large that you cannot realistically download them. If you do, you may find that the act of downloading will use up so much space on your local computer that your quota is reached and the download is aborted. And, even if you do manage to download such files, they will be even larger when you uncompress them.

Thus, be sure to check the size of any file before you download it. It is not uncommon to have someone use up all the free disk space on the local computer by downloading a huge file. If you find yourself in this position, be courteous and remove the file as quickly as you can. (Use the Unix **rm** command.)

Requesting Anonymous FTP Services by Mail

If you do not have access to an **ftp** program, there is a way that you can request Anonymous FTP services by mail. You use a program called Ftpmail (which was written by Paul Vixie, who was then working at the DEC Network Systems Lab).

Even if you regularly use the **ftp** program, you may occasionally find it convenient to make FTP requests by mail, for instance, when you need a long directory listing. (See the last example in this section.) You may also find that it is sometimes easier to send a request by mail than to use the **ftp** program and do all the work yourself.

To use Ftpmail, you mail a message to an Ftpmail server. There are several Ftpmail servers on the Internet, the principal one being a host named **ftpmail.decwrl.dec.com**. In this section, we will discuss how to use this particular Ftpmail server. At the end of the section, we will show you a list of the other Ftpmail servers that you might use.

To use Ftpmail, you mail a message to **ftpmail@decwrl.dec.com**. Within the message, you enclose commands, one per line. The commands are instructions for conducting an Anonymous FTP session. The Ftpmail server will receive the message and carry out the commands on your behalf. The results of the session will be mailed to you automatically. In addition, Ftpmail will mail you a note when it receives your request and another note reporting on the actual Anonymous FTP session. (We discuss the Internet mail facility in Chapters 5 and 6.)

When you mail a message to Ftpmail, it ignores anything in the **Subject** line. However, if you do specify a subject, Ftpmail will use it as part of the **Subject** line when it mails you back a note. Thus, if you are mailing more than one request to Ftpmail, it is handy to use different subjects to help you identify the replies.

To get instructions on how to use Ftpmail, send a message to **ftpmail@decwrl.dec.com**. Here is a short session in which we use the **mail** program (Chapter 6) to send such a message:

```
% mail ftpmail@decwrl.dec.com
Subject: Request for help
help
CTRL-D
EOT
```

Before you use Ftpmail, it is a good idea to read all the instructions. For reference, Figure 16-4 contains a summary of the most important Ftpmail commands. In order for these commands to make sense, you should understand the general principles of Anonymous FTP downloading that we explained earlier in the chapter. Note the Ftpmail command to change directories is **chdir**, not **cd**.

Using Ftpmail is straightforward. To make things easy, use the Ftpmail commands in the order that they are listed in Figure 16-4, leaving out the commands that you don't need. For example, to retrieve a text file, you would normally use the following commands (in this order): **reply**, **connect**, **ascii**, **chdir**, **get**, **quit**.

The following example shows a sample session in which we use commands to connect to the Anonymous FTP host **cathouse.org**, change to the **/misc/fun/humor/british.humour/monty.python/flying.circus** directory, and request the text file named **argument.clinic**.

```
% mail ftpmail@decwrl.dec.com
Subject: Example of requesting a text file
reply harly@fuzzball.ucsb.edu
connect cathouse.org
ascii
chdir/misc/fun/humor/british.humour/monty.python/flying.circus
get argument.clinic
quit
CTRL-D
EOT
```

Specifying Your Mail Address

`reply` *address*   mail response

Connecting to the Host

`connect` *host*   connect to specified Anonymous FTP host

Setting Options

`ascii`   files to be mailed are text files
`binary`   files to be mailed are binary files
`uuencode`   convert binary files to text using `uuencode`
`btoa`   convert binary files to text using `btoa`

Specifying the Directory

`chdir` *directory*   change to the specified directory

Requesting Files

`get` *file*   send a copy of the specified file

Requesting Information

`help`   send a description of a how to use Ftpmail
`dir` [*directory*] send a long directory listing
`ls` [*directory*]   send a short directory listing (names only)

Ending the Session

`quit`   terminate session, ignore rest of message

Figure 16-4.   *Summary of important Ftpmail commands*

If the file you are retrieving is large (more than 64,000 characters), Ftpmail will automatically split it in to pieces and mail each piece separately. It is up to you to take the separate messages and put them back together to re-create the original file. (Don't worry, they will be numbered.)

Remember that the results of your request will be send to you by mail. And, as we explained in Chapter 5, only text files can be sent by regular mail. Thus, if you request a binary file, it must first be converted to a text file. (Remember that files whose names end with **.Z** or **.tar** are binary files.) At the other end, you will have to decode the text and re-create the original binary file.

There are two common systems that are used to perform such conversions, **uuencode/uudecode** and **btoa** (binary to ASCII). We discuss these programs and how to use them in Chapter 18. For now, we will say that if you ask for a binary file, you will have to tell Ftpmail how you want it encoded. Simply use either the **uuencode** or **btoa** command. A good place to do this is right after the **binary** command (which tells Ftpmail that you are requesting a binary file). If you send a binary file and do not specify what type of encoding you want, Ftpmail will use **btoa** by default (which may not be what you want).

Thus, to request a binary file, you would normally use the following commands (in this order): **reply**, **connect**, **binary**, **uuencode** (or **btoa**), **chdir**, **get**, **quit**. The next example shows a sample session in which we use commands to connect to the Anonymous FTP host **ftp.uu.net**, change to the **/doc/literary/obi/DEC/humor** directory, and request the text file names **Lawyer.jokes.Z**. (After we receive this file, we will have to use the **uudecode** program to recover the original binary file, and the uncompress programs to uncompress the file. This is all explained in Chapter 18.)

```
% mail ftpmail@decwrl.dec.com
Subject: example of requesting a binary file
reply harley@fuzzball.ucsb.edu
connect ftp.uu.net
binary
uuencode
chdir /doc/literary/obi/DEC/humor
get Lawyer.jokes.Z
quit
CTRL-D
EOT
```

Our last example requests a directory listing by mail. Requests such as this are handy even when you have access to the **ftp** program. In this example, we will request a large list of humor files. Since we will receive this list as a mail message, it is easy to save it for further reference, or even forward a copy to a friend.

The Anonymous FTP host **cathouse.org** contains a wonderful repository of humor called the Catstyle Archives, compiled by Jason R. Heimbaugh (otherwise known as Catstyle). In the following example, we send Ftpmail commands to connect to **cathouse.org**, change to the **/misc/fun/humor** directory, and request a long listing of all the sub-directories (the command **dir *** specifies that we want to look into all the sub-directories). Once we receive this directory listing, we can scan it for interesting files to request.

```
% mail ftpmail@decwrl.dec.com
Subject: Example of requesting a directory listing
reply harley@fuzzball.ucsb.edu
connect cathouse.org
ascii
chdir /misc/fun/humor
dir *
quit
CTRL-D
EOT
```

As we mentioned earlier, there are several other Ftpmail servers on the Internet. Table 16-1 shows their addresses. In spirit, they work the same way that we have described in this section, although the details and some of the commands are different.

Before you use one of these Ftpmail servers for the first time, send a message asking for instructions. Simply mail a message that contains only one line with the word **help**, just like we did above.

Location	Internet Address	IP Address
France	ftpmail@grasp.insa-lyon.fr	134.214.100.25
Germany	bitftp@vm.gmd.de	192.88.97.13
Ireland	ftpmail@ieunet.ie	192.111.39.1
USA: California	ftpmail@decwrl.dec.com	16.1.0.1
USA: New Jersey	bitftp@pucc.princeton.edu	128.112.129.99

Table 16-1.   *Ftpmail Servers*

## CHAPTER 17

As we explained in Chapter 16, Anonymous FTP provides us with one of the largest collections of public information ever amassed by human beings. Best of all, the Anonymous FTP archives are available for free to anyone on the Internet. Indeed, as we discussed at the end of that chapter, even people who have only Internet mail access can request Anonymous FTP files.

Using Archie to Find Anonymous FTP Files

In order to download a file via Anonymous FTP, you need to know where that file is located. In particular, you need to know the address of an Anonymous FTP host and the name of the directory that contains the file. This is not a lot to ask, but, in the vast world of Anonymous FTP, how do you know where to look?

The answer is you use Archie: the card catalog for the largest library in the world.

What Is Archie?

Throughout the Internet, there are a number of computers, called *Archie servers*, that provide a very important service: to help you find the names of Anonymous FTP hosts that carry a particular file or directory. The name "Archie" was chosen because it sounds like the word "archive". Because of its name, we tend to talk about Archie as if it were a person, or at least an intelligent robot.

When you need to find an Anonymous FTP file or directory, all you have to do is tell Archie what you are looking for. Archie will search its database and display the name of each Anonymous FTP host that has that file or directory. Archie will also show you the exact directory path. Thus, all you need to do is FTP to that host and pick up what you want.

There are three ways to use an Archie server, and we will talk about all of them in this chapter. First, you can telnet to it and log in with a userid of **archie**. (We explain how to telnet to a remote host in Chapter 7.) Once you log in, an Archie program will start automatically. You enter commands, one at a time, and tell Archie what you want to look for. Archie will search its database and display the results. If you are not sure exactly what file you want, Archie offers another facility—called a "whatis" service—that has descriptions of thousands of different programs, data files, and documents.

The second way to utilize Archie is by using a program, called an Archie client, that runs on your own computer. (We explain about clients and servers in Chapter 2.) You tell the Archie client what you are looking for. It automatically connects to an Archie server, asks it to perform the search, accepts the output on your behalf, and then displays the results for you. Once you know how to use Archie, it is a lot easier and a lot faster to use an Archie client. You don't have to initiate a telnet session, nor do you have to remember which Archie commands to use.

Finally, you can mail a request to any Archie server. Archie will carry out your request and mail the results back.

How Does Archie Work?

Conceptually, the workings of Archie are surprisingly simple. At regular intervals, special programs connect to every known Anonymous FTP host and download a full directory listing of all the public files. These lists are stored in what is called the *Internet Archives Database*. When you ask Archie to look for a file, all it needs to do is check the database.

The various Archie server sites around the world each keep track of the Anonymous FTP hosts in a certain portion of the Internet. For example, the Australian Archie host keeps track of all the Australian Anonymous FTP hosts. This information is shared so that all Archie servers are kept up to date as much as possible. On the average, Anonymous FTP hosts are checked about once a week.

Of course, there are a lot of details to take care of, so we should be grateful that someone else is providing this service. At the time we wrote this chapter, Archie servers were keeping track of more than two million files on over 1,500 different Anonymous FTP hosts. In total, these files represent hundreds of gigabytes (billions of characters) of data.

Archie was originally developed as a project by students and volunteer staff at the McGill University School of Computer Science in Montreal, Canada. The software was written by Alan Emtage and Bill Heelan with help from Peter Deutsch. Today, Archie is maintained by Bunyip Information Systems in Montreal, Canada. (You can reach them by sending mail to **info@bunyip.com**.)

Getting Started with an Archie Server

To use an Archie server, telnet to it and log in using a userid of **archie**. (You will not need a password.) Table 17-1 contains a list of the public Archie servers available to Internet users. For example, the following command will telnet to the Archie server at Rutgers University:

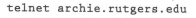

```
telnet archie.rutgers.edu
```

Hint

You can use any Archie server you want, but you will probably find it faster to use one that is relatively close to you.

Archie servers generally have a limit on how many people can telnet to them at the same time. If the closest Archie server is busy, use the next closest one, or wait for a while and try again.

Once you log in as **archie**, you will see some welcoming messages. You will then see the following prompt:

```
archie>
```

Archie is now ready for your commands. To stop Archie, enter the command **quit**. To display help information, use the **help** command. (We will talk more about **help** later, because there are some nuances.)

Location	Internet Address	IP Address
Austria	archie.edvz.uni-linz.ac.at	140.78.3.8
Austria	archie.univie.ac.at	131.130.1.23
Australia	archie.au	139.130.4.6
Canada	archie.uqam.ca	132.208.250.10
England	archie.doc.ic.ac.uk	146.169.11.3
Finland	archie.funet.fi	128.214.6.102
Germany	archie.th-darmstadt.de	130.83.22.60
Israel	archie.cs.huji.ac.il	132.65.6.15
Japan	archie.wide.ad.jp	133.4.3.6
South Korea	archie.sogang.ac.kr	163.239.1.11
Spain	archie.rediris.es	130.206.1.2
Sweden	archie.luth.se	130.240.18.4
Switzerland	archie.switch.ch	130.59.1.40
Taiwan	archie.ncu.edu.tw	140.115.19.24
USA: Maryland	archie.sura.net	128.167.254.179
USA: Nebraska	archie.unl.edu	129.93.1.14
USA: New Jersey	archie.internic.net	198.49.45.10
USA: New Jersey	archie.rutgers.edu	128.6.18.15
USA: New York	archie.ans.net	147.225.1.10

Table 17-1.   *Public Archie Servers*

Using Archie is a three-part process. First, you set things up the way you want, according to your preferences. Then you do the actual work. Then you quit. (Just like life.)

Setting Variables with an Archie Server

To control your working environment, you can change the value of quantities called *variables*. Each variable has a name and a value. You can tell Archie how you want something to work by changing the value of a particular variable. For example, you can specify how you want to sort the output of a search by setting the **sortby** variable. You can examine the current value of all the variables by entering the **show** command:

```
show
```

If you only want to display the setting of a single variable, you can specify that variable. For example:

```
show sortby
```

Figure 17-1 shows typical output from a show command. For now, don't worry about what all the variables mean; we will explain the important ones in a moment.

The best way to begin an Archie session is to start with a **show** command. Take a moment to examine the important variables and see if they are set up to your liking. If not, change them by using the **set** command (described in a moment). Once you have everything just the way you want it, you can enter the commands to tell Archie to search for what you want.

Hint

Before you start an Archie search, make sure that the most important variables are set the way you want. These variables (discussed below) are **maxhits, output_format, pager, search, sortby,** and **status.** Pay particular attention to the **search** variable. In addition, if you plan on asking Archie to mail you the results of a search, set the **mailto** variable.

As you can see from Figure 17-1, there are three types of variables: boolean, numeric, and string.

A *boolean variable* is one that acts as an on/off toggle switch. (The word "boolean" is a programming term named after the nineteenth century English mathematician George Boole.) To turn a boolean variable on, use the **set** command. To turn a boolean variable off, use the **unset** command.

Archie has only two boolean variables: **pager** and **status**. The **pager** variable determines how Archie will display its output. When **pager** is set, the results of a search will be sent to a special program, called a paging program, that will display the output one screenful at a time. When **pager** is unset, all the output will be displayed on your screen without stopping. Most of the time, this would cause all but the last part of the information to scroll off the top of your screen, so normally you will want to make sure that **pager** is set. If it is not already set, you can do so yourself by using the command:

```
set pager
```

There are three common paging programs used in Unix: **more**, **pg**, and **less**. Archie servers display their output using **less**, so that is the paging program you

```
#  'autologout' (type numeric) has the value '15'.
#  'compress' (type string) has the value 'none'.
#  'encode' (type string) has the value 'none'.
#  'language' (type string) has the value 'english'.
#  'mailto' (type string) is not set.
#  'maxhits' (type numeric) has the value '100'.
#  'output_format' (type string) has the value 'verbose'.
#  'pager' (type boolean) is set.
#  'search' (type string) has the value 'exact'.
#  'sortby' (type string) has the value 'time'.
#  'status' (type boolean) is set.
#  'tmpdir' (type string) has the value '/tmp'.
#  'term' (type string) has the value 'vt100 24 80'.
#  'max_split_size' (type numeric) has the value '51200'.
#  'server' (type string) has the value 'archie.rutgers.edu'.
```

Figure 17-1. *Typical output from an archie show command*

will have to learn. Like many Unix programs, **less** has enough commands to choke a horsefly. Fortunately, you can probably get by with knowing only two commands: press SPACE to display the next screen of information, and press **q** (quit) to return to Archie. For reference, we have included a quick summary of the most useful **less** commands later in the chapter.

*FUN TIP: The **more** program gets its name from the fact that, at the bottom of each screen of output, the program displays the word "more". The **less** program is newer and was written to replace **more**. The name **less** was chosen as a wry comment, because the **less** program actually offers more functionality than the **more** program. In other words, "**less** is more".*

The other boolean variable, **status**, is a lot simpler. As Archie performs a search, it can display a status line at the bottom of the screen. When **status** is set, Archie will display this line. When **status** is unset, Archie will not display the line. Normally, you want **status** to be set. If it is not already set, use:

```
set status
```

The second type of variable is a *numeric variable*. As the name implies, you use the **set** command to give these variables a numeric value. There are only two numeric variables that you need to understand: **autologout** and **maxhits**.

The **autologout** variable controls how long Archie will wait for you to enter a command without logging you out. In our example, **autologout** has a value of 15. This means that if you do not enter a command for 15 minutes, Archie will log you out and terminate your connection. To set **autologout** to another value, use a command like:

```
set autologout 25
```

The permissible range is from 1 to 300 minutes.

When you use Archie to search for a file, you will often find that there are many, perhaps hundreds, of Anonymous FTP hosts that contain the file. To speed things up, you can set **maxhits** to tell Archie the maximum number of items that you want to find. When Archie reaches this number, it will stop searching. For example:

```
set maxhits 10
```

Setting **maxhits** is something you will probably want to do because most Archie servers default to a large number (such as 100), and you really only need one good host name.

> ## Hint
>
> To save time, start searching with **maxhits** set to 10. If the results of your search are not adequate, set **maxhits** higher and try again.

The third type of variable is a *string variable*. The name means that these variables store values consisting of a string of characters. The only string variables you need to know about are **mailto**, **output_format**, **search**, and **sortby**. The most important of these is **search**.

The **mailto** variable is used to store a mailing address. As we will explain later, you can use the **mail** command to tell Archie to mail you the results of a search. If you set the **mailto** variable before you issue the **mail** command, Archie will know where to send the output. Otherwise, you will have to specify your address each time you use the **mail** command.

To set the **mailto** variable, use the **set** command and specify your address. For example:

```
set mailto harley@fuzzball.ucsb.edu
```

The **output_format** variable tells Archie in what format you would like the output. You have three choices: **verbose**, **terse**, and **machine**. To make a choice, use the **set** command with the name of the variable, followed by your choice. Be sure to include the _ (underscore) character in the variable name:

```
set output_format verbose
set output_format terse
set output_format machine
```

Normally, you would use either the **verbose** or **terse** format (which we will show you in a moment). Just experiment and pick the one you like. The **machine** format is designed to be used when you will be mailing the result of a search to yourself, and then using a Unix command or a program to manipulate the raw output. You will see an example of this later in the chapter.

Here is an example of one item of output in each of the three formats. We generated this example by using Archie to search for a file named **shoo-fly-pie**. The **verbose** output looked like this:

```
Host mthvax.cs.miami.edu    (129.171.32.5)
Last updated 09:32 17 Jun 1993

   Location: /recipes/ovo
      FILE    -rw-r--r-- 1095 bytes 01:00  4 Dec 1991  shoo-fly-pie
```

Here is the **terse** output:

```
mthvax.cs.miami.edu  01:00  4 Dec 1991 1095 bytes /recipes/ovo/shoo-fly-pie
```

Finally, the **machine** output:

```
19920103010000Z mthvax.cs.miami.edu 1095 bytes -rw-r--r-- /recipes/lacto/shoo-fly-pie
```

The next variable, **search**, is used to tell Archie how you want it to compare patterns as it searches. We will discuss this variable on its own in the next section.

The **sortby** variable indicates in what order you want Archie to display the results of a search. There are several choices:

set sortby none	do not sort
set sortby filename	alphabetical by file name
set sortby hostname	alphabetical by host name
set sortby size	largest to smallest
set sortby time	newest to oldest

You can tell Archie to sort in reverse order by placing an **r** in front of the variable setting:

set sortby rfilename	reverse alphabetical by file name
set sortby rhostname	reverse alphabetical by host name
set sortby rsize	smallest to largest
set sortby rtime	oldest to newest

For example, each file and directory has a time and date that marks when it was last modified. To tell Archie to display its output sorted by time, with the most recently modified items first, enter:

```
set sortby time
```

Setting the search Variable

The main function that Archie performs is to search the Internet Archives Database for a pattern that you specify. To tell Archie to start the search, you use the **find** or **prog** command (discussed later in the chapter). Before you start the search, you can set the **search** variable to control how Archie looks for a match.

There are seven possible settings: **exact**, **sub**, **subcase**, **regex**, **exact_sub**, **exact_subcase**, and **exact_regex**. To set the **search** variable, use the **set** command, followed by **search**, followed by your choice. For example:

```
set search exact
```

The **exact** setting performs a basic, straightforward search. It tells Archie to look for names that are exactly like the one you specified, including upper- and lowercase. For example, if you asked Archie to search for the pattern **IBM-PC**, it would match **IBM-PC**, but not **IBM-pc** or **ibm-pc**. When you know exactly what you want, these types of searches are the fastest and yield the best results.

The **sub** setting tells Archie to search for patterns that contain your specification as a substring. For example, if you asked Archie to search for **PC**, it would match **IBM-PC**, **PC**, or **PC-dos**. This setting does not distinguish between upper- and lowercase, so **PC** will also match **IBM-pc**, **pc**, and **pc-dos**.

The **subcase** setting is the same as **sub** except that upper- and lowercase letters are considered to be different. With this setting, **PC** would match **IBM-PC**, but not **IBM-pc**.

The **regex** setting allows you to use what is called a *regular expression* to tell Archie what you want. Regular expressions are used within Unix as a compact way to specify general patterns. For example, if you tell Archie to search for the regular expression **PC$**, it means to find all names that end with the letters **PC**.

Regular expressions can be complex, and we won't go into all the details here. If you want more information, use the command **help set search**. (We explain the **help** command later in the chapter.)

The last three settings are **exact_sub, exact_subcase**, and **exact_regex**. They tell Archie to first try an **exact** search. Then, if no matches are found, to try again using the second setting. For example:

```
set search exact_sub
```

This tells Archie to start by searching for the specified pattern exactly. If that fails, then try searching for the pattern as a substring.

Performing a Search

To prepare for an Archie search, use the **show** command, check the settings, and change the ones you don't like. If you are not sure what to set, use the following:

```
set mailto your-mail-address
set maxhits 10
set output_format verbose
set pagerset search exact_sub
set sortby time
set status
```

To perform the search, use the **find** command. Enter **find** followed by the pattern you want Archie to locate. Here are some examples:

```
find shoo-fly-pie
find rfc1325.txt
find recipes.tar.Z
```

Another command that you may need to know about is **prog**, which is a synonym for **find**. When Archie was first invented, its database was used primarily to hold information about computer programs. For this reason, the command that started a search was named **prog**.

Today, Archie's database contains information about all kinds of Anonymous FTP resources: not only programs, but documents, electronic magazines, Usenet archives—just about every type of information that you can imagine. Thus, the search command has been renamed from **prog** to **find**.

However, if you are using an Archie server that uses the old version of the Archie software, there may not be a **find** command. In such cases, you will have to use **prog** instead.

Hint

To stop an Archie search before it finishes, you can press CTRL-C.

Mailing the Results of an Archie Search

Once Archie has completed a search, you may find it useful to mail the results to yourself (or to someone else). To do so, all you have to do is enter the **mail** command:

```
mail
```

If you have set the **mailto** variable, Archie will already know where to send the mail. Otherwise, you will have to specify the address as part of the **mail** command:

```
mail harley@fuzzball.ucsb.edu
```

Hint

If you plan on mailing the results of more than one Archie search, it is a lot easier to set the **mailto** variable once, before you start.

Commands to Use While Reading Archie Server Output

We explained earlier that when you set the **pager** variable, it tells Archie to display its output by using a paging program called **less**. When **less** displays output, it shows you one screenful at a time. After each screenful, **less** pauses and waits for a command.

At this point, you can press SPACE to display more data, or **q** to quit **less** and return to Archie. For reference, Table 17-2 contains commands that you can use as you are viewing Archie output. For most of the commands, just press the keys, you do not press RETURN. However, for the **/** and **?** commands, you do have to press RETURN.

Command	Description
SPACE	go forward one screenful
q	quit the program
RETURN	go forward one line
*n*RETURN	go forward *n* lines
b	go backward one screenful
y	go backward one line
*n*y	go backward *n* lines
d	go forward (down) a half screenful
u	go backward (up) a half screenful
g	go to the first line
*n*g	go to line *n*
G	go to the last line
p	go to the line that is *n*% through the output
/*pattern*	search forward for the specified pattern
?*pattern*	search backward for the specified pattern
n	repeat the previous search command

Table 17-2. *Commands to Use When Viewing Output with* **less**

Using the Whatis Database

If you know the name of a file or directory (or part of the name), you can use the **find** command to have Archie search the Internet Archives Database. However, what if you know what you want, but you don't know the name?

To help in such cases, Archie maintains a second collection of information called the *Software Description Database*. This database contains short descriptions of thousands of programs, documents, and data files that are found in the Anonymous FTP archives (a lot more than software, actually).

To search this database, you use the **whatis** command. Type the command followed by any word you want. Archie will search the Software Description Database and display all the entries it finds that contain the word you specified. Each entry will show you a description and a file name. If the description looks like what you want, you will know exactly which file to search for.

Where do the entries come from? Whenever someone makes a file available via Anonymous FTP, he or she can send a short description to the people who maintain Archie, who will place the description in the Software Description Database.

Thus, there is a built-in restriction that limits the effectiveness of the **whatis** command. If the person who created the file you are looking for did not submit a description—which is the case for most files—you will not find it in the Software Description Database. Moreover, unlike the Internet Archives Database, this one is not updated regularly. Thus, you may find the description of a file that does not exist anymore.

Still, when it can find what you want, the **whatis** command can be a great tool, saving you from long searches in the dark. (The name, by the way, comes from the Unix **whatis** command, which performs a similar function for the online Unix manual.)

Here is an example using **whatis**. You would like to find something interesting that has to do with telephones. Enter:

```
whatis phone
```

Here is part of the output:

```
dialup      Maintain a database of phone services and use cu(1)
            to call them
phone       Multi-user real-time "talk" program
phone_kl    Phone another user, typing screen to screen
phoneme     Translate English words into their phonetic spellings
ringback    Implements a ring-back system that allows a phone
```

```
              line that is normally used as a voice line and a
              dial-out data line to be used as a limited dial-in
              data and voice line
 sys5-phone   VAX-like Phone Utility for SysV
 telewords    Telephone number to word conversion
 telno        A telephone number permutation program
```

You decide to try the **telewords** program. All you have to do is enter the
command **find telewords** to display its Anonymous FTP location.

There are several things we would like to point out about this example. First,
notice that **whatis** does a straight letter-by-letter search, ignoring the difference
between upper- and lowercase letters. Second, you will often find items that
have nothing to do with what you are looking for. For example, the **phoneme**
program (although it does look interesting).

Finally, remember that **whatis** does not perform an exhaustive search of the
full Internet Archives Database. It only knows about the items listed in the
Software Description Database. For example, several Anonymous FTP hosts have
directories named **telephone** that contain useful information, but **whatis** cannot
find these directories because they are not in its database.

Displaying Help Information

There are several ways to have Archie display help information. First, you can
use the **help** command to display information about other commands. All you
need to do is enter **help** followed by the name of the command. For example:

```
help find
help set
```

To display a list of all the commands, enter:

```
help ?
```

If the **pager** variable is set, Archie will use the **less** paging program to display the
information, so when you are finished you will have to press **q**. Once you quit
less, you will see a new prompt:

```
help>
```

At this point, you can enter the name of one command after another and receive more help. When you do not need any more help, simply press RETURN and you will return to the regular prompt:

```
archie>
```

You can now enter a regular Archie command.

If you want help for setting a variable, type **help set** followed by the name of the variable. For example:

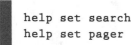
```
help set search
help set pager
```

Again, the information will be displayed using **less**. However, this time, when you quit, you will see the prompt:

```
help set>
```

You can now enter the name of one variable after another and receive more help. When you press RETURN, you will see the **help>** prompt. Pressing RETURN once again returns you to the regular **archie>** prompt.

Hint

In order to go back to the **archie>** prompt after reading **help** information, you will have to press **q** (to quit **less**) and then RETURN (perhaps twice).

Another way to learn about Archie is to read the official documentation. Here is how the Archie documentation is organized.

All Unix systems come with an online manual that contains entries for each command. At any time, you can read the documentation for a particular command by using the **man** command. It is customary to refer to each particular entry in the online manual as a *page* (even though many entries are much longer than a single page). It is common for people who create new software to write a manual page explaining how the software works.

Archie itself has such documentation that explains how to use an Archie server. This documentation, of course, is not available on your local computer. Rather, you can enter the **manpage** command while you are using the Archie server:

```
manpage
```

Hint

Archie will now display its manual page for you.

After you use a **help** or **manpage** command, you can mail the output to yourself by using the **mail** command. This is a good way to get your own copy of the official Archie server documentation.

To finish this section, we will mention three more information commands.

Whenever you log in to an Archie server, you will see a welcoming message. Indeed, many public Internet hosts, such as Anonymous FTP hosts, display general information when you log in. The custom is to call such information the *message of the day*. With an Archie server, you can redisplay this information at any time by entering the **motd** command.

If you would like to find out about other Archie servers, you can display a list by entering the **servers** command.

And finally, you can use the **version** command to display the version of Archie software that your server is using.

A Summary of Archie Server Commands

For reference, Figure 17-2 is a summary of all the important Archie server commands.

Stopping the Archie Session

    `quit`   stop the Archie session and disconnect

Performing a Search

    `find` *pattern*   search the main Anonymous FTP database

    `prog` *pattern*   the old name for `find`

    `whatis` *pattern*   search the Software Description Database

Displaying Information

    `help ?`   display a list of commands

    `help` *command*   display help for the specified command

    `help set` *variable*   display help for the specified variable

    `manpage`   display the Archie manual page

    `motd`   redisplay message of the day

    `servers`   display a list of Archie servers

    `version`   show what version of Archie is being used

Mailing Information

    `set mailto` *address*   specify a mail address

    `mail [`*address*`]`   mail output

Displaying Variable Settings

    `show`   display the value of all the variables

    `show` *variable*   display the value of a specified variable

Figure 17-2.   *Summary of important Archie server commands*
(continued next page)

Setting General Variables

set autologout *minutes*	set maximum idle time before auto logout
set maxhits *number*	set the maximum number of items to find
set pager	display output by using paging program
unset pager	do not display output with paging program
set status	display a status line during search
unset status	do not display status line during search

Setting Output Preferences

set output_format verbose	display output using long format
set output_format terse	display output using short format
set output_format machine	display output using machine format
set sortby none	do not sort output
set sortby filename	sort alphabetically by file name
set sortby hostname	sort alphabetically by host name
set sortby size	sort by size, largest to smallest
set sortby time	sort by time and date, newest to oldest
set sortby rfilename	sort reverse alphabetically by file name
set sortby rhostname	sort reverse alphabetically by host name
set sortby rsize	sort by size, smallest to largest
set sortby rtime	sort by time and date, oldest to newest

Setting Search Preferences

set search exact	search for an exact pattern
set search sub	search for anything that contains pattern
set search subcase	same as sub, but distinguish case
set search regex	search for a regular expression
set search exact_sub	try exact, then sub
set search exact_subcase	try exact, then subcase
set search exact_regex	try exact, then regex

Figure 17-2.   *Summary of important Archie server commands* (continued)

A Typical Session with Archie

It was a dark and stormy night.

Outside, the harsh, driving rain beat a monotonous tattoo against the windowpane. We sat at our desk, hat brim pulled down over our face, grabbing a few quick winks between cases. Things had looked tight for a while in Tangier, but once we had put the screws to the dwarf he had come through. They all do.

We lay back, the old number twelves resting on a desk covered with a month's worth of paperwork, a half-empty bottle of Scotch, and a tarnished .45 that was still smoking. In the corner, the screen of our computer glowed silently.

We didn't hear when she entered, but our nostrils flared at the fragrance of her perfume. "Go away," we said, "we're closed for repairs."

"Well, then," said a sultry Barbara Stanwyck voice, "you'll just have to open early today. I need help, and I'm willing to pay for it."

We looked up and blinked twice at a vision of innocence: five foot three, periwinkle blue eyes, and long blond curls winding over her shoulders like water cascading down a mountain creek. And she was dressed in a skintight outfit that showed more curves than a politician with the bends.

She looked innocent. Too innocent. Something was fishy here, and we didn't mean her perfume.

"Listen," she said, "I need help, and you're the only one who can help me. I'm in an awful jam."

"So what else is new?"

"I've only got seven more hours, or else..."

"Or else what? We already told you, kid, that we weren't going to pull your chestnuts out of the fire anymore. What is it this time? Blackmail? Murder? Or did you have another misunderstanding with Sergeant Rogers?"

She looked at us with large, round, innocent eyes.

"Calculus," she said.

We closed our eyes and sighed.

"Okay, kid, spill it."

Well, we listened to her tale and it wasn't a pretty one. The old, old story. A young, guileless girl from a small country town, still wet behind the ears, full of hope and dreams of tinsel, comes to the big city to make her fortune. But then she falls in with the wrong crowd and gets into trouble.

It starts innocently enough. First, someone takes her to a party where, unknown to her, people in the back room are solving differential equations. Next, she gets invited to a small, private gathering where a few street-wise punks introduce her to basic algebra. At first, the thrill is intoxicating and she goes along with it just for the kicks. Soon, she's into simultaneous equations, conic sections, and fractional exponents. She tells herself that she can take it or leave it, that she can stop whenever she wants. Until one day, she finds herself breaking into a pawn shop to steal a protractor and a Schaums Outline.

And now, calculus.

"You have to help me," she pleaded. "You're my last hope. My first calculus class starts tomorrow and I'm not ready. And you know what The Doctor does to people who are unprepared."

We sighed again.

"Okay," we said, "we'll help you, but it will cost you a hundred bucks. In advance."

"How about a trade," she said. "I know where I can get my hands on a used copy of a complex analysis text."

"That's all we need around here, with the cops breathing down our neck. A hundred bucks, in advance. Think on your feet."

"All right," she pouted and, reaching into the hidden recesses of her outfit, extracted two fifty-dollar bills that had definitely seen better days.

"That's more like it," we said. "Now come over here and sit down."

We pulled her over to the computer which was already logged in to the Internet. "Do you have your own PC at home?"

She nodded.

"Good. We'll get Archie to find a program to help you."

Telneting to a handy Archie server, it was the work of a moment to set the variables and start a search for anything named "calculus". We hit paydirt on the first try: an Anonymous FTP host at Washington University, in St. Louis, Missouri. (The session is shown in Figure 17-3.) We quit Archie, FTP-ed to the Anonymous FTP host, changed to the **calculus** directory, and displayed a directory listing.

"There it is," we said. "This is what you want. A file named **rurc1.zip**. Do you know how to uncompress a zip file?"

She looked at us with scorn. "Does your grandmother know how to suck eggs?"

"Okay, then, all you have to do is unzip this file and you will find a program called *Are You Ready for Calculus?* written by David Lovelock at the University of Arizona. It will help you figure out what you need to review before you start calculus."

She stared at us.

"How do you know all about this?" she asked suspiciously.

We stared back.

"That's what you're paying us for, kid."

"Okay. I guess I know better than to ask questions."

We downloaded the program to our computer and copied it onto a floppy disk for her. She took the disk, slipped it into her blouse, and walked to the door.

She turned and looked at us with those large, blue, innocent eyes.

"I like your style," she said. "Maybe you'd like to come up to my place and uncompress a file sometime." She paused. "You do know how to uncompress, don't you? You just put your fingers on the keyboard—and unzip."

And she was gone, like a wisp of steam on a double-sided razor blade ready to do its work.

We looked outside. It was still a dark and stormy night.

```
% telnet archie.rutgers.edu
Trying 128.6.18.15 ...Connected to dorm.Rutgers.EDU.
Escape character is '^]'.

SunOS UNIX (dorm.rutgers.edu) (ttyq0)
login: archie

Last login: Wed Jun 23 16:33:38 from fuzzball.ucsb.edu
SunOS Release 4.1.3 (TDSERVER-SUN4C-DORM) #1:
Sat May 1 16:46:07 EDT 1993

            Welcome to the Rutgers University Archie Server!

...message of the day deleted...

archie> show
# 'autologout' (type numeric) has the value '15'.
# 'compress' (type string) has the value 'none'.
# 'encode' (type string) has the value 'none'.
# 'language' (type string) has the value 'english'.
# 'mailto' (type string) is not set.
# 'maxhits' (type numeric) has the value '100'.
# 'output_format' (type string) has the value 'verbose'.
# 'pager' (type boolean) is not set.
# 'search' (type string) has the value 'sub'.
# 'sortby' (type string) has the value 'none'.
# 'status' (type boolean) is set.
# 'tmpdir' (type string) has the value '/tmp'.
# 'term' (type string) has the value 'vt100 24 80'.
# 'max_split_size' (type numeric) has the value '51200'.
# 'server' (type string) has the value 'archie.rutgers.edu'.

archie> set maxhits 10
archie> set pager
archie> set search exact_sub
archie> set sortby time

archie> find calculus

...some responses deleted...

Host wuarchive.wustl.edu   (128.252.135.4)
Last updated 05:03 23 May 1993

    Location: /edu/math/msdos
      DIRECTORY rwxr-xr-x        1024  Mar  9 18:02    calculus
```

Figure 17-3. *A typical Archie and Anonymous FTP session*
(continued next page)

```
archie> quit
# Bye.
Connection closed by foreign host.

% ftp wuarchive.wustl.edu
Connected to wuarchive.wustl.edu.
220 wuarchive.wustl.edu FTP server
(Version 2.1WU(2) Wed May 19 07:29:30 CDT 1993) ready.

Name (wuarchive.wustl.edu:harley): anonymous
331 Guest login ok, send your complete e-mail address as password.
Password:

...message of the day deleted...

230 Guest login ok, access restrictions apply.

ftp> cd /edu/math/msdos/calculus
250 CWD command successful.

ftp> dir
200 PORT command successful.
150 Opening ASCII mode data connection for /bin/ls.
total 2344

...some lines deleted...

-rw-r--r--   1 husch     234      171950 Jan 10 18:07 rurc1.zip
-rw-r--r--   1 husch     234       99621 Jan 10 18:08 rurc2.zip
-rw-r--r--   1 husch     234      108018 Jan 11 07:47 rurc3.zip

226 Transfer complete.
2657 bytes received in 5.8 seconds (0.45 Kbytes/s)

ftp> get rurc1.zip
200 PORT command successful.
150 Opening ASCII mode data connection for rurc1.zip (171950 bytes).
226 Transfer complete.
local: rurc1.zip remote: rurc1.zip
172885 bytes received in 1.4e+02 seconds (1.2 Kbytes/s)

ftp> quit
221 Goodbye.
```

Figure 17-3.   *A typical Archie and Anonymous FTP session (continued)*

Understanding Archie Clients

Up to now, we have been talking about using an Archie server directly, by telneting to it, logging in as **archie**, and entering commands. However, once you understand how Archie works, the preferred way to use it is via an Archie client.

An *Archie client* is a program that you run on your own computer. You tell the program what you want to search for. It contacts an Archie server on your behalf, issues the request, accepts the output, and displays it on your screen.

Using an Archie client is not only easier, it can be astonishingly fast. We have often found that with an Archie client, we can get a response in only a few seconds. (Of course, when the server gets busy, your response time can slow down considerably.)

Another consideration is that most Archie servers enforce strict quotas about how many people can telnet in at the same time, and you may not be able to connect whenever you want. Archie clients, being programs that run on your own computer, are always available.

There are various types of Archie clients available. The two most common are **archie** (for regular Unix terminals) and **xarchie** (for X Window terminals, see Chapter 2). There also Archie clients for Next computers, PCs running DOS, and VAX computers running VMS.

The **xarchie** program uses an X Window based graphical user interface and, if you understand how to use Archie, is so easy to use that you won't need our help. So, in this section, we will concentrate on **archie**.

To use **archie**, it must be installed on your system. To find out if this is the case, try performing a search by entering the **archie** command followed by something to search for. For example:

```
archie rfc1325.txt
```

If you have the **archie** program, this command will return some type of response. Otherwise, you will see a message like:

```
archie: Command not found.
```

If this is the case, the only ways for you to access Archie are to either telnet to an Archie server, or to send a request by mail (explained later in the chapter). Since using an Archie client is so much more convenient, you may want to politely ask your system manager if he or she could install **archie** on your system. The program is available by Anonymous FTP. (Just use Archie to search for **archie**.)

If you have the **archie** program on your system, you can read the official documentation from the online manual by using the **man** command:

```
man archie
```

> ## Hint
>
> As you read about Archie, you may see references to Prospero. This is a tool that allows you to access information spread around the Internet. Using Prospero, you can organize and use files that are stored on remote computers as easily as if they were on your own local computer.
>
> The distributed file system that Prospero creates is based on an idea called the Virtual System Model, and is widely used throughout the Internet. Anyone can use Prospero by installing a Prospero client on their computer. Archie clients and servers use the Prospero technology to implement their client/server relationship, although the details are hidden from view.
>
> Prospero was developed by Cliff Neuman of the Information Sciences Institute, a part of the University of Southern California. For more information, you can FTP to **prospero.isi.edu**. You can also subscribe to a general Prospero mailing list by sending a message to **prospero-request@isi.edu**. (We discuss how to use FTP in Chapter 16; mailing lists in Chapter 25.)

Using an Archie Client

In order to learn how to use an Archie client program, you should already understand about Archie and how it works. If you need a refresher, read through the previous sections in this chapter.

The easiest way to use **archie** is to specify the name of a file or directory that you want to find. For example:

```
archie rfc1325.txt
```

The Archie client will connect to an Archie server, have it perform a search, and return the results to you.

One nice thing about using an Archie client is that you can manipulate the output using the standard Unix tools. For example, the following command performs the same Archie search and saves the output in a file named **rfc**:

```
archie rfc1325.txt > rfc
```

If a file named **rfc** does not already exist, Unix will create it for you. If the file does exist, it will be replaced and the original contents will be lost, so be careful.

Hint

In Unix, you can use a **>** (greater than) character followed by a file name at the end of any command to send output to the specified file. This is called "redirecting the standard output". To find out more about such matters, we refer you to any good Unix book. (The best one is *A Student's Guide to Unix*, by Harley Hahn, published by McGraw-Hill.)

As with many Unix commands, **archie** has a number of ways for you to change the behavior of the command. To do so, you include what are called *options* or *switches* directly after the command name. For example, the **-t** option tells Archie to sort the output by the time and date that the file was last modified, from newest to oldest:

```
archie -t rfc1325.txt
```

This is like using **set sortby time** with an Archie server.

Table 17-3 contains a summary of the most important options. Notice that many of them correspond to Archie server commands.

The **-e**, **-s**, **-c**, and **-r** options have the same effect as using the **set search** command with an Archie server. For example, to perform a **sub** search you can use the **-s** option:

```
archie -s rfc1325
```

The **-e** option performs an **exact** search, which is the default. Thus, the following two commands are equivalent:

```
archie rfc1325.txt
archie -e rfc1325.txt
```

If you want to perform a combination search, such as **exact_sub**, you can use **-e** with another search option. For example:

```
archie -e -s rfc1325
```

This command tells Archie to start with an **exact** search. If that doesn't work, Archie will try a **sub** search.

Similarly, you can use the **-t** and **-l** options to control the formatting of the output. The **-t** option sorts by time (like **set sortby time**), and the **-l** option formats the output so that it is suitable to send to another program or Unix command for further processing (like **set output_format machine**):

```
archie -l rfc1325.txt
```

Option	Description
-c	search for substrings, case sensitive (`set search subcase`)
-e	search for an exact match [default] (`set search exact`)
-r	search using a regular expression (`set search regex`)
-s	search for substrings (`set search sub`)
-o*filename*	send output to specified file
-l	list one item per line (`set output_format machine`)
-t	sort output by time and date (`set sortby time`)
-m*number*	set maximum number of items to find (`set maxhits`)
-h*address*	send Archie server requests to the specified host
-L	show list of Archie servers known to the program (`servers`)
-V	(verbose) make comments during a long search

Table 17-3.   *Summary of Important* **archie** *Options*

Here is some sample output from the previous command. Notice that Archie has formatted the output so that each item is on a single line:

```
199205150000002  91885 esel.cosy.sbg.ac.at /pub/mirror/rfc/rfc1325.txt
199205150000002  91885 swdsrv.edvz.univie.ac.at /doc/rfc/rfc1325.txt
199205140000002  91884 plaza.aarnet.edu.au /rfc/rfc1325.txt
199206010000002  91884 sunb.ocs.mq.edu.au /Documents/RFC/rfc1325.txt
199208170000002  91885 sifon.cc.mcgill.ca /pub/ftp_inc/doc/rfc/rfc1325.txt
```

If you know how to use Unix, you can send this type of output to another command. For instance, the following series of commands displays all the items that are stored on Australian computers (those whose addresses end in **.au**):

```
archie -l rfc1325.txt | grep '.au '
```

For the details of how the **grep** command works, check with a good Unix book. (Guess which one we recommend?)

The **-m** option has the same effect as the Archie server command **set maxhits**. For example, to ask for a maximum of 10 items, use a command like the following:

```
archie -m10 rfc1325.txt
```

Hint

To speed up an Archie search, use the **-m** option to limit the amount of output. (We recommend **-m10**.) If you do not find what you want, re-enter the command with a larger value.

As we showed above, you can use the Unix redirection facility to send the output of an Archie search to a file. Alternatively, you can do the same thing with the **-o** option. Thus, the following two commands are equivalent:

```
archie rfc1325.txt -orfc
archie rfc1325.txt > rfc
```

When you use the **archie** program, it connects to the Archie server whose name was specified by your system manager when he or she installed the program. If you want to see the name of this server, enter the following:

```
archie -L
```

This command will also display a list of other Archie servers. If you would like to send a request to a specific Archie server, you can use the **-h** (host) command and specify the server you want. For example, the following command sends a request to the Archie server whose address is **archie.au**:

```
archie -harchie.au rfc1325.txt
```

Finally, the last option that we will discuss is useful when you are requesting a time-consuming search. By using **-V** (verbose), you tell your Archie client to make comments every now and then:

```
archie -V rfc1325.txt
```

Whenever the search seems to be taking a long time, you will see a suitable message:

```
Searching...
```

In this way, you are reassured that something is happening, and your level of anxiety remains at a comfortable level.

Using Archie by Mail

We have already discussed two ways to use Archie. The best way is to use an Archie client program (like **archie** or **xarchie**) that will send requests to an Archie server on your behalf. If this is not possible, you can telnet to an Archie server, log in as **archie**, and enter commands for yourself. In this section, we will describe the third way to access an Archie server: by sending it commands in a mail message.

Using an Archie server by mail is handy if you do not have access to a client and, for some reason, you cannot telnet directly to an Archie server. It is also convenient when you do not need your answer in a hurry or when you need to make multiple searches. You can mail your request and pick up the results later in your mailbox.

To use Archie by mail, send a message to the userid **archie** at the address of one of the Archie servers (which are listed in Table 17-1). When you compose the message, you can leave the subject blank. Within the body of the message, use as many Archie commands as you want, each one on its own line.

Archie will carry out your commands and mail a response back to you. For the most part, the commands that you can use are the same ones that we described earlier in the discussion of Archie servers. (See the summary in Figure 17-2.) The only exception is that you cannot use commands that do not make sense in a mail request (for example, **set pager**).

The first command that you should always use is **set mailto**. This will ensure that Archie has your correct return address. If you leave out this command, Archie will look for your address in the header of your message. Most of the time, this will work, but it is better to specify the address explicitly, so you know that there will not be any ranygazoo. For convenience, you can also use the older version of this command, which is **path**. Thus, the following two commands are equivalent:

```
set mailto harley@fuzzball.ucsb.edu
path harley@fuzzball.ucsb.edu
```

The last command that you should always use is **quit**. This tells Archie that you are finished and to ignore any lines that follow. It is important to include the **quit** command, because some mailing programs allow you to define a signature that will be appended automatically to all your messages. By using **quit**, you ensure that Archie will ignore any extra lines at the end of the message.

Before you use a particular Archie server by mail for the first time, you should send it a message asking for help information. This will show you exactly what commands that particular Archie server will recognize.

Here is a sample session in which we send such a request to the Archie server named **archie.rutgers.edu**:

```
% mail archie@archie.rutgers.edu
Subject:
set mailto harley@fuzzball.ucsb.edu
help
quit
CTRL-D
EOT
```

When you get your response, read it carefully. If your Archie server is using an old version of the Archie software, you may not be able to use all the commands in Figure 17-2. In particular, instead of **set mailto**, you will have to use **path**; and instead of **find**, you will have to use **prog**.

Here is a sample session in which we compose a request for Archie to search for the notorious **rfc1325.txt** file:

```
% mail archie@archie.rutgers.edu
Subject:
set mailto harley@fuzzball.ucsb.edu
set maxhits 25
set output_format verbose
set search exact
set sortby time
find rfc1325.txt
quit
CTRL-D
EOT
```

Once you receive your response from Archie, you can send a message to an Anonymous FTP mail server to retrieve the actual file. (See Chapter 16.)

CHAPTER

18

In this chapter, we will explain what file types you can expect to encounter on Anonymous FTP servers and within Usenet newsgroups. You will find this information essential if you want to access many of the Internet's public files.

We will also answer two questions that are invariably asked by new users: How do I copy a file to or from my personal computer? And, how do I display picture files?

File Types Used on the Internet

Using Archie to Find File Utility Programs

Before we get into the main topics of this chapter, here is a quick prelude.

Within this chapter, we will be using a number of file utility programs. Some of the names are **compress**, **uncompress**, **tar**, **uuencode**, **uudecode**, **btoa**, and **shar**. In our examples, we assume that you are using a Unix computer, in which case, most or all of these programs should be available on your system.

If you are using another type of computer—such as a PC or Macintosh—there are public domain versions of these programs available via Anonymous FTP.

There are two easy ways to find these programs. First, you can check the file utility reference that is maintained as a public service by David Lemson at the University of Illinois. To download this document, FTP to **ftp.cso.uiuc.edu** and look in the directory **/doc/pcnet** for a file named **compression**.

The second way to find a file utility program is to use Archie (see Chapter 17). The easiest method is to use a **regex** search, and look for the name of the program you want. To save time, specify a search pattern that starts with a ^ (circumflex) character. This means that the pattern must come at the beginning of a word. For example, you might search for **^tar**. This would match **tar.exe** but not **game.tar.Z**.

Here are three examples that search for the **tar** program. First, using an Archie client:

```
archie -m25 -r ^tar
```

You will usually find it handy to send the output of an Archie search to a paging program, in order to display the output one screenful at a time:

```
archie -m25 -r ^tar | more
```

Second, using Archie by mail, send a message that contains commands similar to the following ones:

```
set mailto harley@fuzzball.ucsb.edu
set maxhits 25
set sortby time
set search regex
find ^tar
quit
```

Third, after telneting to an Archie server and logging in as **archie**, enter the following commands:

```
set maxhits 25
set pager
set sortby time
set search regex
find ^tar
```

For more details on using Archie, see Chapter 17.

File Types Used with Anonymous FTP

The people who set up Anonymous FTP servers have two important file storage problems to consider. First, many files are large and require a great deal of disk space. Large files also take longer to download than small files. Second, many of the items offered to the public are actually packages and not single files. A single piece of software, for example, might consist of a number of separate programs and documentation files. In such cases, the package itself needs to be easy to store and to download.

For these reasons, Anonymous FTP files are commonly stored in two special kinds of formats. First, large files are usually *compressed* into a form that requires less storage space. After you download such a file, you must use a program to *uncompress* the file before you can use it.

Second, a group of files can be collected into a single package called an *archive*. The archive can be stored and downloaded as a single file. This is analogous to how you might move books from one place to another. You can place a collection of books into a single box, which is easy to store and easy to move. Once the box has been moved, you can open it and take out the books.

Similarly, you can download an archive as a single large file. Once the archive is on your local computer, you use a program to restore the original collection of files. When you do, we say that you are *unpacking* the archive. We also say that you are *extracting* the files.

There are a number of programs that are used to compress and archive files. The most common compression program is named **compress**. When you use this program, it creates a new file that has the same name as the old one with the characters **.Z** added. For example, when you compress a file named **recipe**, it generates a new (smaller) file named **recipe.Z**.

Thus, whenever you see a file whose name ends in **.Z**, you know that this is a compressed file that will have to be uncompressed before you can use it. To do so, you use a program named **uncompress**.

When it comes to archiving, the most commonly used program is **tar**. (The name stands for "tape archiver", from the days when files were archived and stored on magnetic tape.) When you use **tar**, it creates a new file whose name ends in **.tar**. For example, you might archive a group of files that contains all the programs and documentation for a software package named **game** into an archive named **game.tar**. Such archives are commonly called *tarfiles*.

Thus, whenever you see a file whose name ends in **.tar**, you know that this is an archive that must be unpacked before you can use the files. To do so, you use the **tar** program on your local computer.

Hint

A designation like **.Z** or **.tar** that comes at the end of a file name is called an *extension*. Within the world of Unix and the Internet, there are a number of different extensions that are used to indicate file types.

If you have ever used a PC with DOS, you will be used to extensions that must be from one to three letters. In Unix, the rules are more liberal. Extensions are considered to be part of the file name and can be as long as you need. Some files, as you will see below, even have double extensions.

Collecting separate files into an archive does not compress the files. It only makes sense, then, that after you create an archive you might want to compress it. Thus, you will find many files that have extensions of **.tar.Z**, for example, **game.tar.Z**.

This means that a group of files was collected into an archive which was then compressed. When you download such a file, you must first use **uncompress** to restore the archive, and then use **tar** to restore the original files. If you forget what order to perform these operations, just look at the file name and work from right to left.

Later in the chapter, we will explain the details of using both **compress** and **tar**, as well as the other common compression and archiving programs that you may need to use.

Hint

On a DOS-based PC, the name of a file consists of a "filename" and an optional "extension". The filename can be no longer than eight characters, while the extension can be no longer than three characters. For this reason, compressed tarfiles that are meant to be used with DOS are often given an extension of **.taz**. For example, you might see a file named **game.taz**. This is because a name like **game.tar.Z** is not allowed with DOS. When you see such a file, the intention is that you will download it to your PC, and then uncompress and unpack it.

If you are at all involved with the world of PCs, there is another program you should know about. This program, called **pkzip**, both archives and compresses at the same time. To uncompress and unpack, you use a program called **pkunzip**. (The initials **pk** stand for Phil Katz, the program's developer.)

The **pkzip** program creates files with an extension of **.zip**, for example, **view.zip**. Such files are often referred to as *zipfiles*. We will discuss zipfiles and what to do with them later in the chapter.

File Types Used with Usenet and Mail

Broadly speaking, any file can be classified as text or binary. Text files contain ordinary characters: letters, numbers, punctuation, and so on. You can use these files directly as they are. For example, you can display a text file using a paging program (such as **more**, **pg**, or **less**), or edit a text file using a text editor (such as **vi** or Emacs).

Another name for text files is ASCII files. This is because the characters are stored according to a system called the ASCII code. (ASCII stands for the American Standard Code for Information Interchange.)

Binary files contain data that is meant to be used by a particular program. Here is a well-known example. Many people use Anonymous FTP and Usenet to share pictures that can be displayed on a computer screen. These pictures are stored as binary files and must be displayed by a picture-viewing program that understands the file format. If you were to look at a picture file with a paging program—which is designed to display text files—you would see gibberish.

Here is another example. People also share binary files that contain sounds that can be played back on computers with the appropriate hardware. In this case, you need a sound program that understands the binary file format being used. More boring examples are word processors, spreadsheets, and databases that store data in binary files according to their own particular format.

In Chapter 5, we discussed text and binary files in more detail. At the time, we explained that mail messages can contain only text data, not binary data (unless you use a special protocol called MIME: Multipurpose Internet Mail Extensions). Similarly, we saw in Chapter 10 that Usenet articles are also composed of ordinary text.

This poses a problem: How can you mail a binary file to someone? And how can you post a binary file—such as a picture—to a Usenet newsgroup?

The solution is to convert the binary file to ASCII text. Then, you can mail it or post it as a Usenet article. Later, someone else can take the text file and convert it back to binary.

There are two programs that are commonly used to convert files from binary to text. The first is named **uuencode**. We refer to the text file that **uuencode** creates as a *uuencoded file*. To convert a uuencoded file back to binary, we use **uudecode**. (The names were chosen because these two programs were developed as part of UUCP, a system that can be used to copy files from one Unix computer to another. UUCP stands for "Unix to Unix copy".)

When uuencoded files are stored on an Anonymous FTP server, they are usually given an extension of **.uue**, for example, **file.uue**. Such files are sometimes compressed, in which case you will see a file name like **file.uue.Z**.

The other program that is commonly used to convert binary files to text is called **btoa** (binary to ASCII). We refer to the text file that **btoa** creates as a *btoa file*. There is no special extension used to identify such files.

We will discuss the **uuencode/uudecode** and **btoa** programs later in the chapter.

The last program that we will mention here is **shar**. It is used to create archives that are suitable for mailing or posting to Usenet. Files created by **shar** are called *shar files*. When shar files are offered for Anonymous FTP, they are usually given an extension of **.shar**, for example **file.shar**. Sometimes, such files are also compressed, so you will see file names like **file.shar.Z**.

The advantage of using a shar file is it is especially easy to unpack. The shar file itself forms an executable program that, when run, unpacks itself and re-creates the original files. The program is in the form of a Bourne shell script and you can run it by using the **sh** command. (**sh** is the name of the Bourne shell.) There is also a program named **unshar** that you can use for unpacking. We will explain how it all works later in the chapter.

For reference, Table 18-1 contains a summary of the file types we have mentioned so far. As an Internet user, these are the basic file types and programs that you should understand. If you do not use a PC, you can probably ignore zipfiles.

File Type	Example	Principal Programs
compressed file	`file.Z`	compress, uncompress
tarfile	`file.tar`	tar
compressed tarfile	`file.tar.Z`	tar, compress, uncompress
uuencoded file	`file.uue`	uuencode, uudecode
compressed uuencoded	`file.uue.Z`	uuencode, uudecode, compress, uncompress
btoa file	—	btoa
zipfile	`file.zip`	pkzip, pkunzip
shar file	`file.shar`	shar, sh, unshar
compressed shar file	`file.shar.Z`	shar, sh, unshar, compress, uncompress

Table 18-1. *Summary of File Types and Related Programs*

If you would like an exhaustively comprehensive list with every file type known to civilized man, you can download the file utility document we mentioned earlier. FTP to **ftp.cso.uiuc.edu** and look in the directory **/doc/pcnet** for a file named **compression**.

Hint

Before you download a binary file via FTP, don't forget to prepare by entering the **binary** command (see Chapter 16). If you do not use this command, the FTP program may assume that you are transferring an ASCII text file, and your file may be downloaded improperly.

Compressed files (**.Z**), tarfiles (**.tar**), and zipfiles (**.zip**) are always binary.

Uuencoded files (**.uue**), shar files (**.shar**), btoa files, and tarmail files (which we will meet later in the chapter) are always text.

Files That Hold Documents

There are two types of documents that are commonly found on Anonymous FTP hosts. First, there are regular text files. Such files often have an extension of **.txt**, for example, **document.txt**. Once you download a text file, you can view it with any text editor or paging program.

Another common type of document file has an extension of **.ps**, for example, **document.ps**. These are Postscript files and are meant to be printed on a Postscript printer. Postscript files can contain pictures as well as characters.

Technically, Postscript files are text files, and it is possible to look at them with a text editor. However, the data is stored in Postscript format and you won't be able to make any sense out of it. If you need to look at a Postscript file, you either have to print it on a Postscript printer, or use a program that can display Postscript files on your screen.

Compressed Files (.Z)

Say that you have downloaded a file named **recipe.Z** via Anonymous FTP. Since the file name ends with a **.Z** extension, you know that it has been compressed with the **compress** program. To uncompress it, you use the **uncompress** program. Here is what to do.

Simply enter the name of the program followed by the name of the file:

```
uncompress recipe.Z
```

The **uncompress** program always expects file names to end in **.Z**, so you do not actually have to type the **.Z**. For example, the following two commands are equivalent:

```
uncompress recipe
uncompress recipe.Z
```

Either of these commands will uncompress the file named **recipe.Z**. A new uncompressed file, named **recipe**, will be created. If a file with this name already exists, it will be replaced, so be careful. The old file, **recipe.Z**, will be removed automatically.

If you would like **uncompress** to display a message showing you what it did, use the **-v** (verbose) option after the name of the command:

```
uncompress -v recipe
uncompress -v recipe.Z
```

In this case, you will see an informative message like:

```
recipe.Z:  -- replaced with recipe
```

If you would like to display the contents of a compressed file without actually uncompressing it, you can use the **zcat** program:

```
zcat recipe.Z
```

Again, you can leave off the **.Z**:

```
zcat recipe
```

The **zcat** command will show you what is in a compressed file without changing the file itself.

If the uncompressed file is long, most of it will scroll off your screen before you have a chance to read it. To look at the file, one screenful at a time, you can send the output of **zcat** to a paging program, such as **more**, **pg**, or **less**. To do so, follow the **zcat** command with a | (vertical bar) character, followed by the name of your favorite paging program. For example:

```
zcat recipe | more
```

Remember, though, you can only use **zcat** with a compressed text file. If you have a compressed binary file (such as **file.tar.Z**), you must uncompress and then restore the original files before you can view them.

When you use the | character in this way, we say that you *pipe* the output of one program to another. For more details, see a good Unix book. We recommend *A Student's Guide to Unix*, by Harley Hahn (McGraw-Hill).

Hint

The **zcat** program was named after the Unix **cat** command which, in its simplest form, displays the contents of a file with no added processing. The name **cat**, which comes from the archaic word "catenate", to join in a chain, was chosen because the **cat** program is also used to combine files.

To compress a file, you use the **compress** command. Type the command, followed by the name of the file. The **compress** program will create the new **.Z** file for you and automatically remove the original file. For example, to compress a file named **new-recipe**, enter:

```
compress new-recipe
```

When the command finishes, you will have a new file named **new-recipe.Z**, while the original file **new-recipe**, will be removed.

If you would like **compress** to display a message showing you what it did, use the **-v** (verbose) option after the name of the command:

```
compress -v new-recipe
```

You will see an informative message like:

```
new-recipe: Compression: 53.95% -- replaced with new-recipe.Z
```

If you would like more information about using **uncompress**, **zcat**, or **compress**, check with a good Unix book. You can also display the official description of these commands by using the man command to display entries from the online Unix manual. (The online Unix manual is a built-in facility that can display information about any Unix command.) On most systems, these three programs are documented together, so you only need to use the single command:

```
man compress
```

Hint

On many Unix systems, you will find a pair of programs named **gzip** and **gunzip** which were written to be a replacement for the standard **compress** and **uncompress** programs. The **gunzip** program is powerful. It can restore both **.Z** files and **.zip** files. (We discuss **.zip** files later in the chapter.) The **gzip** program has its own type of compression that produces **.gz** files. Thus, if you download a file with an extension of **.gz**, you will need **gunzip** to unpack it.

If your system has **gzip** and **gunzip**, you should be able to display the documentation by using the command:

```
man gzip
```

You will see that there are actually three programs: **gzip**, **gunzip**, and a new version of **zcat**.

Tarfiles (.tar)

When you download a file whose name has an extension of **.tar**, you know that you have a tarfile. This is an archive that must be unpacked by using the **tar** command. When you unpack such a file, you will usually find multiple files, and sometimes directories. (We discuss Unix files and directories in Chapter 16.)

For this reason, it is a good idea to make a new directory to hold the tarfile before you unpack it. This ensures that all the unpacked files and directories will be segregated from your other files. In this section, we can't go into all the details of creating directories—you will need to consult a Unix book; however, we will show you an example.

Suppose you have downloaded a game via Anonymous FTP. The game consists of a number of program and documentation files that are contained in a single tarfile named **game.tar**. Here is what to do, once you have downloaded the file.

Start by creating a new directory for the game:

```
mkdir game
```

Next, move the tarfile to this directory:

```
mv game.tar game
```

Now, change your working directory to be the new directory:

```
cd game
```

Hint

In Unix, you can enter more than one command at a time by separating them with semicolons. Thus, you can perform all three of the previous commands in one line by entering:

```
mkdir game;  mv game.tar game;  cd game
```

At any time, you can display the name of your working directory by using the **pwd** (print working directory) command.

You are now ready to unpack the tarfile. Use the **tar** command, followed by the options **-xvf**, followed by the name of the file. For example:

```
tar -xvf game.tar
```

The **x** tells **tar** that you want to extract files. The **v** (verbose) tells **tar** to display extra information about each file as it is extracted. And the **f** indicates that the name of the tarfile follows directly.

After you unpack an archive, it is a good idea to list the new files and directories. To do so, use the **ls** (file list) command with the **-l** (long listing) option:

```
ls -l
```

The final consideration is that **tar** will not remove the original tarfile. You will have to do this yourself, by using the **rm** command:

```
rm game.tar
```

To summarize, the following series of commands will process an archive named **game.tar**. The commands create a new directory, move the tarfile to that directory, change to the directory, unpack the archive, list all the new files, and then remove the original tarfile:

```
mkdir game
mv game.tar game
cd game
tar -xvf game.tar
ls -l
rm game.tar
```

If you would like to see what is inside a tarfile without actually unpacking it, you can use the **t** (table of contents) option instead of **x** (extract):

```
tar -tvf game.tar
```

Hint

The **tar** command is unusual in that the options do not need to be preceded by a - (hyphen) character, as is the case with most Unix commands. Thus, the following two commands are equivalent:

```
tar -xvf game.tar
tar xvf game.tar
```

You will often see **tar** commands written this way, so don't be confused.

We believe that it is better to get in the habit of using the - character to precede options, because that is the convention that you must follow with virtually all other Unix commands.

To create an archive of your own, use the **tar** command with the **-cvf** options (**c** for create). Type **tar -cvf**, followed by the name you want for the archive, followed by the names of the files you want to include in the archive. For example, let's say that you want to collect three files named **groatcakes**, **chicken-soup**, and **brownies** into an archive named **cookbook.tar**. Use the command:

```
tar -cvf cookbook.tar groatcakes chicken-soup brownies
```

If you would like more information about using **tar**, check with a good Unix book or use the **man** command to display the **tar** entry from the online Unix manual:

```
man tar
```

Compressed Tarfiles (.tar.Z)

It is common to find tarfiles that have been compressed. Such files will have a double extension of **.tar.Z**. To use such files, all you need to do is follow the steps we have already outlined for uncompressing and unpacking. Just remember to uncompress first.

Say, for example, you download a copy of a book via Anonymous FTP. The various chapters of the book have been collected into a tarfile named **book.tar**.

This tarfile has been compressed into a file named **book.tar.Z**, which is the file you download. You need to do the following:

1. Create a new directory for the book.
2. Move the compressed tarfile to this directory.
3. Change your working directory to be this directory.
4. Uncompress the compressed tarfile.
5. Unpack the tarfile.
6. List the new files.
7. If it looks okay, remove the tarfile.

Here are some sample commands that do the job:

```
mkdir book
mv book.tar.Z book
cd book
uncompress book.tar.Z
tar -xvf book.tar
ls -l
rm book.tar
```

Hint

Remember, **uncompress** will remove the compressed file, but **tar -xvf** will not remove the tarfile.

Uuencoded Files (.uue)

A uuencoded file is a binary file that has been encoded as a text file by the **uuencode** program. You will most commonly see uuencoded files posted to Usenet newsgroups or sent as mail messages. This is because both Usenet and mail use text files, and a binary file must be encoded as text before it can be sent.

You will also see such files offered on Anonymous FTP hosts. There is no standard file extension (like **.Z** or **.tar**) that is always used to identify uuencoded files, but you will often see the extension **.uue**.

To decode a uuencoded file, you use the **uudecode** command. All you need to do is type the command followed by the name of the uuencoded file. For example:

```
uudecode picture
```

The **uudecode** program will read the uuencoded file and re-create the original binary file. When a binary file is uuencoded, its name is placed right into the uuencoded file. When the **uudecode** program reads this file, it looks for this name. Thus, **uudecode** always knows what name to give to the newly restored binary file.

Here is an example. You download a file called **harley.uue** via Anonymous FTP. According to some documentation in another file, **harley.uue** contains a binary file named **rick**. To restore the binary file, you enter:

```
uudecode harley.uue
```

Once **uudecode** is finished, you will have a binary file named **rick**. The **uudecode** program does not remove the original uuencoded file, so you must do it yourself. In this case, you would use the command:

```
rm harley.uue
```

When you read Usenet newsgroups that contain pictures, you will find that most of the uuencoded pictures are so large that they are broken into parts. Each part is posted as a separate Usenet article. Later in the chapter, we will show you how to put these parts together and restore the original binary file.

Btoa Files

Like **uuencode**, **btoa** (binary to ASCII) is a program that encodes a binary file as a text file. Btoa files are used mostly to send binary files through the mail. You use **btoa** to encode a file which you mail to a friend. At the other end, your friend saves the message to a file, and then uses **btoa** to restore the original file.

To encode a binary file to text, you use the **btoa** command followed by the name of the binary file, followed by the name you want to use for the text file.

For example, say that you have a binary file named **harley** that you want to mail to a friend. You can use the command:

```
btoa harley temp
```

You can now mail the file **temp** to your friend.

Hint

You can send the output of the **btoa** command directly to the **mail** command, which eliminates the need for a temporary file. For example, say that you want to mail the binary file **harley** to a friend whose address is **addie@nipper.com**. You can use:

```
btoa harley | mail -s 'btoa file of harley' addie@nipper.com
```

The | (vertical bar) character tells Unix to pipe the output of **btoa** directly to the **mail** program. The **-s** option tells **mail** what subject to put on the message. For more details on using **mail**, see Chapter 6.

To decode a btoa file, use the **btoa** command with the **-a** option, followed by the name of the encoded file. You do not need to specify a name for the binary file; it was encoded into the text file and **btoa** will find it automatically.

Thus, suppose you have received the mail message from the previous example. You save this message to a file named **temp**. Now, to recover the binary file, you can use:

```
btoa -a temp
```

When this command is finished, you will have the original file named **harley**.

Hint

If you receive a btoa file as a mail message, you will have to save it as a file. This new file will have header lines at the beginning (see Chapter 5). There is no need to remove these header lines before you decode the file. The **btoa** program will skip over them automatically.

The old version of **btoa** did not use **btoa -a** for decoding. Rather, it used a separate command named **atob** (ASCII to binary). If your system has the old **btoa**, you will have to use the **atob** command to restore the original file. For example:

```
atob temp
```

If this is the case, you might ask your system manager to install the new version of **btoa**.

Mailing Compressed Archives with Tarmail

As we mentioned, one of the most common uses of **btoa** is to encode a binary file as text so it can be mailed. In particular, **btoa** is often used to encode compressed tarfiles. For example, say that you have a number of chapters of a book in separate files that you would like to mail to a friend whose address is **ron@sigstar.com**. These chapters are named **ch1**, **ch2**, **ch3**, and **ch4**. (It's a small book.)

First, use the **tar** command to collect all the files into an archive named **book.tar**:

```
tar -cvf book.tar ch1 ch2 ch3 ch4
```

Next, compress the tarfile:

```
compress book.tar
```

Finally, encode the file from binary to text and mail it:

```
btoa book.tar.Z | mail -s 'Tarfile: Hot new book' ron@sigstar.com
```

At the other end, your recipient can uncompress and unpack the message to recover the original files. (Of course, he will need the **btoa** program.)

If you find this sort of thing handy, there is a program that will do it all for you in one step: collect a group of files into an archive, compress the archive, and then convert it to a text file suitable for mailing. The name of the program is **tarmail**. To use it, you type the name of the program, the address of the recipient, a subject for the message, and the names of all the files you want to send. For example:

```
tarmail ron@sigstar.com 'tarmail book' ch1 ch2 ch3 ch4
```

When you receive a tarmail file through the mail, all you have to do is save it and then use **untarmail** to restore the original files. Simply type **untarmail**, followed by the name of the file.

For example, let's say that our friend receives our message and saves it to a file named **temp**. All he needs to do is enter:

```
untarmail temp
```

The program will restore the original files, in this case, **ch1**, **ch2**, **ch3**, and **ch4**. When **untarmail** is finished, it will delete the tarmail file for him.

Shar Files (.shar)

Shar files are used to collect text files into an archive. The archive is in the form of a shell script so it is easy to unpack.

Now, in order for any of this to make sense, you need to know what a shell script is, so let's take a moment to talk about shells.

In Unix, the program that reads and processes your commands is called the *shell*. When you enter, for example, the **date** command, the shell reads it and figures out what to do. In this case, the shell would execute the **date** program. When you enter the **tar** command, the shell executes the **tar** program, and so on.

Within the world of Unix, there are a number of different shells, the most popular being the C-Shell and the Korn shell. (You are probably using one of these shells, even if you don't know it.)

The original shell was called the Bourne shell. The Bourne shell is considered to be somewhat primitive these days and is not most people's first choice. Still, it is readily available on all Unix systems.

Each shell is itself a program that can be executed, just like any other program. The name of the C-Shell program is **csh**; the name of the Korn shell program is **ksh**; and the name of the Bourne shell program is **sh**. (The Bourne shell was the original shell, so it has the simplest name.)

Hint

You may be wondering, where do the names of the shells come from? The C-Shell was developed by Bill Joy, who named it after the C programming language because they share some features. The Korn shell was named after its developer, David Korn. The Bourne shell was named after its developer, Steven Bourne.

Now, one important aspect of the shell is that it can accept not only commands that you type in at your keyboard, but commands that are stored in a file. For example, you can tell the shell: "I have a file named **harley** that contains shell commands. Read that file, line by line, and execute those commands just as if I had entered them from the keyboard." Such a file is called a *shell script.*

Each shell offers a number of commands that can be used to create complex and powerful shell scripts. Thus, it is possible to write shell scripts that do many different things. For example, the **tarmail** and **untarmail** commands that we discussed earlier are actually shell scripts.

There are several ways to execute a shell script. The simplest way is to type the name of the shell that you want to run the script, followed by the name of the file that holds the script. For example, to have the Bourne shell (whose name is **sh**) run a script named **harley**, you can enter:

```
sh harley
```

Although you can run scripts with the C-Shell or Korn shell, most people use the Bourne shell, because it does a good job of executing scripts, and it is available everywhere. (However, as we mentioned earlier, people generally prefer to use the C-Shell and Korn shell as their primary command processor: the program that reads the commands they type at the keyboard.)

By now, it should make sense when we say that a shar file is actually a shell script. When you execute a shar file, it sends commands to the shell that, when carried out, re-create the original files that were archived. The beauty of this system is that you can archive a group of files into a shar file and mail it to anyone (or post it to Usenet). You are guaranteed that the recipient can unpack the files because everyone who uses a Unix system has access to the Bourne shell.

Thus, you can understand when we tell you that the name "shar" stands for "shell archive".

To create a shar file, you use the **shar** command. Type the name of the command, followed by the names of the files you want to archive, followed by a > (greater than) character, followed by the name of the file you want to hold the archive. For example, say that you want to create an archive named **recipes.shar** to hold three files named **brownies**, **chicken-soup**, and **groatcakes**. You can then mail **recipes.shar** to a friend who appreciates good food.

Use the command:

```
shar brownies chicken-soup groatcakes > recipes.shar
```

The > character tells Unix to store the output of the command in the specified file. When you do this, we say that you are "redirecting the standard output to a file". For more details, see a good Unix book.

An alternate way to specify the name of the output file is to use the **-o** option after the command name. Type **-o** followed by the name of the output file. Thus, the following two commands are equivalent:

```
shar brownies chicken-soup groatcakes > recipes.shar
shar -o recipes.shar brownies chicken-soup groatcakes
```

Either of these commands will produce a shar file named **recipes.shar**.

Hint

One important limitation of shar files is that they can contain only text files. If you want to create a shar archive to hold binary files, you will have to use **uuencode** or **btoa** to encode the binary files as text files. However, in such cases, it is a lot easier to use **tar**, which can handle binary files directly. You can then encode the tarfile as a text file before you mail it.

If you look inside a shar file, you will see a message similar to the following one at the beginning of the file:

```
# This is a shell archive.  Remove anything before this line, then unpack
# it by saving it into a file and typing "sh file".  To overwrite existing
# files, type "sh file -c".  You can also feed this as standard input via
# unshar, or by typing "sh <file", e.g.. If this archive is complete, you
# will see the following message at the end:
#                "End of shell archive."
# Contents:  brownies chicken-soup groatcakes
# Wrapped by harley@fuzzball on Mon Jun 28 11:21:08 1993
```

Here is how to unpack a shar file that you see on Usenet or that you receive through the mail. First, save it to a file. Next, use a text editor and remove all the lines before the top line of the actual archive (as described above). If you have saved a Usenet article or a mail message, there will be header lines that you will have to remove. Finally, enter a command that consists of **sh** followed by the name of the file. The Bourne shell will read the shar file, execute its commands, and unpack the archive.

Here is an example. You are reading a Usenet newsgroup and you come across an article that contains a shar file. This shar file contains a number of recipe files, so you save the article to a file named **recipes.shar**.

Now, using a text editor (such as **vi** or Emacs), you remove all the header lines at the beginning of the file. To unpack the shar file, you enter the command:

```
sh recipes.shar
```

The original files will be unpacked for you.

You may have noticed one problem with this system. What if you don't know how to use a text editor? The solution is that you can use a program named **unshar** to unpack a shar file. The **unshar** program will look for and ignore header lines at the beginning of a file, so you don't have to remove them yourself. Thus, in the last example, we could have used:

```
unshar recipes.shar
```

Some versions of the **unshar** command save the header lines to a file with an extension of **.hdr**. In our example, this file would be called **recipes.shar.hdr**. If you want to tell **unshar** to not save the header lines in this way, you can use the **-n** option:

```
unshar -n recipes.shar
```

Not all Unix systems have the **shar** and **unshar** commands. However, the beauty of the scheme is that, even if you do not have these commands, you can still unpack a shar file, because all Unix systems do have the Bourne shell. Of course, if you do not have **unshar**, you will need to use a text editor. (This is why, in Chapter 1, we recommended that you learn some Unix.)

When you unpack a shar file, the original shar file is not removed. You will have to do this explicitly. For example:

```
rm recipes.shar
```

You may also have to remove the file that contains the header lines:

```
rm recipes.shar.hdr
```

Zipfiles (.zip)

If you use a PC, you should know that the most common programs for compressing and archiving PC files are **pkzip** and its sister program **pkunzip**. The **pkzip** program creates a compressed archive called a *zipfile*. The **pkunzip** program reads a zipfile and restores the original files. Thus, **pkzip** is both an archiver and a compression program (sort of like a combination of **tar** and **compress**). When we use **pkzip** to archive and compress a collection of files, we say that we are *zipping* the files. When we use **pkunzip** to unpack and uncompress, we are *unzipping*.

The PC version of the **pkzip** program is *shareware*. This means that the program can be freely distributed and used by anyone for free. However, if you like the program, you are urged to send a small fee to the company that develops **pkzip** (their name is Pkware). You will then become a registered user and receive certain advantages, such as a printed manual.

Aside from **pkzip** and **pkunzip**, there are programs that run on non-PC computer systems that can create and process zipfiles. Still, the most common zipfile program is the original **pkzip** for PCs. For this reason, you will find that many of the PC files that are available by Anonymous FTP are stored as zipfiles. Thus, as a PC user, it will be a common occurrence for you to download an Anonymous FTP file that needs to be unzipped.

In order to do so, you must follow two steps. First, you must get the file to your PC. (We will discuss how to do this in the next section.) Second, you must use the **pkunzip** program to unzip the file.

This is easy. On your PC, type the command **pkunzip** followed by the name of the zipfile. All zipfiles have an extension of **.zip**, but you can leave this out when you enter the **pkunzip** command. For example, the following two PC commands will both unpack and uncompress a zipfile named **recipes.zip**:

```
pkunzip recipes.zip
pkunzip recipes
```

Although zipfiles can be used to hold any type of data, they are commonly used to hold software packages. For example, later in the chapter, we will talk about software that you can use to view pictures on your PC. Such software involves a whole package of programs and documentation files. To make such a package easy to download, the files are zipped into one tidy zipfile, which is then distributed via Anonymous FTP. As a user, all you need to do is download the zipfile and unzip it.

Hint

Most zipfiles contain a collection of files. Thus, before you unzip, it is a good idea to create a new directory to hold the zipfile. That way, all the unzipped files will be isolated in their own directory.

Among other advantages, this makes it easy to check the new files for viruses before you use them (something you should always do before you use downloaded PC software).

You say that you don't have an anti-virus program? Download one for free using Anonymous FTP. To find such programs, use Archie and search for **virus**.

The **pkzip/pkunzip** programs come with a lot of documentation, and we won't go into all the details here. What we will do is tell you how to get these programs for yourself. There are two ways. First, if your computer has a modem, you can dial the Pkware BBS (bulletin board system) at (414) 354-8670 and download the software (free) directly from the company.

Second, you can use Anonymous FTP and download the software from the Internet. To find an Anonymous FTP host, use Archie with a **regex** search and look for **^zip**. The **^** (circumflex) character tells Archie to match only those items that start with the letters **zip**. (We explain how to use Archie in Chapter 17.)

We mentioned earlier that many PC software packages are distributed as zipfiles. You might ask, is that how **pkzip** itself is distributed? The answer is no. After all, until you have **pkzip** up and running, it is impossible to unzip anything.

For this reason, **pkzip** is distributed as a *self-extracting archive*. This is an executable file that contains a collection of zipped files. When you run the program, it automatically unzips itself to restore the files. For example, at the time we wrote this chapter, the most current version of **pkzip** was distributed as a self-extracting archive named **pkz204g.exe**. When you run this program, it will restore the files that make up the **pkzip** product.

(You might wonder, if the Pkzip software is packaged in a file with a name like **pkz204g.exe**, how can you find it by using Archie to search for **^zip**? The answer is, the Archie search will find you a list of Anonymous FTP hosts that have a directory named **zip**. When you FTP to one of these hosts and look inside the **zip** directory, you will find the **pkzip** program.)

> ## Hint
>
> Program names that have numbers in them usually refer to the version of the program. For example, **pkz204g.exe** contains version 2.04g of **pkzip**. When you have a choice, always download the program that has the highest numbers.
>
> If you see a program name that has the word **beta**, it means that this is a pre-release version that you can use at your own risk.

> ## Hint
>
> We mentioned earlier that there are programs for various different computers that can handle zipfiles. One of the best all round programs for Unix computers is **gzip**. It can not only handle zipfiles, but compressed (**.Z**) files as well.

Uploading and Downloading Files to Your PC or Macintosh

The second most frequently asked question asked by PC and Macintosh users is, how do I get Internet files to my computer? (The most frequently asked question is, how do I look at picture files?)

If you have the right software, downloading and uploading files between your personal computer and the Internet is easy. If you don't have the right software—or if it doesn't work properly—copying files will be impossible.

> ## Hint
>
> Quick review: Downloading refers to copying files from the remote host to your computer. Uploading refers to copying files from your computer to the remote host. To remember, think of the other computer as floating above you in the sky.

Before we start, let us recall that there are two ways to use a PC (or Macintosh) to access the Internet. In the simplest case, your PC might be connected directly to the Internet. However, if you are working at home, you are probably using your PC to run a communications program. This program uses your modem to dial the phone number and connect to a remote Internet host. The program then emulates a terminal. You log in to the host with your userid (login name) and a password. All of this is explained in Chapter 3.

If your PC is connected directly to the Internet, files that you download via Anonymous FTP will come right to the PC. However, if you are using a communications program to connect to a remote host, the FTP files will be sent to that host. You will then have to copy them to your PC. That is what we will discuss in this section.

To copy files from one computer to another, we use what is called a *file transfer protocol*. In the PC and Macintosh world, the most common protocol is called Zmodem. Thus, in order to download and upload files, you need Zmodem software on both your computer and the remote host. All PC and Macintosh communication programs come with Zmodem functionality. What you need to ensure is that such a program exists on the host.

On Unix systems, the most common Zmodem program is named **sz** (the name means "send using Zmodem"). If your system has **sz**, there will be a companion program named **rz** ("receive using Zmodem"). You should be able to display the documentation for both these programs by using the command:

```
man sz
```

If your Unix system does not have a Zmodem program, you may find another file transfer protocol called Kermit. The program will probably be named **kermit**. We will not talk about Kermit here except to say that it is very slow and that you should use it only if you can't use Zmodem.

To download a file from the remote host to your PC or Macintosh, type **sz**, followed by either **-a** or **-b** (we will explain in a moment), followed by the name of the file. For example:

```
sz -a document
sz -b picture.gif
```

If you are downloading a text (ASCII) file, use **-a**. If you are downloading a binary file, use **-b**. If you are not sure what type of file you have, use **-b**.

Extra Important Hint

Never download a binary file without using **-b**. The **sz** program may look like it is working fine, but when the file gets to your PC, the data inside the file will be damaged. This is the most common answer to the question: Why can't I unzip the zipfile that I downloaded to my PC?

As soon as you enter the **sz** command, one of two things will happen. Your communication program may detect that a Zmodem transfer has been initiated and start transferring the file automatically. (Some communication programs can be configured to do this.) Or you will see a message similar to:

```
Sending in Batch Mode
```

At this point, you will have to tell your communication program to start a Zmodem download. How you do this varies from program to program, so you will have to read the documentation that came with your program. As an example, if you use Telix (a popular PC shareware communications program), you would press the PAGE DOWN key and then select "Zmodem" from a menu.

To upload a file from your PC or Macintosh to the remote host, start by entering one of the following commands (on the host):

```
rz -a
rz -b
```

(Remember to use **-b** if you are uploading a binary file.)

After you enter this command, you will see a message similar to:

```
ready. To begin transfer, type "sz file ..." to your modem program
```

You must now tell your communications program to begin a Zmodem upload. Again, how you do this depends on your program. With Telix, you would press the PAGE UP key, select "Zmodem" from a menu, and then enter the name of the file you want to upload.

As we said before, when it all works, it's simple. When it doesn't work, it will make you tear your hair out and wonder if Man is really Nature's last word.

Downloading and Viewing Picture Files

We don't know if you have ever read a Perry Mason book. If you have, you will know that every now and then, while Perry Mason is in court defending a client, the district attorney will ask a question and Perry Mason will jump up and say, "I object, the question is incompetent, irrelevant, and immaterial. The district attorney has not laid a proper foundation."

Lest we be accused of just such a transgression, we will now lay a proper foundation for the viewing of picture files.

Pictures are stored as binary files. There are various formats for storing such files, the two most common being GIF and JPEG. To display a picture, you use a program called a *viewer* that knows how to read the picture file and turn it into an image. To view a GIF file, you need a GIF viewer; to view a JPEG file, you need a JPEG viewer. You can obtain a viewer for free via Anonymous FTP.

GIF pictures are stored in files that have an extension of **.gif**. The name GIF stands for Graphics Interchange Format. JPEG pictures are stored in files that have an extension of **.jpg**. The name JPEG refers to the people that devised the standard: the Joint Photographic Experts Group.

GIF and JPEG files hold regular pictures. There is another common type of file, a GL file, that contains animated (moving) pictures. Such files have an extension of **.gl**. To view such pictures, you need a GL viewer.

Of all the Usenet newsgroups, the ones that use the most network resources are those devoted to swapping pictures. Near the end of Chapter 9, we showed a list of the most popular Usenet newsgroups. If you look at this list (Table 9-4), you see that the sixth most popular newsgroup is **alt.binaries.pictures.erotica** (right in between **alt.sex.stories** and **rec.arts.erotica**). As you can see, sex is one of the most popular subjects on Usenet, and, as you might guess, most of the pictures that are posted are erotic. However, there are newsgroups that contain other types of pictures, for example, **alt.binaries.pictures.misc**.

To view a picture, you must go through the following steps:

1. Use your newsreader program (see Chapter 11) to read the newsgroup that contains the type of pictures you like. For a list of newsgroups, see Appendix G. Look at the groups whose names start with **alt.binaries.pictures**.

2. Pictures are stored in uuencoded format. Most pictures are so large that they will be posted in several parts. Find a picture that appeals to you and save each part to a separate file.

3. Using a text editor, combine the files (in the correct order) into a single uuencoded file. As you do, you must strip out all the superfluous header lines. If you find yourself working with a lot of picture files, you can get a

utility program that will do all the combining and stripping automatically. For more details, read the frequently asked question list mentioned below.

4. Now, use **uudecode** to decode the uuencoded file. You should end up with a single **.gif**, **.jpg**, or **.gl** file

5. If necessary, download this file to the computer on which you want to view the picture. For example, you may want to download the file to your PC or Macintosh.

6. Use your viewer to look at the picture.

Hint

Remember that pictures are binary files. If you are downloading with **sz**, be sure to use the **-b** option.

There is a frequently asked question (FAQ) list that explains all about viewing pictures. (We described FAQ lists in Chapter 9.) The best way to introduce yourself to picture viewing is to read the FAQ list. It comes in three parts and is posted regularly to all of the picture newsgroups as well as **news.answers**.

If you would like to download the FAQ list by Anonymous FTP, use Archie to search for **pictures-faq**. This will be a directory that contains the three parts of the FAQ list. One place from which you can FTP these files is **rtfm.mit.edu**. Change to the directory named **/pub/usenet/news.answers/pictures-faq**, and download the files named **part1**, **part2**, and **part3**.

The last point that we will cover is where to find a picture viewer. There are several places to look. The best way to start is by reading **part3** of FAQ list. This article contains a large list of all different kinds of picture-oriented software and where to get them via Anonymous FTP. Look for the section that pertains to your computer and then look for the subsection called "Picture Viewers".

Along with the description of each viewer, you will also see the location of an Anonymous FTP host that carries the file. In addition, once you know the name of the file, you can use Archie to search for alternate sites.

Another place to look is the Usenet newsgroup **alt.binaries.pictures.utilities**. You will often see PC picture viewing programs posted here. Typically, the programs will be split into several postings. When you put together the postings and decode them, you will have a zipfile, which you can download to your PC.

One good place to know about is the Anonymous FTP host **bongo.cc.utexas.edu**. FTP to this host and change to the **/gifstuff** directory. You will see a number of sub-directories that contain files relating to various types of computers. For example, if you are a PC user, you would change to the **msdos** sub-directory. Within **/gifstuff**, there is also a file named **ftpsites**. This file contains a list of a large number of Anonymous FTP hosts that carry picture-related files and programs.

Hint

Erotic pictures are available only in the Usenet newsgroups. There is no Anonymous FTP host that offers erotic pictures, so don't waste your time looking for one.

CHAPTER 19

Talking

In this chapter, we will show you how to have a computer-mediated conversation with another Internet user. You will be able to connect your computer to another person's computer and type messages back and forth. He will see what you type, and you will see what he types.

A handy and convenient communication tool... when it works. For those times when it doesn't work, we will show what might be wrong and give you hints on what to do.

Finally, we will tell you about a program that allows several people to have a group conversation, sort of like an Internet conference call.

Introducing the talk Program

The **talk** program allows you to connect your computer to someone else's computer, and then type messages back and forth. (Of course, both computers must be on the Internet.)

When you communicate in this way, we say that you are *talking* to that person. Of course, you are not talking in the usual sense—by speaking and listening—but you do see what the other person types as soon as he or she presses the key.

For this reason, the convention is to refer to such connections in an informal sense. For example, you might say that you had a "conversation" with someone yesterday. Or you might send a mail message that says, "If you have any problems, just talk to me."

The great thing about the **talk** program is that you can talk with people all over the world for free. For example, as we worked on this book in California, we often had conversations with a well-known computer expert in Austria.

As you talk to someone, the actual communication is handled by a program called a *talk daemon*. As we explained in Chapter 5, a daemon is a program that executes in the background, in order to provide a service of general interest. In this case, the daemon provides a talk service. In Chapter 8, we saw that a different daemon provides the Finger service.

To start a conversation, type **talk**, followed by the address of the person to whom you would like to talk. For example, say that your userid is **harley** and you are logged in to the computer named **fuzzball.ucsb.edu**. You want to talk to a friend whose userid is **rick**, who is logged in to the computer named **tsi.com**. Enter:

```
talk rick@tsi.com
```

The program will send a message to your friend's computer and tell him that you want to talk with him. Out of nowhere, he will see the following message on his screen and hear a beep.

```
Message from Talk_Daemon@tsi at 13:56 ...
talk: connection requested by harley@fuzzball.ucsb.edu.
talk: respond with:  talk harley@fuzzball.ucsb.edu
```

At this point, all he has to do is follow the instructions. In this case, he would enter the command:

```
talk harley@fuzzball.ucsb.edu
```

Once he does, the connection is complete and you can start talking.

If the person you are trying to reach is logged in, but is not responding, the **talk** daemon will keep sending him a message, about every ten seconds. As it does, you will see:

```
[Ringing your party again]
```

If you decide to give up, just press CTRL-C to stop the **talk** program.

Of course, you yourself may receive a **talk** request unexpectedly. If you do, just remember to enter the **talk** command exactly as you are requested in order to make the connection.

When you have finished your conversation, either person can quit by pressing CTRL-C. (If you are not using a Unix system, just press whatever key is used with your computer to stop a program.)

If you want to talk to someone who is logged in to the same computer as you, you need specify only a userid and not the full address. For example, say that you want to talk to someone on your computer whose userid is **mike**. You can use the command:

```
talk mike
```

The talk Screen

Once your **talk** daemon has made the connection with another person's **talk** daemon, you will see the message:

```
[Connection established]
```

and hear a beep.

The **talk** program will draw a horizontal line in the middle of your screen, dividing it into an upper half and a lower half. Whatever you type will be displayed in the upper half, whatever the other person types will be displayed in the lower half. Figure 19-1 shows a typical **talk** screen.

Hint

If you are using Unix and you notice that the output is displayed in a garbled fashion, it may be that you have not initialized your terminal type: that is, you have not told Unix what type of terminal you are using. Typically, you will see a line of characters going down the left-hand side of the screen.

We discuss terminal initialization in Chapter 3. If you are having this problem, check with a good Unix book or ask someone for help.

You can both type at the same time, and whatever you type will be displayed immediately on both screens. As you type, you can press BACKSPACE to correct the previous character. You can also use CTRL-W to erase an entire word and CTRL-U to erase an entire line. At any time, you can tell **talk** to re-display the entire screen by pressing CTRL-L. This can come in handy if an unexpected message—like someone else trying to talk with you—has appeared on your screen.

(You may have to experiment a little. With some computers, you have to use DELETE instead of BACKSPACE to erase a single character. Similarly, you may have to use CTRL-X instead of CTRL-U to erase an entire line.)

```
What you type comes out here, above the dividing line.
Each time you type a character, it appears immediately on your
screen and on the other person's screen.
If you want a blank line, simply press the Return key twice.

The lines are displayed one after another.
When the last line on the bottom is filled, the next line
will wrap around to the top.


----------------------------------------------------------------
What the other person types is displayed here, on the
bottom half of your screen.

Since your messages and his are displayed in different places,
you can both type at the same time.
```

Figure 19-1. *A typical* **talk** *screen*

Hint

Because the other person can see each character as you type it, he will also see you backspace when you make corrections. Thus, do not type something offensive thinking that you can change it. The other person will see exactly what you type, as you type it.

Hint

One of the biggest mistakes that people make is pressing BACKSPACE repeatedly to correct a few words or the entire line. It is much faster to use CTRL-W and CTRL-U (or CTRL-X).

Talk as long as you like. When you are finished, either person can sever the connection by pressing CTRL-C. It is considered good manners to say goodbye and to wait for an acknowledgment before ending a conversation. When the **talk** program is finished, you will see:

```
[Connection closing. Exiting]
```

Guidelines for Talking

Having a conversation via **talk** combines the immediacy of a face to face encounter with the verbosity of written language. For this reason, you will notice two significant differences between talking by computer and talking by mouth.

First, talking by computer is a lot slower, because almost everybody can speak faster than they can type (not that this is always an advantage). Second, because you are not looking at the other person, the customary nuances and inflections that comprise so much of the non-verbal aspects of communication are missing.

We make up for these limitations in two ways. First, there are a number of abbreviations that are commonly used to speed things up. Some of these are shown in Table 19-1. These abbreviations are the spiritual descendants of the Morse code conventions used by ham radio operators since the 1920s.

Abbreviation	Meaning
BCNU	be seeing you
BRB	be right back
BTW	by the way
BYE	goodbye (I am ready to stop talking)
BYE?	goodbye? (are you ready to stop talking?)
CU	see you
CUL	see you later
FYI	for your information
FWIW	for what it's worth
GA	go ahead and type (I will wait)
IMHO	in my humble opinion
IMO	in my opinion
JAM	just a minute
O	over (your turn to speak)
OO	over and out (goodbye)
OBTW	oh, by the way
ROTFL	rolling on the floor laughing
R U THERE?	are you there?
SEC...	wait a second
WRT	with respect to

Table 19-1. *Common Communication Abbreviations*

The second important convention recognizes that it is easy to offend someone (or at least have your words misconstrued) when you are not talking in person. Over the computer, we are unable to use subtle body language to indicate that what we just said was debonair and ironic, and not really as dumb and insensitive as it seems.

Thus, when there is any doubt as to whether or not the other person might be offended, it is a good idea to use a smiley to show that you are just kidding. As we explained in Chapter 10, a smiley is a little face that we draw with punctuation characters. Here is the basic smiley:

```
:-)
```

(To see the face, tilt your head to the left.) You can use a smiley whenever you say
something that might bother the other person, for example:

```
I'd feel the same way as you, if I had a
head that looked like a ripe eggplant :-)
```

If you want to see a few more smileys, look at Figure 10-3 in Chapter 10.

FUN TIP: For possibly the first time in your life, spelling does not count.

*Because typing is so much slower than talking, it is perfectly acceptable to type
as fast as you can and not worry about spelling mistakes. Nothing is worse than
watching someone laboriously backspacing and correcting each word. Just go
ahead and type and don't worry. Almost everthigg is undrstanable with kust a
few worng lettrs and. no doubt,m the other pweosn will get the g9ist of thwta
you are syaing.*

Problems with Addresses and What to Do

There are several problems that might keep you from making a **talk** connection.
    First, the person you want to talk to may not be logged in. If this is the case,
you may see a message like:

```
[Your party is not logged on]
```

The solution, of course, is to try again later. You may also want to mail the person
a message, asking him to talk to you when he logs in.
    Second, it may be that the computer you are trying to reach is not on the
Internet. Remember, there are some computers that can exchange mail with the
Internet, but are not on the Internet itself. If you specify the address of a
computer that is not on the Internet, you will see a message similar to:

```
elmer.com is an unknown host
```

You may also see either of these messages if you make a typing mistake when you enter the **talk** command. For example, say that you want to talk to a friend whose userid is **rick** on the computer named **tsi.com**. However, by mistake, you type:

```
talk ric@tsi.com
```

You will see the message:

```
[Your party is not logged on]
```

Do not misinterpret this message. It does not imply that the userid you typed is valid, but the person is not logged in. All it means is that no one with that userid is currently logged in. Thus, when you get this message, be sure to check your spelling and, if necessary, retype the command.

Similarly, if you enter:

```
talk rick@tsii.com
```

you will see:

```
tsii.com is an unknown host
```

At least this time, the message is more informative and your spelling mistake is obvious.

The next reason that your connection might not work is that you may be using a mail address that is an alias and not the person's actual address.

As we explained in Chapter 4, some organizations allow their users to use a simplified mailing address. For example, say that you have a friend named Melissa who has userid **wx1523** on the computer named **sdcc99.ucsd.edu**. For convenience, her organization allows her to receive mail at the address **melissa@ ucsd.edu**. When a message arrives, the local mail program looks up the name in a master list and sends the mail to the correct local address.

However, if you want to talk to your friend, you cannot use **melissa@ucsd .edu**, as this is only a mail alias. You must use the exact address of **wx1523@ sdcc99.ucsd.edu**. If you are not sure of the correct address, you can always mail her a message and ask.

The "Checking for Invitation" Problem and What to Do

So far, we have discussed problems that stem from an incorrect address, or from a person not being logged in. However, there is a problem that commonly arises even when you have the correct address and the person is logged in. The reason is that not all Internet computers can use **talk** to connect to one another.

There may be a time when you enter a **talk** command and see the following message:

```
[Checking for invitation on caller's machine]
```

This looks like a nice, benign message. It seems to imply that the other person has been notified that you want to talk, and you are just waiting for him or her to enter the appropriate **talk** command.

In reality, this message is highly misleading. What it usually means is that your **talk** program and the other **talk** program are incompatible. (More precisely, your daemons are incompatible.) The reason is a little technical, but here it is.

As you may know, computer data is stored in bytes, each byte holding a single character. Inside the computer, the bytes are collected into words. There may be two bytes per word, four bytes per word, or whatever. The question is, within each word, what order are the bytes stored in?

For example, say that you have a word that contains two bytes, call them A and B. Some computers store the bytes as AB, while other computers store the bytes as BA. Normally, the hardware takes care of all the details, so you don't have to worry about it. The trouble comes when one **talk** program tries to communicate with another one.

If you are using the original version of **talk**, communication can only be established between two computers that store bytes in the same order. Unfortunately, many Internet computers still use this version of **talk**, and when two incompatible computers try to connect, you get stuck with the "checking for invitation" message. In this situation, the only thing you can do is press CTRL-C and abort the command. (But see below for some more ideas.)

Technical Hint

The two schemes that we described are called *big-endian* ("big end first") and *little-endian* ("little end first"). Sun computers and Macintoshes are big-endian. VAX computers and PCs are little-endian.

This incompatibility is sometimes referred to as the *NUXI problem*, because the word "UNIX" on one computer would be stored as "NUXI" on another.

What this all means is that if you are using the original version of **talk**, you can only talk to people who are using compatible computers. To solve this problem, another version of **talk**, called **ntalk** (new **talk**), was developed. The **ntalk** program can talk to any other **ntalk** program, regardless of what computer it is running on. (In case you are interested, **ntalk** communicates via a facility called "sockets", which is not sensitive to byte order.)

Many system managers have installed **ntalk** on their machines, so if you are having trouble with **talk**, try entering the same command again using **ntalk**. For example:

```
ntalk rick@tsi.com
```

Still, there is a problem. Some system managers install **ntalk** under that name. When you want the new version, you use **ntalk**, when you want the old version, you use **talk**.

Other system managers install **ntalk** under the name **talk** (to make it more convenient!) and rename the original version to **otalk** (old **talk**). On those systems, you use **talk** to get the new version and **otalk** to get the old version.

Don't be confused. If you have problems using **talk** to connect to a particular computer, just experiment with **ntalk**, **talk**, and **otalk** until you find the combination that works.

The real solution to all of this silliness is to use a much better program called **ytalk**, which we will discuss later in the chapter. Not only will **ytalk** work with any other computer—whether it is running **talk**, **ntalk**, or **ytalk**—it allows more than one person to talk at the same time. The **ytalk** program is readily available via Anonymous FTP and, if you are having **talk** problems, you might prevail upon your system manager to install **ytalk**.

FUN TIP: What can you do if you and your friend do not have two compatible versions of **talk**? *You will have to find another way to get together.*

For example, you could agree to meet on Internet Relay Chat (IRC), which we discuss in Chapter 20. You could also use one of the Internet-accessible bulletin board systems (BBS), which we mention in the catalog. And, of course, you can always send messages by mail.

Refusing to Talk

There may be times when it is inconvenient to receive invitations to talk. For example, you may be editing an important file and you do not want your screen cluttered up with messages from the **talk** daemon. Or you may be a private person who simply does not want to talk to anyone else.

By using the **mesg** (message) command, you can tell your system to refuse or accept **talk** requests on your behalf. To refuse all such invitations, type **mesg** followed by **n** (no):

```
mesg n
```

To ask your **talk** daemon to show you such invitations, use **mesg** followed by **y** (yes):

```
mesg y
```

Normally, **mesg** is set to **y**, so you do not have to worry about it unless you want to cut yourself off from the outside world. To check your status, just enter the command by itself:

```
mesg
```

and **mesg** will display the current setting.

If you try to talk to a person who has set **mesg** to n, you will see a message like:

```
[Your party is refusing messages]
```

In such cases, about all you can do is mail a message.

One way to check if a person is accepting messages is to finger him (see Chapter 8). Some Finger servers will tell you whether or not a userid is accepting messages. Others will not; you will just have to try.

Here is an example. You have a friend whose address is **tln@nipper.com** and who has set **mesg** to **n**. You enter:

```
finger tln@nipper.com
```

As part of the output, you see the line:

```
Login name: tln  (messages off)  In real life: The Little Nipper
```

This shows you that this person is not accepting **talk** requests. You can also finger the **nipper.com** computer to show a summary of all the users who are currently logged in:

```
finger @nipper.com
```

When you do, one of the lines of output is:

```
tln     The Little Nipper   *p4  7 Tue 21:48
```

The details of reading such output are explained in Chapter 8. What we want you to notice is the * character. This tells you that this person is not accepting **talk** requests.

Hint

Before you try to talk to someone, you may want to finger their userid. Aside from checking if the person is accepting **talk** requests, you may also see if the person is logged in.

If the person is logged in, but you see that he has been idle for a long time (some Finger servers tell you this), it is likely that he will not respond to your **talk** request.

The ytalk Program

The **ytalk** program was written by Britt Yenne as a replacement for the standard Unix **talk** programs. The name **ytalk** stands variously for either "Yenne **talk**", or "Why talk?" depending on your state of mind.

The nice thing about **ytalk** is that it will work with both the old and new versions of **talk** that we discussed earlier in the chapter. If you are using **ytalk**, you never have to worry about what type of **talk** your friends are using.

Moreover, if you know other people who use **ytalk**, you can use it to hold a group conversation. At the end of this section, we will tell you how to get the **ytalk** program if it is not already installed on your computer.

The essential form of the **ytalk** command is the same as the **talk** command. Enter the command, followed by the address of the person you want to talk to:

```
ytalk rick@tsi.com
```

When you talk with one person, the screen divides into two parts, just like with **talk**. One important difference is that **ytalk** will show you the userid of the person to whom you are talking. Figure 19-2 shows such a screen. What you type is displayed on the top half. What the other person types is displayed in the bottom half.

If you want to talk with more than one person, all you have to do is specify more than one address as part of the command. For example, here is a command that sets up a five-way conversation:

```
ytalk  rick@tsi.com  mike  addie@nipper.com  tln@nipper.com
```

```
----------------------= YTalk Version 2.0 =--------------------
What you type is displayed here.

Just as with the regular talk command,
one can type at the same time as
the other person.




----------------------------= rick =----------------------------
What the other person types is displayed here.

He can type at the same time as you and
both parts of the screen will be updated
simultaneously.
```

Figure 19-2. *A **ytalk** screen with a two-way conversation*

In such cases, **ytalk** will divide the screen into as many partitions as necessary. In this example, the screen would look like Figure 19-3.

At any time, you can change the participants by calling up a special menu. To do so, press the ESC (Escape) key. You will see:

```
############################################################
# a) Add a new user to session                            #
# d) Delete a user from session                           #
# o) Output a user to a file                               #
# Your choice:                                             #
############################################################
```

You can now press either **a**, **d**, or **o**.

If you press **a**, you will be asked to enter the address of the person you want to add. Once you enter this address, **ytalk** will send them a message. When they respond, they will be incorporated into the conversation.

If you press **d**, you will be asked to enter the userid of the person you want to delete. In this case, you need only enter the userid and not the full address. Once you do, the person will be removed from the conversation.

If you press **o**, you will be asked to enter the name of a userid and then the name of a file. From then on, everything that is typed by that person will not only be displayed on your screen, but will be copied to the file as well. This option is useful when you want to save what someone is typing.

```
---------------------= YTalk Version 2.0 =--------------------
What you type is displayed here.

---------------------------= rick =---------------------------
What Rick types is displayed here.

---------------------------= mike =---------------------------
What Mike types is displayed here.

---------------------------= addie =--------------------------
What Addie types is displayed here.

---------------------------= tln =----------------------------
What The Little Nipper types is displayed here.
```

Figure 19-3.   *A* **ytalk** *screen with a five-way conversation*

In order to have a group conversation, all the people need to be using **ytalk**. However, if you have **ytalk**, you can still talk to people who are using the regular **talk** program. It's just that they will only be able to talk to you, and not to anyone else.

For example, say that you have **ytalk** and you want to talk with two friends who only have **talk**. If you connect to both of them, you will see a three-person conversation. Each of them, however, will only see the standard two-person **talk** screen. They will be able to talk with you, but not with each other.

What happens if you are in the middle of a conversation and a new person tries to talk to you? You will see a message like the following:

```
###############################################################
# Talk with mschuster@netsys.com?                             #
###############################################################
```

You can now press **y** (yes) to add the person to the conversation, or **n** (no) to turn them down. If you press **y**, **ytalk** will redraw your screen to add another partition.

When this happens, each person who is using **ytalk** will be asked if he wants to add the new person to his screen. He will see a message like:

```
###############################################################
# Import mschuster@netsys.com?                                #
###############################################################
```

He can now answer **y** or **n** as he wishes.

As you can see, **ytalk** is an especially handy program. However, it is not a standard part of Unix, so there is a good chance that it may not be on your system. (Indeed, it will only be on your system if someone has deliberately installed it.)

If this is the case, you can download **ytalk** via Anonymous FTP. To find the program, you can use Archie. (We discuss Anonymous FTP in Chapter 16 and Archie in Chapter 17.)

When you use Archie, specify a **regex** search and look for the pattern **^ytalk**. For example, with an Archie client you would use the command:

```
archie -r ^ytalk
```

(The **^** character tells Archie to look for names that start with **ytalk**. If you do not do this, you will find some superfluous names that have **ytalk** in the middle.)

Once you download the program, you can install it if you know how to install programs. Otherwise, you will have to ask someone for help. (Show this page of the book to your system manager and ask nicely. He or she may agree to install **ytalk** as a public service.)

Hint

Whenever you look for a program on Anonymous FTP, make sure you get the newest version, because not all FTP hosts are up to date. You can often tell the version by looking for a number in the file name. For example, if you see a file named **ytalk-2.0.tar.Z**, you know that it holds version 2.0 of the program. (BTW, we explain how to deal with a **.tar.Z** file in Chapter 18.)

CHAPTER 20

Internet Relay Chat

In this chapter, we will introduce you to one of the most wonderful facilities on the entire Internet. You will learn how to join a permanent, 24-hours-a-day gathering, where people from all over the world carry on a multitude of conversations about everything and anything.

What Is Internet Relay Chat?

In Chapter 19, we discussed two ways for you to talk directly with other Internet users. You can use the **talk** program and talk to one person at a time, or you can use **ytalk** and talk to one or more people at the same time. What these two programs have in common is that your conversations are private, like a telephone call. You control who you talk to and only those people can take part in the conversation. A *chat facility* affords a different type of talking experience in which you talk to many people from all over the Internet.

Imagine a vast number of people mingling with one another, perhaps at a large party. You are standing with a group of people who are having several conversations. Whatever you say can be heard by everyone else in your group. You can also eavesdrop on other people's conversations. If you'd like, you can walk from one group to another, join in one of their conversations or just listen. You can also invite someone (or several people) over into a corner for a more private talk. And, if the need arises, you can whisper a private message in someone's ear.

This is the *Internet Relay Chat* or *IRC*.

IRC was developed in 1988 by Jarkko Oikarinen in Finland. The current version of IRC is called IRC II. Since its inception, IRC has become one of the most popular Internet resources and is frequented by people in many different countries. Here is how it works.

You use a client program to act as your interface. (We discuss the client/server relationship in Chapter 2.) Your client connects to an IRC server. You can then enter IRC commands, of which there are many. You can join a group of people or move from one group to another, talking and listening as you please.

Each IRC server is connected to nearby servers. Thus, all the IRC servers are connected (at least indirectly) with one another. So when you make contact, you are connected to a global web of IRC users, all of them talking. Twenty-four hours a day, you will find a vast number of people from many different countries, all connected to IRC, talking, talking, talking.

To bring order to what otherwise would be conversational chaos, IRC maintains a number of different *channels*. When you first connect to IRC, you choose which channel you want to join. You can now talk to all the other people on the same channel. When you feel like a switch, you can leave that channel and join another one. If you like, you can even join more than one channel at the same time.

At any particular time, you will find several thousand people, participating in hundreds of different channels. As you move from one channel to another, you will find people from all over the world. Most of the conversations will be in English, but many are in other languages.

To keep things from being too confused, IRC uses two conventions. First, every channel has its own name. Most channel names begin with a # character, for example, **#hottub**. Some channels are for discussing specific topics. Others are just for talking about whatever comes up. There are public channels, private channels, and secret "invisible" channels.

How does a channel get created? Very simply. Whenever you enter the command to join a channel, IRC checks to see if that channel already exists. If so, you will join it. Otherwise, IRC will create a new channel using whatever name you specified. At first, of course, you will be the only one on the channel, but you can wait for other people to join you. When the last person leaves a channel, IRC removes it.

The first person who joins (and thus, creates) a channel is given the status of *channel operator*. The operator has control over various aspects of the channel. For example, he or she can make the channel private, so that only people who are invited are allowed to join. An operator can also extend operator status for that channel to another person and set the topic for discussion.

Thus, IRC is more than people talking on public channels. You can set up your own channels whenever you want. For example, the President of the United States might send you a mail message and ask, "Would you like to join Hillary and me on channel **#algore** tonight at 8:30 PM, Eastern time?"

FUN TIP: If you would like to mail a message to the President, asking him to join you on IRC, his address is **president@whitehouse.gov**. *If he can't make it, Rush Limbaugh's address is* **70277.2502@compuserve.com**.

The other convention is that every IRC participant has a nickname. When you start a new IRC session, you specify a nickname, up to nine characters, which you can also change at any time. The only limitation is that you cannot choose a nickname that is currently in use. If you want a permanent nickname, you can register it with a special IRC database, so you can always be known by the same name.

As you might imagine, the world of IRC is filled with all manner of imaginative nicknames. However, nicknames offer more than whimsy: they allow people to participate freely and openly while still retaining their anonymity.

At all times, your IRC client will act as your window, showing you what everyone in your channel is saying. Your screen display will vary, depending on what client program you use. Figure 20-1 shows a typical screen.

The bottom line of the screen is where you type your input. In our example, we have typed:

Here is where you type, on the bottom line...

What you type is shown on this line, but is not actually transmitted until you press RETURN. This allows you to make corrections as you type.

Hint

IRC does not distinguish between upper- and lowercase letters when you type a command, a nickname, or a channel name.

```
<Dizney> So of course I told her to go ahead
<Davidg> So then what happened?
<Oinker> How can you say that, Sharkface?
<Dizney> How about you, Harley?
> I have never had that exact same experience
<Dizney> How do you know?
*** Nipper (tln@nipper.com) has joined channel #hottub
> Well, I'm not exactly sure
*** Addie (addie@fuzzball.ucsb.edu) has joined channel #hottub
<Nipper> Hi everyone
<Mindblow> Hello
<Addie> Hi Nipper, how are you?
<Davidg> Dizney, you never told us what happened
<Nipper> Addie, I am doing just fine, just ate dinner but
<Nipper> I accidentally knocked down a picture frame
<Nipper> and broke the glass.
<StreetWise> Where is everyone tonight?
*** Oinker has left channel #hottub
*** Guessme (whoami@silly.com) has joined channel #hottub
[1] 23:06 Harley (+i) on #hottub (+nt) * type /help for help
Here is where you type, on the bottom line...
```

Figure 20-1.   *A typical conversation with Internet Relay Chat*

The second line from the bottom is a status line. In our example, it is:

```
[1] 23:06 Harley (+i) on #hottub (+nt) * type /help for help
```

In this case, we are using the nickname **Harley** and have joined the **#hottub** channel. We can also see the time (23:06, which is 11:06 PM), some technical information, and a reminder as to how to display help information.

The rest of the screen is given over to the actual conversations. As you watch the screen, you will see what is being typed by the people in your channel. Line after line will appear at the bottom of the display area. As new lines appear, the old ones scroll up. Thus, an IRC experience is one of watching many messages, one after the other.

You will sometimes notice that, if parts of the network slow down, people's replies are delayed and become unsynchronized, leading to various amusing and incoherent situations. In such cases, we refer to the delay as *lag*. Thus, you might see someone say, "There was so much lag, I thought you had forgotten all about me."

In our example, you can see several people talking at once. Each line is prefaced by the nickname of the person who typed the message. For example, the line:

```
<Nipper> I accidentally knocked down a picture frame
```

was typed by the person whose nickname is **Nipper**. The lines that you type do not have a nickname when they are displayed on your screen. Rather they have a single > character. For example, we typed the line:

```
> Well, I'm not exactly sure
```

Of course, on everyone else's screen, that same line will appear with your nickname. For example, if your nickname is **Harley**, everyone else would see:

```
<Harley> Well, I'm not exactly sure
```

Lines that begin with *** show informative system messages. For example:

```
*** Nipper (tln@nipper.com) has joined channel #hottub
```

You will often see people carrying on more than one discussion at the same time, and when things get busy it can be somewhat disorienting. Still, you will find that in a short time, you will synchronize with the IRC rhythm and it will all make sense. So much so, that unless you develop some self-discipline, you will find yourself spending more and more time chatting and less and less time participating in regular life (paying your bills, studying, washing dishes, buying computer books, and so on). Thus, before you get started, let us warn you:

> ## Hint
>
> IRC is addictive.

Learning About IRC

The best way to learn how to use IRC is to read something about it. Start with the IRC primer and the IRC tutorial, both of which are available by Anonymous FTP. You should also read the IRC frequently asked question (FAQ) list. If you want more technical information about IRC and its protocol, look at RFC #1459. (We discuss protocols in Chapter 2 and RFCs in Chapter 16.)

You can download the IRC primer by Anonymous FTP. Look in the following places for a file that begins with the name **IRCprimer**.

Host Name	Directory
nic.funet.fi	/pub/unix/irc/docs
cs.bu.edu	/irc/support
coombs.anu.edu.au	/pub/irc/docs

There are three versions of the file, as you will see when you look in one of these directories. The file with an extension of **.txt** is a regular text file. The one with an extension of **.ps** is a Postscript file. And the one with an extension of **.tex** is a Tex file. (Tex is a typesetting system.) If you download a file whose name ends in **.Z** you will have to uncompress it. (We discuss Anonymous FTP in Chapter 16 and file types in Chapter 18.)

You can download the IRC tutorial by Anonymous FTP. FTP to **cs.bu.edu** and change to the directory named **/irc/support**. Download the files that begin with the name **tutorial** (**tutorial.1, tutorial.2**, and so on). You can use the command **mget tutorial.***.

The IRC frequently asked question list is posted regularly to several Usenet newsgroups, including **alt.irc** and **news.answers**. You can check these groups, or you can retrieve the document via Anonymous FTP. FTP to **rtfm.mit.edu** and look in the directory **/pub/usenet/news.answers** for a file named **irc-faq**. (We discuss FAQ lists in Chapter 9.)

More technical information about IRC can be found in RFC #1459. You can find it by using Archie to search for **rfc1459**. (We discuss RFCs in Chapter 16 and Archie in Chapter 17.)

Starting Your IRC Client

To use IRC, you start an IRC client program. The client will automatically connect you to an IRC server and the rest of the world.

It may be that you already have an IRC client (probably named **irc**) on your computer. If so, you should be able to connect to IRC just by entering **irc** followed by the nickname you want to use. For example:

```
irc Harley
```

If you have a Unix system, you can display the documentation for this command by using the **man** (online manual) command at the Unix prompt (not from within IRC):

```
man irc
```

If you do not have an IRC client, you have three choices. First, you can ask your system manager to install one. (Ask nicely. Some system managers feel that IRC is a waste of time.)

Second, you can download the client program from an Anonymous FTP host. The FAQ list will show you where to look. There are various clients available for different types of computer systems. If you download a client program, you will have to install it yourself. If you have any trouble, you may need to ask for help from your system administrator or a friendly computer nerd.

Hint

If you decide to download an IRC client, make sure you get the program for the client and not for the server.

Finally, there are public IRC clients that you can use via Telnet. Look in the IRC FAQ list for an address and then telnet to that address. You may have to specify a port number. (We discuss Telnet and port numbers in Chapter 7.) Once you connect, an IRC client program will start automatically. Conceptually, this is like telneting to an Archie server when you do not have an Archie client of your own.

Hint

The least preferred way to use IRC is by telneting to a public client. Such systems limit the number of users they will accept, and they are often busy. If you get at all serious about IRC, you will find it much more convenient to use a client program on your own computer.

Getting Started

Once you are connected to IRC, there are a great many commands that you can use. All IRC commands start with a / (slash) character. The / must be the first character on the line. As you type a command, it will appear on the bottom line of your screen. When you press RETURN, the command will be sent to IRC. Whenever you type a line that does not start with a / it is considered to be part of a conversation (a regular line) and will be sent to your channel.

The first command you should learn is the one to quit. It is:

```
/quit
```

To display a list of all IRC commands, enter:

```
/help
```

To display help information about a specific command, use **/help** followed by that command. For example:

```
/help join
```

As a new user, there are two special **/help** commands to use. To display information of interest to beginners, enter:

```
/help newuser
```

To display a short introduction to IRC, use:

```
/help intro
```

When you first connect to IRC, you will not be in a channel, and any conversation that you type will be discarded. The first thing you will want to do is decide which channel to join. To display a list of all the channels, enter:

```
/list
```

For each channel, you will see the name, the number of users, and the topic (if one is set).

Hint

Some commands—such as **/list**, **/who ***, or **/whois *** (both described later)—can display a long list of information that seems to go on and on. To stop the flow, you can enter the command:

```
/flush
```

However, have patience. This command does not always take effect right away. The best thing is to avoid using commands that produce so much output.

To make the channel list more manageable, you can ask to see only those channels that have a minimum number of participants, by using **-min** followed by a number. For example, to list only those channels that have at least fifteen people, use:

```
/list -min 15
```

You can also use **-max** to display channels that have no more than a specified number of participants:

```
/list -max 5
```

To display information about a particular channel, you can specify its name. For example:

```
/list #hottub
```

Aside from these variations, there are other options that you can use with **/list**. For more information, use **/help list**.

Once you decide what channel looks good, you can join by using **/join** followed by the name of the channel. Remember, most channel names start with a # character, for instance **#hottub**. Thus, to join this channel, you would enter:

```
/join #hottub
```

From now on, everything you type (except commands) will be sent to that channel. If you are composing a long message, it is not necessary to keep pressing RETURN. IRC will allow you to type up to 256 characters in a single message.

Hint

If you are not sure which channel to join, **#hottub** is a good place to start. Think of it as a large group of people, sitting around in a hot tub, talking about anything that enters their minds. (You will need to supply your own water.)

To leave a channel, type **/leave** followed by the name of the channel. For example:

```
/leave #hottub
```

When you enter a **/join** command to join a new channel, IRC will automatically remove you from your current channel. Once you are more experienced, you can issue the command:

```
/set novice off
```

This changes several things about your IRC session. In particular, it allows you to join more than one channel at the same time. (For more information, use **/help set novice**.)

FUN TIP: Once you use **/set novice off***, you can join as many channels as you want. However, if you are trying to follow too many conversations at once, you will probably fry your brain.*

From time to time, you will see a message, called an *action*, that says that someone has done something. For example, you might see:

```
* Nipper is disgusted.
```

Actions are used to imply feelings and ideas where regular messages would be inadequate. To display an action, you use the **/me** command. Type **/me** followed by your action. Everyone on the channel will see your nickname, followed by the action. For example, if your nickname is **Nipper** and you enter:

```
/me breaks the glass.
```

Everyone (including you) will see:

```
* Nipper breaks the glass.
```

Nicknames

When you first connect to IRC, you can specify a nickname (which can be up to nine characters). At any time, you can change your nickname by typing **/nick**, followed by the new nickname. For example, if you nickname is **Irishboy** and you want to change it to **MikeP**, use:

```
/nick MikeP
```

When you do, IRC will notify everybody in your channel. You will all see a message like:

```
*** Irishboy is now known as MikeP
```

When you specify a nickname, you can type upper- or lowercase letters, and IRC will remember what you use. Whenever IRC displays your nickname, it will be just as you specified it. However, when you type a nickname as part of a command, you can use all lowercase letters, which is a lot easier to type.

When you specify a new nickname, IRC checks it against all the nicknames currently in use (ignoring any differences due to upper- and lowercase). If someone is already using the same name, you will be asked to select another one. If the nickname is registered to someone else, but is not currently in use, you will be allowed to use it, but you will see a warning. In such cases, the polite thing to do is to change to another nickname, so the person to whom the nickname is registered can use it if he or she connects.

If you enter the command by itself:

```
/nick
```

you will see your current nickname. If you would like to see the last person who used your current nickname, enter:

```
/whowas
```

To find out who was the last person to use a particular nickname, use **/whowas** followed by that name. For example:

```
/whowas nipper
```

IRC nicknames are maintained by an automated service called *Nickserv*. To learn how to use Nickserv to register your own nickname, enter:

```
/msg nickserv@service.de intro
```

This command sends a private message to the Nickserv program, asking it to display the introductory information.

Commands for Talking to Specific People

To send a private message to someone, use **/msg** followed by his or her nickname, followed by the message. For example:

```
/msg dizney What's up guy?
```

The message will be seen only on that person's screen.

When you send private messages, you do not have to be on the same channel as the recipient. Wherever the person is, he or she will get the message. (That is one reason why no two people can use the same nickname at the same time.)

As you look at your screen, normal messages are prefaced by the nickname of the originator, enclosed in **< >** characters. For example:

```
<Addie> Hi Nipper, how are you?
```

When someone sends you a private message, the nickname will be marked with ***** characters instead. For example:

```
*Addie* What are you working on?
```

When you send a private message to someone, it will show up on your screen with his or her nickname preceded by the characters **->**. For example, if you send a message to the person whose nickname is **Addie**, what you see on your screen looks like this:

```
-> *Addie* I am working on Chapter 20.
```

Hint

Because the world of IRC is so open and public, it is necessary to maintain a protective aura of anonymity. At any time, there may be thousands of people talking to one another, but each particular person is known only by a nickname. For this reason, many people consider it good manners to refer to a person only by his or her nickname, even if you know their real name.

When you use **/msg**, there are two shortcuts available. First, if you use a **,** (comma) character instead of a nickname, the message will go to the last person who sent you a private message. For example, if the last person who sent you a private message was **Nipper**, the following two commands would be equivalent:

```
/msg nipper What's up guy?
/msg , What's up guy?
```

Alternatively, if you use a **.** (period) character instead of a nickname, the message will go to the last person to whom you sent a private message. For example:

```
/msg . Did you get my last message?
```

If you want to send a private message to more than one person, you can specify more than one nickname. However, you must separate the names with a comma and not put in any spaces:

```
/msg nipper,addie  Hello you two.
```

Remember that any line that starts with a **/** (slash) character is interpreted as a command, and is not sent to the channel. Similarly, any line that does not begin with a **/** is considered to be part of the general conversation and is sent to the channel.

Thus, if you type a **/msg** command, but accidentally put a space in front of the **/** character (or omit the **/** character), the line would be broadcast for everyone to see.

*FUN TIP: To make new friends, join a channel with a lot of people and then type the following line. Be sure to put a space in front of the **/** character.*

```
/msg . sounds great!  Shall I bring my whip with me too?
```

If you find yourself sending more than a few **/msg** lines to the same person, you can simplify things by using the **/query** command followed by one or more nicknames. As with **/msg**, you must separate the nicknames with a comma and not put in extra spaces.

This command tells IRC that, until further notice, everything you type (except commands, of course) should be sent to the specified people. Here are two examples:

```
/query addie
/query addie,nipper
```

The first command says that all our messages should go only to **addie**. The second command sends messages to both **addie** and **nipper**.

When you are finished with your private conversation, you can turn it off by using the command without a nickname:

```
/query
```

Now all your messages will go to your channel.

Hint

Even though you may be sending private messages with **/msg** or **/query**, your screen will still show all the other messages that are being sent to the channel, which can be distracting. If you want a quiet conversation with someone, ask the person to join you on another channel. For example, you might type:

```
/msg addie Let's talk more quietly. Join me on '#foobar'.
```

If you specify a channel name that is unused, you will have your own private channel. After you are both on the channel, you can keep it private by entering the command:

```
/mode * +pi
```

For more information about the **/mode** command, use **/help mode**.

Displaying Information About People

The last few commands that we will mention show you information about people. First, to display information about a particular person, use **/whois** followed by the nickname, for example:

```
/whois nipper
```

You also can display information about a person by using the **/who** command:

```
/who nipper
```

However, **/whois** is better as it shows you more information.

Where **/who** comes in handy is in showing you information about all the people in a particular channel. For example, to find out who has joined channel **#hottub**, use:

```
/who #hottub
```

If you want to see a list of everybody who is joined to your current channel, you can enter:

```
/who *
```

Hint

If you have not joined a channel, some IRC clients will interpret the command **/who *** as meaning that you want a list of everybody who is using IRC. Similarly, if you type **/whois ***, IRC will display information about everybody who is connected.

Do not do this, for the list will go on forever (or until you die, whichever comes first). If you do find yourself in this situation, remember that you can use **/flush** to stop the rest of the output (eventually).

Summary of Basic IRC Commands

For reference, Table 20-1 contains a summary of the basic IRC commands that we have covered in this chapter. These commands are enough to get you started. However, there are a lot more commands to learn and a lot of surprises waiting for you (like robots and the flower server).

Command	Description
/flush	throw away remaining output for current command
/help	display a list of all IRC commands
/help *command*	display help regarding specified command
/help intro	display an introduction to IRC
/help newuser	display information for new users
/join *channel*	join the specified channel
/leave *channel*	leave the specified channel
/list	display information about all channels
/list *channel*	display information about specified channel
/list -min *n*	display channels that have at least *n* people
/list -max *n*	display channels that have no more than *n* people
/me *action*	display specified action
/mode * +pi	make current channel completely private
/msg *nicknames text*	send private message to specified people
/msg , *text*	send message to last person who sent you a message
/msg . *text*	send message to last person you sent a message to
/nick	display your current nickname
/nick *nickname*	change your nickname to specified name
/query *nicknames*	send all your messages to specified people
/query	stop sending private messages
/quit	quit IRC
/set novice off	allow certain actions (like joining multiple channels)
/who *channel*	show who is joined to specified channel
/who *nickname*	show information about the specified person
/who *	show who is joined to the current channel
/whois *nickname*	show all information about the specified person
/whois *	show all information about everybody

Table 20-1. *Summary of Basic IRC Commands*

CHAPTER 21

The Gopher, Veronica, and Jughead

If we could use only one Internet resource, it would be the Gopher. There are two reasons. First, the Gopher is easier to use than anything else on the Internet. Second, the Gopher allows you to access a larger variety of information and services than anything else on the Internet.

Does this mean that the Gopher is the ultimate Internet information tool? No, because there are many ways to think about information, and we need a variety of tools to serve a variety of people. We recommend that you learn to use all the basic Internet services so that, whatever your needs, you are always able to use the best tool for the job. For example, it is possible to access Anonymous FTP files using the Gopher, but most of the time it is easier to use FTP directly.

Still, the Gopher is the closest thing we have to a one-size-fits-all service bureau and, as such, it provides a unique window into the Internet. Moreover, when you have spare time, the Gopher is probably the best tool for browsing for those unexpected and exotic items that make life on the Internet so interesting.

What Is the Gopher?

The Gopher is a powerful system that allows you to access many of the resources of the Internet in a simple, consistent manner. To use the Gopher, all you need to do is make selections from a menu. Each time you make a selection, the Gopher does whatever is necessary to carry out your request. For example, if you select a menu item that represents a text file, the Gopher will get that file—wherever it happens to be—and display it for you.

Some menu items represent other menus. If you choose one of these, the Gopher will obtain the new menu and display it for you. Thus, you can move from menu to menu, using only a handful of keys (or a mouse) to navigate.

The power of the Gopher lies in the fact that the resources listed in a menu may be anywhere on the Internet. When you select an item, the Gopher will get or do whatever is necessary to carry out that request. Much of the time, the Gopher will have to connect to another computer, but it will all be transparent to you. All you will notice is that your request has been fulfilled simply and easily.

How Does the Gopher Work?

The Gopher is a client/server system. (We discuss clients and servers in Chapter 2.) To use the Gopher, you run a program called a *Gopher client*. This program displays the menus for you and carries out your requests. From time to time, your Gopher client contacts a *Gopher server* to ask for information on your behalf. If it becomes necessary to contact another type of service—say, to set up a Telnet session, or to download a file—your Gopher client will take care of that as well.

There are several thousand Gopher servers around the Internet, and they store all manner of information. You will find Gopher servers in many universities, companies, and other organizations. Indeed, within a university, you will often find separate departments that have their own Gopher server.

Each Gopher server carries information that is of interest to the local users. For example, the Gopher server for a university department would have information that is of interest to the members of that department. The Gopher server at a company would have information for its customers and employees.

What is great about this system is that, for the most part, all the Gopher servers on the Internet are public. Although most Gopher servers are set up for a particular set of people, you will find that much of the information you encounter is of general interest.

Travelling Through Gopherspace

The Gopher offers more information than any single person can comprehend. We refer to the sum total of all this information—everything that is available via the Gopher—as *gopherspace*. To help you get a feeling for what travelling through gopherspace is like, here is a typical scenario that describes an everyday excursion.

In our example, you are a student at the Unix Studies Department at the University of Foobar. Your department, as well as others at the university, has its own Gopher server. To use the Gopher, you start your Gopher client program (we will explain how later). Your Gopher client connects to the local Gopher server and asks for the initial menu. The Gopher server sends this menu, which your Gopher client displays on your screen.

```
        Internet Gopher Information Client v1.30


   University of Foobar, Unix Studies Department Gopher Server


  -->   1.   About the Unix Studies Department.
        2.   Where to buy the book "A Student's Guide to Unix".
        3.   Search the Online Unix Manual <?>
        4.   The Internet Studies Department Gopher/
        5.   Gopher Servers at the University of Foobar/
        6.   Other Gopher Servers Around the World/
        7.   University of Foobar Directory <CSO>
        8.   The Unix Daemon <Picture>
        9.   University of Foobar Library Catalog <TEL>
       10.   The Sound of Unix <)
       11.   Fun and Games/


   Press ? for Help, q to Quit, u to go up a menu         Page: 1/1
```

You decide that you want to know more about the Unix Studies department, so you select item number 1. Your Gopher client contacts the Gopher server and asks for the information. As you wait, you see the message:

```
Receiving Information..
```

A moment later, the Gopher server has sent back a short description of the department, which your Gopher client displays for you. After you are finished reading, you see the message:

```
Press <RETURN> to continue, <m> to mail, <s> to save, or <p> to print:
```

You press RETURN, and your Gopher client re-displays the previous menu.

This time, you decide to see what another Gopher server might have to offer, so you select item number 6. Your Gopher client sends the request to the Gopher server, which sends back a new menu.

This menu contains a long list of many different Gopher servers. Your Gopher client displays the first screenful, which it calls a page. You move through the list, page by page. On the 37th page, you see the following:

```
660. West Virginia Network for Educational Telecomputing/
661. Western Illinois University, IL USA/
662. Wheaton College, Wheaton, IL/
663. Whole Earth 'Lectronic Magazine--The WELL/
664. Wisconsin Interlibrary Services/
665. Wittenberg University/
666. Worcester Foundation for Experimental Biology/
```

You decide to check out The Well, so you select item number 663. Your Gopher client connects to the Well's Gopher server—which happens to be in Sausalito, California, just north of San Francisco—and The Well's Gopher server sends its main menu to your Gopher client. Your Gopher client displays this menu. As you look at it, you notice the following item:

```
11. Publications (includes Zines)/
```

You happen to know that a "zine" is a small, personal, opinionated publication, devoted to a single group or topic, usually far from the center of mainstream culture. This item looks intriguing. You select it and your Gopher client sends the request to The Well's Gopher server, which responds by sending back another menu. Your Gopher client displays this menu. As you expected, you see a lot of weird-looking items, among which you notice:

> 10. Online Zines/

This looks even more intriguing: zines in electronic form that you can read right on your screen. You select this item and your Gopher client again passes the request to The Well's Gopher server. Once again, the Gopher server sends back a menu which your Gopher client displays. This menu contains a list of online zines. You select item number 16:

> 16. Obscure Electronic #4

Your Gopher client sends in the request. The Gopher server sends back its response: a copy of a zine called *Obscure Electronic* edited by Jim Romenesko of Milwaukee, Wisconsin. This issue consists of a long profile of Dan Kelly from Chicago, Illinois, who publishes his own zine called *EVIL: The Newsletter for True Crime Book Fanatics*. Your Gopher client displays the zine, one page at a time. As you read, you are fascinated to learn all about Kelly's long-time interest in serial killers. When you are finished, your Gopher client displays the message:

> Press <RETURN> to continue, <m> to mail, <s> to save, or <p> to print:

Since the article was so interesting, you decide to mail a copy to the President of the United States (his address is **president@whitehouse.gov**). You press the **m** key and your Gopher client asks you to enter a mail address. You enter the address and press RETURN. Your Gopher client mails a copy of the article as you requested. On second thought, you decide that you would like to have a copy for yourself, so you repeat the procedure and enter your own mail address.

You now move back up one level and display the previous menu (the one about online zines). This menu proved to be so useful that you use a special command to tell your Gopher client to remember the menu, so you can get to it quickly whenever you want.

Your Gopher client makes a note called a *bookmark*. Any time you want, you can ask your Gopher client to display your personal *bookmark list*. You can then select this item and have your Gopher client go to it directly, without having to make your way through so many servers and menus like you did the first time.

And now, feeling pleasantly exhausted from your travels through the idyllic islands of gopherspace, you bid a fond farewell to the gracious and friendly natives at The Well in Sausalito, California, and to the two wise men in Milwaukee and Chicago. As the sun sinks slowly over your gopherspace paradise, you press the **Q** (quit) key and stop your Gopher client.

Musing on the Gopher

As you will have noticed from the last section, your Gopher client allows you to move from one Gopher server to another, smoothly and easily. For this reason, we like to consider the hundreds of interconnected Gopher servers to be one large distributed entity; thus, we speak of "the Gopher", with its many small parts, spread throughout the Internet. The Gopher is always growing, always changing. It's everywhere, but you can't locate it; it's always there, but you can't see it.

FUN TIP: *The Gopher is the largest and most practical example of applied pantheism in the history of mankind.*

Thus, the existence of gopherspace depends upon two completely different facilities. First, there is the Gopher itself: a vast information-based lifeform consisting of many interconnected Gopher servers.

Second, there is your Gopher client: the program you use to visit gopherspace and to interact with the Gopher. For this reason, when we talk about "your Gopher", we mean your client program.

You might be wondering, where did the name "Gopher" come from? Certainly one can imagine an electronic version of a furry little creature who burrows through gopherspace, seeking out treasures on your behalf. Still, the analogy is not apt, for it is neither the Gopher nor your Gopher client that does the burrowing: it is your mind, jumping from topic to topic as you read the menus and make your choices. No, the actual origin of the name "Gopher" is more prosaic, but not without its points of interest.

The original Gopher was developed at the University of Minnesota in April of 1991 by a team consisting of Bob Alberti, Farhad Anklesaria, Paul Lindler, Mark McCahill, and Daniel Torrey. The work was done within the Department of Computer and Information Systems in order to provide a cheap and easy way for various campus departments to make information available to the campus at large. The idea was—and still is—that each interested organization can maintain their own Gopher server and put whatever they want on it. Thus, every organization has control over its own section of gopherspace, which it shares with anyone who has a Gopher client.

And what about the name "Gopher"? It happens that Minnesota is known as the Gopher State. For example, the sports teams at the main campus of the University of Minnesota are named the Golden Gophers. No one knows exactly why Minnesota is the Gopher State—the most widely accepted theory is that the inventor of Minnesota (Phineas T. Bushbottom) actually resembled a rodent—still, there it is.

Moreover, there is a serendipitous association between the name "Gopher" and the American slang expression "gofer". This epithet describes a person whose position in life is not unlike that of a Gopher client: to go for this and go for that.

In the charming argot of the American midwest, "go for" becomes "gofer" which becomes "Gopher". Indeed, this is probably why so many tourists are astonished to find, upon visiting Minnesota, that the natives are so cooperative when it comes to running errands for other people.

Starting Your Gopher Client

There are a variety of Gopher clients available for different systems. The most widely used Gopher client is the one that runs on Unix systems with regular text-based terminals. The name of this program is **gopher**. There is also a Gopher client, named **xgopher**, for X Window systems. (We discuss X Window in Chapter 2.)

If you have a choice, use a Gopher client that takes advantage of the special features of your computer. For example, if your system uses pull-down menus, scroll bars, and a mouse, a Gopher client tailored to your computer will make use of these features.

If your system has a Gopher client installed, all you need to do is enter the name of the program to start it. For most Unix users, this program will probably be **gopher**. So to start your Gopher client, all you need to enter is:

```
gopher
```

If you are using an X Window system, you can certainly run **gopher**, but you may prefer **xgopher** if it is available.

You will remember that in the last section we explained that you can save locations in gopherspace to a menu of your own. Each location you save is called a bookmark. If you would like to start your Gopher client with the menu that holds your personal bookmarks—rather than with the main menu of the default Gopher server—specify the **-b** (bookmark) option when you start the program:

```
gopher -b
```

From time to time, you may have the address of a particular Gopher server that you want to use. If so, you can enter the address as part of the **gopher** command. This tells your Gopher to connect directly to the specified server instead of using whatever Gopher server is the default.

For example, in the catalog in the back of this book, there is a Gopher at Northwestern University that specializes in aviation. The address of the Gopher is **av.eecs.nwu.edu**. If you want to connect directly to this particular Gopher server, you can enter:

```
gopher av.eecs.nwu.edu
```

Some Gopher servers require that you specify a particular port number (see Chapter 7). For example, the catalog mentions a Gopher at the University of Tuebingen in Germany that contains documents about MUD (Multiple User Dimension role-playing games). The address is listed as **nova.tat.physik.uni-tuebingen.de 4242**. The **4242** is the port number. Simply type it as part of the command:

```
gopher nova.tat.physik.uni-tuebingen.de 4242
```

If you enter the **gopher** command and you see the following message:

```
gopher: Command not found.
```

it means that your system does not have a Gopher client installed. You have several choices, which we will discuss in the next section.

If you are using a Unix system and you do have a Gopher client, you can display the official documentation about the **gopher** command by using the **man** command to access the online Unix manual:

```
man gopher
```

(Note: This will display documentation about the **gopher** command, not about the Gopher system in general.) You can also use **man** to display documentation about the **xgopher** command, if it is available on your system:

```
man xgopher
```

Once you start your Gopher, it will contact whatever Gopher server is designated as the place to start and ask for the main menu. If your organization has its own Gopher server, this, no doubt, will be where you will start. Otherwise, your system manager will have chosen some other Gopher server, somewhere on the Internet, as the starting place. Once you see the main menu, you are ready to begin: simply make your first selection.

If you want to learn more about the Gopher, there are several places to look. First, many Gopher servers have menu items that display information about the Gopher itself.

Second, there are two Usenet newsgroups devoted to Gopher-related discussions: **comp.infosystems.gopher** and **alt.gopher**.

Finally, you can read the Gopher frequently asked question (FAQ) list. You can obtain this document via Anonymous FTP from **rtfm.mit.edu**. Change to the directory **/pub/usenet/news.answers** and download the file named **gopher-faq**. (We discuss Usenet newsgroups and FAQ lists in Chapter 9. We discuss Anonymous FTP in Chapter 16.)

Public Gopher Clients

If your system does not have a Gopher client installed, you have two choices. In the long run, the best solution is to install a Gopher client. Ask your system manager if he or she would be so kind as to download the program via Anonymous FTP and install it for everybody. The location of where to find such clients is given in the Gopher FAQ list.

The second choice is to use a public Gopher client. All you need to do is telnet to one of the hosts in Table 21-1. Once you log in using the specified userid, a Gopher client will start automatically. (We discuss Telnet in Chapter 7.)

Hint

Using a public Gopher client will somewhat restrict what you can do. For example, after your Gopher has displayed a text file, you will see the following message:

```
Press <RETURN> to continue, <m> to mail, <s> to save, or <p> to print:
```

If you are using a Gopher client on your own computer, you can use all three options. However, if you have telneted to a remote Gopher client, you can only use mail (because you cannot save or print files on a remote computer).

You may also find that, as a remote user, you are denied other privileges, such as being able to initiate a Telnet session.

Location	Internet Address	IP Address	Log in as...
Australia	info.anu.edu.au	150.203.84.20	info
Chile	gopher.puc.cl	146.155.1.16	gopher
Denmark	gopher.denet.dk	129.142.6.66	gopher
Ecuador	ecnet.ec	157.100.45.2	gopher
England	gopher.brad.ac.uk	143.53.2.5	info
Germany	gopher.th-darmstadt.de	130.83.55.75	gopher
Japan	gopher.ncc.go.jp	160.190.10.1	gopher
Spain	gopher.uv.es	147.156.1.12	gopher
Sweden	gopher.chalmers.se	129.16.221.40	gopher
Sweden	gopher.sunet.se	192.36.125.2	gopher
USA: California	infopath.ucsd.edu	132.239.50.100	infopath
USA: California	scilibx.ucsc.edu	128.114.143.4	gopher
USA: Georgia	grits.valdosta.peachnet.edu	131.144.8.206	gopher
USA: Illinois	gopher.uiuc.edu	128.174.5.61	gopher
USA: Iowa	panda.uiowa.edu	128.255.40.201	—
USA: Michigan	gopher.msu.edu	35.8.2.61	gopher
USA: Minnesota	consultant.micro.umn.edu	134.84.132.4	gopher
USA: North Carolina	gopher.unc.edu	152.2.22.81	gopher
USA: North Carolina	twosocks.ces.ncsu.edu	152.1.45.21	gopher
USA: Ohio	gopher.ohiolink.edu	130.108.120.25	gopher
USA: Virginia	ecosys.drdr.virginia.edu	128.143.96.10	gopher
USA: Virginia	gopher.virginia.edu	128.143.22.36	gwis
USA: Washington	wsuaix.csc.wsu.edu	134.121.1.40	wsuinfo

Table 21-1. *Public Gopher Clients to Which You Can Telnet*

Basic Gopher Commands

There are a variety of commands that you can use to negotiate gopherspace. Most of the time, you can go wherever you want by using just six keys: the four cursor control keys (with the arrows) RIGHT, LEFT, DOWN, and UP; the SPACE key; and the **B** key. We will go into the details in a moment.

There are three basic commands that you should always remember. First, to display help information about all the commands, press **?** (question mark).

Second, to stop your Gopher client, press **q** (quit). You will be asked if you really want to quit, to which you can answer **y** (yes) or **n** (no).

Finally, to quit immediately, without being asked for confirmation, press **Q** (uppercase "Q").

Your Gopher works in what is called *cbreak mode*, which means that, for single character commands, you do not have to press the RETURN key. For example, to stop your Gopher, you need only press the **q** key. Do not press RETURN.

Now, let's take a look at a typical Gopher menu:

```
              Internet Gopher Information Client v1.30

     University of Foobar, Unix Studies Department Gopher Server

  -->  1.  About the Unix Studies Department.
       2.  Where to buy the book "A Student's Guide to Unix".
       3.  Search the Online Unix Manual <?>
       4.  The Internet Studies Department Gopher/
       5.  Gopher Servers at the University of Foobar/
       6.  Other Gopher Servers Around the World/
       7.  University of Foobar Directory <CSO>
       8.  The Unix Daemon <Picture>
       9.  University of Foobar Library Catalog <TEL>
      10.  The Sound of Unix <)
      11.  Fun and Games/


  Press ? for Help, q to Quit, u to go up a menu          Page: 1/1
```

At all times a pointer, shown as -->, will be beside one of the items. In this case, it is pointing to item number 1. The idea is to move the pointer to the item you want and then select it. To select an item, you can press either RIGHT or RETURN.

If you select another menu, your Gopher will get it for you and display it. If you select a text file, your Gopher will get a copy and display it one page at a time. For other types of menu items, your Gopher will do whatever is appropriate. (We will have more to say about this later in the chapter.)

Moving Around a Gopher Menu

As we mentioned, your current position within a menu will be marked by a pointer. There are several ways to move this pointer. The easiest way is to use the UP and DOWN keys. If your keyboard does not have these keys, or if you prefer not to use them, you can use **k** or CTRL-P to move up, and **j** or CTRL-N to move down. Although these choices may seem odd, they are used within the **vi** text editor and are second nature to a great many Unix users.

If you would like to jump to and select a particular item, simply enter its number. For example, to select item number 10, type **10** and then press RETURN.

Some menus are so long that they will not fit onto a single page. In this case, the message at the bottom right will tell you that there is more than one page. For example, you might see:

```
Press ? for Help, q to Quit, u to go up a menu          Page: 1/7
```

To move to the next page, press SPACE. To move to the previous page, press **b** (back). Again, there are alternative keys to use. To move to the next page, you can press > (greater than) or + (plus). To move to the previous page, you can press < (less than) or − (minus).

If a menu is so long as to require more than one page, you can jump to and select an item, even if it is not on the current page. For example, say that you are looking at page 7 of a long menu. You can select item number 1 directly. Simply type **1** and press RETURN.

Another way to jump is to tell your Gopher to search the current menu for an item that contains a particular pattern. To do so, type a / (slash) character, followed by the pattern, and then press RETURN. When it searches, your Gopher does not distinguish between either upper- or lowercase letters.

For example, say you are looking at a very long list of Gopher servers, and you would like to jump directly to the item for the University of Foobar. Enter:

```
/foobar
```

If the item your Gopher finds is not the one you wanted, you can repeat the same search by pressing the **n** (next) key.

As you know, many of the items you select will themselves be menus. When you select such an item, your Gopher will request the menu and then display it for you. Thus, travelling through gopherspace entails moving up and down a tree of menus.

There are a few commands that you can use to move from one menu to the next. As we mentioned above, to select a new menu, point to that item and press RIGHT or RETURN. To return to the previous menu, press LEFT or **u** (for up).

Now you can see the beauty of the system. Although there are many Gopher commands, all you really need to navigate gopherspace are UP and DOWN (to move within a menu), and RIGHT and LEFT (to move from one menu to another). To this basic command set, you can add SPACE and **b** to page forward and backward through long menus.

At any time, you can press **m** to jump to the main menu (the one your Gopher started with). This is handy when you are far down a series of menus and you want to jump back to the beginning.

Saving a File

There are several ways to save a file. As we mentioned earlier, once your Gopher has displayed a file, you will see:

```
Press <RETURN> to continue, <m> to mail, <s> to save, or <p> to print:
```

You can now mail a copy of the file to someone, save it, or print it.

There is a shortcut for saving a file that does not require you to display it first. Simply move to the menu item and press **s** (save). You will then be asked to supply a file name. Your Gopher will suggest a file name that you can accept or modify as you wish. For example, say that the current menu item is:

```
--> 16. Obscure Electronic #4
```

You decide to save it to a file. To do so, you press **s**. You now see:

```
+-----------------------------------------------------------------+
|                                                                 |
| Save in file:    Obscure-Electronic-#4                          |
|                                                                 |
|                        [Cancel ^G] [Accept - Enter]             |
|                                                                 |
+-----------------------------------------------------------------+
```

You can now either press RETURN to save the file, or press CTRL-G to forget the whole thing. Before you save, you can change the name by typing a new one.

Saving has two limitations. First, the current item must be something that would be meaningful if it were saved in a personal file. For example, you can save a text file, but you cannot save a menu. Second, you must be using a Gopher client that is running on your own computer. This makes sense as you do not have permission to save files on a remote host.

Another way to save a file is by downloading. This is handy when you are using a PC or Macintosh to connect to a remote Internet host. You can use the **D** (uppercase "D") command to download an item directly to your computer. (We discuss the basic PC-to-host connection in Chapter 3. We discuss downloading in Chapter 18.)

When you press **D**, you will see a menu asking you to choose which file transfer protocol (download method) you want to use:

```
+---------Obscure Electronic #4------------+
|                                          |
| 1. Zmodem                                |
| 2. Ymodem                                |
| 3. Xmodem-1K                             |
| 4. Xmodem-CRC                            |
| 5. Kermit                                |
| 6. Text                                  |
|                                          |
| Choose a download method:                |
|                                          |
| [Cancel ^G]   [Choose 1-6]               |
|                                          |
+------------------------------------------+
```

After you make a choice, your Gopher will initiate the file transfer. Depending on how your communication program is set up, you may then see a message telling you to start your download. (The whole thing works just like we described in Chapter 18.)

Customizing Your Gopher Environment

As we explained earlier, you can use a bookmark to save a menu item. Your Gopher stores all your bookmarks in a bookmark list that you can jump to whenever you want. This allows you to remember and visit any location in gopherspace. If you are using a Gopher client on your own machine, your bookmark list will be saved automatically, so you can use it each time you explore gopherspace.

There are several commands that you can use to create your personal bookmark list. To add a menu item to your list, move to it and press **a** (add). To add the entire menu to your bookmark list, press **A** (uppercase "A").

At any time, you can jump to your bookmark list (which is itself a menu) by pressing **v** (view). To return to the previous menu, simply press LEFT or **u**.

While you are looking at your bookmark list, you can delete an item by moving to it and pressing **d** (delete).

> ## Hint
>
> The art of using the Gopher well is to build up a personalized bookmark list. Many people forget to use this facility and end up retracing their steps time and time again. Moreover, it is all too easy to forget where in gopherspace you found a particular item. If you do not save interesting items to your bookmark list, you may never see them again.

The next command displays technical information about any menu item. Simply move to an item and press = (equal sign). You will see the technical description of the item as it appears in the Gopher server's database.

For example, say that you are pointing to the following item:

```
  --> 1.  CIA World Fact Book/
```

If you press =, you will see something similar to:

```
Type=1
Name=CIA World Fact Book
Path=1/info/Government/Factbook
Host=info.umd.edu
Port=901
```

You don't really need to worry about the details. What is interesting, though, is that you can see where in gopherspace the item is located. In this case, the item is stored on **info.umd.edu**, a computer at the University of Maryland in the United States.

The last command that we will mention is **O** (uppercase "O"). This tells your Gopher to display a list of options. You can examine the list and make changes as you wish. Most of the time, you will probably not want to make any changes, so if you don't understand what something does, you are best off leaving it alone.

Summary of Gopher Commands

For reference, Figure 21-1 contains a summary of the Gopher commands that we discussed. Remember, you can display a command summary at any time by pressing **?**.

Basic Commands

Q	quit Gopher immediately
q	quit Gopher, but ask for confirmation
?	display a help summary
=	display technical information about an item
O	examine and change Gopher options

Fundamental Commands to Move Through Gopherspace

RIGHT	select the current item
LEFT	back up one level to the previous menu
UP	move pointer up one item
DOWN	move pointer down one item
SPACE	move to next page of the menu
b	move back to previous page of the menu
number	jump to and select specified item
/pattern	search for next menu item containing pattern
n	search for next menu item using same pattern
m	jump to the main menu

Alternate Commands for Moving the Pointer

RETURN	select the current item
u	back up one level to the previous menu
k	move pointer up one item
j	move pointer down one item
CTRL-P	move pointer up one item (previous)
CTRL-N	move pointer down one item (next)
>	move to next page of the menu
<	move to previous page of the menu
+	move to next page of the menu
-	move to previous page of the menu

Saving Information

s	save current item to a file
D	download current item to a file

Using Bookmarks

a	add current item to the bookmark list
A	add current menu or search to the bookmark list
d	delete a bookmark
v	jump to the bookmark list (view)

Figure 21-1.   *Summary of Gopher commands*

Types of Gopher Resources

Let's take another look at a typical Gopher menu. Notice that each item ends with a symbol. The symbol tells you what type of resource is represented. In this section, we will discuss the various resources that you will find and explain how to use them.

```
          Internet Gopher Information Client v1.30


    University of Foobar, Unix Studies Department Gopher Server


  --> 1.   About the Unix Studies Department.
       2.   Where to buy the book "A Student's Guide to Unix".
       3.   Search the Online Unix Manual <?>
       4.   The Internet Studies Department Gopher/
       5.   Gopher Servers at the University of Foobar/
       6.   Other Gopher Servers Around the World/
       7.   University of Foobar Directory <CSO>
       8.   The Unix Daemon <Picture>
       9.   University of Foobar Library Catalog <TEL>
      10.   The Sound of Unix <)
      11.   Fun and Games/


  Press ? for Help, q to Quit, u to go up a menu          Page: 1/1
```

The most common symbol that you will see at the end of an item is a / (slash) character. This indicates that the item represents another menu. For example, look at item number 4.

```
       4.   The Internet Studies Department Gopher/
```

If you choose this item, you will get the main menu for the Internet Studies Department Gopher server.

One thing you may have noticed is that the Gopher menu structure resembles the Unix file system. Menus and sub-menus correspond to directories and sub-directories. For this reason, you will sometimes see Gopher menus referred to as directories. For example, when you are waiting for a menu to be fetched from a remote Gopher server, you will see the message:

```
   Retrieving Directory...
```

(We discuss the Unix file system in Chapter 16.)

Hint

From time to time, you will notice that the Gopher is slow. For example, you may have to wait while your Gopher client connects to a remote Gopher server. Or you may initiate a database search (see below) that seems to go on and on.

In such cases, there is not much that you can do except wait. If you are a Unix user, you may be tempted to press CTRL-C, the key that is usually used to abort something. Unfortunately, all that will happen is that you will see a message asking if you want to abort the Gopher client itself. There is no simple way to stop a Gopher activity that is in progress.

The next most common symbol is a . (period) character. This indicates that the item is a text file: something to be displayed. For example, item number 1 represents a text file that contains information about the Unix Studies Department:

```
1.   About the Unix Studies Department.
```

When you select such an item, your Gopher will fetch it and display it for you, one page at a time.

Hint

If you are a Unix user using your own Gopher client, you can specify the paging program you prefer to use. To do so, set the **PAGER** environment variable to the name of your favorite paging program. For example, if you use the C-Shell, you could place the following command in your **.login** initialization file:

```
setenv PAGER more
```

If you use the Korn shell or Bourne shell, you could place the following two commands in your **.profile** initialization file:

```
PAGER=more
export PAGER
```

This will ensure that the **PAGER** environment variable is defined properly each time you log in. Moreover, this variable will be used by a number of programs (such as your mail program) and not only by your Gopher client.

The discussion of environment variables and initialization files is beyond the scope of this book. For more information, see a good Unix book. We recommend *A Student's Guide to Unix*, by Harley Hahn (McGraw-Hill).

The <TEL> symbol indicates a completely different type of resource: a Telnet session. In our example, we see:

```
    9.   University of Foobar Library Catalog <TEL>
```

If we select this item, our Gopher will initiate a Telnet connection to a remote host. In this case, it is a host that offers access to the University of Foobar library catalog. Just before the connection is made, you will see a warning telling you that, in order to make the connection, you are going to have to go outside the Gopher system:

```
+---------University of Foobar Library Catalog----------+
|                                                       |
| Warning!!!!!, you are about to leave the Internet     |
| Gopher program and connect to another host. If        |
| you get stuck press the control key and the ] key,    |
| and then type quit                                    |
|                                                       |
|                                                       |
|                        [Cancel - ^G] [OK - Enter]     |
|                                                       |
+-------------------------------------------------------+
```

What this message is telling you is that you are about to embark on a regular Telnet connection. To continue, press RETURN. To forget the whole thing, press CTRL-G.

Once you finish with the Telnet connection, you will be sent back to your Gopher client. However, during the session, whatever you type will be under the auspices of Telnet. In particular, you can use the Telnet escape key, CTRL-]. For more details on how to conduct a Telnet session and a discussion of the Telnet commands, see Chapter 7.

The next Gopher symbol is **<CSO>**. This indicates a facility called a *CSO name server* that contains information about the people at a particular organization. For example:

```
    7.   University of Foobar Directory <CSO>
```

In this case, you can select this item to search for information about someone at the University of Foobar. For example, you may need to find out someone's electronic mail address.

The name "CSO" refers to the Computing Services Office at the University of Illinois, Urbana, where the software was first developed. There are many CSO name servers on the Internet, and it is common to see one on the main Gopher menu of a university or other large organization. CSO name servers belong to a family of programs called "white pages directories" that are used to find someone on the Internet. (We discuss such services in Chapter 22.)

If you select a **<CSO>** item, you will see a screen asking you what to search for. The following example shows one such screen. If you use a number of CSO name servers, you will encounter different variations. However, in principle, they all work the same way.

```
+---------------University of Foobar Directory---------------+
|                                                            |
|name         _____                |
|email        _____                |
|department _____                  |
|                                                            |
|  [Switch Fields - TAB]           [Cancel ^G] [Accept - Enter]  |
|                                                            |
+------------------------------------------------------------+
```

Each of the categories—**name**, **email**, and **department**—is called a *field*. All you have to do is fill in whatever information you know and press RETURN. The CSO server will check its database and display what it finds as a single text file.

As you are typing, you can move from one field to another by pressing the TAB key. If you decide to forget about searching, simply press CTRL-G to return to the last Gopher menu.

The next symbol to discuss is **<?>**. This indicates a database that you can search by entering one or more keywords. For example:

```
    3.   Search the Online Unix Manual <?>
```

When you select this item, you will see a screen that asks you to specify what you want to search for. For example:

```
+---------------Search the Online Unix Manual---------------+
|                                                           |
|Words to search for _____             |
|                                                           |
|                    [Cancel ^G] [Accept - Enter]           |
|                                                           |
+-----------------------------------------------------------+
```

You fill in one or more keywords and press RETURN. Your Gopher will initiate the search. Once the results come back, your Gopher will display them for you in the form of a menu. Simply select whatever looks good.

Of course, the results of your search are only as good as the database you are searching and only as exact as the keywords you specify. One problem is that, unless the description of the item tells you, you have no way to know what type of search program and database you are using. If you are looking for files, you may be using Archie (Chapter 17). If you are searching plain text, you may be using Wais (Chapter 23).

Hint

You can use the = command to display technical information about a <?> menu item. This will sometimes give you a hint about what type of database is being searched. For example, you may see the word "wais" within the information.

The best thing to do is experiment. If you enter a keyword that doesn't get you what you want, try something else. Remember that Archie can only search for a single word, but Wais will search for more than one word.

Hint

It is often better to use Wais directly than through a Gopher client. This is because the Gopher client limits you as to what types of searches you can perform. When you use Wais directly, you can take advantage of all the Wais resources. For more information about Wais and how to use it, see Chapter 23.

If you like how a search turned out, you may want to save it to use again on another day. To do so, wait until the search is finished and use the **A** command. This will save the search (not its results) to your bookmark list. Whenever you want, you can jump to your bookmark list and select this exact search.

For example, some Gopher servers allow you to search news articles that arrive via Clarinet (see Chapter 9). You can create a specific search that checks the news for your favorite topic, and then save the search to your bookmark list. Now, each day when you start work, you can call up your bookmark list and use your preconfigured search to look for new articles.

The last two menu symbols we will discuss are **<Picture>** and **<)**. The **<Picture>** symbol indicates a binary file that contains a picture. The **<)** symbol indicates a binary file that contains a sound. (The **<)** is supposed to look like a stereo speaker.) Here are two examples showing these symbols:

```
 8.   The Unix Daemon <Picture>
10.   The Sound of Unix <)
```

You can only access such items if your computer has the appropriate hardware and software. If this is the case, your system manager will have specified which programs to use at the time your Gopher client was installed. All you have to do is select the item.

Hint

If you are connecting to an Internet host by using a PC or Macintosh that is able to display pictures or play sounds, you can use the **D** command (described earlier) to download such files to your own computer. Of course, your computer must be able to understand the particular file format that is being used. For example, you can't play sound files from a Next machine on your PC.

Using Veronica to Search Gopherspace

Veronica is a Gopher-based resource that you can use to search gopherspace for all the menu items that contain specified words.

For example, say that you are interested in math jokes. You know that somewhere out in gopherspace, there must be some menu items that contain math jokes. But where are they? Veronica can find out.

Accessing Veronica is easy. It is simply a menu item that you can select like any other. Look for an item similar to the following:

```
 2.   Search titles in Gopherspace using veronica/
```

You will often find Veronica under the same heading as a list of Gopher servers. For example, a good place to look would be within an item like:

```
 8.   Other Gopher and Information Servers/
```

> ## Hint
>
> You only need to find Veronica once. Once you do, you can place it in your personal bookmark list by using the **a** (add) command. At any time, you can use the **v** (view) command to display this list and then select Veronica.

Once you select a menu item that mentions Veronica, you will see a list of Veronica-oriented items. Here is a typical example:

```
         Search titles in Gopherspace using veronica

-->  1.   Search gopherspace using veronica at NYSERNet <?>
     2.   Search gopherspace using veronica at University of Pisa <?>
     3.   Search gopherspace for GOPHER DIRECTORIES  (NYSERNet) <?>
     4.   Search gopherspace for GOPHER DIRECTORIES  (U. Pisa) <?>
     5.   How to compose veronica queries.
     6.   FAQ:  Frequently-Asked Questions about veronica.
```

In this example, you would choose item 1 or 2 to make a general search of gopherspace. (It is usually best to choose the location that is closest to you.) Veronica will ask you to specify what you are looking for:

```
+--------Search gopherspace using veronica at NYSERNet---------+
|                                                              |
|Words to search for _____                     |
|                                                              |
|                    [Cancel ^G] [Accept - Enter]              |
|                                                              |
+-------------------------------------------------------------+
```

We happen to be looking for math jokes, so we type **math** and press RETURN.

Once you press RETURN, Veronica will search all of known gopherspace for menu items that contain the word math. These items will be selected and placed in a menu of their own which will be presented to you. Here is an example of a few such items:

```
-->  1.   sci-math-faq.

     2.   space_math.

     3.   AMATH1.ZIP - Animated Math: counting/addition/subtraction.

     4.   UCALC21.ZIP math expression evaluator.
```

The nice thing about Veronica is that it shows you its results in the same form as a regular Gopher menu. Thus, you can use the standard Gopher commands to examine the results of a Veronica search. If you like what you see, you can use the **a** command to add an item to your personal bookmark list.

When Veronica performs a search, it does not distinguish between upper- and lowercase letters. Thus, you get the same results searching for **math**, **Math**, or **MATH**. Indeed, you will notice that item number 3 above contains the word **Math**.

In our example, we are looking for math jokes, but our search was too general. Veronica has found all kinds of items that have to do with math, but are not funny. The best idea is to perform another search, this time looking for two words:

```
math joke
```

This search yields better results. Here are the first few items:

```
--> 1.  misc.math.jokes.
    2.  sci.math #6242 - Re: Math Jokes Needed
    3.  sci.math #6272 - Re: Math Jokes Needed
    4.  Re: Math jokes wanted ....
```

Evidently, we have found a number of Usenet articles (see Chapters 9 and 10). We are getting closer, but it would be nice to narrow down the search criteria even further.

To do so, Veronica allows us to specify certain qualifiers: **and**, **or**, and **not**. We can also group words in parentheses. Here are some examples.

When you use **and**, it tells Veronica to search for items that contain more than one word. By default, Veronica considers that you are using **and** whenever you specify more than one word. For example, the following two search lists are equivalent:

```
math jokes
math and jokes
```

You can use **or** to tell Veronica to search for items that contain any of the specified words. For example, to find all the items that contain either the word **jokes** or the word **humor** (or both), search for:

```
jokes or humor
```

You can use parentheses to indicate that a particular combination should be treated as a single word. For example, to find all the items that contain the word **math** and at least one of the words **jokes** or **humor**, search for:

```
math and (jokes or humor)
```

As we mentioned, Veronica will assume that you are using **and** by default. Thus, the last search is equivalent to:

```
math (jokes or humor)
```

Finally, you can use **not** in indicate that the item must not contain a specific word. For example, to find all the items that contain the words **math** and **jokes**, but not the word **wanted**, you would search for:

```
math jokes not wanted
```

There will be times when you want to search for (or not search for) words that are similar to one another. For example, you may want to search for items that contain **mathematics** as well as **math**, or **joke** as well as **jokes**. To do so, you can use a * (asterisk) character at the end of a word to represent any number of extra characters.

For example, to find all the items that contain any word that begins with **math** and any word that begins with **joke**, you would search for:

```
math* joke*
```

To find all the items that contain these same words but do not contain any word that begins with either **want** or **need**, search for:

```
math* joke* not (want* or need*)
```

The last thing that we will mention about searching is to remind you that we had two choices as to what type of Veronica search we wanted to use:

```
1.   Search gopherspace using veronica at NYSERNet <?>
3.   Search gopherspace for GOPHER DIRECTORIES  (NYSERNet) <?>
```

The regular search (number 1 in our example) finds all kinds of menu items. The directory search (number 3) only finds items that are themselves menus. If you are looking for something that you know is a menu, use a directory search. Not only will it be faster, but the results will contain fewer irrelevant items.

Hint

You may find that a Veronica search comes up with nothing when you know there must be something. For example, if you search for **math*** and Veronica cannot find anything, you should be suspicious.

There are several reasons why this may happen. The most common reason is that you have specified a bad search pattern. Ask Veronica to search again and, this time, re-check your spelling and your logic.

The second most common reason is that the remote Veronica host may not be working. If you suspect this is the case, try using a different Veronica host or try again later.

Learning More About Veronica

To learn more about Veronica, there are two places to look, both of which are accessible from most Veronica menus. First, you can read the article on "How to compose Veronica queries". This will explain all the details of how to tell Veronica what you want.

Second, you can read the Veronica frequently asked question list. This contains many questions and answers about Veronica.

To satisfy your curiosity, we will answer one interesting question right now: how does Veronica work?

The first version of Veronica was developed in November 1992. The work was done by Steven Foster and Fred Barrie of the University of Nevada at Reno, System Computing Department. Foster and Barrie were Gopher users who realized that the potential of the Gopher could never be reached unless there was an easy way to find items in gopherspace. They saw how frustrating and unproductive it was to hunt for specific items when you didn't know where to look.

For inspiration, they turned to Archie, which had been developed to provide a searchable database for the Internet's Anonymous FTP archives. They developed what they called a *Veronica server* to provide a similar function for the Gopher.

On a regular basis (every one to two weeks), their Veronica server would contact every known Gopher server and ask for a copy of all the menus. These menus were stored in a database. At any time, a user could use a Veronica search program to look for items in the database.

It wasn't long before Veronica became an integral part of the Gopher culture. Today, there are a number of Veronica servers around the Internet, and new work on Veronica is coordinated with mainstream Gopher development.

Of course, Veronica is not perfect. It is too easy to come up with a lot of irrelevant items or even no items at all. Some of this has to do with how Veronica works; some of this reflects the nature of gopherspace itself. After all, each Gopher server is maintained locally and there are no global standards as to how menu items should be named or organized. Nor should there be.

The strength of the Gopher—and the Internet as a whole—lies in the fact that no one person or organization is in charge. This makes for a fertile environment in which tools such as Anonymous FTP, Archie, the Gopher, and Veronica can develop. Of course, it also makes for a certain amount of disorganization and chaos.

Using Jughead to Search Gopherspace

As you know, Veronica is a tool that allows you to search all of gopherspace. Jughead is a similar tool, except that it searches only a confined area of gopherspace. Jughead was developed by Rhett (Jonzy) Jones at the University of Utah Computer Center and was first released on March 25, 1993.

The reason that Jughead is so important is that there are times when you will want to search only one small area of gopherspace. For example, if you are a user at a university, you may want to search only those Gopher servers within the university. Similarly, if you use a public access Internet service that maintains a large Gopher server, it is handy to be able to search only the menu items on that server.

The same thing applies to outside Gopher users. If you are interested in finding something, say, at a particular university, you would like to be able to connect to one of their Gopher servers and search all the gopherspace at the university.

When a system manager decides that his users would benefit from some type of limited gopherspace search, he sets up a *Jughead server*. The role of the Jughead server is to maintain a database of all the menu items within the part of gopherspace specified by the system manager. At any time, a user can use Jughead to search for whatever he wants within that limited area.

For example, let's say that someone is interested in books and libraries at a particular university. He might think about using Veronica to search for **book* librar***. However, there are three problems with this approach. First, it would find menu items throughout all of gopherspace, which would be too much. Second, Veronica has a built-in limit as to how many items it will find on a single search, and this limit may be exceeded before Veronica finds anything at the particular university. Third, it will take a while for Veronica to perform such a general search.

However, if the person uses Jughead, he can confine his search to the university. Moreover, the search will be faster than a Veronica search and will yield complete results. The only limitation is that someone must have already set up a Jughead server for that part of gopherspace.

How do you access Jughead? The same way you access Veronica: by selecting an item on a Gopher menu. The difference is that, where there are only a relatively few Veronica servers in the world, there are many Jugheads, each offering a service for a particular part of gopherspace. Here are some typical Jughead menu items:

```
4.   Use Jughead to search menus of University of Utah <?>
5.   Use Jughead to search menus of State of Utah <?>
```

Performing a Jughead search is a lot like searching with Veronica. You enter one or more words and Jughead looks for them. The results are returned to you as Gopher menu items, so you can select them in the regular manner. When you compose a Jughead search, you can use the special words **and**, **or**, and **not**. (For more information and some examples, see the previous discussion about Veronica.)

If you want to display help, all you need to do is use Jughead to search for **?** (question mark). You will then see an item for Jughead help, which you can display.

The Names Veronica and Jughead

The name Archie (Chapter 17) was used for the Anonymous FTP searching tool because it sounds like the word "archive". As Archie became popular, it became common to refer to it as if it were a person.

When Steven Foster and Fred Barrie needed a name for their Gopher search tool, they thought immediately of Archie Andrews: the perennial teenager who is the mainstay of a whole family of American comic books, comic strips, and cartoons.

In the comics, Archie has several friends, including Veronica and Jughead. Foster and Barrie chose Veronica. Later, Rhett (Jonzy) Jones chose Jughead.

If you are an Archie aficionado, you will know that Archie has other well-known friends, among them Betty, Reggie, and Moose. You might ask, will we soon see their names on the Internet? We think so.

Hint

As you now know, the names Veronica and Jughead were chosen because of their affinity to the Archie Andrews comics. Thus, there is no truth to the rumor that the name Veronica really stands for "Very Easy Rodent-Oriented Netwide Index to Computerized Archives". Nor does the name Jughead really mean "Jonzy's Universal Gopher Hierarchy Excavation And Display".

You will see these long names, but they were contrived after the fact. This may be disappointing to those people who love cryptic acronyms, but as one of our readers, we thought you deserve to know the real truth.

FUN TIP: If you are at all inclined to metaphysical speculation, think about this: in the comics, Jughead's last name is Jones. On the Internet, the Jughead program was developed by Rhett Jones. Cosmic or what?

CHAPTER 22

Finding Someone on the Internet

If you know someone's name, how do you find his or her Internet address? When you see a userid and address in a Usenet article, is there a way to find out who that person is? Or let's say you have a friend at a particular university, how do you send him mail?

These are some of the most common questions that Internet users ask. There are many different situations and variations, but it all boils down to the same thing: How do you find someone on the Internet?

In this chapter, we will explain what Internet facilities exist for locating people and show you how to use them. We will also show you how to find the location of a particular computer, when all you know is its name.

An Approach to Finding Someone

As you know, there is no central authority to organize the Internet. Indeed, there are so many computers on the Internet that no one can even keep track of all of them. Thus, it is not surprising that there is no such thing as a central Internet users directory.

This means that if you are looking for someone on the Internet, there is no one place to look. What you need to do is develop a strategy.

In this chapter, we will show you what facilities exist for finding people. We suggest that you become familiar with the entire chapter. Then, when you need to track down someone, you can decide where is the best place to start.

Although there is no single Internet directory, there are many services that will help you find someone. We can divide these services into three categories. If you know the organization to which the person belongs, you can try to find a white pages directory that contains information about the people in that organization. If the person sends articles to Usenet, you can use the Usenet address server. Or, if you have some idea where the person is located, you can use Netfind.

Now, before we start, here is a general hint that pertains to all the services we will discuss in this chapter.

Hint

In general, programs that search for names do not distinguish between upper- and lowercase letters. For example, if you want to find "Harley Hahn" you can search for "harley hahn", or "Harley Hahn", or "HARLEY HAHN".

White Pages Directories

A *white pages directory* is a service that allows you to search a database for someone's name or electronic mail address. White pages directories often contain other information as well, such as a postal address, a department name, a telephone number, and so on. For the most part, using a white pages directory is simple: you enter a name and, if it is in the directory, you will be shown whatever information is known about that name.

The name "white pages directory" was coined to suggest the image of a computerized telephone book. However, the analogy is misleading. All telephone books work pretty much the same way and contain the same standard information. On the Internet, there are a variety of white pages directories and they do not all work the same. In this section, we will discuss the most well-known white pages directories and how you can use them.

The Gopher

If you know someone's organization, especially if he or she is at a university, the best place to start is the Gopher. Many organizations have CSO name servers or some other type of white pages directory. (We explain all about the Gopher and CSO name servers in Chapter 21.)

Thus, start by looking for a Gopher server for that organization. If you find one, look for a menu item that contains some type of directory, such as a CSO name server or an item that refers to a "phone book". For example, if you look at the main menu for the Gopher at the University of Wisconsin at Madison, you will see the item:

> `Phone books/`

Selecting this menu leads you to a number of white pages directories.

Another alternative is to connect to the Gopher at Notre Dame University, in the state of Indiana. (The address is **gopher.nd.edu**.) From the main menu, select:

> `Non-Notre Dame Information Sources/`

From the next menu, select:

> `Phone Books--Other Institutions/`

You will now see a menu that offers several categories of white pages directories (which they call "directory servers"). From here you have access to hundreds of different white pages directories, not only for universities, but for all kinds of organizations.

Whois Servers

A *Whois server* is another type of white pages directory that allows you to query a database of names and electronic mail addresses, usually for a particular organization. There are a large number of Whois servers on the Internet.

You can get a list of Whois servers via Anonymous FTP. To find which Anonymous FTP hosts carry this list, use Archie to perform a **sub** search for **whois-servers**. (We discuss Anonymous FTP in Chapter 16 and Archie in Chapter 17.) The most well-known Whois server is **whois.internic.net**, which contains a lot of Internet names and addresses.

There are two main ways to access a Whois server. First, your system may have a command called **whois**. If so, you can type the name of the command, followed by **-h**, followed by the name of a particular Whois server, followed by the name of the person you are looking for. For example, to look for someone whose last name is Helliwell using the Whois server at **whois.internic.net**, enter:

```
whois -h whois.internic.net helliwell
```

(The **-h** stands for "host name".)

If you want to search for a first and last name, for example, Catherine Helliwell, use the following format:

```
whois -h whois.internic.net 'helliwell, catherine'
```

Be sure not to leave out the quotes. Notice that Whois servers do not distinguish between upper- and lowercase letters.

If you use the **whois** command without the **-h** and a host name, it will use a default Whois server. If your organization has its own Whois server, this may be the default. Otherwise, it will probably be a well-known server like **whois.internic.net**.

For example, if your organization has a Whois server, you should be able to search for all the people whose last name is Smith by using:

```
whois smith
```

If you are using a Unix system that has a **whois** command, you can read the documentation by using the **man** command to display the **whois** entry in the online Unix manual:

```
man whois
```

The second way to use a Whois server is to telnet to it. If the server provides public access, you should be able to log in with a userid of **whois**. For example, to use **whois.internic.net**, enter the command:

```
telnet whois.internic.net
```

Log in as **whois** (or follow whatever instructions you see). When you see the **Whois:** prompt, enter whatever name you want to search for. If you need help, enter **help**. (We explain how to use Telnet in Chapter 7.)

The Knowbot

A *knowbot* is an automated robot-like program that intelligently searches for information on your behalf. For instance, if you had a knowbot, you could tell it that you were looking for a recipe for vegetarian meatloaf with no radishes, and it would search the appropriate places on the Internet for you.

People have been talking about knowbots for some time now, but, as of yet, they are a dream unrealized. However, there is a primitive knowbot-like program that you can use to search many white pages directories at once, including one that contains people who use MCI Mail.

To use this knowbot, telnet to **nri.reston.va.us** using port number **185**:

```
telnet nri.reston.va.us 185
```

Once you are connected, enter **?** to display a summary of commands. It may take you a little while to learn how to use the knowbot effectively, but it is worth knowing. The basic command to use is **query**. For example, to search for someone whose name is Catherine Helliwell, enter:

```
query catherine helliwell
```

If you have an idea where the person may be, you can use some of the other knowbot commands to narrow down the search.

Fred and the X.500 Directories

X.500 is an international white pages directory that was developed to provide a universally accepted system for keeping track of people. Well, we are still waiting for X.500 to be universally accepted. In the meantime, a number of organizations have adopted the scheme, and you can search some of their directories by using a program called Fred.

To do so, telnet to either **wp.psi.com** or **wp2.psi.com** and log in as **fred**. (You will not need a password.) For example:

```
telnet wp.psi.com
```

Once you log in, enter **help** for a list of commands. For more detailed documentation, enter **manual**.

By the way, the name "Fred" stands for "Front End to the Directory".

The Usenet Address Server

You may already know the computer **rtfm.mit.edu** as the main repository of Usenet archives. For example, it contains copies of all the frequently asked question (FAQ) lists (see Chapter 9).

This computer is administered by Jonathan Kamens, who has set up an important white pages directory called the *Usenet address server*. A program on **rtfm.mit.edu** routinely scans each article that is submitted to Usenet looking for the **From** header line. (As we explained in Chapter 10, the **From** header line shows the name and address of the person who sent the article.) The program extracts all these **From** lines, and saves them in a database. Thus, if the person you are looking for has posted a Usenet article within the last year, there is a good chance that **rtfm.mit.edu** will have his or her name and mail address.

To use this service, mail a message to:

```
mail-server@rtfm.mit.edu
```

Leave the **Subject** line blank. In the body of the letter, type a line with the following format:

```
send usenet-addresses/name
```

where *name* describes the person you are looking for. The program on **rtfm.mit.edu** will search the Usenet address database and send back all the lines that contain the pattern you specified. Thus, you can use a first name, last name, or userid. Moreover, you do not need to use capital letters as the search program does not distinguish between upper- and lowercase.

As an example, we decided to search for Michael Peirce in Ireland by sending the following message:

```
To: mail-server@rtfm.mit.edu
Subject:
send usenet-addresses/peirce
```

In a short time, we received mail that contained 21 different lines containing **peirce** or **Peirce**, two of which referred to a Michael Peirce. By searching for the line with an Irish address (one that ends with the domain **ie**), we were able to find the person we wanted.

Hint

You can send more than one request per message. If you are not sure how to spell someone's name, send more than one request with alternate spellings. For example:

```
mail mail-server@rtfm.mit.edu
Subject:
send usenet-addresses/peirce
send usenet-addresses/pierce
send usenet-addresses/pirce
```

Each request will be answered in a separate message.

Another way to use the Usenet Address Server is via Wais. This allows you to search the database directly. All you need to do is start Wais and select the database called **usenet-addresses**. The nice thing about using Wais is that your searching can be more complex. For example, you can specify more than one word to search for and Wais will rank the responses in order of relevancy. (We explain how to use Wais in Chapter 23.)

Netfind: The Program

Netfind is a program that will actively search through the Internet, looking for a computer that knows about the person you are trying to find. To use Netfind, you need to have some idea of where the person is.

Netfind will not only look for a name and mail address, it will try to find Finger information about the person (see Chapter 8). If the person seems to have more than one mail address, Netfind will make an educated recommendation as to which is the best one to use.

There are two ways to use Netfind. First, you may have a Netfind program on your computer. If not, you can telnet to a public Netfind host and use the program there. We will explain how to do this in the next section. (If this is the case for you, you should still read this section to see how Netfind works.)

If you have a Netfind program on your computer, you will be able to display the documentation from the online Unix manual—Netfind only runs on Unix systems—by using the command:

```
man netfind
```

You will see that the **netfind** program is complex and has all kinds of options. We will describe how to use it in its simplest sense, which will work fine most of the time.

Enter the **netfind** command, followed by a single name, followed by one or more locations. The name can be a first name, last name, or userid. The locations can be geographical or domains from an Internet address. Netfind does not distinguish between upper- and lowercase letters.

For example, let's say that you are looking for John Navarra, at Northwestern University in Evanston, Illinois. Here are several commands that you could try:

```
netfind navarra evanston
netfind navarra evanston illinois
netfind navarra northwestern
netfind navarra northwestern university
```

You could, of course, look for **john**, but there are probably a lot of Johns, and you are only allowed to specify one name.

Netfind works especially well if you know all or part of the person's address. For example, say that you happen to know that John Navarra has a computer account with Academic Computing and Network Services at the university. You also know that such accounts have mail addresses that end with **acns.nwu.edu**. You can use the following command:

```
netfind navarra acns nwu edu
```

Notice that you type each part of the address separately, leaving out the . characters.

As Netfind works, it will display a lot of technical messages showing its progress. Most of the time, these messages just get in the way. You can turn them off by using the option **-dfmn** just after the command name. For example:

```
netfind -dfmn navarra northwestern
netfind -dfmn navarra acns nwu edu
```

(The meaning of this option is described in the documentation. Essentially, what it does is turn off certain debugging information.)

When Netfind starts a search, it looks at the part of the Internet that you described and checks how many computers there are. If there are more than a few, Netfind will tell you that there are too many "domains". It will then list the computers that look promising and ask you to choose which ones you would like to search. For example, we enter the command:

```
netfind -dfmn navarra northwestern
```

Here are the first few lines of the output:

```
There are too many domains in the list.
Please select at most 3 of the following:

0. acns.nwu.edu (academic computing and network services,
   northwestern university, evanston, illinois)
1. astro.nwu.edu (astronomy department, northwestern university,
   evanston, illinois)
2. biochem.nwu.edu (biochemistry department, northwestern
   university, evanston, illinois)
3. cas.nwu.edu (northwestern university, evanston, illinois)
```

In this case, we know enough to choose number 0. Otherwise, we would have to guess.

Hint

There are two problems that you will encounter with Netfind's output. First, Netfind does not break up long lines as nicely as we did in our example, so the output usually looks ugly. Unfortunately, there is nothing you can do about it.

Second, Netfind will often generate so much output that it scrolls off your screen before you can read it. The best thing to do is use whatever facilities exist on your system to handle fast output. For example, on a Unix system, you can press CTRL-S to pause the output and CTRL-Q to continue. On a PC, you can press the PAUSE key to pause and any other key to continue. If you are using Netfind within a window under X Window, you can use the scroll bar to look back at the previous output.

Once you tell Netfind what it needs to know, it will proceed with an intelligent search of all the computers in the appropriate part of the Internet. In our example, Netfind will search all the computers that fall under the domain of Academic Computing and Network Services at Northwestern University. (That is, all the computers whose addresses end with **acns.nwu.edu**.) Netfind will attempt to connect to each computer looking for information about a user whose name or userid is **navarra**.

When it finds such a person, it will request the Finger information and display it for you. You will frequently find that a person has Finger information on more than one computer—say, on every computer in a departmental network. In such cases, Netfind will display a plethora of information, and you will have to figure out which parts are the most valuable.

If possible, Netfind will look at the times that the person has last logged in to each computer and deduce which of several mail addresses would be the best to use. (For an explanation about domains and addresses, see Chapter 4. For information about interpreting Finger information, see Chapter 8.)

Hint

You can interrupt a Netfind search by pressing CTRL-C.

Netfind: The Server

If your computer does not have a Netfind program, you can telnet to one of the public Netfind servers listed in Table 22-1 and log in as **netfind**. You will not need a password.

When you use the Netfind server, you make selections from a menu to tell Netfind what to do. When you get to the point where you must tell Netfind what to search for, enter the same type of information that we discussed in the previous section.

After logging in as **netfind**, you will see the following main menu:

```
Top level choices:
        1. Help
        2. Search
        3. Seed database lookup
        4. Options
        5. Quit (exit server)
```

Select number 2. You will now see:

```
Enter person and keys (blank to exit) -->
```

You can now tell Netfind what to search for. For example, you might enter:

```
navarra acns nwu edu
```

Location	Internet Address	IP Address
Australia	archie.au	139.130.4.6
Canada	macs.ee.mcgill.ca	132.206.61.15
Chile	malloco.ing.puc.cl	146.155.1.43
Czech Republic	netfind.vslib.cz	147.230.16.1
England	monolith.cc.ic.ac.uk	155.198.5.3
Singapore	lincoln.technet.sg	192.169.33.6
Slovakia	nic.uakom.sk	192.108.131.12
South Korea	nic.nm.kr	143.248.1.100
USA: Alabama	redmont.cis.uab.edu	138.26.64.4
USA: Colorado	bruno.cs.colorado.edu	128.138.243.151
USA: Minnesota	mudhoney.micro.umn.edu	134.84.132.7
USA: Texas	netfind.oc.com	192.82.215.92
USA: Virginia	ds.internic.net	198.49.45.10
Venezuela	dino.conicit.ve	150.188.1.10

Log in with a userid of **netfind**.

Table 22-1.   *Public* **netfind** *Servers*

Netfind will perform its search and display its output in the same manner that we have already discussed.

Hint

You can interrupt a Netfind search by pressing CTRL-C.

When your search is finished, you will once again see the message:

```
Enter person and keys (blank to exit) -->
```

Press RETURN and you will be back at the main menu. You can now select number 5 to quit.

Using Netfind to Find the Location of a Computer

There will be times when you know the address of a computer and you want to find out where it is. Or you may want to find all the computers in a certain area. You can use a Netfind server for both of these tasks.

Here is how it works. Telnet to one of the public Netfind servers in Table 22-1 and log in as **netfind**. You will see the main menu:

```
Top level choices:
        1. Help
        2. Search
        3. Seed database lookup
        4. Options
        5. Quit (exit server)
```

Select number 3 (the *seed database* is the list of computers and locations that Netfind uses for its searches). You will see another menu:

```
Seed database choices:
        1. Seed database help
        2. Seed database search
        3. Toggle seed database search output format
        4. Quit menu (back to top level)
```

Select number 2. You will see the following:

```
Keys (blank to exit):
```

You can now enter the address of a computer or a geographical location. If you enter the address, use only the rightmost two domains and leave out the **.** character. Remember, Netfind does not distinguish between upper- and lowercase letters.

For example, say that you want to find out the location of the computer **rtfm.mit.edu**. Enter:

```
mit edu
```

In this case, you would see a list of computers in that part of the Internet. In particular, you will see:

```
mit.edu massachusetts institute of technology, cambridge
```

This is enough to show you that **rtfm.mit.edu** is probably in Cambridge, Massachusetts.

Here is another example. Say that you want to look for computers in Dublin, Ireland. Tell Netfind to search for:

```
dublin ireland
```

You will see a long list of computers. Be aware that these are not all the computers in Dublin, but only the ones Netfind knows about.

When you are finished, press RETURN to get back to the menu, and then work your way back out by choosing "Quit". (It will be obvious.)

What to Do When All Else Fails

If you have tried everything you can think of to find someone and you still can't get anywhere, there are several alternatives.

If you know a person's userid and address—or if you can make a good guess—you can use the Finger service to display whatever public information is offered about that person. (See Chapter 8.)

Second, you can post an article to the Usenet newsgroup named **soc.net-people**. One of the purposes of this group is for people to ask questions like, "Does anyone know where to find Chuck Wagon, who graduated from the University of Foobar in 1983 from the Unix Studies department?" When you send the article, make sure that the **Subject** line has the person's name and location (if you know it).

If you are looking for someone at a university, there is an article by David Lamb of Queens University in Canada that you will find helpful. The article has a summary of various methods of finding people, followed by specific information about many different universities.

Lamb's article is posted regularly to the **soc.net-people** newsgroup in three parts. You can also obtain it via Anonymous FTP from **rtfm.mit.edu**. Change to the directory named **/pub/usenet/news.answers/mail/college-email** and download the three files named **part1**, **part2**, and **part3**. The command **mget part*** will do the job. (We explain how to use Anonymous FTP in Chapter 16.)

Another article to read is called "Finding Addresses". It was written by Jonathan Kamens (the keeper of the Usenet Archive at **rtfm.mit.edu**). This article describes many different ways to find someone, including some esoteric techniques that we did not cover in this chapter. The article is also posted regularly to **soc.net-people**. Or, you can get it via Anonymous FTP from **rtfm.mit.edu**. Change to the directory named **/pub/usenet/news.answers** and download the file named **finding-addresses**. (In passing, we would like to mention that the same directory contains an article called **finding-sources** that will show you how to find computer programs in source code form.)

If you know the name of the computer on which someone has an account, you can send mail to the userid named **postmaster** at that address and ask for help. For example, if you were looking for someone who uses the computer named **fuzzball.ucsb.edu**, you could send mail to:

```
postmaster@fuzzball.ucsb.edu.
```

Most Internet hosts have someone who answers such mail. Remember, though, this will require a busy person to take the time to answer you, so make sure you have tried the other alternatives first.

Hint

When nothing else has worked, try the telephone.

CHAPTER 23

Wais

Wais is a tool that allows you to search large amounts of information quickly and thoroughly.

Of course, such tools are only as good as the quality and variety of the information, as well as the power of the searching program. As you will see, Wais can search any of several hundred information sources on a large variety of topics.

Nevertheless, Wais is far from perfect, and we will discuss its limitations. But, within these limitations, Wais is an amazing tool. Once you learn how to use it, Wais is like having a team of simple-minded, but efficient librarians, ready to search for whatever it is you need, whenever you need it.

The Motivation for Wais

The name Wais (pronounced "Wayz") stands for "Wide Area Information Service". As an Internet service, Wais is unique in that it grew out of a project started by three commercial companies: Apple, Thinking Machines, and Dow Jones.

The original idea behind Wais was an ambitious one. In a world of too much information, a computer could keep track of a vast amount of data, sift it for you, and present you with only the information that is relevant to your needs. The idea was to create a program that would act as a personal reference librarian, saving you time and making it possible to access information that you might never even know existed.

For example, say that your daily newspaper was delivered to you in electronic form, as information that you could display on your computer screen. Wouldn't it be nice if you could tell a program what you wanted to see and have it do all the work?

For example: "Each day, I want you to show me sports news, but not business. I want to see anything that has to do with baseball, but only if it mentions the New York Mets. And I don't want to see any football stuff, except when it's time for the Super Bowl. Oh yes, I also want to see the world news, but not too much politics, but if there is anything about airplanes, be sure to include it..."

You get the idea: let a computer program do the work. Moreover, when the program finds an article that really interests you, you should be able to tell it to keep an eye out for other articles on the same subject.

Of course, once you have such a system, you do not have to confine yourself to news. Any type of data that can be accessed by computer can, in principle, be searched on your behalf by a discriminating computer program.

You can see the motivation here for Wais: Apple sells personal computers with an easy-to-use graphical interface; Thinking Machines makes computers that contain many processors, suitable for searching large amounts of data quickly; and Dow Jones runs a service that sells news and information.

Obviously there are some wonderful aspects to such a system. You can let a computer program handle all of the details of searching for information. All you have to do is sit back like an aristocrat, reading your daily personalized newspaper.

Still, there are important limitations. First, a series of articles on your computer screen, no matter how personalized, will never take the place of an actual newspaper or book that you can lay flat on the table and scan with your eye, page by page.

Second, whatever data you receive is only as good as the questions that you ask, plus whatever the computer program can infer about your likes and dislikes. All too often, it is impossible to ask exactly the right questions, even though you may know what you want when you see it.

Finally, the computer program can only find data that is computerized, which limits things enormously. For example, think about how much information you can access in a regular library. You can walk into a library, take any book off the shelf, open the book and read it. Although you may sometimes have trouble finding exactly what you want, you do have access to every book on the shelves, as well as a catalog and, possibly, a human reference librarian.

Hint

One mistake that people often make is thinking that just because they can search for a book in a computerized library catalog, they should be able to read the text of that book using the computer. In most cases, even if you use a computer to help you find a book, you will still have to get the real thing in order to read it.

Compared to a real library, there are very, very few books and periodicals whose text is stored on a computer.

The original idea behind Wais was to develop a generalized system of information retrieval that would be able to access collections of data all around the world. Some of the data would be available for free over the Internet, but it wouldn't be long before people would start to pay for information.

What happened to Wais is somewhat predictable, at least in hindsight. It did not turn out to be the information utility of the future. Yes, there are some people who do use Wais-like computer systems for which they pay real money, but most of us still depend on old-fashioned paper.

So, with respect to the Internet, Thinking Machines has stopped supporting the publicly distributed Wais. A free version of Wais, called *Freewais*, is now maintained by an organization called the CNIDR (Clearinghouse for Networked Information Discovery and Retrieval).

Nevertheless, Wais is important to the Internet. Why? Because it evolved into a system that can perform one vital service extremely well.

What Is Wais?

Wais is an Internet service that can search any of hundreds of collections of data. Each such collection is called a *source*. You tell Wais which sources to use and what you would like to find. To do so, you specify one or more words, called *keywords*, for which you want Wais to search. Wais will search the entire text of each item in the sources you specify, and find the ones that meet your criteria. This is called a *full-text search*.

For example, you can point Wais at a large set of recipes and tell it to find all the recipes that contain the words **garlic**, **chicken**, and **rice**. Or you can search a collection of Bill Clinton's speeches from the 1992 U.S. presidential campaign and find all the speeches that mention **tax**, **cut**, and **promise**. Or you can search a reference of protein sites and patterns, and select all the citations that contain the words **dihydrofolate reductase**. Or you can search the Bible, the Quran, and the Book of Mormon, looking for all the passages that contain the word **adultery**. (Think how much time this could save you.)

Using Wais is reasonably simple. All you do is select one or more sources and then tell Wais what to search for. Wais will connect to the computers that contain the sources and ask them to carry out the search. The output will be a list of articles or citations.

Wais will display the list for you, showing the items from most to least relevant. What Wais thinks is relevant is based on how often the keywords you specified showed up in each article. Wais considers the most relevant articles to be the ones with the most occurrences of the keywords.

Once Wais shows you what it has found, you go through the list and pick out which items you want to look at. Wais will then retrieve the actual text and display it on your screen. If you would like to keep the item, you can tell Wais to save it to a file or mail it to yourself (or to someone else).

When you find an article that is especially germane to your interests, you can tell Wais to scan all or part of the article and to use those words for future searches. This is called *relevance feedback*.

The service that Wais provides can be extremely useful. Still there are limitations. First, Wais can only help you if it has access to the information you want. Although several hundred information sources sounds like a lot, they do not begin to cover all the questions a human being might have.

Second, Wais searches virtually every word contained in a source—ignoring small, common words like "a" and "the"—but only in a simple, non-contextual manner. For instance, you can tell Wais to search a collection of speeches for the words **tax**, **cut**, and **promise**, but those words might be anywhere in the text. For example, the person giving the speech might have said, "We must tax our ingenuity to cut down on our inability to fulfill the promise of the future." There is no way to tell Wais to find references to "promises about tax cuts".

Still, when it works well, Wais is a fabulous tool that can save you many hours of research. So let's take a look at how it works and how to use it.

Wais Clients

Like many Internet services, Wais uses a client/server system (see Chapter 2). You interact directly with a *Wais client* program. It is this program that displays information and carries out your commands. Each Wais information source is maintained by a *Wais server* program. There are many public Wais servers, all over the Internet, and all you need to be able to access them is a Wais client.

Whenever you make a request, your Wais client connects to the appropriate Wais server and asks it to perform a search. The server carries out the search, and returns the results to the client, which displays the results on your screen. When you indicate that you would like to view a particular item, your client again sends the request to the server. The server sends back the item, which your client displays, one screenful at a time.

There are a variety of Wais clients available for free, via Anonymous FTP. It may be that your system already has a Wais client. If so, all you have to do is start the program. Two of the most common Wais clients are **swais** and **waissearch** for regular Unix systems, and **xwais** for X Window systems. (We discuss X Window in Chapter 2.) There are also Wais clients for a large number of other systems.

If you do not have a Wais client on your computer, you can telnet to one of the hosts listed in Table 23-1 and log in as indicated. Each of these computers provides a public Wais client that anyone can use. Once you log in, the client will start automatically. No password is necessary. (We discuss Telnet in Chapter 7.)

Hint

The public Wais hosts do not all offer the same set of sources. You may want to try the various hosts to see what is available.

When you telnet to a public Wais client, you will be using the **swais** client. (The name means "simple Wais".) For the rest of the chapter, we will assume that this is the client you are using. If you are using a different client, you will have to read the appropriate documentation. However, the basic concepts as to how to use Wais will be the same as what we discuss.

For more information about Wais, there are two Usenet discussion groups you can subscribe to: **comp.infosystems.wais** and **alt.wais**. There is also a Wais frequently asked question (FAQ) list that you will find helpful, especially if you are looking for a Wais client for your computer.

Location	Internet Address	IP Address	Log in as...
Finland	`info.funet.fi`	`128.214.6.102`	`wais`
USA: California	`swais.cwis.uci.edu`	`128.200.15.2`	`swais`
USA: Massachusetts	`nnsc.nsf.net`	`128.89.1.178`	`wais`
USA: Massachusetts	`quake.think.com`	`192.31.181.1`	`wais`
USA: North Carolina	`kudzu.cnidr.org`	`128.109.130.57`	`wais`
USA: North Carolina	`sunsite.unc.edu`	`152.2.22.81`	`swais`

Table 23-1.   *Public Wais Clients to Which You Can Telnet*

The FAQ list is posted regularly to **comp.infosystems.wais**. It can also be downloaded via Anonymous FTP from **rtfm.mit.edu**. Look in the directory named **/pub/usenet/news.answers/wais-faq** for a file called **getting-started**. (We discuss Usenet newsgroups and FAQ lists in Chapter 9. We discuss Anonymous FTP in Chapter 16.)

Musing on Wais

In the following sections, we will show you how to use a public access Wais client to perform a search. Before we do, we have a few words of advice.

It can take a while to get good at using Wais. Your results depend very much on how good you are at choosing which sources to search, and how good you are at formulating keywords. Moreover, the Wais interface is kind of funky and takes some getting used to. (For those of you who do not live in Southern California, funky means "low quality, but still cool".)

There are nuances about Wais that you must come to understand over a period of time, akin to those that a good librarian develops when it comes to knowing where to look for just the right book. Don't be frustrated if, at the beginning, you find it difficult to come up with anything useful. Once you get some experience, it becomes a lot easier to use Wais effectively.

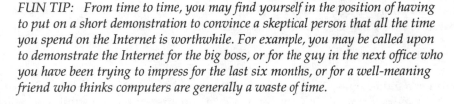

FUN TIP: From time to time, you may find yourself in the position of having to put on a short demonstration to convince a skeptical person that all the time you spend on the Internet is worthwhile. For example, you may be called upon to demonstrate the Internet for the big boss, or for the guy in the next office who you have been trying to impress for the last six months, or for a well-meaning friend who thinks computers are generally a waste of time.

Our advice is to show them the Gopher, which is a lot more reliable and easier to use than Wais. If you decide to show off by using Wais, that will be the day that the client program is slow and uncooperative, and that every source you search will be temporarily not working. Instead of showing them how well Wais can find data, you will probably end up with nothing but an embarrassed look on your face.

Okay, having offered the preceding caveat, let's see what it looks like to use Wais.

Starting to Use Wais

To start Wais, you can telnet to one of the public clients. For example:

```
telnet quake.think.com
```

As the Telnet connection is made, you will see the following messages. (For an explanation of how to use Telnet, see Chapter 7.)

```
Trying 192.31.181.1...
Connected to quake.think.com.
Escape character is '^]'.

SunOS UNIX (quake)
login:
```

Log in as **wais** (you will not need a password). You will then see:

```
Last login: Sat Jul 10 15:32:36 from jazz.ucc.uno.edu
SunOS Release 4.1.1 (QUAKE) #3: Tue Jul 7 11:09:01 PDT 1992

Welcome to swais.
Please type user identifier (optional, i.e user@host):
```

At this point, you should enter your Internet mail address and press RETURN. The remote host now asks what type of terminal you are using:

```
TERM = (vt100)
```

You can either enter the type of terminal or press RETURN to indicate a VT-100. (We discuss terminals in Chapter 3.) Once you answer the question, you will have to wait a moment for the **swais** client to start. When it starts, you will see a list of sources. You are now ready to start work.

```
SWAIS                       Source Selection           Sources: 463
  #         Server                  Source                    Cost
001:  [          archie.au]  aarnet-resource-guide            Free
002:  [    munin.ub2.lu.se]  academic_email_conf              Free
003:  [wraith.cs.uow.edu.au] acronyms                         Free
004:  [     archive.orst.edu] aeronautics                     Free
005:  [ ftp.cs.colorado.edu] aftp-cs-colorado-edu             Free
006:  [nostromo.oes.orst.ed] agricultural-market-news         Free
007:  [    archive.orst.edu] alt.drugs                        Free
008:  [    wais.oit.unc.edu] alt.gopher                       Free
009:  [sun-wais.oit.unc.edu] alt.sys.sun                      Free
010:  [    wais.oit.unc.edu] alt.wais                         Free
011:  [alfred.ccs.carleton.] amiga-slip                       Free
012:  [    munin.ub2.lu.se]  amiga_fish_contents              Free
013:  [   coombs.anu.edu.au] ANU-Aboriginal-Studies    $0.00/minute
014:  [   coombs.anu.edu.au] ANU-Asian-Computing       $0.00/minute
015:  [   coombs.anu.edu.au] ANU-Asian-Religions       $0.00/minute
016:  [        150.203.76.2] ANU-CAUT-Academics        $0.00/minute
017:  [   coombs.anu.edu.au] ANU-CAUT-Projects         $0.00/minute
018:  [   coombs.anu.edu.au] ANU-Coombspapers-Index    $0.00/minute

Keywords:

<space> selects, w for keywords, arrows move, <return> searches, q quits, or ?
```

Let's take a look at the various parts of the screen. First, the top line tells us that this is the Wais source selection screen. (Remember, we have to select our sources before we can search.) On the top right, we see that this client knows about 463 different sources.

Following this, we see a summary of the first 18 sources. Each line shows the address of the host that contains the source, the name of the source itself, and the price of performing a search. Notice that the first source is highlighted. As you use this list, you will move from one source to another. The highlighting will move also, to indicate the *current source*.

> ## Hint
>
> Don't worry about the price. All the sources are free. You cannot spend money by accident.

On the second to last line we see **Keywords:**. This is where we will type our keywords when we are ready to begin searching.

Finally, the last line summarizes the basic **swais** commands. These will make more sense once you learn how to use **swais**.

In the next few sections, we will show you how to select sources and how to perform a search. Before we do, we would like to introduce you to the two most basic **swais** commands.

First, you can display a summary of commands by pressing **h** (help) or **?**. The **swais** client works in what is called *cbreak mode*. This means that you do not have to press RETURN when you use a single character command. For example, when you want help, you only need press the **h** key. Do not press RETURN.

Second, to stop **swais** and disconnect from the remote host, press **q** (quit).

> ## Hint
>
> Be careful with the **q** key. It is all too easy to press **q** inadvertently—say when you are reading the list of sources—and irrevocably abort your session with Wais.

Selecting a Source

The general plan for using Wais is to select one or more sources, and then perform a search. In this section, we will go over the commands that you use to select a source. In the next section, we will show you how to perform a search. We will then present an example that illustrates how it all works.

To select a source, you move to it and then press SPACE. Wais will show you that a source is selected by marking it with a * character. For example, in the following list, source number 10 is selected:

```
008:    [    wais.oit.unc.edu]   alt.gopher          Free
009:    [sun-wais.oit.unc.edu]   alt.sys.sun         Free
010: *  [    wais.oit.unc.edu]   alt.wais            Free
011:    [alfred.ccs.carleton.]   amiga-slip          Free
```

You can select as many sources as you want before you start to search. At any time, you can select a new source by moving to it and pressing SPACE. If you decide to stop using a source that is already selected, move to it and press SPACE once again. This will unselect the source. To unselect all the sources and start fresh, press = (equal sign).

Hint

Before you start a new search, press = to clear any sources that may have been selected from the previous search.

There are several ways to move from one line to another. The easiest way is to use the UP and DOWN keys (the cursor control keys, with the arrows). If these do not work on your terminal, or if you prefer not to use them, you can use **j** or CTRL-N to move down to the next line, and **k** or CTRL-P to move up to the previous line. (Although this choice of keys may seem odd, they are similar to what is used with the **vi** editor and are second nature to many Unix users.)

To move down to the next screenful of sources, you can use either **J** or CTRL-D. To move up to the previous screenful of sources, you can use either **K** or CTRL-U. (Note: these are the uppercase "J" and "K" letters.)

To jump directly to a particular source, simply enter its number. For example, to jump to source number 15, enter:

    15

(You do not have to type the leading **0** in **015**.)

If you know the name of a source, you can jump to it by entering a / (slash) character, followed by the first few letters of its name. For example, to jump to the source named **amiga_fish_contents**, it is enough to enter:

    /amiga_f

If you would like to see the technical information that Wais knows about a source, move to it and press **v** (view). For example, if you move to **amiga_fish_contents** and press **v**, you will see the following:

```
Name:           amiga_fish_contents.src
Directory:      /sources/
Maintainer:     hakan@hera.dit.lth.se
Selected:       No
Cost:           Free
Server:         munin.ub2.lu.se
Service:        210
Database:       amiga_fish_contents
Description:
Server created with WAIS release 8 b5 on Aug 29 16:24:34 1992
   by anders@munin

This is an index of the contents of Fred Fish's disks #1-current
with a freely distributable AMIGA software library containing an
extensive collection of PD, shareware and demo programs.

Search for 'disknr' to see the current (latest) disk number.
Search for 'ftp' to see some ftp sites where you can get the
   disks.
```

Performing a Search

Once you have selected one or more sources, you are ready to perform a search. To do so, press RETURN. The cursor will move to the line near the bottom of the screen.

```
Keywords:
```

You can now type whatever keywords you want. As you type, there are several ways for you to make corrections.

To erase a single character, press BACKSPACE. If this does not work, you can also try CTRL-H, DELETE, or CTRL-BACKSPACE. For reasons we won't go into, one of these keys will work.

To erase an entire word, press CTRL-W. To erase the entire line, press CTRL-U.

Here is an example of what the line looks like after we have entered several keywords:

```
Keywords: tax cut promise
```

Once you have specified your keywords, press RETURN to begin the search. Wais will check each source that you specified and display a list of possibilities. Wais will organize this list so that the best matches come first. Here is an example that we generated by searching the source named **clinton-speeches** for the above-mentioned keywords:

```
SWAIS                              Search Results              Items: 40
  #    Score    Source                        Title              Lines
001:   [1000]  (clinton-speeche)    VP DEBATE ANALYSIS: Encyclopedi   697
002:   [ 928]  (clinton-speeche)    THE ECONOMY: Statement            791
003:   [ 928]  (clinton-speeche)    THE ECONOMY: Speech - Detroit,    559
004:   [ 819]  (clinton-speeche)    LABOR: Speech - Washington, DC    525
005:   [ 782]  (clinton-speeche)    ECONOMIC STRATEGY: 6/21/92        877
006:   [ 728]  (clinton-speeche)    ECONOMIC PLAN: Position Paper -   773
007:   [ 691]  (clinton-speeche)    EDUCATION: Speech - Rockville,    658
008:   [ 673]  (clinton-speeche)    VARIOUS TOPICS: Interview - Atl   830
009:   [ 655]  (clinton-speeche)    ECONOMICS: "New Covenant" Speec   634
010:   [ 637]  (clinton-speeche)    SMALL BUSINESS PLAN: Position P   598
011:   [ 601]  (clinton-speeche)    MANUFACTURING: Speech - Washing   706
012:   [ 600]  (clinton-speeche)    ON RESPONSIBILITY (BUSH'S): Spe   237
013:   [ 600]  (clinton-speeche)    THE ECONOMY: Speech - New Orlea   227
014:   [ 564]  (clinton-speeche)    HEALTH CARE: Speech - Macon, GA   440
015:   [ 546]  (clinton-speeche)    LABOR: Speech - San Diego, CA     507
016:   [ 509]  (clinton-speeche)    ECONOMICS: Speech - Flint, MI -   283
017:   [ 509]  (clinton-speeche)    JOB RETRAINING: Speech - Dayton   359
018:   [ 509]  (clinton-speeche)    EDUCATION: Speech - Los Angeles   638

<space> selects, arrows move, w for keywords, s for sources, ? for help
```

Now that you have seen what Wais has found, you can decide if you want to look at anything. Notice the score in the second column. This gives you an indication of how well each item matched your keywords. The item that had the most matches is always given a score of 1000. All the other items are graded relative to the best match.

Hint

A score of 1000 does not mean that this is the best item to meet your needs. It simply means that—according to Wais' criteria—this particular item had the most keyword matches. You will still have to examine the results. In many cases, you will want to specify better keywords and try again.

At this point, you have two choices. You can look at some of the items, or you can try another search. To look at an item, move to it and press RETURN. Wais will connect to the appropriate server and ask for a copy of the actual file. When it arrives, Wais will use a paging program to display the file, one screenful at a time.

As you read a file, your screen is controlled by the paging program. At this point, there are two basic commands that you can use. To display the next screenful of text, press SPACE. To stop displaying the article, press **q** (quit).

Hint

While you are reading an article, pressing **q** will quit the paging program. At all other times, pressing **q** will quit Wais. You will have to get used to this.

If you would like to have a permanent copy of what you have just read, you can use the **S** (uppercase "S") command to save the text to a file, or the **m** command to mail the text. Of course, you can only use the **S** command if you are using a Wais client on your own computer. (You cannot save files on a public Wais host.) However, you can always use the **m** command to mail a copy of the text to yourself, or to someone else.

If you are not satisfied with the results of the search, you can try again by pressing **w** (keyword). This will bring back the keyword entry line. You can now change your keywords and start a new search.

When you decide that you are finished searching this particular set of sources, press **s**. This will bring back the sources screen. You can now select or unselect whatever sources you want. Remember, if you decide to start fresh with new sources, you can press = to unselect everything.

Until you perform a new search, you can re-display the results of the previous search by pressing **r**.

At this point, we would like to say a few words about relevance feedback. With some Wais clients, you can select all or part of an item as being representative of the type of thing you are looking for. You can then tell Wais to save all the words in the item to help with future searches. This is called relevance feedback.

When you telnet to a public Wais client, relevance feedback is not much of a consideration. However, if you are using a sophisticated Wais client on your own computer, you can use relevance feedback to help you create personalized searches of great specificity.

Hint

If you get serious about Wais, get yourself a good Wais client and spend some time with it.

FUN TIP: Wais is like a sewer. What you get out of it depends on what you put into it.

A Summary of Wais Commands

The summary in Figure 23-1 shows the basic commands for using Wais with the **swais** client program. If you are using another client, the commands will be different (and possibly more powerful), but the basic concepts will be the same.

A Real-Life Example of Using Wais

To complete this chapter, we will use Wais to search for some important information that will be of immediate practical benefit in your everyday life.

You have decided that you are tired of the pace of modern life and you want to go live on a small island, somewhere in the tropics. The only stipulation is that there must be plenty of coconuts and guano so that you can support yourself. The question is: where should you go? To find the answer, you use Wais.

To begin, you telnet to a public Wais client and log in. For this example, you use:

```
telnet nnsc.nsf.net
```

and log in as **wais**. Once you have specified your terminal type, the Wais client starts by displaying the list of sources.

To start, you need to know which source (or sources) to search. If you are intimately familiar with all the hundreds of different sources, this should not be much of a problem. However, if you are not sure which one is best, you can use Wais to help you find it.

There happens to be one particular source that contains information about all the other sources. It is called **directory-of-servers**. You can search this source to help you find other sources.

First, jump to this source by entering:

```
/directory-of-servers
```

Next, select this source by pressing SPACE. To tell Wais that you are ready to enter keywords, press RETURN. The cursor will now move to the keyword line.

Basic Commands

h	display summary of commands [help]
?	same as h
q	quit Wais

Moving the Current Line

DOWN	move down one line
j	same as DOWN
CTRL-N	same as DOWN
UP	move up one line
k	same as UP
CTRL-P	same as UP
number	jump to specified line number
/*pattern*	jump to source that begins with specified pattern
J	move down to next screenful
CTRL-D	same as J
K	move up to previous screenful
CTRL-U	same as K

Selecting a Source

SPACE	select or unselect a source
=	unselect all sources
RETURN	(after selecting sources) ask for new keyword
r	re-display results of the previous search
v	display technical information about source [view]

Performing a Search

RETURN	(after entering keywords) start search
RETURN	(while examining results of a search) display item
w	ask for new keywords
s	re-display the source screen

Reading an Article

SPACE	display next screenful
q	stop reading article [quit]

Figure 23-1. *Commands for using Wais with the **swais** client program*

Now, what keywords can you use to choose a source for finding some place on the Earth? Try the following:

```
world geography
```

Press RETURN to start the search. After a few moments, Wais displays the results of your search:

```
SWAIS                           Search Results              Items: 24
         Score      Source                 Title
 --> 001 [1000] (directory-of-se)  world-factbook.src           21
     002 [ 526] (directory-of-se)  eros-data-center.src         94
     003 [ 474] (directory-of-se)  CCINFO.src                  106
     004 [ 368] (directory-of-se)  unced-agenda.src             71
     005 [ 368] (directory-of-se)  world91a.src                 26
     006 [ 316] (directory-of-se)  ANU-Pacific-Relations.src    90
     007 [ 316] (directory-of-se)  ANU-Thai-Yunnan.src          83
     008 [ 316] (directory-of-se)  ASK-SISY-Software-Information.src  34
     009 [ 316] (directory-of-se)  Arabidopsis_thaliana_Genome.src    37
     010 [ 316] (directory-of-se)  Connection-Machine.src       25
     011 [ 316] (directory-of-se)  Func-Prog-Abstracts.src      23
     012 [ 316] (directory-of-se)  POETRY-index.src             28
     013 [ 316] (directory-of-se)  Queer-Resources.src          21
     014 [ 316] (directory-of-se)  US-State-Department-Travel-Advisorie  89
     015 [ 316] (directory-of-se)  comp.doc.techreports.src     18

 <space> selects, arrows move, w for keywords, s for sources, ? for help
```

Wais has found a number of possible sources for you, but only number 1 even comes close to a score of 1000. To find out something about it, press RETURN. You see:

```
(:source
  :version  3
  :ip-address "131.239.2.100"
  :ip-name "cmns-moon.think.com"
  :tcp-port 210
  :database-name "CIA"
  :cost 0.00
  :cost-unit :free
  :maintainer "bug-public@think.com"
  :subjects "social sciences demographics politics CIA worldfact
book population economics imports exports business"
  :description "Connection Machine WAIS server.  The WorldFact
book by the CIA which contains a good description of every
country. The entry for WORLD is also particularly good.

Descriptions of 249 nations, dependent areas, and other entities
with information on population, economic condition, imports/exports,
conflicts and wars, and politics.  Produced annually by the CIA.
Search 'World Factbook' for table of contents.")
```

Well, isn't that nice of the CIA to provide you with such useful information for free? This sounds like just what you need. Press **q** (to end the paging program). You are now back at the results screen.

To prepare for your next search, you need to select a new source. Press **s** to return to the sources screen. Next, press = to unselect all selected sources. (In this case, we only had one source already selected, but pressing = before starting a new search is a good habit.)

Now we need to select the CIA World Factbook. To jump to this source, enter:

```
/world-factbook
```

To select it, press SPACE. To specify new keywords, press RETURN.

You will be back at the keyword line, but the previous keywords will still show. To erase them, press CTRL-U. Now, type the new keywords:

```
coconut guano
```

(Remember, we are looking for a tropical island with coconuts and guano.) To start the search, press RETURN. Here are the results:

```
SWAIS                         Search Results                  Items: 25
             Score _____Source_____  _____Title_____
 --> 001    [1000] (cmns-moon.think)  Glorioso Islands (French possession)    79
     002    [ 543] (cmns-moon.think)  Juan de Nova Island (French possessi    83
     003    [ 533] (cmns-moon.think)  Navassa Island (territory of the US)    80
     004    [ 515] (cmns-moon.think)  Johnston Atoll (territory of the US)   107
     005    [ 515] (cmns-moon.think)  Jarvis Island (territory of the US)     90
     006    [ 515] (cmns-moon.think)  Howland Island (territory of the US)    99
     007    [ 515] (cmns-moon.think)  Baker Island (territory of the US)      95
     008    [ 477] (cmns-moon.think)  Soviet Union  Geography Total area:    467
     009    [ 440] (cmns-moon.think)  Palmyra Atoll (territory of the US)     79
     010    [ 421] (cmns-moon.think)  Tonga  Geography Total area: 748 km2   286
     011    [ 421] (cmns-moon.think)  Niue (free association with New Zeal   271
     012    [ 412] (cmns-moon.think)  Western Samoa  Geography Total area:   288
     013    [ 412] (cmns-moon.think)  Saint Lucia  Geography Total area: 6   290
     014    [ 412] (cmns-moon.think)  Philippines  Geography Total area: 3   381
     015    [ 412] (cmns-moon.think)  French Polynesia (overseas territory   288

 <space> selects, arrows move, w for keywords, s for sources, ? for help
```

Again, Wais has found a number of items, but only one has a high score. To display this item, press RETURN. Here is what you see (with some of the blank lines omitted):

```
CIA World Factbook      Glorioso Islands (French possession)

Geography
Total area: 5 km2; land area: 5 km2; includes Ile Glorieuse,
Ile du Lys, Verte Rocks, Wreck Rock, and South Rock

Comparative area: about 8.5 times the size of The Mall in
Washington, DC

Land boundaries: none
Coastline: 35.2 km
Maritime claims:
Contiguous zone: 12 nm;
Exclusive economic zone: 200 nm;
Territorial sea: 12 nm
Disputes: claimed by Madagascar
Climate: tropical
Terrain: undetermined
Natural resources: guano, coconuts
Land use: arable land 0%; permanent crops 0%;
        meadows and pastures 0%; forest and woodland 0%;
        other--lush vegetation and coconut palms 100%
Environment: subject to periodic cyclones

Note: located in the Indian Ocean just north of the Mozambique
Channel between Africa and Madagascar

People
Population: uninhabited

Government
Long-form name: none
Type: French possession administered by Commissioner of the
Republic Daniel CONSTANTIN, resident in Reunion

Economy
Overview: no economic activity

Communications
Airports: 1 with runway 1,220-2,439 m
Ports: none; offshore anchorage only


Defense Forces
Note: defense is the responsibility of France
```

Judging by this information, it looks like you have found your ideal home away from home (as long as you can arrange for Internet access). To quit, you press **q** twice: once to stop the paging program and a second time to stop Wais.

Tips for Choosing a Source

Sometimes, the hardest part of using Wais is knowing what sources to use. To help you, here are two tips.

The first tip is to make a preliminary search of the **directory-of-servers** source using keywords that describe the source you think will help you. Remember, though, in this preliminary search, you should be more general than when you are looking for a specific item. Ask yourself, what kind of reference is likely to contain the information I want?

For instance, in our example, we wanted to find a place that was associated with both coconuts and guano, so we needed some type of reference that contained geographical facts about the world. This is why we were successful in searching the **directory-of-servers** with the keywords **geography** and **world**.

Hint

With Wais, being able to think on a higher level of abstraction than your ultimate goal is often the prerequisite to a successful search.

The second tip is a real time saver. You can download a summary of Wais sources by Anonymous FTP. This summary contains short descriptions of all the Wais sources, collected into categories.

A quick and easy way to choose a source is to look in this file, find the category that meets your needs, and use whichever sources look good. Moreover, the file is great for browsing, as it gives you a good overview of everything that Wais can access. The only caveat is that the summary is often out of date; there will probably be newer sources that Wais knows about that are not yet categorized.

You can download the Wais source summary via Anonymous FTP from **kirk.Bond.edu.au**. Look in the directory named **/pub/Bond_Uni/doc/wais** for a file named **src-list.txt**. An alternate Anonymous FTP host is **archive.orst.edu**, where the same file is stored in the directory named **/pub/doc/wais**. (We discuss Anonymous FTP in Chapter 16.)

CHAPTER 24

T he *World Wide Web*, or more simply, the *Web*, is an ambitious project whose goal is to offer a simple, consistent interface to the vast resources of the Internet.

When you use the Web, you follow your nose: that is, you start anywhere you want, and you jump from one place to another pursuing whatever strikes your fancy. The amazing thing is that with only a few simple commands, you can jump your way around the Internet like a hyperactive flea at a dog convention.

Intrigued? Read on.

The World Wide Web

What Is the Web?

The World Wide Web has more names than any other Internet resource. We like to call it the Web, but you will also see it referred to as *WWW* or *W3*. To understand the Web, we need to start with the idea of hypertext.

Hypertext is data that contains *links* to other data. A simple example of hypertext is an encyclopedia. Say that you are reading the entry on "Trees". At the end of the article, you see a reference that says, "For related information, see Plants". This last line is a link, from the "Trees" article to the "Plants" article.

Of course, this is a simple example. The Web is based on hypertext that is a lot more complex. In particular, there may be links anywhere within a document, not just at the end.

Here is an imaginary example. Say that you are using the Web to read a hypertext article about trees. Every time the name of a new tree is mentioned there is a link. Each link is marked in some way so that it stands out. For example, a word that has a link may be highlighted or underlined, or it may be identified by a number.

If you follow that link, you will jump to an article about that particular type of tree. Within the main article, there are also links to other related topics such as "rain forests" or "wood". These links lead to complete articles. You will also find links to technical terms such as "deciduous" and "coniferous". When you follow one of these links, you will find a definition.

In the language of the Web, a hypertext *document* is something that contains data and, possibly, links to other documents. The program that you use to read a hypertext document is called a *browser*. As you follow one link to another, we say that you are *navigating* the Web.

In theory, there is no reason to confine ourselves to plain textual data. For example, when we follow a link within the tree document, we might find a picture of a specific type of tree. Or we might find a video clip of an aerial view of a forest. Perhaps we might even find a sound (say, of a tree falling when there is no one around to hear it).

In the world of hypertext, we use the word *hypermedia* to refer to documents that might contain a variety of data types and not just plain text. The Web does contain some hypermedia, but most of what you see will be plain vanilla hypertext: textual material that you can display on your screen.

What makes the Web so powerful is that a link might go to any type of Internet resource: a text file, a Telnet session, a Gopher, a Usenet newsgroup, and so on. The job of your browser is to act as a window into the Internet by following whatever links you desire, and by accessing each document using an appropriate method.

For example, if you follow a link to a text file, your browser will fetch it and display it for you, one screenful at a time. If you follow a link to a Telnet session (see Chapter 7), your browser will initiate the session for you. You won't have to know anything technical like the address or port number. If you follow a link that leads to a Usenet newsgroup (see Chapter 9), your browser will present the articles using a simple, hypertext format.

Thus, we can characterize the Web a little more precisely. The Web is an attempt to organize all the information on the Internet (plus any local information that you would like to add on your own) as a set of hypertext documents. Although this dream may be somewhat unrealistic, the Web does allow you to access all kinds of Internet resources, just by using a browser to "read" the appropriate document.

Why Would You Use the Web?

The Web was originally developed in Switzerland, at the CERN research center. The idea was to create a way for the CERN physicists to share their work and to use community information. Before long, the idea of the Web expanded and was embraced within the Internet as a general mechanism for accessing information and services.

In theory, the idea of an Internet-wide hypertext system seems to offer the holy grail of information retrieval: a simple, easy method for finding and using just about any type of data that exists. However, in practice, you will find that the Web's usefulness is generally limited to two main functions: reading hypertext articles and accessing Internet resources.

The idea of hypertext is not a new one. Perhaps you have already used such a system. For example, you may have used the Hypercard program on a Macintosh, or the InfoExplorer facility on an IBM RS/6000. Hypertext can be a great way to read information, but there are several unavoidable problems.

First, what makes hypertext more valuable than regular text is having all the links. The problem is that putting links into hypertext is a time-consuming process. It takes so much effort, that there is still not all that much hypertext available. There has been some work done to try to automate the process: to have a program read regular text and try to figure out where to put links. However, there is really no substitute for the judgment of a human being.

The second problem is that the value added to regular text by hypertext links depends on how useful the links really are. Each link is supposed to represent a mental leap that a person would want to make as he or she reads the text. Unfortunately, many types of text do not lend themselves to jumping. Moreover, even when jumps would be useful, it is not always obvious where the links should go. What you personally get out of a hypertext document very much depends on how close your thinking is to that of the person who put the links in the document.

Still, there are some types of information for which it is more or less obvious where the links should go. For example, a scholarly document might contain a link to each footnote. When you use the Web to read such documents, you will find that hypertext can save you a lot of time. Not only can you jump to a related topic as the need arises, you can also skip a lot of the extraneous details by ignoring the links that look boring (to you, anyway).

The second important use for the Web is accessing Internet resources. The best way for you to see how convenient this can be is to try it for yourself. Indeed, in many cases, the Web will offer you resources that you would never even know existed. We will show you a typical example later in this chapter.

Hint

Set aside a few hours to explore the Web. Once you have a rough idea of what is available, you will be able to call on the Web whenever necessary.

The Web isn't perfect. There will be times when you are better off using a tool that was designed specifically for a particular type of data. For example, although you can use the Web to read Usenet articles, you are usually better off with a regular newsreader. Still, the Web is definitely the most flexible tool you have for exploring the Internet.

Using the Web

Like many Internet resources, the Web uses a client/server system (see Chapter 2). You use a client program called a browser to act as your window into the Web. From the point of view of the Web, everything in the universe consists of either documents or links. Thus, the job of your browser is to read documents and to follow whatever links you select.

A browser knows how to access just about every service and resource on the Internet. It can set up a Wais search, read Usenet articles, access Gopher items, initiate Telnet sessions, and so on. What is most important is that a browser also knows how to connect to *WWW servers* that offer public hypertext documents.

There are dozens of WWW servers around the Internet and most of them specialize in a particular area. For example, the WWW server at the Cornell University Law School contains a great many documents dealing with U.S. law.

In general, you will encounter two types of documents. Text, that can be read, and *indexes*, that can be searched. When your browser finds that a link points to a text document, it retrieves the document and displays it for you, one screenful at a time.

When your browser finds an index, it shows you a short description and invites you to enter one or more keywords. The browser then searches the index and returns whatever data best matches the keywords. A Web search is a lot like a Wais search (see Chapter 23), and, in fact, many indexes are actually Wais sources.

This means that there are only three basic skills you need to use the Web. First, you need to be able to control the display of text. Second, you must know how to tell your browser to follow a link. Third, you must be able to specify how you want to search an index.

The exact way in which these operations work depends on which browser you are using. For example, if your browser uses a graphical user interface with a mouse, the links are highlighted and you can select one simply by clicking on it. With a character-based browser, each link is given a number; to select a link, you enter the number. Still, no matter what type of browser you have, using the Web is easy and intuitive.

Choosing a Browser

More so than most Internet services, your experience working with the Web depends very much on what client program (browser) you use. The best way to access the Web is to use a browser that is designed to take advantage of the special features of your system. For example, if you use X Window (see Chapter 2), you can use an X Window browser that will make use of the mouse and the graphical user interface.

There are browsers available for many different systems: X Window, PCs with Microsoft Windows, Macintosh, VMS, various types of Unix, and so on. It may be that your system already has a browser installed. (You will have to ask around and find out.) If not, perhaps you can prevail upon your system manager to install one. Later in the chapter, we will show you where you can find browser software.

If you do not have a browser on your own computer, you can telnet to one of the hosts listed in Table 24-1. Each of these hosts offers a public browser that can

Location	Internet Address	IP Address
Finland	info.funet.fi	128.214.6.102
Hungary	fserv.kfki.hu	148.6.0.3
Israel	vms.huji.ac.il	128.139.4.3
Slovakia	sun.uakom.cs	192.108.131.11
Switzerland	info.cern.ch	128.141.201.74
USA: Kansas	ukanaix.cc.ukans.edu	129.237.1.30
USA: New Jersey	www.njit.edu	128.235.163.2
USA: New York	fatty.law.cornell.edu	132.236.108.5

If you are asked to log in, use a userid of **www**.

Table 24-1.   *Public World Wide Web Browsers to Which You Can Telnet*

be used by anyone. With some hosts, the browser will start automatically. With others you must log in as **www**. (We discuss how to use Telnet in Chapter 7.)

All of these hosts offer the type of browser that can be used remotely by people all over the Internet. That means the browser must assume that you are using a plain, ordinary computer terminal with no special features. There are two types of browsers that are used for this purpose. When you telnet to a public Web host, you will be using one of these browsers.

A *line mode browser* is the simplest type. It displays output by writing one line after another to your terminal. To you, it will look as if everything appears at the bottom of your screen and scrolls up. A line mode browser does not care what type of terminal you are using, since all it is doing is sending one line of text after another.

Hint

You may already have a line mode browser on your system. If you do, you may be able to start it by entering the command:

    www

Try it and see what happens.

A *screen mode browser* makes use of your entire screen. Like the line mode browser, a screen mode browser uses only characters. However, it can write those characters anywhere on the screen. This makes the browser faster, and allows it to use a more complex interface. However, it also means that the browser needs to know what type of terminal you are using.

When you log in to a system that uses a screen mode browser, you will be asked to specify your terminal type. Since most systems emulate a VT-100 for remote Telnet sessions, this will be the default.

At the time we wrote this chapter, the hosts in New Jersey, Israel, and Kansas all used (different) screen mode browsers. The other hosts used a simple line mode browser that we will describe in the next section. All of the browsers use English. The Israeli browser also offers Hebrew, while the one in Slovakia also offers Slovak.

An Example of Using the Web

To give you a feeling of what it is like to use the Web, we will show you an example. However, the Web is vast and our example is a small one. To really appreciate how the Web works, you should spend some time exploring it on your own.

In the example, we will telnet to one of the public browsers, the one in Switzerland. We chose this host because it offers a line mode browser (which is easy to use), and because it is oriented towards the general user.

To start, initiate a Telnet session:

```
telnet info.cern.ch
```

Hint

If the line mode browser is already installed on your system, you can use the Web without having to telnet. Just enter the **www** command.

Once you make the connection, the line mode browser starts automatically and you see the welcoming screen:

```
                                     Overview of the Web (23/27)
                     GENERAL OVERVIEW

There is no "top" to the World-Wide Web. You can look at it from many points
of view. If you have no other bias, here are some places to start:

by Subject[1]          A classification by subject of interest. Incomplete
                       but easiest to use.

by Type[2]             Looking by type of service (access protocol, etc) may
                       allow to find things if you know what you are looking
                       for.

About WWW[3]           About the World-Wide Web global information sharing
                       project

Starting somewhere else

   To use a different default page, perhaps one representing your field of
   interest, see  "customizing your home page"[4].

What happened to CERN?

1-6, Up, <RETURN> for more, Quit, or Help:
```

At the top right of the screen, you see **(23/27)**. This tells you that you are looking at the first 23 lines of a 27-line document. (That is why the last part of the text appears to be cut off.) On the bottom line, you see a reminder of the basic commands that you might want to use. (We will explain them later.)

Although what you see looks simple enough, it is actually hypertext. The numbers in square brackets, such as **[1]** or **[2]**, are the links. To follow one of these links, all you have to do is enter the appropriate number. For example, to jump to the document that explains about the World Wide Web (link **[3]**), all you need to do is type **3** and press RETURN:

3

Hint

When you use a browser that supports a mouse, you can select a link by clicking on it.

If you want to display the rest of this document (remember, you are looking at only the first 23 lines), you can press RETURN. If you want to display a summary of all the available commands, as well as other useful information, you can enter:

```
help
```

When you are finished working with the Web, you quit by entering:

```
quit
```

You can enter this command at any time, no matter what document you happen to be reading.

Now, to continue with our example, let's say that you want to use the Web to find an interesting fact that most people don't know. Since that's somewhat of a general goal, you can start by exploring.

By looking at the starting document, you see that there are two ways to begin looking for information, by **Subject[1]** or by **Type[2]**. Start with the second choice by entering:

```
2
```

Your browser follows this hypertext link and you now see the beginning of a new document:

```
Data sources classified by access protocol (23/43)
             DATA SOURCES CLASSIFIED BY TYPE OF SERVICE

See also categorization exist by subject[1] .

World-Wide Web[2]        List of W3 servers . See also: about the WWW
                         initiative[3] .

WAIS[4]                  Find WAIS index servers using the directory of
                         servers[5]. or lists by name[6] or domain[7].
                         See also: about WAIS[8] .

Network News[9]          Available directly in all www browsers. See also
                         this list of FAQs[10].

Gopher[11]               Campus-wide information systems. etc. listed
                         geographically. See also: about Gopher[12].
```

```
Telnet access[13]        Hypertext catalogues by Peter Scott. See also:
                         list by Scott Yanoff[14] . Also, Art St George's
                         index[15] (yet to be hyperized) etc.

VAX/VMS HELP[16]         Available using the help gateway[17] to WWW.
```

`1-26, Back, Up, <RETURN> for more, Quit, or Help:`

Take a look at link **[10]**. It shows you that you can use the Web to access the frequently asked question (FAQ) lists. (We discuss FAQ lists in Chapter 9.) This should be a good place to look for something interesting. Enter:

 10

Your browser follows the link and displays the first part of another document:

```
                         List of USENET FAQs
                         USENET FAQS

This document contains a list of all USENET FAQs found in
news.answers. The document is alphabetized by topic (more or
less). Many of the FAQs in this list are presented in the
same format as they appear in the newsgroup, while others
have been further processed and split into additional documents.
For more information on all aspects of this project, see the
technical notes.[1]

Please send comments and complaints to fine@cis.ohio-state.edu.

   3b1 FAQ[2]
   3b2 FAQ[3]
   Acorn[4]
   Active Newsgroups[5]
   Address Book[6]
   Ai FAQ[7]
```

`1-282, Back, Up, <RETURN> for more, Quit, or Help:`

Notice the numbers in the bottom left-hand corner: **1-282**. This shows you that there are a total of 282 links in this document (of which you can see **1** through 7). At this point, there are several commands you can use to examine the rest of the document.

To look at the next screenful of information, press RETURN. To re-display the previous screenful of the current document, enter the command **up**. To start again at the beginning of the document, enter **top**. To go to the end of the document, enter **bottom**.

Hint

Most browser commands can be abbreviated. In the next section, you will find a table that summarizes all the commands and shows the abbreviations.

Remember, your goal is to find something interesting that most people don't know. Press RETURN a few times and look at the next few screenfuls of the document. Eventually, you will see the following:

```
Canadian Football League (CFL) - Frequently Asked Questions [38]
```

Maybe you can find something here. Tell your browser to follow this link, by entering:

 38

This particular link points to an article in the form of a text file, which your browser obligingly fetches. The browser displays the article one screenful at a time. As you read, you can use the same commands that we mentioned earlier: RETURN, **up**, **top**, and **bottom**. So, pressing RETURN, you work your way through the article and come upon the following paragraph:

```
    The Montreal Alouettes folded shortly before the start of the
1987 season, forcing the league to eight teams.  In 1992, the
league announced a plan to first go to twelve, then eventually
sixteen teams, including U.S. teams.  However, these plans were
then reduced to ten, with the two teams being the Sacramento Gold
Miners and a team in San Antonio.  However, only Sacramento will
be joining the league for 1993.
```

Now, here's something interesting that most people don't know: the Canadian Football League has a team in Sacramento (California), with a new team planned for San Antonio (Texas).

Having fulfilled your quota of new knowledge for the day, you decide to quit the Web. Enter:

```
quit
```

The connection is broken and your browser stops.

Commands to Use with the Line Mode Browser

In the previous section, we used the Web's line mode browser for a bit of exploring. In this section, we will go over all the commands, including a few that we have not yet mentioned. If you use a different browser, the commands will, of course, be different. Still, the basic concepts will be similar to what you have already seen.

One nice thing about the line mode browser's commands is that they can all be abbreviated. Most of the commands require only a single letter (such as **h** for **help**); a few require two letters (such as **bo** for **bottom**). The only exception is **quit**, which you must spell out in full. This prevents you from inadvertently aborting your Web session prematurely.

At any time, you can display a quick command summary by using the **help** command. If you would like more detailed assistance, use the **manual** command. This will jump to a hypertext document that contains links you can follow for specific information.

When you enter the **help** command, you will also see some technical information about the current document. For example, if we entered **help** while we were reading the list of FAQs (in the previous section), we would have seen:

```
You are reading
 "List of USENET FAQs"
whose address is
 http://www.cis.ohio-state.edu/hypertext/faq/usenet/top.html
```

The first part of this information reminds us what we are reading. The second part shows us the official technical description of the document's Internet location. This information is called a *Uniform Resource Locator* or *URL*.

Within the Internet, every resource can be described by a URL. We won't go into all the details of URLs, but, for the most part, they are not too hard to understand. The first part of a URL specifies what type of resource is being described, the next part shows the address of the resource.

In this case, the first part is **http**. This indicates we are looking at a hypertext document. (The name **http** refers to Hypertext Transport Protocol, the protocol used by the Web to move data from place to place.) After this, we see the name of a computer at the Computer and Information Services Department of Ohio State University (**www.cis.ohio-state.edu**). Finally, we see a path name on this computer (**hypertext/faq/usenet/top.html**).

FUN TIP: *The Web really shows off the power of the Internet. In our example, we use a computer in Switzerland to find a document in the United States (at Ohio State University) that contains information about the Canadian Football League.*

If the technical details of the URL don't mean anything to you, don't worry about it. The reason we mention it is because, one day, you may encounter a URL that you want to access using the Web. To do so, you can use the **go** command. Simply type this command, followed by the entire URL.

For example, let's say that one of your friends tells you about a cool document that you should check out. All you have to do is ask him for the URL. You can then use the **go** command to access the document. For example:

```
go http://www.cis.ohio-state.edu/hypertext/faq/usenet/alt-sex/faq/part2/faq.html
```

To continue, there are several commands that you can use to move within a document. To display the next page, just press RETURN. To re-display the previous page, use the **up** command. To move to the beginning of the document, use **top**. To move to the end of the document, use **bottom**.

The next set of commands allow you to use links. The simplest command is to follow a link. To do so, all you need to do is enter its number. For example, to jump to a link shown as **[15]**, enter:

15

If you want a quick list of all the links in the current document (without any of the text), use the **list** command. This command will also show you the URLs of all the links.

Hint

If there are a lot of links in the current document, the output from the **list** command may scroll by so fast that you cannot read most of it. To slow things down, you will have to use whatever pause key is used with your system. For example, with a Unix system, you can press CTRL-S to pause the output and CTRL-Q to continue. With a PC, you can press the PAUSE key to pause and any other key to continue.

Once you have been exploring for a while, you may want to return to a document that you looked at earlier in the session. Use the **recall** command to display a list of all such documents. Here is a sample list:

```
Documents you have visited:-

R   1) in Overview of the Web
R   2) in Data sources classified by access protocol
R   3) in List of USENET FAQs
R   4) Canadian Football League (CFL) - Frequently Asked Questions
```

The word "in" means that you have passed through this document to get to where you are.

To jump to one of these documents, use the **recall** command followed by the appropriate number. For example:

```
recall 3
```

There are several other commands that you can use to jump to other documents. To return to the very first document that you saw when you started, use the **home** command. To move back to the previous document in the recall list (the one you just came from), use the **back** command.

Hint

When using the Web, the two most common things to do are:

- enter a number to follow a link forward
- use the **back** command to move one step backward

In other words, entering a number moves you "in"; the **back** command moves you "out". If you use **back** repeatedly, you will follow your path backwards and, eventually, end up at the home document (the very first document you saw).

The next two commands allow you move within a sequence of documents. To jump to the next link from the last document, use **next**. To jump to the previous link from the last document, use **previous**.

At first, these two commands can be a little confusing, so here is an example. Remember that in the last section we looked at a document that had links to FAQ lists. We saw the following list:

```
3b1 FAQ[2]
3b2 FAQ[3]
Acorn[4]
Active Newsgroups[5]
Address Book[6]
Ai FAQ[7]
```

Let's say that you want to jump to the **Acorn** document. Simply enter **4**. While you are reading the **Acorn** document, you can use the **next** command to jump directly to the **Active Newsgroups** document; if you use **next** again, you will jump to the **Address Book** document; and so on.

The **previous** command makes the same type of jump in the opposite direction. For example, if you are reading the **Address Book** document and you use the **previous** command, you will jump to the **Active Newsgroups** document.

The last command we want to mention is used to perform a search. From time to time, you will follow a link that leads to an index. (Remember, "index" is Web terminology for something that can be searched.) Many indexes are actually Wais sources. When you jump to an index, you will see the following on the bottom line of your screen:

```
FIND <keywords>, Back, Up, Quit, or Help:
```

When you see "FIND" it means you are within an index.

You can now type **find**, followed by as many keywords as you want. Your browser will send the search request to the appropriate server and display the results when they arrive. For example, you might enter:

```
find history eighteenth century
```

When you perform a search, there is a shortcut you can use. If the first keyword is not the name of a browser command, you do not have to type **find**. For example, the following two commands would be equivalent:

```
find history eighteenth century
history eighteenth century
```

However, in the following case, you would have to specify **find** as the first keyword happens to be the name of a command:

```
find top performer
```

For reference, Table 24-2 contains a summary of all the basic commands that we discussed in this section.

Finding a Browser

We mentioned earlier that, if you get serious about the Web, you should definitely use a browser on your own computer. If one is not already installed, ask your system manager if he or she could install one for everybody.

There are browser programs for a wide variety of different systems available via Anonymous FTP from **info.cern.ch**. For more information, look in the directory named **/pub/www** for a file named **README.txt**.

If you have an X Window system, we recommend that you try the Mosaic browser. Mosaic was written by Marc Andreessen of the National Center for Supercomputer Applications (NCSA) at the University of Illinois at Urbana.

Mosaic has a large number of advanced features, such as embedded graphics and the launching of multimedia applications. It also uses colors to help you keep track of links. For example, Mosaic remembers every document you have ever read. When a link points to something you have already seen, Mosaic will draw that link using a special color. This helps you to avoid wasting your time

Command	Full Name	Description
—	`number`	follow the specified link
`b`	`back`	jump to previous document in recall list
`bo`	`bottom`	jump to the end of the current document
`f` *words*	`find` *words*	use specified keywords to search index
`g` *UDL*	`go` *UDL*	jump to specified document or resource
`h`	`help`	display command summary and technical info
`ho`	`home`	jump to the starting document
`l`	`list`	display a list of links within current document
`m`	`manual`	jump to the browser reference document
`n`	`next`	jump to the next link within the last document
`p`	`previous`	jump to previous link within the last document
—	`quit`	quit the Web
`r`	`recall`	display list of previously-visited documents
`r` *number*	`recall` *number*	jump to specified document from recall list
—	RETURN	display next screenful of the current document
`t`	`top`	jump to the beginning of the current document
`u`	`up`	display previous screenful of current document

Table 24-2. *Summary of Line Mode Browser Commands*

on documents you have already seen. It also makes it easy to follow your previous path through unknown territory.

Another nice feature is that Mosaic allows you to attach personal notes within any document. These notes will appear each time you read the document. (Of course, the notes are private; only you can see them.)

Mosaic looks and feels like a Motif application, although you don't need to have Motif on your computer. (Motif is a graphical user interface that runs on top of X Window.)

Installing Mosaic is easy, as it comes all ready to run. All you need to do is download the program, uncompress it (using the **uncompress** command) and make it executable (using the **chmod** command). There is nothing to set up.

Mosaic is available via Anonymous FTP from **ftp.ncsa.uiuc.edu**. The actual name of the program is **xmosaic**. Change to the directory **/Web/xmosaic-binaries** and look for the version for your system.

If you get interested in the World Wide Web, there is a Usenet newsgroup that you can subscribe to called **comp.infosystems.www**. For a discussion of hypertext, subscribe to **alt.hypertext**.

(We discuss X Window in Chapter 2, Anonymous FTP in Chapter 16, uncompressing files in Chapter 18, and Usenet in Chapter 9.)

CHAPTER 25

I n this chapter, we will show you how to access a huge network of discussion groups: one that is carried entirely by mail. To participate in these discussions, you do not need to use Usenet or any other Internet services. All you need to know is how to send and read mail (which we discuss in Chapters 5 and 6).

There are thousands of discussions going on right now on every topic imaginable. And to participate, all you need is an electronic mail address.

Mailing Lists

What Is a Mailing List?

Like all successful Internet services, mailing lists are based on a simple idea. When you mail a message to someone, you specify an address. If you want to mail a message to more than one person, you can set up a special name, called an *alias*, that represents a group of people.

For example, let's say you set up an alias called **executives** to represent the addresses of three people named Curly, Larry, and Moe. Whenever you mail a message to **executives**, the mail program will automatically send it to each of these three users.

Imagine how these three people could use this alias to have a discussion group. Say that Curly gets an idea that he wants to share with the others. All he has to do is mail a message to **executives** and everybody gets a copy. Now let's say that Moe wants to comment on something in Curly's message. He sends his own message to **executives**. Again, the message is automatically sent to everybody on the list.

Now, think about the same sort of thing on a larger scale. Imagine an alias that contains the mail addresses of tens or even hundreds of users, scattered all around the Internet. Any message sent to the alias will be automatically sent to everyone in the group. People can talk, argue, help one another, discuss problems, share information, and so on. Everything that anyone says goes to all the people in the group.

This is a *mailing list*. As an Internet user, you have access to several thousand such lists, each of which is devoted to a specific topic.

How Are Mailing Lists Different from Usenet Newsgroups?

In Chapters 9 and 10, we discussed Usenet, the worldwide collection of discussion groups. (In Usenet, we refer to the discussion groups as newsgroups even though they do not contain actual news.) We mentioned that there are over 5,000 different Usenet newsgroups.

Many of these newsgroups are of local or regional interest, and even large computer systems will usually carry no more than a couple of thousand newsgroups. Still, that is a huge variety. To see for yourself, take a moment to glance at Appendix G, which contains descriptions of many of these newsgroups.

Since Usenet offers such a large variety of discussion topics, it is natural to ask how Usenet newsgroups differ from mailing lists.

The first big difference is that you have to learn a lot more to participate in Usenet. In particular, you have to learn how to use a newsreader program. Such programs are complex, and it may take a while for you to feel at home with Usenet. With a mailing list, all you need to know is how to send and read mail, something which you should probably learn anyway. This means that people who do not have access to Usenet—or don't want to learn how to use it—can still participate in discussion groups.

The next difference is that the discussions in a mailing list come to you in the form of messages sent to your personal mailbox. Some lists have only a few messages a day. However, it is not uncommon for a busy list to generate dozens of messages a day. Since all these messages show up in your mailbox, you have to do something with them. Usenet articles, on the other hand, are stored in a central location on your network and are administered by a system manager. When you participate in mailing lists, it is not uncommon to return from a two week vacation and find hundreds of messages in your mailbox. (However, as we will explain later in the chapter, it is usually possible to tell a mailing list to stop sending you messages temporarily.)

Hint

Mailing lists can generate a lot of messages. Although you can subscribe to as many mailing lists as you want, it is best to confine yourself to no more than five. Otherwise, you are guaranteed to find your mailbox constantly filled with unread messages.

One of the nice things about mailing lists is that you can choose a few that interest you and count on getting the messages automatically. There is nothing for you to do but read your mail. With Usenet, you have to start your newsreader program and check your favorite newsgroups every time you want to see what has arrived. Moreover, most system managers automatically delete Usenet articles after they have been around for a fixed period (anywhere from 1-2 days to several weeks).

However, Usenet is more convenient in other ways. As a Usenet user, you only need to participate when you want. This means that you can drop in and out of discussions as the mood strikes you. Moreover, it is easy to sample a variety of newsgroups quickly. With a mailing list (as you will see), you have to send a special mail message to be put on the list and another message to be taken off.

FUN TIP: *The best part about both Usenet and mailing lists is that they are free.*

Hint

To use a mailing list, there are two main ideas you need to understand. First, you should understand the basic concepts behind the Internet mail system and how to use a mail program. Second, you need to know how to read and understand Internet addresses.

If you feel like you need a quick review, you might want to take a few moments to skim Chapters 4 and 5 where we discuss these topics. If you use the standard Unix mail program, you can also read Chapter 6. Otherwise, you will have to find some documentation (or a friendly person) to help you with whatever mail program you are using.

Moderated and Unmoderated Mailing Lists

Each mailing list has an administrator. In most cases, this is one person, variously referred to as the list manager, administrator, or coordinator. The main job of the administrator is to keep the list of mail addresses up to date.

All mailing lists have an official address. With most lists, every message that is sent to that address is automatically passed on to everyone on the list. This means that anyone can contribute to a mailing list just by sending a message to the appropriate address.

When you receive a message, it will be sent to you from that same address. Thus, if you use the built-in feature of your mailing program to reply to the sender, your reply will go to everyone on this list.

Hint

Before you reply to any message, always ask yourself if it would be better to send a private response to the person who sent the message. (This person's name will be in the message.) In many cases, it is more appropriate to send a private response than a message that will be sent to everyone in the group.

Some mailing lists are *moderated*. This means that all the messages go to one person called the *moderator* (which may or may not be the same person as the administrator). The moderator decides which articles should be sent out to the members of the list. Most moderators also perform some basic editing and organization on the raw material.

Some moderators will organize messages into a collection called a *digest*. A digest is like an issue of an electronic magazine: a whole set of messages and articles in one easy to read package. Some moderators will include a table of contents with each digest so it is easy to find the messages that interest you.

The advantage of a moderated mailing list is that you see only the best messages (in someone's opinion, anyway). Many unmoderated mailing lists have a lot of boring and redundant messages that you will have to wade through to find the jewels.

The main disadvantage of a moderated list is that maintaining it is a lot of work. Moreover, the only compensation that moderators receive is that warm feeling that comes to those who help their fellows. (Thus, most mailing lists are not moderated.)

How Mailing Lists Are Administered

As we explained in the last section, the most important part of administering a mailing list is keeping track of the people on the list. When you request to be put on a mailing list, we say that you *subscribe* to that list. When you ask to be taken off the list, you *unsubscribe*. (Remember, though, even though we talk about "subscribing", there is no charge for the service.)

There are two basic ways in which mailing lists are administered. Some lists are maintained by a program. To subscribe, you send a message to a special address. All mail to this address is automatically processed by the list administration program. Since the message will be read by a program, you must use a particular format (which we explain later).

Other lists are maintained by a person. With these lists, you send a message that is read by the administrator, who manually adds or deletes people from the list. Although most lists are public, there are some private lists that you cannot join without the permission of the administrator.

The most common mailing list administration system is called *Listserv*. (The name stands for "list server".) The convention is to speak of a computer that provides this service as "a Listserv". For example, you might read, "To get basic information about subscribing to a mailing list, send a message with the word **help** to any Listserv."

The Listserv system was developed to coordinate mailing lists on the large Bitnet network (described in the next section). We will discuss Listserv and the commands that it uses later in the chapter.

Bitnet

Bitnet is a worldwide network—separate from the Internet—that connects well over a thousand academic and research institutions in more than 40 countries. Many of the Bitnet sites are IBM mainframe computers running the VM operating system and supporting hundreds of users. Thus, Bitnet serves a very large number of people. In this section, we will explain a little about Bitnet because, as you will see, it is the source of many of the mailing lists to which you have access.

Bitnet began in 1981 as a small network of IBM mainframe computers at the City University of New York (CUNY). The name Bitnet was chosen to stand for the "Because It's Time Network". (No doubt, this name would make sense to you if you were using an IBM mainframe computer in New York in 1981.)

In the United States, Bitnet is associated with Educom, a non-profit consortium of educational institutions, and is administered by the Corporation for Research and Educational Networking (CREN). Educom supports the Bitnet Network Information Center (Bitnic). The job of Bitnic is to promote the use of Bitnet in higher education.

Outside the U.S., Bitnet is known by different names. In Canada, it is called Netnorth. In Europe, it is the European Academic Research Network (EARN). In Latin America and Asia, you will find other names for Bitnet.

In Chapter 2, we discussed protocols (technical specifications) and explained how the Internet is based on a family of protocols called TCP/IP. Bitnet is based on a family of IBM protocols called RSCS (Remote Spooling Communications Subsystem) and NJE (Network Job Entry).

Historically, Bitnet developed within a technology that did not allow for systems like Usenet. Thus, an elaborate mailing list system, based on Listserv, developed.

Hint

If you hang around Bitnet, you will see many common abbreviations in the names of computers and mailing lists: **vm** (the name of the principle IBM mainframe operating system), **cuny** (City University of New York, where Bitnet was developed), **bitnic** (Bitnet Network Information Center, a main site of Bitnet in the U.S.), **earn** (European Academic and Research Network, the European part of Bitnet), and so on.

You will also find that Bitnet people tend to use acronyms and commands that are all uppercase letters, like BITNET and LISTSERV. This is part of the IBM mainframe culture. Compare this to our more genteel Internet traditions, which encourage the use of terminology that is almost exclusively lowercase.

FUN TIP: You, can learn a lot about a culture by observing how it uses upper- and lowercase letters.

Bitnet Mailing Lists and Usenet

Many (but not all) of the Bitnet lists are available via Usenet as well as through the mail. Each message that is sent to one of these lists is also sent to a Usenet newsgroup which you can read in the regular fashion. All these Bitnet newsgroups are in the **bit** hierarchy and have names that start with **bit.listserv**. (We discuss newsgroup names in Chapter 9.)

For example, the Bitnet mailing list named **film-l** is devoted to film and the cinema. Messages to this list can also be read as Usenet articles in the newsgroup **bit.listserv.film-l**. Moreover, when you send an article to the Usenet group, it will be forwarded to the list itself and, from there, to all the subscribers. (Although, this doesn't always work properly.)

If you would like to see a list of the Usenet newsgroups that contain Bitnet mailing lists, look at Appendix G and read the descriptions of the newsgroups in the **bit** hierarchy.

> ## Hint
>
> Many Bitnet mailing lists have names that end in **-l** (the "l" stands for list).

The actual work of sending messages between Usenet and Bitnet is done by a computer called an *Bitnet/Usenet gateway*. There are a number of such computers that act as gateways for the various lists. If you would like more information about the gateways or about the Bitnet/Usenet newsgroups, there are three articles you can read.

The first article, named **bitlist**, contains a master list of all the Bitnet-oriented newsgroups, with a one-line description of each group. The second article, named **gatelist**, shows each Bitnet mailing list, its corresponding newsgroup, and the name of the gateway. The third article is named **policy**. It provides a general explanation of Usenet, Bitnet, and how they are connected.

All three articles are posted regularly to the newsgroups **bit.admin** and **news.answers**. They are also available via Anonymous FTP from **rtfm.mit.edu** in the directory **/pub/usenet/news.answers/bit**.

Sending Mail to Bitnet

To participate in a mailing list, all you have to do is find the list and send a message asking to subscribe. We will explain how to do all of that later in the chapter. First, though, we need to take a moment to discuss how to mail a message to Bitnet. That is because many of the mailing lists that you will encounter are Bitnet lists. To subscribe (and unsubscribe), you will need to send a message to a Listserv. And once you subscribe, you will need to send messages to the list itself in order to participate. In both cases, you may find yourself having to deal with a Bitnet address.

Before we start, we would like to remind you that we discuss Internet addresses in Chapter 4. At the end of that chapter, we explain how to mail a message to Bitnet. You might want to take a moment to review that material before you go on.

Within Bitnet, addresses are simple. All you need is a user name and a computer name. For example, say that a Bitnet user has a friend whose user name is **lunaea** on the computer known as **psuvm**. He or she could mail a message to the friend by using the address **lunaea@psuvm** (or as a Bitnet person would put it, **LUNAEA@PSUVM**).

From the Internet, we cannot send mail directly to Bitnet. It must first go through a *Bitnet/Internet gateway*: a computer that is connected to both networks. If we wanted to send a message to the same person we mentioned above, we would have to send a message to a gateway and tell it: "Please take this message and pass it to the Bitnet computer named **psuvm** and, once there, have that computer deliver the message to **lunaea**."

On many Internet systems, this is easy. All you have to do is use the following address format:

```
username@computer.bitnet
```

For example:

```
lunaea@psuvm.bitnet
```

What happens is that the mail program on your system recognizes the name **bitnet** (called a pseudo domain) and knows how to send your message to a Bitnet/Internet gateway.

Some Internet systems, however, are not set up to recognize the **bitnet** pseudo domain. If your system is like this, your mail will come back marked "Host unknown" (like in the Elvis Presley song). In such cases, you will have to send your Bitnet messages directly to a gateway. These days, many Bitnet computers are also on the Internet and can act as gateways between the two networks. For reference, the list below shows a number of such gateways that you can use.

cornellc.cit.cornell.edu
cunyvm.cuny.edu
mitvma.mit.edu
pucc.princeton.edu
vm1.nodak.edu

When you send mail via a gateway, there are two addressing formats that you can use. The preferable one uses the UUCP bang path notation (that we discussed in Chapter 4):

```
gateway!computer.bitnet!userid
```

For example, to send a message to **lunaea** on the Bitnet computer named **psuvm**, you might use the following address:

```
cornellc.cit.cornell.edu!psuvm.bitnet!lunaea
```

Hint

If you are using a Unix system with the C-Shell, the ! character will have a special meaning. Consequently, when you enter an address that contains ! characters, you will see the following error message:

```
Event not found.
```

If this happens, you must place a \ (backslash) character in front of each ! character. For example:

```
cornellc.cit.cornell.edu\!psuvm.bitnet\!lunaea
```

The \ character tells the C-Shell that the next character is to be taken literally.

The second format uses the % notation (also discussed in Chapter 4):

userid%*computer*.bitnet@*gateway*

For example:

```
lunaea%psuvm.bitnet@cornellc.cit.cornell.edu
```

Hint

When sending mail via a Bitnet/Internet gateway, it is better to use the bang path format. The % format is not officially supported anymore (although it usually works). The bang path format will always work and is easier to understand.

Here are two real life examples of how to mail messages to Bitnet. To send a message to a Listserv, you use a user name of **listserv**. Thus, here are three addresses that you might use to send a message to the Listserv at the Bitnet computer named **templevm**:

```
listserv@templevm.bitnet
cunyvm.cuny.edu!templevm.bitnet!listserv
listserv%templevm.bitnet@cunyvm.cuny.edu
```

In the earlier examples, we used the gateway at Cornell University. In these examples, we use the gateway at CUNY. It really doesn't matter which one you use, although you may get faster service if you use the gateway closest to you.

Here is our second real-life example. Most Bitnet lists are administered by a single Bitnet computer. When you read a description of a mailing list, you will typically see the name of the list and the name of the computer. For example, the list named **help-net** resides at **templevm**. When you read about it, it will be called **help-net@templevm**. (By the way, the **help-net** list is devoted to general questions and answers regarding Bitnet and the Internet.)

Let's say that you have already subscribed to this list. To mail a message to the list, you will have to send it to user name **help-net** at the computer **templevm**. As an Internet user, here are three addresses you could use:

```
help-net@templevm.bitnet
cunyvm.cuny.edu!templevm.bitnet!help-net
help-net%templevm.bitnet@cunyvm.cuny.edu
```

When a Bitnet computer is also on the Internet, it will have an Internet address as well as a Bitnet address. In such cases, you can use the Internet address (which will be faster). For example, if you wanted to send mail to the Listserv at CUNY, you could send it directly to:

```
listserv@cunyvm.cuny.edu
```

The last thing we would like to mention is that some lists can be administered by more than one Bitnet computer. These computers are called *peers*. In such a case, you can send mail regarding the list to any one of the peers.

Finding Mailing Lists

By now, you are probably wondering what mailing lists are available. Well, there are a lot of them, and they cover just about any topic you can imagine. Of course, there are new lists being formed all the time, and (less often) old lists disappearing.

If you want to find out what is available, there are a number of sources from which you can obtain a summary of mailing lists. Such a summary is usually referred to as a *list-of-lists*. In this section, we will show you where to obtain various lists-of-lists. Each one is put together by different people (usually volunteers) and covers a different territory.

Hint

Most of the lists-of-lists are large and take up a lot of disk space. If you copy a list-of-lists to your computer, we suggest that you delete the file when you are finished looking at it. There is no real reason to save a list-of-lists: it would just go out of date and you can get a newer version for free whenever you want.

The SRI List-of-Lists

One of the two largest list-of-lists is maintained by SRI International at their Network Information Systems Center (NISC) in Menlo Park, California. The purpose of this list-of-lists is to offer a description of every public mailing list—both Internet and Bitnet—along with instructions on how to subscribe.

This list is a useful, albeit very long reference. Unfortunately, the instructions are somewhat confusing because they were written by different people at different times. For example, there is really only one way to subscribe to a Bitnet list (as you will see later in the chapter). However, in the SRI list-of-lists, each Bitnet list has a separate description that shows you how to subscribe. Actually, all you really need to do is find out where a mailing list is, and then follow the instructions in this chapter.

There are two ways to get the SRI list-of-lists. First, you can download it (copy it to your computer) via Anonymous FTP from **ftp.nisc.sri.com**. Look in the directory named **/netinfo** for a file named **interest-groups.Z**. You will have to uncompress this file once you download it. (We discuss Anonymous FTP in Chapter 16. We discuss uncompressing files in Chapter 18.)

The second way to get this list is to send mail to **mail-server@nisc.sri.com**. In the body of the letter, put the message **send /netinfo/interest-groups**. (Note: The list is huge. It will arrive in the form of many separate messages that you will have to save and join together.)

The Dartmouth List-of-Lists

The other very large list-of-lists is maintained at Dartmouth University in New Hampshire. Like the SRI list, the Dartmouth list also attempts to keep track of every Internet and Bitnet mailing list in existence. The Dartmouth people edit their list to remove duplications and mailing lists of only local interest.

Before you download this list (which comes in parts), you should read a file that contains instructions. Use Anonymous FTP to connect to **dartcms1.dartmouth.edu**. This system is an IBM mainframe and it may look a little strange if you are used to Unix computers; still, the regular FTP commands will work.

Change to the directory named **siglists** (use **cd siglists**). Download the instruction file named **read.me** (use **get read.me**). Within this file, you will see several ways to obtain the list-of-lists. If you want to download it using regular FTP, use the command **get listserv.lists**.

Wais

When you are interested in a particular subject and you want to find a mailing list in that area, it is a bother to have to download and examine a large set of files. An easier way is to use Wais. All you have to do is search the source (Wais database) named **mailing-lists**. (We explain how to use Wais in Chapter 23.) This source contains several Bitnet and Internet lists-of-lists as well as a list of Usenet newsgroups.

Listserv Mailing Lists

You can get a summary of all the Listserv mailing lists on Bitnet by mailing a message to any Listserv. In the body of the message, put the command **list global**. The Listserv will mail you back a list in which each mailing list is summarized on a single line.

If you are interested in a particular subject, use the command **list global/***subject*, specifying whatever subject you want. For example, if you want a summary of all the lists that have something to do with poetry, use **list global/poetry**.

Here is an example of a message that requests the overall list. Notice that when you send commands to a Listserv you do not need to specify a Subject. (Remember, you can send this request to any Listserv.)

```
mail listserv@cunyvm.cuny.edu
Subject:
list global
```

The Internet List-of-Lists

There is a list-of-lists that keeps track of public Internet mailing lists. This list is updated regularly and posted to the Usenet newsgroups **news.lists**, **news.announce.newusers**, and **news.answers**.

You can download the list via Anonymous FTP from **rtfm.mit.edu**. Look for the files named **part1**, **part2**, and so on, in the directory **/pub/usenet/news.answers/mail/mailing-lists**.

Academic Mailing Lists for Scholars

A group at Kent State University in Ohio maintains a list-of-lists that describes mailing lists devoted to scholarly and academic subjects. If you are a student or teacher, this is a good place to look for a mailing list for your discipline.

You can download this list-of-lists via Anonymous FTP from **ksuvxa.kent.edu**. Once you log in, change to the **library** directory. (Use the command **cd library**.) For instructions and a description of the list-of-lists, download the file named **acadlist.readme**. The actual list is divided into parts and stored in the files named **acadlist.file1**, **acadlist.file2**, and so on.

Reviews of Mailing Lists

If you are interested in reviews of mailing lists, you can subscribe to the Bitnet mailing list named **lstrev-l@umslvma**. This is a service provided by Raleigh C. Muns. He subscribes to a mailing list for a while, then offers his comments to the world.

New Bitnet Mailing Lists

There is a Bitnet mailing list that announces all new lists. You can subscribe to this list in order to keep up on what is new and exciting. The name of the mailing list is **new-list@vm1.nodak.edu**. This list is also available on Usenet as the newsgroup **bit.listserv.new-list**.

Subscribing to a Bitnet Mailing List

To subscribe to a mailing list, all you have to do is mail a simple message. If the list is administered by a program, such as Listserv, you send the message to that program. If the list is administered by a person, you send a message to that person.

When you subscribe to a mailing list, there are only two important guidelines to remember. First, you must make sure that you send your requests to the address of the person or program that administers the mailing list. Do not send requests to the address of the list itself. Second, when you send a message to a program, you must use commands in a particular format.

Hint

Remember, many Bitnet mailing lists are also available on Usenet.

We can divide the whole world of mailing lists into two groups: Listserv-administered lists and manually administered Internet lists. When you subscribe to a mailing list, the first thing you have to do is figure out if it is administered by a Listserv. You can usually do so by reading the description of the list.

If you are dealing with a Listserv-administered list, there will be several clues. First, the description may tell you outright that to subscribe you must send a message to a particular Listserv. Second, the description may tell you that this is a Bitnet mailing list. (Virtually all Bitnet lists use Listserv.) Third, you may see instructions telling you to send a subscription message to an address that starts with the name **listserv**. Finally, you may recognize that address of the list as a Bitnet address.

Here is a typical example. Let's say you see the following description of a mailing list:

```
HELP-NET on LISTSERV@TEMPLEVM.Bitnet

Help-Net is a discussion list for the purposes of solving user
problems with utilities and software related to the Internet and
Bitnet networks.  In addition, LISTSERV at TEMPLEVM maintains a
set of low-level help files intended to help the beginning user
acclimatize himself to the network systems.
```

You can tell that you would use Listserv to subscribe to this mailing list: the top line contains the address of a Listserv. To subscribe to the mailing list, you would send mail to the Bitnet computer named **templevm**. Use the address:

```
listserv@templevm.bitnet
```

(If this address doesn't work, you can use one of the variations we discussed earlier.)

The mailing list itself will have an address at the same computer as the Listserv. Thus, once you have subscribed to the list, you can participate by sending mail to:

```
help-net@templevm.bitnet
```

Sometimes you will see a much briefer description of a mailing list. For example:

```
HELP-NET          HELP-NET@TEMPLEVM Bitnet/Internet Help Resource
```

In this case, all you really see is the address of the list itself, **help-net@templevm**. You have to be able to look at this address and realize that it refers to a Bitnet computer. (You can tell because the address looks like a Bitnet address and not like an Internet address.) Once you realize that you are dealing with a Bitnet mailing list, you know that you can subscribe by sending a message to the Listserv at the same address.

To subscribe to a Bitnet mailing list, mail a short message to the Listserv at the same address. You do not need to put a subject in the message. In the body of the message, put a single line that contains the word **subscribe**, followed by the name of the mailing list, followed by your full name. The following is a typical subscription message:

```
mail listserv@templevm.bitnet
Subject:
subscribe help-net Chuck Wagon
```

This message asks the Listserv at **templevm** to add the person named Chuck Wagon to the **help-net** mailing list. Notice that you do not have to specify your mailing address: the Listserv will figure it out from the header of your message.

To unsubscribe to a Bitnet mailing list, send a message to the Listserv that administers the list. In the body of the message, put a single line that contains the word **unsubscribe** followed by the name of the mailing list. For example:

```
mail listserv@templevm.bitnet
Subject:
unsubscribe help-net
```

Hint

What do you do if you accidentally spelled your name wrong when you subscribed to a Bitnet mailing list? Not to worry. Simply send in another **subscribe** message with your name spelled correctly. You do not need to unsubscribe just to change your name.

Aside from **subscribe** and **unsubscribe**, there is a whole set of commands that you can send to a Listserv. All you have to do is send a message to the Listserv, with a single line that contains the command. The Listserv will read your message, carry out the command, and mail you the results.

The first command you might want to use is **help**. This will send you a summary of all the basic commands. You can also use the **info** command to request information on a particular topic. Use **info ?** to ask for a list of topics.

One interesting command is **review**. Type this command followed by the name of a mailing list. For example, to ask for information about the **help-net** mailing list, use:

```
review help-net
```

The Listserv will send you information about the list, including a description of the list, and the names and addresses of all the subscribers.

The Bitnet Listservs have an interesting characteristic. Whenever you send a request for information, the Listserv will send you two messages. The first is a confirmation message; the second is the actual information. Here is a typical confirmation message that we received after sending the command **list global** to a Listserv:

```
From: BITNET list server at BITNIC (1.7f)
<LISTSERV@BITNIC.EDUCOM.EDU>
Subject:      Output of your job "harley"


> list global


Summary of resource utilization
─────────────────────────────────────  -

  CPU time:        4.997 sec         Device I/O:    158
  Overhead CPU:    1.358 sec         Paging I/O:    213
  CPU model:       9377              DASD model:   9335
```

The name **harley** that you see in the **Subject:** line is the userid that sent the mail to the Listserv. This name is inserted automatically.

A short time after we received this message, another message arrived with the information we requested. You will receive similar confirmation messages whenever you subscribe or unsubscribe to a mailing list, and (if you choose) whenever you send a message to a list.

To control certain aspects of a mailing list subscription, you can use the **set** command. This command tells the Listserv that administers a mailing list that you request certain options regarding that list. Here is how it works.

Send a message to the appropriate Listserv. Within the body of the message, put a single line that contains **set**, followed by the name of the mailing list, followed by the option you want. For example:

```
set help-net noack
```

This command tells the Listserv to set the **noack** option (explained below) for you with respect to the **help-net** mailing list. There are a number of options you can use, of which we will discuss the most important.

To tell the Listserv that you do not want to receive confirmation messages whenever you send mail to a list, use **noack** (no acknowledgment). To turn this service back on, use **ack**. The default is **noack**.

If you do not want to receive mail temporarily (say, while you are away on vacation), you can use the **nomail** option to tell the Listserv to stop sending you **mail**. When you want to start receiving mail again, use the **mail** option. When you use **nomail**, you are still subscribed to the mailing list. Thus, if you want to forget about the list completely, you should unsubscribe.

When you send a message to a mailing list, the Listserv that administers the list does not send you a copy of the message. If you would like to receive copies of your own messages, you can use the **set** command with the **repro** option. To turn this off again, use the **norepro** option.

For reference, Table 25-1 contains a summary of these basic Listserv commands. Notice that **signup** is a synonym for **subscribe**, and **signoff** is a synonym for **unsubscribe**.

Subscribing to an Internet Mailing List

You will find that some Internet mailing lists are administered by Listserv-like software running on an Internet computer. Still, from our point of view, they

Command	Function
`help`	send summary of basic commands
`info ?`	send list of information topics
`info topic`	send information about specified topic
`list global`	send summary of all Bitnet mailing lists
`review list`	send information about specified mailing list
`set list ack`	send confirmation messages for specified list
`set list noack`	do not send confirmation messages for specified list
`set list mail`	start sending mail from specified list
`set list no mail`	stop sending mail from specified list
`set list repro`	send copies of your own messages
`set list norepro`	do not send copies of your own messages
`subscribe list yourname`	subscribe to specified mailing list
`signup list yourname`	same as `subscribe`
`unsubscribe list`	unsubscribe to specified mailing list
`signoff list`	same as `unsubscribe`

Table 25-1.   *Summary of Important Listserv Commands*

both work the same way and you can subscribe and unsubscribe by using the same commands. (To be very technical, some people refer to the Bitnet service as "LISTSERV", and to a similar Internet service as "listserv".)

Most Internet mailing lists, however, are administered by a person. In such cases, you need to send that person a short message asking him or her to add (or delete) you from the list. The main thing you have to know is where to send your request.

Each Internet mailing list has an address in the standard Internet format (see Chapter 4). For example, there is an Internet mailing list named **musicals**, devoted to musical theater. The address of this list is:

> `musicals@world.std.com.`

By convention, administrative requests for Internet lists should be sent to an address similar to that of the list, but with the word **-request** added to the userid. For example, to subscribe to the **musicals** mailing list, you would send a message to:

> `musicals-request@world.std.com.`

Once you have subscribed, you can send messages to the list using the regular address. If you want to unsubscribe, you would again send mail to the **-request** address.

Here is another example. There is a list called **disney-comics** devoted to discussion of Disney-based comic books and characters. The address of the list is:

> `disney-comics@student.docs.uu.se`

But to subscribe or unsubscribe, you would send a message to:

> `disney-comics-request@student.docs.uu.se`

When you read about a mailing list, you can usually tell from the description if it is a standard Internet list. For example, the following is a description of the **musicals** mailing list. Notice the word **-request** within the "contact" address:

```
Contact:  musicals-request@world.std.com
Purpose:  This forum is intended for the general discussion of
musical theater, in whatever form it may take, but related non-
musical theater topics are welcome too.
```

When you subscribe to an Internet mailing list, remember that, in most cases, your message will be read by a human volunteer, not a program. Thus, be sure to be polite. Use a subject that says what you want, and include your full name and mailing address in the body of the message. Here is an example of how you might subscribe to the **musicals** mailing list:

```
mail musicals-request@world.std.com
Subject: subscription request
Please subscribe me to the musicals mailing list.
Thank you.
Harley Hahn    harley@fuzzball.ucsb.edu
```

Hint

Remember, you should never send a subscribe or unsubscribe request to the actual mailing list. If you do, everyone on the list will receive your request. Be sure to send all such requests to the special administrative address.

Catalog of Internet Resources

You will enjoy this catalog. We have spent a long, long time exploring the Internet and tracking down hundreds of interesting and useful resources. Moreover, the items in the catalog are really only starting places. Almost all of them will lead you to something else: there are many hours of enjoyment and fascination ahead of you.

It is, of course, our intention to maintain and enhance this catalog in further editions of this book and, towards that end, we would like your help. If you have a favorite Internet resource that does not appear in our list, we want to know about it.

All you have to do is describe the resource and mail it to us. If possible, use a format similar to what you see in the catalog. Mail your contributions to:

catalog@rain.org

Similarly, if you find an omission or inaccuracy, and you can take the time to drop us a note, we would appreciate it. Thanks for your help.

Within the catalog, you will see a variety of ways to access information: Usenet, the Gopher, Anonymous FTP, and so on. If you are not completely familiar within one of these, you can refer to the appropriate chapter for more information.

Chapter	Topic
4	Internet Addressing
5-6	Mail
7	Telnet
8	Finger
9-11	Usenet
12-15	Usenet Newsreaders
16	Anonymous FTP
17	Archie
21	Gopher
23	Wais
24	World Wide Web
25	Mailing Lists

To help you find your way around the catalog, here is a list of all the major categories that we cover. Within the catalog, each of these categories is in alphabetical order. In addition, each major category is indexed and appears in the general index at the end of the book.

Agriculture
Archeology
Art
Astronomy
Automobiles
Aviation
BBSs
Biology
Bizarre
Books
Business and Finance
Calculators
Chemistry
Children
Computer Culture
Computer Hardware
Computer Literature
Computer Networks
Computer Technology
Computer Vendors
Consumer Services
Cyberpunk
Economics
Education
Environment
Food and Drink
Fun
Games
Games (Video)
Geoculture
Geography
Geology
Government
Health
Historical Documents
History
Hobbies
Humanities
Humor
Humor Archives
Internet
Intrigue
Language

Languages
Law
Libraries
Literature: Authors
Literature: Collection
Literature: Titles
Locks and Keys
Magazines
Mathematics
Medicine
Military
Movies
MUDs
Music
News
Oceanography
Operating Systems
Organizations
Pets
Physics
Pictures and Sound
Politics
Programming
Publications
Religion
Science
Sex
Software
Software: Archives
Software: Internet
Software: Macintosh
Software: Utilities
Space
Sports and Athletics
Technology
Telephones
Television
Travel
Trivia
Unix
Usenet
Weather

Agriculture

Advanced Technology Information Network

Current market information, weather forecasts, events, and news relating to farming and agriculture.

How to access   Telnet: **caticsuf.csufresno.edu**
Login: **super**

Bee Biology

Mailing list information, details on honeybees and bumble bees, pollination, honey and related material.

How to access   1. Anonymous FTP: **sunsite.unc.edu**
Path: **/pub/academic/agriculture/sustainable_agriculture/ beekeeping/***

2. Listserv mailing list: **bee-l@albnyvm1.bitnet**

Commodity Market Reports

This server contains the agricultural commodity market reports compiled by the Agricultural Market News Service of the United States Department of Agriculture. There are over a thousand reports from all over the United States. Most of these reports are updated daily. Try searching for "portland grain."

How to access   Wais: **agricultural-market-news**

Iowa State University SCHOLAR System

Various publication databases, including agriculture, applied science and technology, biology, general business, and humanities.

How to access   Telnet: **isn.iastate.edu**
Login: **scholar**
Choose: **agri**

Not Just Cows

A guide to agricultural resources on the Net.

How to access   Anonymous FTP: **ftp.sura.net**
Path: **/pub/nic/agricultural.list**

PENpages

International Food & Nutrition Database, National Family Database, The 4-H Youth Development Database, agricultural and weather statistics, market news, newsletters, and drought information. This resource is provided by the Penn State College of Agricultural Sciences.

How to access	Telnet: **psupen.psu.edu**
	Login: Two-letter identifier for your state (i.e., **CA**)

Archaeology

Archaeological Computing

A bibliography of archaeological computing in BibTeX format.

How to access	Wais: **archaeological_computing**

Art

Arts Online

A bibliography of arts-related resources, available on the Internet and other networks.

How to access	Anonymous FTP: **nic.funet.fi**
	Path: **/pub/doc/library/artbase.txt.Z**

ASCII Cartoons

A selection of ASCII art, including the Simpsons, cows, smileys, spaceships, dragons, and Slimer.

How to access	Gopher: Universitaet des Saarlandes
	Address: **pfsparc02.phil15.uni-sb.de**
	Choose: **INFO-SYSTEM BENUTZEN** | **Fun** | **Cartoons**

OTIS Project

OTIS (Operative Word Is Stimulate) distributes original artwork and photographs over the network for public perusal, scrutiny, and distribution. OTIS also offers a forum for critique and exhibition of your works. A virtual art gallery that never closes and exists in an information dimension where your submissions will hang as wallpaper on thousands of glowing monitors.

How to access	Anonymous FTP: **sunsite.unc.edu**
	Path: **/pub/multimedia/pictures/OTIS**

Astronomy

Astronomy

Programs for all popular systems, texts, documents, pictures, news and equipment information about astronomy and star gazing.

How to access   Anonymous FTP: **nic.funet.fi**
Path: **/pub/astro/***

Automobiles

Automobile Discussion Groups

Everything to do with automobiles, including design, construction, service, tires, karting, competitions, driving, manufacturers, antique cars, and much more.

How to access   Usenet: **rec.autos.tech, rec.autos.sport, rec.autos.driving, rec.autos.vw, rec.autos.antique, rec.autos, alt.hotrod, alt.autos.rod-n-custom, alt.autos.karting, rec.audio.car**.

Automotive Archives

Mailing lists, archive and new user guides, consumer automotive FAQs, and other material about automobiles, automotive products and laws.

How to access   Anonymous FTP: **rtfm.mit.edu**
Path: **/pub/usenet/rec.autos/***

Driving

Material concerning driving, parking tickets, traffic laws, insurance, highway patrol and other aspects of driving in California.

How to access   Anonymous FTP: **rtfm.mit.edu**
Path: **/pub/usenet/news.answers/ca-driving-faq**

Motorcycles

Great roads list and survey, picture guide, and other material all about motorcycles.

How to access   Anonymous FTP: **ftp.cs.dal.ca**
Path: **/comp.archives/alt.surfing/***

Aviation

Aviation Archives

In-flight cockpit visits to aviation jokes, humor, trivia, and so on.

How to access　　Anonymous FTP: **rascal.ics.utexas.edu**
Path: **/misc/av**

Aviation Gopher

This gopher has bundles of information about aviation, and contains most of the items that appear on or are covered in the Usenet newsgroup **rec.aviation**. It offers numerous articles, pictures, stories, and also has weather and fly-in information.

How to access　　Gopher: Northwestern University
Address: **av.eecs.nwu.edu**

DUATS

Aviation weather, PIREPS, and file flight plans.

How to access　1. Telnet: **duat.gtefsd.com23** (for pilots only)

2. Telnet: **duat.contel.com** (pilots only)

3. Telnet: **duats.contel.com** (for non-pilots)

Flight Planning

Public domain flight-planning software and data, written in C and complete with source.

How to access　1. Anonymous FTP: **eecs.nwu.edu**
Path: **/pub/aviation/***

2. Anonymous FTP: **lifshitz.ph.utexas.edu**
Path: **/pub/aviation/***

Flight Simulators

Flight simulation theory, products, reviews, scenery for specific software, and other material related to air and spacecraft simulators.

How to access　1. Anonymous FTP: **rtfm.mit.edu**
Path: **/pub/usenet/news.answers/aviation/
flight-simulators**

2. Anonymous FTP: **ftp.ulowell.edu**
Path: **/msdos/Games/FltSim/***

3. Anonymous FTP: **ftp.iup.edu**
Path: **/flight-sim/***

4. Anonymous FTP: **onion.rain.com**
   Path: **/pub/falcon3/***

5. Anonymous FTP: **kth.se**
   Path: **/kth/misc/fltsim.tar.Z 6.**

6. Anonymous FTP: **nic.funet.fi**
   Path: **/pub/X11/contrib/acm2.4.tar.Z**

7. Wais: **flight_sim**

Flying

Material about learning how to fly, technical information, ownership costs, equipment guides, aviation policies, and many more related FAQs.

How to access   Anonymous FTP: **rtfm.mit.edu**
                       Path: **/pub/usenet/news.answers/aviation/***

Jumbo Jets

Archives and information about airliners from the Usenet newsgroup **sci.aeronautics.airliners**. Includes airliner specifications and Boeing codes.

How to access   Anonymous FTP: **ftp.eff.org**
                       Path: **/pub/airliners/***

BBSs

AfterFive

Both a BBS and a MUD chat-like system in one. The BBS has lots of active topics while the chat is based on the colorful adults-only Bourbon St. in New Orleans. Open 5pm until 8am CST weekdays, and 24 hours on weekends.

How to access   Telnet: **af.itd.com 9999**

Auggie BBS

A very varied BBS, with a wide spread of discussion boards, public files, chat and talk facilities. Friendly people are always ready to chat, day or night, through the numerous online communication programs.

How to access   Telnet: **bbs.augsburg.edu**
                       Login: **bbs**

CetysBBS

A Mexican BBS, with many Spanish language discussion groups. Interesting place to practice your Spanish, but English speakers are welcome also.

| **How to access** | Telnet: **infux.mxl.cetys.mx** |
| | Login: **cetysbbs** |

Csb/Sju BBS

This easy to use menu-driven BBS, with its simple and unique message navigating commands and use of graphics, allows for quick scanning of all the different topics it has to offer.

| **How to access** | Telnet: **tiny.computing.csbsju.edu** |
| | Login: **bbs** |

CueCosy

A conferencing system in Canada allowing posting and reading of messages on many topics. There is a special education section called TIX, the Teachers Information Exchange.

| **How to access** | Telnet: **cue.bc.ca** |
| | Login: **cosy** |

Cybernet

Offers many interesting and varied resources, including Internet mail, limited Usenet, Hytelnet, finger, talk, multi-user chat, file downloading, games, and even a match-making program for lonely hearts.

| **How to access** | Telnet: **cybernet.cse.fau.edu** |
| | Login: **bbs** |

DUBBS

A quiet little BBS in the Netherlands, offering message and bulletin systems, and lots of downloadable computer files. If you have a flair for Dutch, check it out.

| **How to access** | Telnet: **tudrwa.tudelft.nl** |
| | Login: **bbs** |

Eagles' Nest BBS

Offers lots of variety in its discussion groups, two public chat rooms, and an IRC link with Auggie BBS.

How to access	Telnet: **seabass.st.usm.edu**
	Login: **bbs**
	Password: **bbs**

Endless Forest BBS

Some say alternate space/time continuums exist for the known reality—the Endless Forest is one. Here the Forest dwellers roam, purely for the exchange of technical information, controversial debate, inane babble, and general fun.

How to access   Telnet: **forest.unomaha.edu**
Login: **ef**

European Southern Observatory Bulletin Board

A bulletin board system for people involved in, or interested in, the European Southern Observatory. Discussion and information regarding astronomy and telescopes.

How to access   Telnet: **bbhost.hq.eso.org**
Login: **esobb**

Foothills Multi-User Chat

A very popular chat system, and a great place to relax and talk. Foothills provides a nice secure environment in which you can converse, with a multitude of features that make life easier.

How to access   Telnet: **vulture.dcs.king.ac.uk 2010**

Freenets via Gopher

Access the Freenets via Gopher menu. Simply choose the Freenet you wish to reach from the menu and the connection will be made.

How to access   Gopher: Yale University
Address: **yaleinfo.yale.edu**
Choose: **The Internet | FREE-NET systems**

ISCA BBS

The biggest and most popular BBS on the Internet, allowing a massive 310 users online at one time. There are discussion groups to fit all tastes, especially some of a more esoteric nature that seem to be lacking from Usenet. ISCA is often full, with users from all over the globe, busily rambling away.

How to access   1. Telnet: **bbs.isca.uiowa.edu**
Login: **guest**

2. Telnet: **whip.isca.uiowa.edu**
Login: **guest**

Internet BBS Access via Gopher

Access more BBSs than you can imagine through Gopher. Just choose the BBS you wish to use from the large list generated, and you will be connected. Get rid of those long cumbersome BBS lists.

How to access   Gopher: Use veronica to search for **Bulletin Board**.

Internet BBS Lists

CC's list, Zamfield's BBS list, and Yanoff's Internet list together form the most complete updated source of all the Internet BBS addresses and like-minded services that you will ever need. Endless hours of roaming the Net await amidst their pages.

How to access   1. Mail: **bbslist@aug3.augsburg.edu**

2. Anonymous FTP: **aug3.augsburg.edu**
   Path: **files/bbs_lists**

Launchpad BBS

Much more than a BBS, this Internet Service Mediator welcomes all new users with open arms. Offering complete Network News, local mail, Wais, Gopher, and access to many other information systems, it is well worth investigating.

How to access   Telnet: **launchpad.unc.edu**
Login: **launch**

Monochrome

Monochrome is a sophisticated multi-user messaging system. This includes its local messages, multitudes of files, a multi-user talker, and a unique presentation capability which takes full advantage of your terminal type, throwing delightful quotes and scrolling messages at you constantly.

How to access   Telnet: **mono.city.ac.uk**
Login: **mono**
Password: **mono**

NSYSU BBS

A friendly BBS, located in the heart of Taiwan. It offers an amazing selection of resources, including local discussion boards, games, Usenet news, Gopher, and great chat facilities. Can display everything in Chinese, if you download the necessary client, which is an interesting touch.

How to access   Telnet: **cc.nsysu.edu.tw**
Login: **bbs**

OuluBox

A small BBS located in Finland, offering bulletins, a selection of discussion conferences, and downloadable files. A great place to practice your Finnish.

    How to access   Telnet: **tolsun.oulu.fi**
                                   Login: **box**

Prism Hotel BBS

The Prism Hotel is divided into multiple floors, each with its own subject area. On each floor there are numerous rooms which can be entered, to view the discussion posts therein. All this makes for interesting BBS navigation.

    How to access   Telnet: **bbs.fdu.edu**
                                   Login: **bbs**

Quartz BBS

One of the oldest BBSs on the Internet. Age has not withered it, as there are many interesting discussion topics here, and also a useful Internet information system.

    How to access   Telnet: **quartz.rutgers.edu**
                                   Login: **bbs**

Radford University CS BBS

Offers an amazing number of services, including general discussion groups, IRC, public files, local mail, access to libraries, and games.

    How to access   Telnet: **muselab-gw.runet.edu**
                                   Login: **bbs**

Skynet BBS

A friendly BBS, based in Norway, with many varied and interesting discussion groups. The policy of no censorship has lead to many diverse and informative topics, covering a wide spectrum of life's mysteries.

    How to access   Telnet: **hpx6.aid.no**
                                   Login: **skynet**

Softwords COSY

A friendly conferencing system with a variety of discussion groups, including many technical and business issues. Also offers Internet mail.

    How to access   Telnet: **softwords.bc.ca**
                                   Login: **cosy**

Sunset BBS

Lots of varied discussion groups, local mail, and a scenic login screen.

How to access   Telnet: **paladine.hacks.arizona.edu**
Login: **bbs**

UTBBS

Based in Holland, with both English and Dutch-speaking users, UTBBS offers public and personal messaging, online chat, and a very large file selection covering many areas of computing and more.

How to access   Telnet: **utbbs.civ.utwente.nl**
Login: **bbs**

Zamfield's Internet BBS List

Zamfield's Wonderfully Incomplete, Complete Internet BBS List. A list of BBSs and miscellaneous Internet services.

How to access   Anonymous FTP: **lilac.berkeley.edu**
Path: **/help/cat/network/information/zamfield_list**

Biology

Biologist's Guide to Internet Resources

This document explains how to find everything of use to a biologist on the Internet.

How to access   Gopher: Center for Scientific Computing
Address: **finsun.csc.fi**
Choose: **Finnish EMBnet.. | FAQ Files
| A Biologist's Guide..**

Biology Journals

Periodical references to various biology journals.

How to access   Wais: **biology-journal-contents**

Biology Newsletter

Selection of newsletters about agriculture, botany, ecosystems, genetics and general biology, including publications such as Starnet, Tiempo, and Flora online.

How to access   Anonymous FTP: **nigel.msen.com**
Path: **/pub/newsletters/Bio/***

Biology Resources

Explore the many biology-related resources available on the Internet, via the Gopher, from here. Contains FAQs, FTP sites, book lists, and access to many bio servers around the globe.

How to access   Gopher: Center for Scientific Computing
Address: **finsun.csc.fi**
Choose: **Finnish EMBnet BioBox**

Taxacom FTP Server

A Listserv, technical information, software, and information on standards, and workshops in the field of systematic biology.

How to access   Anonymous FTP: **huh.harvard.edu**
Path: **/pub/README.TAX**

Taxacom Listserv Lists

A BITNET Listserv mailing list which focuses on systematic biology and accepts mail messages from any network source, and then distributes those messages to list subscribers. The Listserv is capable of delivering mail to users on BITNET, the Internet, or to any mail service on a "gatewayed" network. There are no restrictions on participation, and there is no cost.

How to access   Anonymous FTP: **huh.harvard.edu**
Path: **/pub/taxacom/taxacom.txt**

Bizarre

Aleister Crowley

Documents and texts by Aleister Crowley, including Book 4, Magick Without Tears, Book of the Law, Magick in Theory and Practice, and Equinox.

How to access   Anonymous FTP: **slopoke.mlb.semi.harris.com**
Path: **/pub/magick/magick/Crowley/***

Future Culture Digest Archives

Collection of archives for the Future Culture mailing list, which brings tomorrow's reality today, including the Future Culture and **alt.cyberpunk** FAQs.

How to access   Anonymous FTP: **ftp.rahul.net**
Path: **/pub/atman/UTLCD-preview/future-culture/***

High Weirdness

Incredible collection of information and lists, telling you where to locate the more esoteric and weird resources on the Internet.

How to access   Anonymous FTP: **slopoke.nlb.semi.harris.com**
Path: **/pub/weirdness/weird2_1.doc**
Path: **/pub/weirdness/weird2_1.sup**

Illegal Recreational Drug Information

Lots of information about drugs, including drug tests, FAQs, statistics, growing methods, chemical notes, marijuana-brownie recipes, and much more.

How to access   Anonymous FTP: **ftp.u.washington.edu**
Path: **/public/alt.drugs/***

Necromicon

FAQ and material about this near-legendary text written in Damascus in 730 A.D. Includes documents about the Voymich Manuscript.

How to access   Anonymous FTP: **nic.funet.fi**
Path: **/pub/doc/occult/necronomicon/***

Occult

Collection of occult material and pictures, covering such subjects as astrology, druidism, herbs, magick, rituals, tarot, and wicca.

How to access   1. Anonymous FTP: **slopoke.mlb.semi.harris.com**
Path: **/pub/magick/***

2. Anonymous FTP: **ftp.funet.fi**
Path: **/pub/doc/occult/***

3. Anonymous FTP: **seismo.soar.cs.cmu.edu**
Path: **/occult/***

Paganism and Magick, Occultism, Satanism

Open discussions about everything related to paganism, magick, satanism, and the occult.

How to access   Usenet: **alt.pagan**, **alt.magick**, **alt.satanism**, **alt.horror.cthulhu**, **alt.necromicon**, **alt.religion.sabaean**

UFOs

Digests all about UFOs from the Paranet Information Service.

How to access   1. Telnet: **grind.isca.uiowa.edu**
Login: **iscabbs**

2. Anonymous FTP: **grind.isca.uiowa.edu**
Path: **/info/paranet/***

UFOs and Paranormal Phenomena

Open discussion of unidentified flying objects, and occurrences of the paranormal and unusual.

How to access   Usenet: **alt.alien.visitors, alt.paranormal, alt.dreams, alt.out-of-body, sci.skeptic**

Unplastic News

An electronic magazine devoted to the aberrant, bizarre, and preposterous, containing weird and humorous quotes from the computer underground.

How to access   Anonymous FTP: **ftp.eff.org**
Path: **/pub/journals/Unplastic_News**

Wicca

Documents detailing the Wiccan tradition branch of the occult, including spells, beliefs, exercises, and much more.

How to access   Anonymous FTP: **nic.funet.fi**
Path: **/pub/doc/occult/wicca/***

Books

Book Reviews

Book reviews from the Whole Earth Review magazine.

How to access   Gopher: Whole Earth Lectronic Link
Address: **gopher.well.sf.ca.us**
Choose: **Art and Culture | Book Reviews**

Book Talk Mailing List

A discussion list for soon-to-be published books, CDs, and videos.

How to access   Listserv mailing list: **book-talk@columbia.ilc.com**

Bookstore Reviews

Reviews and interesting information about bookstores. From the motif of the store to ratings on salespeople. These reviews are done by book buffs, and their works are compiled and categorized by region and city on this Gopher server.

How to access   1. Gopher: University of Minnesota
Address: **consultant.micro.umn.edu**
Choose: **Fun & Games | Games | Bookstores**

2. Anonymous FTP: **rtfm.mit.edu**
Path: **/pub/usenet/news.answers/books/stores/***

Computer Books

Offers news, book descriptions, lists, and ordering information, for publications from O'Reilly & Associates.

How to access   Gopher: O'Reilly & Associates
Address: **ora.com**

O'Reilly and Associates

News, book descriptions and information, a complete listing of book titles, online indexes, instructions on obtaining book samples and archives.

How to access   1. Gopher: O'Reilly & Associates
Address: **ora.com**
Choose: **Book Descriptions and Information**

2. Anonymous FTP: **ftp.ora.com**
Path: **/pub**

O'Reilly Book Samples

Samples from many of the computer books published by O'Reilly & Associate0s.

How to access   Anonymous FTP: **ftp.uu.net**
Path: **/published/oreilly**

Online Bookstore

An online bookstore with computer and other books. Mail for more information. Order extra copies of this book for all your friends.

How to access   Mail: **obs@tic.com**

Unix Book Bibliography

A bibliography of some of the best books and documentation on Unix and related areas, based on recommendations from netnews readers and sales figures.

How to access   1. Gopher: O'Reilly & Associates
Address: **ora.com**
Choose: **Unix Bibliography**

2. Anonymous FTP: **rtfm.mit.edu**
Path: **/pub/usenet/news.answers/books**

The Unix Book List

A compilation of titles and other pertinent information on books about the Unix operating system and the C programming language. This list was organized and is maintained by Mitch Wright.

How to access   1. Anonymous FTP: **ftp.rahul.net**
Path: **/pub/mitch/YABL/**

2. Anonymous FTP: **ucselx.sdsu.edu**
Path: **/pub/doc/general/Unix-C-Booklist**

Business and Finance

Business Statistics

General business indicators, commodity prices, construction and real estate stats, and many more indicators and statistics. Also includes industry statistics for such business types as the food industry, leather, lumber, metals, manufacturers, and so on.

How to access   Gopher: University of California San Diego
Address: **infopath.ucsd.edu**
Choose: **News & Services | Economic..**
| Current Business..

Eastern Europe Trade Leads

A repository of requests from entrepreneurs in eastern European countries seeking business partners and trade leads in the U.S.

How to access   Gopher: University of California San Diego
Address: **infopath.ucsd.edu**
Choose: **News & Services | Economic..**
| Eastern Europe trade..

EBB and Agency Information

The Economic Bulletin Board (EBB) is an electronic bulletin board run by the U.S. Department of Commerce, Office of Business Analysis, that provides a one-stop source of current economic information. The EBB contains press releases and statistical information from the Bureau of Economic Analysis, the Bureau of the Census, the Federal Reserve Board, the Bureau of Labor Statistics, the Department of Treasury, and several other Federal Government agencies. A menu-driven system.

How to access   Gopher: University of California San Diego
Address: **infopath.ucsd.edu**
Choose: **News & Services | Economic..
| EBB and Agency Info..**

Economic Indicators

The raw data for the leading (and lesser) economic indicators in the U.S.

How to access   Gopher: University of California San Diego
Address: **infopath.ucsd.edu**
Choose: **News & Services | Economic..
| Economic Indicators**

Financial Ratios for Manufacturing Corporations

Supporting data and computation of FRMCs.

How to access   Gopher: University of California San Diego
Address: **infopath.ucsd.edu**
Choose: **News & Services | Economic..
| Special Studies and Reports**

Industry Statistics

Benchmark and periodic statistics for a number of industry segments. Includes quarterly financial reports and technical documents.

How to access   Gopher: University of California San Diego
Address: **infopath.ucsd.edu**
Choose: **News & Services | Economic..
| Industry Statistics**

International Market Insight (IMI) reports

Market briefs on opportunities and news in international and foreign markets.

How to access   Gopher: University of California San Diego
Address: **infopath.ucsd.edu**
Choose: **News & Services | Economic..
| International Market...**

Office Automation

A paper discussing the methods necessary for an office to thrive by having a hierarchy of well-defined realms which are controlled by and support the needs of each group.

How to access   Gopher: Whole Earth Lectronic Link
Address: **gopher.well.sf.ca.us**
Choose: **Whole Systems | Good Office Patterns**

Stock Market

View stock market closing quotes and comments for specific recent dates.

How to access   Gopher: Colorado State University
Address: **lobo.rmhs.colorado.edu**
Choose: **Other Information Services
| Stock Market Closing Quotes**

Stock Market Report

Daily stock market summary report. Provided as a free service of a2i.

How to access   Telnet: **a2i.rahul.net**
Login: **guest**
Choose: **n** (new screen-oriented guest menu)
Choose: **Current system information | Market report**

Calculators

HP Calculator BBS

A BBS just for users of HP calculators. Chat or message with other users and enthusiasts of HP scientific and business calculators. The HP Calculator Bulletin Board is a free service to allow for the exchange of software and information between HP calculator users, software developers, and distributors.

How to access   Telnet: **hpcvbbs.cv.hp.com**

Chemistry

Chemistry Information

Chemistry Information is an electronic reference source. It provides answers to frequently asked chemistry questions through library resources. It covers nomenclature, compound identification, properties, structure determination, toxicity, synthesis, registry numbers, and synthesis. For each component it lists the most appropriate reference resources (online catalog, indexes, journals, etc.).

How to access   Anonymous FTP: **ucssun1.sdsu.edu**
Path: **pub/chemras/***

Periodic Table of the Elements

Download a text or a graphic version of the periodic table of the elements.

How to access   1. Telnet: **kufacts.cc.ukans.edu**
Login: **kufacts**
Choose: **Reference Shelf**

                         2. Telnet: **camms2.caos.kun.nl 2034**

Children

Childcare Newsletters

Selection of newsletters about kids and childcare in general.

How to access   Anonymous FTP: **nigel.msen.com**
Path: **/pub/newsletters/Kids/***

KIDS

A spin-off of the KIDSPHERE mailing list (see under *Education*), KIDS exists for children to post messages to other children around the world. Send a mail message to the address below to join.

How to access   Mail: **joinkids@vms.cis.pitt.edu**

Computer Culture

Art of Technology Digest

Set of journals dedicated to sharing information among computerists and to the presentation and debate of diverse views.

How to access   Anonymous FTP: **wuarchive.wustl.edu**
Path: **/doc/misc/aot/***

Computer Underground Digest

The complete collection of the Computer Underground Digests, the weekly electronic publication covering matters concerning the computer underground.

How to access   Anonymous FTP: **ftp.eff.org**
Path: **/pub/cud/cud/***

Computers and Academic Freedom (CAF)

Discussion about everything to do with computers and academic freedom, and how it should be applied to university computers and networks.

How to access   1. Gopher: Electronic Frontier Foundation
Address: **gopher.eff.org**
Choose: **Computers & Academic Freedom mailing list archives & info**

2. Anonymous FTP: **ftp.eff.org**
Path: **/pub/academic/***

3. Usenet: **alt.comp.acad-freedom.talk,
alt.comp.acad-freedom.news**

Ethics

Dissertations on the computer ethics policies of many universities and organizations.

How to access   Anonymous FTP: **ariel.unm.edu**
Path: **/ethics/***

Hack Chat

Chat about computer hacking and learn more about the Net.

How to access   Internet Relay Chat: **#hack, #mindvox**

Computer Hardware

Hard Disk Guide

Comprehensive dictionary of hard drives, floppy drives, optical drives, drive controllers and host adapters. Designed to help the novice and pro alike with integration problems and system setups.

How to access   Anonymous FTP: **nic.funet.fi**
Path: **/pub/doc/harddisks/***

Hardware Architectures

Technical information and tutorials about some of the large IBM systems, including the ES/9000, RS600, and the Scalable Powerparallel systems.

How to access	Gopher: Cornell University
	Address: **gopher.tc.cornell.edu**
	Choose: **Hardware Architectures**

A List of EPROM Models and Manufacturers

This file is a list of manufacturers, models, and statistics for many EPROM (Erasable/Programmable Read Only Memory) chips.

| **How to access** | Anonymous FTP: **oak.oakland.edu** |
| | Path: **/pub/misc/eprom/eprom-types.list** |

Modem News

Recent editions of the online Modem News publication.

| **How to access** | Anonymous FTP: **wuarchive.wustl.edu** |
| | Path: **/pub/modemnews/*** |

Modems

Discussion, reviews, comparative statistics, program source code, and much more for and about modems.

| **How to access** | Anonymous FTP: **oak.oakland.edu** |
| | Path: **/pub/misc/modems** |

Powerful Computer List

A list of the world's most powerful computing sites.

| **How to access** | Mail: **gunter@yarrow.wt.uwa.edu.au** |

RS232 Pinouts

Detailed technical information about the connections of the RS232 terminal lead.

How to access	Gopher: University of Surrey
	Address: **gopher.cpe.surrey.ac.uk**
	Choose: **Misc Technical notes	RS232 pinouts**

SFI BBS

A research BBS devoted to Complex Systems.

| **How to access** | Telnet: **bbs.santafe.edu** |
| | Login: **bbs** |

Supercomputer Documentation

Collection of information about applications packages, graphics software, languages and compilers, and scientific libraries available for supercomputers.

How to access   Gopher: Texas A&M University
Address: **gopher.tamu.edu**
Choose: **Texas A&M Gophers**
 | **Supercomputer Center gopher**
 | **Available Applications & Software**

Ultrasound

An area devoted to information and utilities supporting the Gravis Ultrasound card for ISA-based computers, including IBM-PCs. Includes demos, digests, games, bulletins, sounds, and utilities.

How to access   Anonymous FTP: **wuarchive.wustl.edu**
Path: **/systems/ibmpc/ultrasound/***

Vaxbook

A guide made by users for users of VAX/VMS, available in postscript and TeX formats.

How to access   Anonymous FTP: **nic.funet.fi**
Path: **/pub/doc/vaxbook/***

Computer Literature

Amateur Computerist

Complete run of the Amateur Computerist Newsletter, and archives from the **alt.amateur-comp** Usenet group, dedicated to informing people of developments in an effort to advance computer education.

How to access   Anonymous FTP: **wuarchive.wustl.edu**
Path: **/doc/misc/acn/***

BBS Issues

Items of interest to BBS operators, including the text and analysis of lawsuits, FCC regulations, opinions, and more.

How to access   Anonymous FTP: **oak.oakland.edu**
Path: **/pub/misc/bbs**

CERT Archive

Lots of technical and advisory documents about specific computer security problems and bugs from the Computer Emergency Response Team (CERT).

How to access   Anonymous FTP: **cert.org**
Path: **/pub/***

Computer Science Technical Reports

Large archive of technical computer science reports, arranged by institution.

How to access   1. Gopher: Simon Fraser University
Address: **fas.sfu.ca**
Choose: **Internet Resource Projects | EPiCS
| Technical Report Archives in Computer Science**

2. Anonymous FTP: **fas.sfu.ca**
Path: **/projects/EPiCS/CS-TechReports/***

Computer Underground

Largest collection of hacker, phreaker, anarchist, cyberpunk, and underground material to be found anywhere on the Internet. Includes such fabled publications as Phrack, Magik, Phantasy, and the Legion of Doom Technical Journals.

How to access   Anonymous FTP: **ftp.eff.org**
Path: **/pub/cud/***

See also   Computer Culture

Computer Virus Technical Information

Detailed technical information about many of the known viruses, including how to detect them, the damage they cause, and eradicating them. Includes entries for MS-DOS, Amiga, and Macintosh viruses.

How to access   Anonymous FTP: **oak.oakland.edu**
Path: **/pub/misc/virus**

Computing Across America

The tales of adventure from Nomad and his electronic cottage, on a techno-gizmo encumbered recumbent bicycle, as he travelled across America.

How to access   1. Gopher: Mount Holyoke College
Address: **orixa.mtholyoke.edu**
Choose: **Document Library | Misc & Fun Stuff
| The NOMAD Papers**

2. Anonymous FTP: **ftp.telebit.com**
Path: **/pub/nomad**

Computing Newsletters

Selection of newsletters and articles about operating systems, software, and other technical computer topics.

How to access   Anonymous FTP: **nigel.msen.com**
Path: **/pub/newsletters/Computing/***

Desktop Publishing

FAQs, current job opportunities, accounting database and cashbook programs, for Pagemaker users and desktop publishers in general.

How to access   Anonymous FTP: **wuarchive.wustl.edu**
Path: **/doc/misc/pagemaker/***

Guide to PC Downloading

A guide to downloading Internet files to a PC, using the Procomm or Kermit communications programs and protocols.

How to access   Anonymous FTP: **nic.funet.fi**
Path: **/pub/doc/library/download.txt**

The Hacker's Dictionary

A comprehensive compendium of hacker slang illuminating many aspects of hackish tradition, folklore, and humor. Also known as "the jargon file".

How to access   Gopher: Mount Holyoke College
Address: **orixa.mtholyoke.edu**
Choose: **Document Library | The Hackers Dictionary**

How to Steal Code

Also known as "Inventing the Wheel Only Once". A guide by Henry Spencer on the merits of using the wealth of existing software and libraries instead of rewriting programs from scratch.

How to access   Anonymous FTP: **ucselx.sdsu.edu**
Path: **/pub/doc/general/steal.doc**

HPCwire

HPCwire publishes a weekly news bulletin on high-performance computing, distributed to thousands of users on the Internet. Topic matter ranges from workstations through supercomputers, with news briefs, feature stories, and in-depth exclusive interviews.

How to access	Telnet: **supernet.ans.net**
	Login: **hpcwire**

Kermit Manual

Sixth edition of the manual for the Kermit file transfer protocol, as often used by communications programs to transfer files using a modem.

How to access	Anonymous FTP: **nic.funet.fi**
	Path: **/pub/doc/kermit-protocol-manual/***

Network Bibliography

A list of many computer networks.

How to access	Mail: **comserve@rpiecs.bitnet**
	Body: **send compunet biblio**

Network Newsletters

Selection of newsletters related to networks, including their social aspect and impact.

How to access	Anonymous FTP: **nigel.msen.com**
	Path: **/pub/newsletters/Networker/***

Networking Computers

Lots of information on computer networking, school computing, hardware for networking, Unix networking, and programming languages.

How to access	Gopher: Pacific Systems Group
	Address: **gopher.psg.com**
	Choose: **Networking...**

PC/MS-DOS: The Essentials

A brief guide for beginners with MS-DOS/PC-DOS computers. Written by George Campbell, this guide starts at the very beginning and takes it one step at a time.

How to access	Anonymous FTP: **ucselx.sdsu.edu**
	Path: **/pub/doc/general/msdos.txt**

Tao of Programming

A complete text.

How to access	Anonymous FTP: **ucselx.sdsu.edu**
	Path: **/pub/doc/etext/tao-of-programming**

Computer Networks

Matrix News

Samples and information about Matrix News, a newsletter about cross-network issues.

How to access   1. Gopher: Texas Internet Consulting
Address: **tic.com**
Choose: **Matrix Information...** | **Matrix News, the..**

2. Anonymous FTP: **tic.com**
Path: **/matrix/news/***

Network Maps

A color geographical map of the world, showing the main global networks in the matrix of computer networks that exchange electronic mail.

How to access   1. Gopher: Texas Internet Consulting
Address: **tic.com**
Choose: **Matrix Information..** | **Maps of Networks**

2. Anonymous FTP: **tic.com**
Path: **/matrix/maps/***

Computer Technology

Computer Speech

Archives and information about computer speech technology and speech science.

How to access   Anonymous FTP: **svr-ftp.eng.cam.ac.uk**
Path: **/comp.speech/***

Virtual Reality

This menu contains items relevant to the concept of a single-user virtual reality. It includes some interesting articles and access to the Interactive Fiction archive.

How to access   Anonymous FTP: **ftp.u.washington.edu**
Path: **/public/VirtualReality/***

Computer Vendors

Toll-free Phone Numbers for Computer Companies

800 numbers for many computer hardware and software companies.

How to access	Anonymous FTP: **oak.oakland.edu** Path: **/pub/misc/telephone/tollfree.num**
See also	Telephones

Consumer Services

FAXNET

Send and receive faxes to and from anywhere in the world from your computer or terminal. All you need is to be able to send mail on the Internet. Send a mail message to the address below for a FAQ list about this service. (This is a service of AnyWare Associates and they do charge fees.)

How to access	Mail: **info@awa.com**

Infinity Link Corporation Consumer Access Services

Bookstores, music CDs, Unix software, and movies on VHS video cassettes and laserdiscs. This service provides you with a window into a wide variety of unique and hard to find products. Some sections of ILC/CAS provide you with information on these products, and some sections allow you to actually purchase these products online with only a few keystrokes.

How to access	Telnet: **columbia.ilc.com** Login: **cas**

Online Bookstore

An online bookstore with computer and other books. Mail for more information. Order extra copies of this book for all your friends.

How to access	Mail: **obs@tic.com**
See also	Books

Your Complete Guide to Credit

A consumer guide by Mark J. Allen.

How to access   Anonymous FTP: **oak.oakland.edu**
Path: **/pub/misc/consumers/credit**

Cyberpunk

Agrippa: A Book of the Dead

The complete text of this cyberpunk tale by William Gibson.

How to access   Anonymous FTP: **ftp.rahul.net**
Path: **/pub/atman/UTLCD-preview/assorted-text/agrippa.arj**

Bruce Sterling Articles

A selection of essays and articles by Bruce Sterling, the renowned cyberpunk author, covering a variety of topics about cyberspace and its surrounding culture.

How to access   1. Gopher: Whole Earth Lectronic Link
Address: **gopher.well.sf.ca.us**
Choose: **Cyberpunk**
 | **Essays and articles by Bruce Sterling**

2. Anonymous FTP: **ftp.eff.org**
Path: **/pub/agitprop/***

Cyberpunk News

Collection of cyberpunk related material including the Locus magazine, Bruce Sterling articles, and the latest information about cyperpunk conventions, those mind-blowing meetings of the computer underground.

How to access   Gopher: Whole Earth Lectronic Link
Address: **gopher.well.sf.ca.us**
Choose: **Cyberpunk**

Cypherpunks

Clipper documents, cryptanalysis tools, crypt FTP site list, FAQs, PGP, and other cryptography-related material can be found at this Cypherpunk mailing list archive.

How to access   1. Anonymous FTP: **soda.berkeley.edu**
Path: **/pub/cypherpunks/***

2. Anonymous FTP: **nic.funet.fi**
Path: **/pub/doc/cypherpunks/***

Future Culture and Cyberpunks Mailing List

Discussion and news about tomorrow's reality today. Become a dreamer of the dreams.

How to access   Listserv mailing list: **futurec@uafsysb.uark.edu**

Economics

Employment Statistics

Civilian and government labor force statistics and unemployment information by state.

How to access   Gopher: University of California San Diego
Address: **infopath.ucsd.edu**
Choose: **News & Services** | **Economic..**
| **Employment Statistics**

Energy Statistics

Compilations of statistics on energy consumption, requirements, and reserves in the U.S., including specific information on coal, crude oil, natural gas, and other sources.

How to access   Gopher: University of California San Diego
Address: **infopath.ucsd.edu**
Choose: **News & Services** | **Economic. .** | **Energy Statistics**

See also   Government, Science

Foreign Trade

Statistics and data on U.S. foreign trade of merchandise and textiles. Includes import and export data, as well as foreign spending, capital expenditures, and summaries.

How to access   Gopher: University of California San Diego
Address: **infopath.ucsd.edu**
Choose: **News & Services** | **Economic..** | **Foreign Trade**

Monetary Statistics

Foreign exchange rates, aggregate reserves, daily bond rates, consumer credit, flow of funds, savings bond rates, treasury yields, and much more.

How to access   Gopher: University of California San Diego
Address: **infopath.ucsd.edu**
Choose: **News & Services** | **Economic..**
| **Monetary Statistics**

E

National Income and Products Accounts

Annual income statistics for past years, estimates, gross domestic product tables, quarterly and annual NIPA reports, and source statistics used to calculate GDP.

How to access   Gopher: University of California San Diego
Address: **infopath.ucsd.edu**
Choose: **News & Services | Economic..
| National Income..**

Press Releases from the U.S. Trade Representative

A collection of the press releases from the U.S. Trade Representative. Releases are named by subject.

How to access   Gopher: University of California San Diego
Address: **infopath.ucsd.edu**
Choose: **News & Services | Economic.. | Press releases...**

Price and Productivity Statistics

Supporting data and statistics used in computation of the Consumer Price Index, import and export price indexes, the producer price index, and productivity and cost statistics.

How to access   Gopher: University of California San Diego
Address: **infopath.ucsd.edu**
Choose: **News & Services | Economic..
| Price and Productivity...**

Regional Economic Statistics

Supporting data and computations of disposable per capita income by state, total personal incomes, metropolitan, state, and regional incomes. Also includes wage and salary by industry data by state, region and U.S. total.

How to access   Gopher: University of California San Diego
Address: **infopath.ucsd.edu**
Choose: **News & Services | Economic..
| Regional Economic Stat...**

Summaries of Current Economic Conditions

Government and industry statistical summaries on trade, the balance of payments, the consumer price index, retail sales, inventories, construction, durable goods, employment cost index, housing starts, production and capacity, plant and equipment spending, and much more.

How to access   Gopher: University of California San Diego
Address: **infopath.ucsd.edu**
Choose: **News & Services | Economic...
| Summaries of current...**

Education

Academic BBSs

Allows access to a bundle of academic bulletin boards and online information systems.

How to access   Gopher: Yale University
Address: **yaleinfo.yale.edu**
Choose: **Internet Resources | Miscellaneous resources**

Academic Dialogs

Offers articles and comments on several academic disciplines that use computers, to which readers can respond, using the unique posting capability of this Gopher.

How to access   Gopher: Apple Computer, Inc.
Address: **info.hed.apple.com**
Choose: **Computing in Higher Ed Academics**

Apple Computer Higher Education Gopher

Offers product information, news, publications, and support from Apple. The Gopher has the unique capability of being able to "post" to it using email.

How to access   Gopher: Apple Computer, Inc.
Address: **info.hed.apple.com**

Chronicle of Higher Education

Offers information about job openings, best-selling books on campuses, and news articles from the Chronicle of Higher Education in its free online service "Academe This Week".

How to access   Gopher: Merit Computer Network
Address: **chronicle.merit.edu**

EDUPAGE

EDUPAGE is an education-related news service provided by EDUCOM, a consortium of leading colleges and universities seeking to transform education through the use of information technology. The vehicle for news delivery is periodical mail messages with summary news briefs.

How to access   Mail: **edupage@educom.edu**

E

FEDIX/MOLIS

Information on scholarships, minority assistance.

How to access   Telnet: **fedix.fie.com**
Login: **fedix**

Health Sciences Libraries Consortium

The Health Sciences Libraries Consortium (HSLC) Computer Based Learning Software Database contains listings of PC compatible and Macintosh programs used in health sciences education.

How to access   Telnet: **shrsys.hslc.org**
Login: **cbl**

See also   Medicine

Incomplete Guide to the Internet

The Incomplete Guide to the Internet and Other Telecommunications Opportunities Especially for Teachers and Students, K-12 is a resource guide and how-to manual for beginning and intermediate Internet users as well as an excellent reference for advanced users.

How to access   1. Anonymous FTP: **zaphod.ncsa.uiuc.edu**
Path: **Education/Education_Resources/Incomplete_Guide/***

2. Anonymous FTP: **ftp.ncsa.uiuc.edu**
Path: **Education/Education_Resources/Incomplete_Guide/***

Information System for Advanced Academic Computing

An service for promoting the use of high-end computers for research and education.

How to access   Telnet: **isaac.engr.washington.edu**

JANET Network

Long- and short-term information on the JANET network, the UK's joint academic network.

How to access   Gopher: Joint Academic Network
Address: **news.janet.ac.uk**

KIDSPHERE

Kidsphere is a mailing list for teachers (grades K-12), school administrators, scientists, and others around the world. Send a mail message to join the list.

How to access   Mail: **kidsphere-request@vms.cis.pitt.edu**

MicroMUSE

Educational multi-user simulated environment.

How to access   Telnet: **michael.ai.mit.edu**
Login: **guest**

National Education BBS

A system sponsored by the National Education Supercomputing Program (NESP). Full access is only available to NESP members, but limited access is granted to guests. Includes files, news, talk, and chat.

How to access   Telnet: **nebbs.nersc.gov**
Login: **guest**

Newton

A BBS for anyone teaching or studying science, math, or computer science.

How to access   Telnet: **newton.dep.anl.gov**
Login: **cocotext**

QUERRI

Questions on University Extension Regional Resource Information (QUERRI) is a database maintained by the North Central Region Educational Materials Project. QUERRI offers online access to bibliographic information and thousands of North Central Region Extension resources.

How to access   Telnet: **isn.rdns.iastate.edu**
Login: **querri**

TECHNET

Technical Support for Education and Research (TECHNET) is an open, unmoderated discussion list for technical support staff at universities and other non-profit educational/research institutions worldwide. Discussions may cover such topics as electronic and software design, interfacing of laboratory equipment to computers, construction of unique laboratory equipment, data collection methods, etc.

How to access   Listserv mailing list: **technet@acadvm1.uottawa.ca**

Technical Reports and Publication Archives

Numerous technical reports covering a large variety of academic fields, collected from hundreds of institutions, companies, and universities.

E

How to access   Gopher: University of Nottingham
Address: **trellis.cs.nott.ac.uk**
Choose: **Technical Reports and Publications Archives**

Usenet University

FAQs, history, ideas, papers, and newsgroup lists for the Usenet University, a society of people interested in learning, teaching, and tutoring.

How to access   Anonymous FTP: **nic.funet.fi**
Path: **/pub/doc/uu/***

Environment

EnviroGopher

EnviroGopher is a large resource of environmental information from the EnviroLink Network. It covers all aspects of the environment, including environmental action, issues, media, networks, organization and politics, and gives easy access to other environmental Gopher servers.

How to access   Gopher: Carnegie Mellon University
Address: **envirolink.hss.cmu.edu**

Environmental Issues

A variety of stories, essays, and book reviews covering various environmental issues and ideas.

How to access   Gopher: Whole Earth Lectronic Link
Address: **gopher.well.sf.ca.us**
Choose: **Environmental Issues and Ideas**

FireNet

FireNet as an information service for everyone interested in any aspect of rural and landscape fires. The information concerns all aspects of fire science and management—including fire behavior, fire weather, fire prevention, mitigation and suppression, plant and animal responses to fire, and all aspects of fire effects.

How to access   1. Gopher: Australian National University
Address: **life.anu.edu.au**
Choose: **Landscape Ecology, Fire,.. | firenet**

2. Anonymous FTP: **life.anu.edu.au**
Path: **pub/landscape_ecology/firenet**

3. Listserv mailing list: **firenet@life.anu.edu.au**

Food and Drink

Beer Archive

Offers a beer FAQ from the Usenet **alt.beer** newsgroup, lists of beer magazines, and information about CAMRA, the Campaign for Real Ale.

How to access   Anonymous FTP: **ftp.cwru.edu**
Path: **/pub/alt.beer/***

Big Drink List

Mixing hints, tips, and recipes for hundreds of mixed drinks.

How to access   Anonymous FTP: **ocf.berkeley.edu**
Path: **/pub/Library/Recreation/big-drink-list**

Booze Cookbook

More drink recipes, including Hemorrhaging Brain and the Vulcan Death Grip.

How to access   Anonymous FTP: **ocf.berkeley.edu**
Path: **/pub/Library/Recreation/Booze_Cookbook**

Cooking

International conversion helper, food terms, liquid measures, ingredient guides, and other material related to cooking.

How to access   Anonymous FTP: **rtfm.mit.edu**
Path: **/pub/usenet/news.answers/cooking-faq**

Drinking Games

Magnificent collection of nearly one hundred beer drinking games, including such gems as Beat the Barman, Kings and Blood, and Viking.

How to access   1. Anonymous FTP: **sauna.cs.hut.fi**
Path: **/pub/drinking_games/***

2. Anonymous FTP: **ocf.berkeley.edu**
Path: **/pub/Library/Recreation**

Good Food

Discussion of cooking techniques, equipment, recipes, vegetarianism, restaurants, and food and drink in general.

How to access   Usenet: **rec.food.cooking, rec.food.recipes, rec.food.veg, rec.food.drink, rec.food.restaurants, alt.food**

F

Homebrewing

All kinds of material related to beer and homebrewing, including recipes, Homebrew Digest archives, color images of various beer labels and coasters, and much more.

How to access   Anonymous FTP: **nic.funet.fi**
Path: **/pub/culture/beer/***

The Recipe Archives

Search this large database of food recipes for that special something.

How to access   1. Anonymous FTP: **gatekeeper.dec.com**
Path: **/pub/recipes**

2. Anonymous FTP: **mthvax.cs.miami.edu**
Path: **/pub/recipes**

3. Gopher: Albert Einstein College of Medicine
Address: **gopher.aecom.yu.edu**
Choose: **Internet Resources | Miscellaneous
| Search the Food Recipes Database**

4. Anonymous FTP: **wuarchive.wustl.edu**
Path: **/usenet/rec.food.recipes/recipes**

5. Wais: **usenet-cookbook**

6. Wais: **recipes**

Vegetarian

General information on all aspects of vegetarianism, answers to common questions, vegetarian recipes, and a world guide to vegetarianism.

How to access   1. Anonymous FTP: **rtfm.mit.edu**
Path: **/pub/usenet/rec.answers/vegetarian/***

2. Anonymous FTP: **flubber.cs.umd.edu**
Path: **/other/tms/veg/***

Fun

Conversational Hypertext Access Technology

CHAT (Conversational Hypertext Access Technology) is a natural language database query engine. Take part in a simulated conversation with a dragon or a woman named Alice. They correctly understand just about anything you say to them. Be careful though, it takes skill and cunning to talk the dragon out of flaming you on the spot.

How to access   Telnet: **debra.dgbt.doc.ca**
Login: **chat**

Crosswords

Dictionary, word-list books, technical paper guides, software pointers, solution tips, and other material about crosswords.

How to access   Anonymous FTP: **rtfm.mit.edu**
Path: **/pub/usenet/news.answers/crossword-faq**

Disney

Interesting facts, animated feature film guides, character notes, event news, and more about Walt Disney and the magic kingdom he created.

How to access   Anonymous FTP: **rtfm.mit.edu**
Path: **/pub/usenet/news.answers/disney-faq/***

Disneyland

Prices, hours, entertainment guide, attraction information and ratings, and a list of things not to be missed, such as the Matterhorn bobsleds, Fantasmic, and Pirates of the Caribbean.

How to access   Anonymous FTP: **rtfm.mit.edu**
Path: **/pub/usenet/news.answers/disney-faq/disneyland**

Historical Costuming

Pattern and supplier lists, bibliography, information relevant to SCA periods and Civil War, and some pointers to historical reenactment groups.

How to access   Anonymous FTP: **rtfm.mit.edu**
Path: **/pub/usenet/news.answers/crafts-historical-costuming**

Juggling

Large collection of information, news, FAQs, animations, publications, help, programs, and archives to do with juggling. Learn how to juggle the cascade, fountain, or shower patterns in no time.

How to access   Anonymous FTP: **moocow.cogsci.indiana.edu**
Path: **/pub/juggling**

Nude Beaches

Lists of some magnificent beaches, hot springs, and parks around the world where clothing is optional.

F

How to access   Anonymous FTP: **rtfm.mit.edu**
Path: **/pub/usenet/rec.answers/nude-faq/beaches/***

Puzzles

Hundreds of puzzles to open your mind, complete with solutions. If 6 cats can kill 6 rats in 6 minutes, how many cats does it take to kill one rat in one minute?

How to access   Anonymous FTP: **rtfm.mit.edu**
Path: **/pub/usenet/news.answers/puzzles-faq/***

Roller Coaster

All sorts of goodies about roller coasters, including images, a FAQ, and descriptions and reviews of parks and coasters.

How to access   Anonymous FTP: **gboro.rowan.edu**
Path: **/pub/Coasters/***

Games

See also   MUDs

Advanced Dungeons and Dragons

Lots of Advanced Dungeons and Dragons role-playing material, including spell and priest books.

How to access   1. Anonymous FTP: **ccosun.caltech.edu**
Path: **/pub/adnd/***

2. Anonymous FTP: **greyhawk.stanford.edu**
Path: **D_D/***

3. Anonymous FTP: **ftp.cs.pdx.edu**
Path: **/pub/frp/***

Backgammon Servers

Play backgammon.

How to access   Telnet: **ouzo.rog.rwth-aachen.de 8765**
Login: **guest**

Chess

Meet and play chess with other chess enthusiasts throughout the world. Watch others play, or join in and play a game of your own. You can save a game and return to it later.

How to access   1. Mail: **shaheen@eve.assumption.edu**

2. Telnet: **cirrus.gp.cs.cmu.edu**

Core War

Offers documents, tutorials, source code, and system information for a variety of formats for Core War, the system where programs battle against programs to try to destroy each other in a virtual memory.

How to access   Anonymous FTP: **soda.berkeley.edu**
Path: **/pub/corewar/***

Diplomacy

Diplomacy is a game you play by mail. The setting is pre-WWI Europe. Players send in their moves every week and results are distributed each week. Conspire with your neighbors to conquer the old world.

How to access   1. Mail: **judge@morrolan.eff.org**

2. Mail: **judge@dipvax.dsto.gov.au**

3. Mail: **judge@shrike.und.ac.za**

4. Mail: **judge@u.washington.edu**

Fantasy Role-Playing Games

This site has a number of role-playing games to choose from, such as "Empire of the Petal Throne".

How to access   Gopher: University of Minnesota
Address: **consultant.micro.umn.edu**
Choose: **Fun & Games | Games
| Fantasy Role-Playing Games**

Flame BBS Gopher

Offers lots of information from Flame BBS including information boards and many documents. It even allows exploration of rooms and is a neat place "to flop around and zip and burble".

How to access   Gopher: University of Western Australia
Address: **mackerel.gu.uwa.edu.au 3452**

Gambling

Collection of gambling information, including a FAQ list, blackjack card-counting information, poker statistics, and much more.

G

How to access   Anonymous FTP: **soda.berkeley.edu**
Path: **/pub/rec.gambling/***

Game Solutions

Hints and solutions to more games than you will ever have time to play. Covers hundreds of popular adventure games.

How to access   Anonymous FTP: **nic.funet.fi**
Path: **/pub/doc/games/solutions/***

Go

Play go with other admirers of this popular oriental game. Watch others play, join in yourself, or discuss strategy with the masters.

How to access   1. Telnet: **lacerta.unm.edu 6969**

2. Telnet: **hellspark.wharton.upenn.edu 6969**

3. Telnet: **ftp.pasteur.fr 6969**

Role-playing

A very large collection of information connected to role-playing, including convention reports, FTP site lists, list of games and companies, and game-system specifics.

How to access   Gopher: University of San Diego
Address: **teetot.acusd.edu**
Choose: **Everything.. | Entertainment and Food
| Role-Playing**

Role-playing Archives

Large archive of role-playing materials—pictures, programs, game sheets, stories, specific systems, and texts on all aspects of role-playing. Includes AD&D, Traveller, Champions, Navero, Runequest, and others.

How to access   1. Anonymous FTP: **ics.uci.edu**
Path: **/usenet/rec.games.frp/***

2. **Anonymous FTP: ftp.cs.pdx.edu**
Path: **/pub/frp/***

3. Anonymous FTP: **ccosun.caltech.edu**
Path: **/pub/adnd/***

4. Anonymous FTP: **greyhawk.stanford.edu**
Path: **D_D/***

5. Anonymous FTP: **soda.berkeley.edu**
Path: **pub/***

6. Anonymous FTP: **ocf.berkeley.edu**
   Path: **/pub/Traveller/***

7. Anonymous FTP: **ftp.white.toronto.edu**
   Path: **/pub/frp/***

8. Anonymous FTP: **nic.funet.fi**
   Path: **/pub/doc/games/roleplay/***

ScrabbleMOO

Play the popular Scrabble board game.

How to access Telnet: **next2.cas.muohio.edu 7777**

Source code to OMEGA (a ROGUE-style game)

Complete C source code to OMEGA, a popular role-playing game. This source also exists in a number of other locations. To find other sites, do an Archie search for "**omega**".

How to access Anonymous FTP: **sun.soe.clarkson.edu**
Path: **/src/games/omega/***

Tiddlywinks

Learn the rules, history, and terminology of this complex game of strategy and tactics.

How to access Anonymous FTP: **rtfm.mit.edu**
Path: **/pub/usenet/alt.games.tiddlywinks/***

Trade Wars

Play this multi-user game that simulates international trade.

How to access Telnet: **harmony.cern.ch 2002**

Games (Video)

Mortal Kombat

Discuss strategy with the masters of this popular arcade game.

How to access Gopher: University of Minnesota
Address: **consultant.micro.umn.edu**
Choose: **Fun & Games | Games | Arcade Games
| Mortal Kombat**

G

Video Game Discussions

Discussion groups relating to many popular video and arcade games.

How to access   Usenet: **rec.games.video.arcade**

Geoculture

Japan

Japanese language and culture-related material, including programs and documentation for processing Japanese language documents with computers.

How to access   Anonymous FTP: **nic.funet.fi**
Path: **/pub/culture/japan/***

New Zealand

Everything you could want to know about beautiful New Zealand. Statistics on population, economy, politics, geography, climate, shopping, and other tourist information.

How to access   Gopher: Wellington City Council
Address: **gopher.wcc.govt.nz**
Choose: **New Zealand information**

Russia

About the Russian language and culture, including a literature guide, computer-related material, glossaries and word lists, recipes, jokes, lyrics, politics, and translation information.

How to access   Anonymous FTP: **nic.funet.fi**
Path: **/pub/culture/russian/***

Usenet Cultural Groups

About the languages, culture, people, customs, and many other aspects of countries around the world.

How to access   Usenet: **soc.culture.***

See also   Appendix G

Geography

CIA World Factbook

The complete text. Detailed information about every country and territory in the world. Includes geographic, climate, economic, and political information.

How to access   1. Gopher: University of Minnesota
Address: **consultant.micro.umn.edu**
Choose: **Libraries | Reference Works
| CIA World Fact Book**

2. Anonymous FTP: **ucselx.sdsu.edu**
Path: **/pub/doc/etext/world.text.Z**

3. Wais: **world-factbook**

Earth

A world fact book, world map, and topological relief database about the Earth.

How to access   Anonymous FTP: **nic.funet.fi**
Path: **/pub/doc/world/***

European Postal Codes

List of examples of postal country codes used by most European and Mediterranean countries.

How to access   Anonymous FTP: **nic.funet.fi**
Path: **/pub/doc/mail/stamps/***

Geography Server

Get information about cities, regions, countries, etc., including population, latitude and longitude, elevation, and so on.

How to access   Telnet: **martini.eecs.umich.edu 3000**

Global Land Information System

Download land use and geological survey maps of the United States.

How to access   Telnet: **glis.cr.usgs.gov**
Login: **guest**

U.S. Geographic Name Server

Obtain geographic position, population, elevation, state, and other statistics for a U.S. city or zip code.

G

How to access	Gopher: NASA Goddard Space Flight Center
	Address: **gopher.gsfc.nasa.gov**
	Choose: **Other Resources	US geographic name server**

Zip Codes of the U.S.

A file with all the zip codes in the U.S. and the city and state of their location.

How to access	1. Gopher: University of California San Diego
	Address: **infopath.ucsd.edu**
	Choose: **Reference Services	ZIP Codes**
	2. Anonymous FTP: **oes.orst.edu**
	Path: **/pub/data/zipcode**

Geology

Earthquake Information

Get up-to-date news about earthquakes around the world.

How to access	1. Finger: **quake@geophys.washington.edu**
	2. Telnet: **geophys.washington.edu**
	Login: **quake**

World Paleomagnetic Database

This program allows remote users to search the "Abase" ASCII version of the World Paleomagnetic Database. The Search program is very simple to use and will search the Soviet, non-Soviet, rock unit, and reference databases and create output files which can be downloaded back to a researcher's local system via Anonymous FTP.

How to access	Telnet: **earth.eps.pitt.edu**
	Login: **Search**

Government

1990 USA Census Information

The full text from Project Gutenberg.

How to access	Gopher: University of Minnesota
	Address: **consultant.micro.umn.edu**
	Choose: **Libraries	Electronic Books	By Title**

Clinton's Inaugural Address

The full text from Project Gutenberg.

How to access Gopher: University of Minnesota
Address: **consultant.micro.umn.edu**
Choose: **Libraries** | **Electronic Books** | **By Title**

Congress Members

A database containing the names, addresses, and phone numbers of members of Congress for all fifty states.

How to access Gopher: Albert Einstein College of Medicine
Address: **gopher.aecom.yu.edu**
Choose: **Internet Resources** | **Miscellaneous**
| **Search the Members of Congress Database**

Constitution of the United States of America

The complete text of the Constitution.

How to access 1. Anonymous FTP: **ucselx.sdsu.edu**
Path: **/pub/doc/etext/USConstitution.txt**

2. Anonymous FTP: **ocf.berkeley.edu**
Path: **/pub/Library/Politics**

3. Anonymous FTP: **nic.funet.fi**
Path: **/pub/doc/misc/us.independence.Z**

See also Historical Documents

Daily Summary of White House Press Releases

Receive by mail daily summaries of the activities and goings-on in the White House and government. Send a mail message to start the service with the line 'subscribe wh-summary' in the body of your message.

How to access Mail: **almanac@esusda.gov**
Body: **subscribe wh-summary**

Declaration of Independence

The complete text of the Declaration of Independence.

How to access 1. Anonymous FTP: **ucselx.sdsu.edu**
Path: **/pub/doc/etext/Declaration.txt**

2. Anonymous FTP: **ocf.berkeley.edu**
Path: **/pub/Library/Politics**

See also Historical Documents

G

Federal Information Exchange (FEDIX)

Federal Information Exchange, Inc. (FIE) is a diversified information services company providing a full range of database services, software development, and technical support to the government, private sector, and academic communities.

How to access   Gopher: University of California San Diego
Address: **infopath.ucsd.edu**
Choose: **The World | Misc Special.. | Federal Gov..
| FEDIX**

Federal Information Processing Standards

A publication of the U.S. Department of Commerce describing standards for information processing.

How to access   Anonymous FTP: **oak.oakland.edu**
Path: **/pub/misc/standards**

Federal Register

A sample access, and information on how to gain full access, to the daily U.S. Federal Register via the Internet.

How to access   Gopher: Texas A&M University
Address: **gopher.tamu.edu**
Choose: **Hot.Topics & Whats New in Gopher
| Federal Register..**

NATODATA

A mailing list that distributes public information about the North Atlantic Treaty Organization, including press releases, communiques, articles, factsheets, programs, speeches, and the NATO handbook.

How to access   Listserv mailing list: **natodata@cc1.kuleuven.ac.be**

The President of the United States

E-mail directly to the President.

How to access   Mail: **president@whitehouse.gov**

President's Daily Schedule

Ever wonder what the President of the United States is doing today? Examine the public schedule for the current month or previous months.

How to access   Gopher: Texas A&M University
Address: **gopher.tamu.edu**
Choose: **Hot.Topics & Whats New in Gopher**
  | Information from the White House
  | President's Daily Schedule

President's Economic Plan

The real report.

How to access   Gopher: University of California San Diego
Address: **infopath.ucsd.edu**
Choose: **News & Services | Economic..**
  | Summaries of current..
  | President Clinton's Economic Plan

Presidential Documents

Lots of information from the White House, including domestic, international, business, and economic affairs, and the President's daily schedule.

How to access   Gopher: Texas A&M University
Address: **gopher.tamu.edu**
Choose: **Hot.Topics & Whats New...**
  | Information...White House

Social Security Administration

Information about the Social Security Administration, Social Security numbers, and obtaining information on others for genealogical purposes.

How to access   Anonymous FTP: **oak.oakland.edu**
Path: **/pub/misc/ss-info**

United Nations Gopher

Contains lots of information about the United Nations, including what it is, what it does, its organizations and resources, press releases, conference news, and much more.

How to access   Gopher: United Nations
Address: **nywork1.undp.org**

U.S. Environmental Protection Agency

A menu-driven system that provides information on the EPA and what they're up to.

How to access   Telnet: **epaibm.rtpnc.epa.gov**

G

U.S. Federal Government Information

Don't take the politicians' word. Look up government data yourself and make your own decisions. Information is categorized by source (i.e., Government Accounting Office). Choose a source, then a document.

How to access   Gopher: University of Minnesota
Address: **consultant.micro.umn.edu**
Choose: **Libraries**
| Information from the US Federal Government

U.S. Government BBS List

List of dial-up U.S. Government Department bulletin boards, including BBSs for the Commerce, Customs, Defense, Energy, and Justice Departments.

How to access   Gopher: Panix
Address: **gopher.panix.com**
Choose: **Society for Electronic Access | List of U.S. Gov..**

U.S. Government Gophers

Access the Gophers of agencies directly connected with the U.S. Government from this menu, including NASA, National Science Foundation, FEDIX, and many more.

How to access   Gopher: National Science Foundation
Address: **stis.nsf.gov**
Choose: **Other U.S. Government Gopher Servers**

The Vice President of the United States

E-mail directly to the Vice President.

How to access   Mail: **vice.president@whitehouse.gov**

The White House

Details the White House electronic mail system, and presents such connected information as press conferences, information policy, and the federal budget.

How to access   Gopher: Whole Earth Lectronic Link
Address: **gopher.well.sf.ca.us**
Choose: **Politics | The White House**

White House Electronic Publications

Daily distribution of electronic publications via e-mail, search for and retrieve White House documents, and get information on sending e-mail to the White House.

How to access   1. Mail: **clinton-hq@campaign92.org**

2. Mail: **clinton-info@campaign92.org**

The White House Papers

White House press briefings and other papers dealing with the President, Vice President, First Lady, the Cabinet, Socks (the First Cat), and other important White House personalities.

How to access   Wais: **White-House-Papers**

Health

Health Newsletters

Selection of newsletters covering medicine, medical research, disease, and therapy.

How to access   Anonymous FTP: **nigel.msen.com**
Path: **/pub/newsletters/Health/***

See also   Medicine

Health Sciences Libraries Consortium (HSLC)

The Health Sciences Libraries Consortium (HSLC) Computer Based Learning Software Database, begun in 1987, contains listings of PC compatible and Macintosh programs used in health sciences education. This project has been endorsed, and funded, by the American Medical Informatics Association's Education Working Group. Records have also been contributed by the University of Michigan's Software for Health Sciences Education (supported by a grant from Sandoz Pharmaceuticals).

How to access   Telnet: **shrwyw.hslc.org**
Login: **cbl**

See also   Medicine

National Institute of Health

Announcements, information for researchers, a molecular biology database, library and literature resources, the NIH phone book, and more.

How to access   Gopher: University of California San Diego
Address: **infopath.ucsd.edu**
Choose: **The World** | **Misc Special..** | **Federal Gov..**
| **National...**

See also   Medicine

H

Software and Information for the Handicapped

This FTP site has many directories of informational files of interest to the handicapped.

How to access   Anonymous FTP: **handicap.shel.isc-br.com**

Typing Injuries

All the information about typing injuries, and their solutions, available on the Internet. Includes a long list of keyboard alternatives and some related gif pictures.

How to access   Anonymous FTP: **soda.berkeley.edu**
Path: **/pub/typing-injury/***

See also   Medicine

Historical Documents

American Historical Documents

Amendments to the Constitution, Annapolis Convention, Articles of Confederation, Bill of Rights, Charlotte Town Resolves, the Constitution, Continental Congress Resolves, Japanese and German surrenders, "I Have a Dream", Inaugural addresses, the Monroe Doctrine, Rights of Man, treaties, and more.

How to access   Gopher: University of Minnesota
Address: **consultant.micro.umn.edu**
Choose: **Libraries | Electronic Books | By Title | Historical**

See also   Government

The Federalist Papers

The full text from Project Gutenberg.

How to access   Gopher: University of Minnesota
Address: **consultant.micro.umn.edu**
Choose: **Libraries | Electronic Books | By Title**

See also   Government

History

History Databases

American history topics.

How to access   Telnet: **ukanaix.cc.ukans.edu**
Login: **history**

History Mailing List

A mailing list devoted to people interested in any aspect of history. Discussions range from trivial to very serious. Newcomers are welcome.

How to access   Listserv mailing list: **history@psuvm.psu.edu**

I Have a Dream

Martin Luther King, Jr.'s "I have a dream" speech.

How to access   Anonymous FTP: **ocf.berkeley.edu**
Path: **/pub/Library/Politics**

Hobbies

Amateur Radio

Interested in amateur radio? This site has all the information, including how to get started in this exciting hobby. ARRL offers information and files by e-mail.

How to access   Mail: **info@arrl.org**
Body: **help**

Bonsai

Dictionary of terms, soil information, FAQs, reading guides, supplier lists, and other material relevant to this ancient art.

How to access   1. Anonymous FTP: **rtfm.mit.edu**
Path: **/pub/usenet/news.answers/bonsai-faq/***

2. Anonymous FTP: **bonsai.pass.wayne.edu**
Paths: **/pub/GIFS/***; **/pub/Information/***

Collecting

Buy, sell, discuss, and collect stamps, coins, comics, cards, autographs, magazines, watches, and many other things.

How to access   Usenet: **rec.collecting**

H

Dance

Material related to different types of dance, including a large list of ballroom dance CDs.

How to access   1. Anonymous FTP: **ftp.fuzzy.ucsc.edu**
Path: **/pub/bds93.Z 2.**

2. Anonymous FTP: **ftp.cs.dal.ca**
Paths: **/comp.archives/rec.arts.dance/***;
/comp.archives/rec.folk-dancing/*

3. Usenet: **rec.arts.dance, rec.folk-dancing**

Gardening

Large collection of material about fertilizers, herbs, peppers, ivy, poisonous plants, pruning, roses, seeds, and much more about gardening in general.

How to access   Anonymous FTP: **sunsite.unc.edu**
Path: **/pub/academic/agriculture/sustainable_agriculture/
recgardens/***

See also   Agriculture

Guns

Discussions and material about shooting sports, reloading, training, personal defense, gun laws, weaponry, and other topics related to firearms in general.

How to access   1. Anonymous FTP: **flubber.cs.umd.edu**
Path: **/rec/***

2. Anonymous FTP: **ftp.vmars.tuwien.ac.at**
Path: **/pub/misc/guns/***

3. Anonymous FTP: **ftp.uu.net**
Path: **/usenet/rec.guns/***

4. Usenet: **rec.guns**

Ham-Radio Archives

Archives of interest to ham-radio buffs.

How to access   1. Anonymous FTP: **oak.oakland.edu**
Path: **/pub/misc/hamradio**

2. Gopher: Texas A&M University
Address: **gopher.cs.tamu.edu**
Choose: **Access to TAMU CS Anonymous FTP Files
| ham-radio**

3. Anonymous FTP: **ftp.cs.tamu.edu**
   Path: **/pub/ham-radio/***

4. Anonymous FTP: **ftp.cs.buffalo.edu**
   Path: **/pub/ham-radio/***

Ham-Radio Callbooks

The national ham-radio call-sign book.

How to access   1. Telnet: **callsign.cs.buffalo.edu 2000**

2. Telnet: **ham.njit.edu 2000**

Kites

All the information you need to know about kites. Offers kite reviews, stories, tips on flying, general information, event guides, and much more.

How to access   Gopher: University of Surrey
Address: **gopher.cpe.surrey.ac.uk**
Choose: **Kites**

Model Railroads

Modeling techniques, railroading information sources, operation guides, and notes on real railroads.

How to access   Anonymous FTP: **rtfm.mit.edu**
Path: **/pub/usenet/rec.models.railroad/***

Photography

Lexicon of terms, FAQs, equipment reviews, lens information, useful addresses and phone numbers, and archives having to do with photos and cameras.

How to access   1. Anonymous FTP: **rtfm.mit.edu**
   Path: **/pub/usenet/news.answers/rec-photo-faq**

2. Anonymous FTP: **moink.nmsu.edu**
   Path: **/rec.photo/***

Radio-Controlled Models

Lots of information about all types of radio-controlled models for hobbyists.

How to access   Anonymous FTP: **rtfm.mit.edu**
Path: **/pub/usenet/rec.models.rc/***

H

Sewing

Supply information, antique sewing machine guide, book list and much more material about sewing, fitting, and pattern drafting.

How to access   Anonymous FTP: **rtfm.mit.edu**
Paths: **/pub/usenet/news.answers/crafts-textiles**;
/pub/usenet/news.answers/crafts-textiles-books/*

Stamp Collecting

Collection of stamp pictures and related material.

How to access   Anonymous FTP: **nic.funet.fi**
Path: **/pub/doc/mail/stamps/***

Usenet Hobby Groups

Discussion of pastimes, hobbies, crafts, and many other recreational activities.

How to access   Usenet: **rec.***
See also   Appendix G

Woodworking

Discussions of tools, safety notes, FAQs, Woodsmith plans, electric woodworking motor information, supplier addresses, and more.

How to access   1. Anonymous FTP: **ftp.cs.rochester.edu**
Path: **/pub/rec.woodworking/***

2. Anonymous FTP: **ftp.cs.purdue.edu**
Path: **/pub/sjc/woodworking/***

3. Anonymous FTP: **rtfm.mit.edu**
Path: **/pub/usenet/news.answers/woodworking/***

Humanities

American Philosophical Association BBS

A BBS offering information about philosophical societies, grants, fellowships, seminars, and institutes. Also a mail directory of the APA membership, and bibliographical and journal information.

How to access   Telnet: **atl.calstate.edu**
Login: **apa**

Cognition

Selection of publications from the Indiana-based Center for Research on Concepts and Cognition, about cognition, the mind, and related subjects.

    How to access   Anonymous FTP: **moocow.cogsci.indiana.edu**
                                Path: **/pub/***

Coombspapers Social Sciences Research Data Bank

Electronic repository of social science and humanities papers, offprints, departmental publications, bibliographies, directories, abstracts of theses, and other material.

    How to access   1. Gopher: Australian National University
                                  Address: **coombs.anu.edu.au**
                                  Choose: **Coombspapers Soc.Sci.Research Data Bank**

                             2. Anonymous FTP: **wuarchive.wustl.edu**
                                  Path: **/doc/coombspapers/***

Objectivism Mailing List

Students of Objectivism and Ayn Rand discuss their ideas, concrete issues, and exchange news. Any issue of relevance to Objectivists is welcomed.

    How to access   Mail: **objectivism-request@vix.com**
                                  Body: **SUBSCRIBE** [your name]

H

Humor

Catstyle Archives

Hundreds of humorous and amusing files. Animal jokes, ASCII art, why a * is better than a *, British humor, holiday jokes, Dave Barry, geography, sex, jobs, life, politics, Murphy's laws, political correctness, quotes, religion, sports, and much more.

    How to access   Anonymous FTP: **cathouse.org**
                                  Path: **/pub/cathouse**

Comics

Discussions about the comic "Watchmen", and a bibliography, episode guide, and glossary for the comic book "Cerebus".

    How to access   Anonymous FTP: **ftp.white.toronto.edu**
                                  Path: **/pub/comics/***

Cookie Server

Get a humorous quote or rhyme.

How to access   Telnet: **astro.temple.edu 12345**

Hacker Test

Find out if you are a computer illiterate, nerd, hacker, guru, or wizard, with this set of questions.

How to access   Gopher: Universitaet des Saarlandes
Address: **pfsparc02.phil15.uni-sb.de**
Choose: **INFO-SYSTEM BENUTZEN | Fun | Hacker**

Jive Server

Send a mail message to this system and within minutes you'll receive your message translated into jive. Example: "Ask not what your country can do for you, but what you can do for your country", becomes "Ask not whut yo' country kin do fo' ya', but whut ya' kin do fo' yo' country. Slap mah fro!"

How to access   Mail: **jive@ifi.unizh.ch**

Jokes and Stories

A collection of popular jokes and humorous stories, such as "A Role-player's Famous Last Words" and "The Vaxorcist".

How to access   Gopher: Technische Universitaet Muenchen
Address: **gopher.informatik.tu-muenchen.de**
Choose: **Vershiedenes | Fun and Jokes | English**

Manly Men's Ten Commandments

It is hoped that all men obey these sacred laws, for any breach of these written rules will be considered a sin against womanhood, and may result in the loss of manly privileges such as Monday Night Football.

How to access   Gopher: Manchester Computing Centre
Address: **uts.mcc.ac.uk**
Choose: **Experimental..**
| Manly-Men's Ten Commandments

Monty Python

Collection of all the popular Monty Python sketches and screenplays, including "The Holy Grail", and "Life of Brian".

How to access   1. Anonymous FTP: **ocf.berkeley.edu**
Path: **/pub/Library/Monty_Python**

2. Anonymous FTP: **nic.funet.fi**
Path: **/pub/culture/tv+film/MontyPython/***

More Comics

Graphics and album lists from a selection of comics, including Tintin, and the French Valerian.

How to access   Anonymous FTP: **nic.funet.fi**
Path: **/pub/culture/comics/***

Netwit Mailing List

Netwit is a mailing list of jokes and net-humor. To subscribe, send a mail message.

How to access   1. Mail: **netwit@netwit.cmhnet.org**
Body: **ADD** [your address]

2. Mail: **help@netwit.cmhnet.org**
Body: **ADD** [your address]

Parodies

Parodies on popular movies and TV shows.

How to access   Anonymous FTP: **ocf.berkeley.edu**
Path: **/pub/Library/Parodies**

Streamlined Bill O' Rights

The new, streamlined Bill O' Rights, as amended by the recent federal and state decisions. (Humorous)

How to access   Gopher: NASA Goddard Space Flight Center
Address: **gopher.gsfc.nasa.gov**
Choose: **Other Resources | Bill O' Rights**

The Ten Commandments for C Programmers

Example: "Thou shalt check the array bounds of all strings (indeed, all arrays), for surely where thou typest 'foo' someone someday shall type 'supercalifragilisticexpialidocious'".

How to access   Anonymous FTP: **ucselx.sdsu.edu**
Path: **/pub/doc/general/ten_commandments**

H

Tintin

Extensive list of Tintin comic albums, and scanned images of album covers.

How to access   Anonymous FTP: **nic.funet.fi**
Path: **/pub/culture/comics/Tintin/***

Toxic Custard Workshop

Complete archives for this mildly warped and totally entertaining set of brain drippings from Australia.

How to access   Anonymous FTP: **ftp.cs.widener.edu**
Path: **/pub/tcwf/***

Usenet Humor

Usenet news groups are a seemingly endless source of jokes and off-color humor.

How to access   1. Usenet: **alt.tasteless.jokes**

2. Usenet: **rec.humor**

3. Usenet: **rec.humor.d**

4. Usenet: **rec.humor.funny**

5. Usenet: **rec.humor.oracle**

6. Usenet: **rec.humor.oracle.d**

Virus Jokes

A small collection of amusing computer virus jokes.

How to access   Gopher: North Dakota State University
Address: **chiphead.cc.ndsu.nodak.edu**
Choose: **Other Stuff | Fun Stuff | Virus Alert**

Humor Archives

Massive collections of jokes, anecdotes, humorous stories, one-liners, and riddles.

How to access   1. Anonymous FTP: **donau.et.tudelft.nl**
Path: **/pub/humor**

2. Anonymous FTP: **ftp.uu.net**
Path: **/doc/literary/obi/DEC/humor**

3. Anonymous FTP: **gatekeeper.dec.com**
Path: **/pub/misc/humour**

4. Anonymous FTP: **jerico.usc.edu**
   Path: **/pub/jamin/sciina**

5. Anonymous FTP: **mc.lcs.mit.edu**
   Path: **/its/ai/humor**

6. Anonymous FTP: **mintaka.lcs.mit.edu**
   Path: **/humor**

7. Anonymous FTP: **nic.funet.fi**
   Path: **/pub/doc/humour**

8. Anonymous FTP: **ocf.berkeley.edu**
   Path: **/pub/Library/Humor**

9. Anonymous FTP: **pit-manager.mit.edu**
   Path: **/pub/humor**

10. Anonymous FTP: **puffin.doc.ic.ac.uk**
    Path: **/doc/humour**

11. Anonymous FTP: **quartz.rutgers.edu**
    Path: **/pub/humor**

12. Anonymous FTP: **rascal.ics.utexas.edu**
    Path: **/misc/funny**

13. Anonymous FTP: **shape.mps.ohio-state.edu**
    Path: **/pub/jokes**

14. Anonymous FTP: **sifon.cc.macgill.ca**
    Path: **/pub/docs/misc/dave_barry**

15. Anonymous FTP: **slopoke.mlb.semi.harris.com**
    Path: **/pub/doc/humor**

16. Anonymous FTP: **srawgw.sra.co.jp**
    Paths:
    /.a/sranha-bp/arch/arch/comp.archives/auto/alt.humor.oracle;
    /.a/sranha-bp/arch/arch/comp.archives/auto/rec.humor;
    /.a/sranha-bp/arch/arch/comp.archives/auto/rec.humor.d;

17. Anonymous FTP: **theta.iis.u-tokyo.ac.jp**
    Path: **/JUNET-DB/jokes**

18. Anonymous FTP: **toklab.ics.osaka-u.ac.jp**
    Path: **/JUNET-DB/jokes**

19. Anonymous FTP: **tolsun.oulu.fi**
    Path: **/pub/humor**

20. Anonymous FTP: **trantor.ee.msstate.edu**
    Path: **/files/Text**

21. Anonymous FTP: **tybalt.caltech.edu**
    Path: **/pub/bjmccall/non-political/Funny**

H

22. Anonymous FTP: **ftp.cs.dal.ca**
    Paths: **/pub/comp.archives/alt.humor.oracle;**
    /pub/comp.archives/rec.humor;
    /pub/comp.archives/rec.humor.d

Internet

Anonymous FTP Site List

A list of all of the Internet's Anonymous FTP sites.

How to access   Anonymous FTP: **ftp.shsu.edu**
Path: **/pub/ftp-list/sites.Z**

Answers for Experienced Internet Users

An Internet RFC (1207).

How to access   Anonymous FTP: **ds.internic.net**
Path: **rfc/rfc1207.txt**

Answers for New Internet Users

An Internet RFC (1325).

How to access   Anonymous FTP: **ds.internic.net**
Path: **rfc/rfc1325.txt**

BBS Lists

Lists of BBSs compiled from a number of sources, including Internet BBSs, government lists, law BBSs, and so on.

How to access   Anonymous FTP: **oak.oakland.edu**
Path: **/pub/misc/bbslists**

Bay Area Frequently Asked Questions

This newsgroup and FAQ are focused primarily on the San Francisco Bay Area, but most of the information is applicable to any area. The FAQ contains information on publications, Usenet, FAQ archives, Gopher, mail, uucp, and much more.

How to access   1. Usenet: **ba.internet**

2. Anonymous FTP: **wiretap.spies.com**
   Path: **/ba.internet/FAQ**

A Bibliography of Internetworking Information

An Internet RFC (1175).

How to access   Anonymous FTP: **ds.internic.net**
Path: **rfc/rfc1175.txt**

Big Fun List

Big Fun in the Internet with Uncle Bert. A list of fun resources on the Net.

How to access   Anonymous FTP: **cerberus.cor.epa.gov**
Path: **/pub/bigfun/bigfun.txt.z**

Computer Security Gopher

The DFN-Cert Gopher in Germany offers lots of information about computer security through their archives, including information on firewalls, worm-attacks, and Unix security.

How to access   1. Gopher: University of Hamburg
Address: **gopher.informatik.uni-hamburg.de**

2. Anonymous FTP: **ftp.informatik.uni-hamburg.de**
Path: **/pub/security/***

Connecting to the Internet

An Internet RFC (1359).

How to access   Anonymous FTP: **ds.internic.net**
Path: **rfc/rfc1359.txt**

Cyberspace Communications

A collection of articles and papers covering communications in cyberspace and the Internet, and what we have to gain or lose therein.

How to access   Gopher: Whole Earth Lectronic Link
Address: **gopher.well.sf.ca.us**
Choose: **Communications**

Data Explorer

Tutorials, news, examples, extensions, and other information for the IBM Visualization Data Explorer software.

How to access   1. Gopher: Cornell University
Address: **gopher.tc.cornell.edu**
Choose: **Anonymous FTP | Data Explorer Repository**

2. Anonymous FTP: **info.tc.cornell.edu**
Path: **/pub/vis/Data.Explorer/***

Database via finger

Several databases provided for public use by the University of Sydney in Australia. Finger for more information and instructions.

How to access   finger: **help@dir.su.oz.au**

Databases

Major commercial and free databases around the world, including Nicolas, Penpages, Dow Jones, Epic/Firstsearch, and Orbit.

How to access   Gopher: Swedish University Network
Address: **sunic.sunet.se**
Choose: **Databases via telnet**

December List

Information sources on the Internet.

How to access   Anonymous FTP: **ftp.rpi.edu**
Path: **/pub/communications/internet-cmc**

FAQ Archive

Very large collection of FAQs from all the popular newsgroups, as posted to the Usenet **group news.answers**.

How to access   Anonymous FTP: **rtfm.mit.edu**
Path: **/pub/usenet/news.answers/***

Fileserver via E-mail

Humorous and other files and BBS services

How to access   Mail: **mailserv@uiuc.edu**

Find an E-mail Address

To find someone's mail address when you only know their userid or real name, try this service.

How to access   Mail: **mail-server@pit-manager.mit.edu**
Body: **send usenet-addresses/**[the name you know]

FineArt Forum's Directory of Online Resources

An excellent directory of Internet resources.

How to access   Anonymous FTP: **ra.msstate.edu**
Path: **/pub/archives/fineart_online/Online_Directory**

Frequently Asked Questions (the Master List)

The master list of FAQ lists. Includes cross references to Usenet news groups and Internet resources.

How to access   Anonymous FTP: **rtfm.mit.edu**
Path: **/pub/usenet/news.answers/index**

Frequently Asked Questions re: Internet BBSs

Frequently asked questions (FAQ) (and answers) about Internet bulletin boards.

How to access   1. Usenet: **alt.bbs.internet**

2. Anonymous FTP: **pit-manager.mit.edu**
Path: **/pub/usenet/news.answers/inet-bbs-faq.Z**

FTP Services for Non-FTP Users

A guide to FTP-by-mail and other FTP services for those without FTP.

How to access   Mail: **ftpmail@decwrl.dec.com**
Body: **HELP**

FTP Tutorial

A quick tutorial on using FTP (file transfer protocol). Lists and explains common FTP commands, and outlines what files are available at these sites.

How to access   Finger: **ftp@piggy.ucsb.edu**
Finger: **ftp@ferkel.ucsb.edu**

Future of the Internet

Information and technical proposals for the future of the Internet.

How to access   Mail: **Internet-drafts@nri.reston.va.us**
Body: **help**

Gold in the Network

An Internet RFC (1402).

How to access   Anonymous FTP: **ds.internic.net**
Path: **rfc/rfc1402.txt**

Hitchhiker's Guide to the Internet

Detailed hints to allow new network participants to understand how the direction of the Internet is set, how to acquire online information and how to be a good Internet neighbor. (RFC-1118)

How to access   1. Anonymous FTP: **wuarchive.wustl.edu**
Path: **/doc/internet-info/hitchhikers.guide**

2. Anonymous FTP: **nic.ddn.mil**
Path: **rfc/rfc1118.txt**

3. Anonymous FTP: **nis.nsf.net**
Path: **rfc/rfc1118.txt**

Inter-Network Mail Guide

A publication by John Chew and Scott Yanoff that documents methods for sending mail from one network to another. If you're not sure how to e-mail someone on CompuServe, America On-Line, or to someone on Compuserve from Prodigy, or any of many different networks, this document has the information with detailed instructions.

How to access   1. Finger: **yanoff@csd4.csd.uwm.edu**

2. Usenet: **alt.internet.services**

3. Anonymous FTP: **csd4.csd.uwm.edu**
Path: **/pub/internetwork-mail-guide**

InterNIC Information Services

Find information about people, organizations, and resources on the Internet. Also find and retrieve documents from all over the world with lookups by name or key word.

How to access   1. Telnet: **ds.internic.net** and follow the instructions

2. Telnet: **rs.internic.net** and follow the instructions

Internet Engineering

Charters, technical documents, service and up-to-date activity information about the Internet Engineering Task Force (IETF).

How to access   Anonymous FTP: **wuarchive.wustl.edu**
Path: **/doc/ietf/***

Internet Help

Collection of documents, guides, and publications to help you find your way around the vastness of the Internet. It includes a FAQ, Hitchhiker's Guide to the Internet, Zen and the Art of the Internet, and much more.

How to access   Gopher: Yale University
Address: **yaleinfo.yale.edu**
Choose: **Internet Resources | Help about the Internet**

Internet Hunt

A monthly scavenger hunt for facts and trivia on and about the Net. Be the first to submit the correct answers to the questions and win fame and notoriety. Participate in the individual category, or work with friends in the team category. Look for the hunt questions on usenet.

How to access   1. Gopher: CICnet Inc.
Address: **gopher.cic.net**
Choose: **The Internet Hunt**

2. Anonymous FTP: **ftp.cic.net**
Path: **pub/internet-hunt/***

3. Usenet: **alt.bbs.internet**, **alt.internet.services**

Internet Information

Over fifty documents and resources detailing the Internet, its resources, culture, technical features, and much more.

How to access   Gopher: Swedish University Network
Address: **sunic.sunet.se**
Choose: **Internet Information**

Internet Overview

An interesting history and overview of the Internet, written by cyberpunk author Bruce Sterling.

How to access   Gopher: University of Texas at Austin
Address: **actlab.rtf.utexas.edu**
Choose: **Networks | The Internet
| An Article on the Internet...**

Internet Resource Guide

A comprehensive document listing Internet resources and how to access them.

How to access   Mail: **info-server@nnsc.nsf.net**

Internet Services

A list of documents on how to find indexes to Internet services.

How to access   Mail: **fileserv@shsu.edu**
                      Body: **SENDME MaasInfo.TopIndex***

Internet Services and Resources

The LIBS system is a comprehensive collection of Internet resources presented in an easy-to-use menu-driven interface. The system operates like a bulletin board, but offers direct access to remote resources.

How to access   1. Telnet: **nessie.cc.wwu.edu** (in the U.S.)
                         Login: **LIBS**

                      2. Telnet: **info.anu.edu.au** (in Australia)

                      3. Telnet: **garam.kreonet.re.kr** (in South Korea)

Internet Society Gopher Server

Contains detailed information about the Internet Society Organization, including conference news, charts and graphics, newsletters, and other general information related to the technical side of the Internet.

How to access   Gopher: Corporation for National Research Initiatives
                      Address: **ietf.cnri.reston.va.us**

Internet User's Glossary

A glossary for all the technical terms and phrases encountered on the Internet. An Internet RFC (1392).

How to access   1. Gopher: Swedish University Network
                         Address: **sunic.sunet.se**
                         Choose: **Internet Users' Glossary**

                      2. Anonymous FTP: **ds.internic.net**
                         Path: **rfc/rfc1392.txt**

See also   Jargon File, Computer Literature

Internet Worm

Technical papers and reports about the Internet Worm, the program that crippled the Internet.

How to access   Anonymous FTP: **nic.funet.fi**
                      Path: **/pub/doc/security/worm/***

IP Address Resolver

This resource will determine the IP address of an Internet site and send you a mail message with the address. This could be useful for people who don't have access to **host** or **nslookup** commands.

How to access   1. Mail: **resolve@cs.widener.edu**
Body: **site** [site name]

2. Mail: **dns@grasp.insa-lyon.fr**
Body: **site** [site name]

IRC II Client

IRC II help files and clients for Unix and VMS.

How to access   Anonymous FTP: **slopoke.mlb.semi.harris.com**
Path: **/pub/irc/***

IRC Thesis

An honors thesis entitled "Electropolis: Communication and Community on IRC" by E.M. Reid, detailing the culture and ways of Internet Relay Chat.

How to access   Gopher: University of Tuebingen
Address: **nova.tat.physik.uni-tuebingen.de 4242**
Choose: **Electropolis: Communication and Community on IRC**

Jargon File

Pronunciation, definitions, and examples of computer and Internet terms, acronyms, and abbreviations. (Humorous, but informative.) This file is available via Anonymous FTP from many sites. One is listed below. Search for "jargon" with Archie for other sites.

How to access   1. Anonymous FTP: **world.std.com**
Path: **/obi/Nerd.Humor/webster/jargon**

2. Wais: **jargon**

See also   Internet User's Glossary, Computer Literature, Humor

Knowbot Information Service

Knowbot, or netaddress, is an information service that provides a uniform user interface to the various Internet information services. By learning the Knowbot interface, you can query a variety of remote information services and see the results of your search in a uniform format.

How to access   Telnet: **nri.reston.va.us 185**

List of Internet Mailing Lists

A comprehensive list of mailing lists you can join on the Internet. An electronic version of the Prentice Hall book *Internet: Mailing Lists*. Nearly 500 pages of detailed descriptions and instructions. Topics range from children's rights and other political issues to sailing and yachting.

How to access   Anonymous FTP: **ftp.nisc.sri.com**
Path: **/netinfo/interest-groups**

List of Mailing Lists

A very large list of all the mailing lists in operation on the Internet.

How to access   Mail: **mail-server@nisc.sri.com**
Body: **send netinfo/interest-groups**

Loopback Service

Use this Internet address and you'll bounce back to your own system.

How to access   Telnet: **127.0.0.1**
FTP: **127.0.0.1**
Finger: **@127.0.0.1**

MaasInfo Files

A particularly comprehensive collection of information about what is available on the Net.

How to access   Anonymous FTP: **niord.shsu.edu**
Path: **maasinfo/***

Mining the Internet

A tutorial on using the Internet from the University of California. Request this document by mailing Gee Lee.

How to access   Mail: **gblee@ucdavis.edu**

Net Happenings

A mailing list of everything interesting going on with the Net. Includes interesting events such as The Geek of the Week contest.

How to access   Listserv mailing list: **net-happenings@is.internic.net**

Network Information Services

Easy access to information about important networks and information services around the globe, including ASK, DDN, Hytelnet, Janet, MichNet, and others.

How to access	Gopher: Yale University
	Address: **yaleinfo.yale.edu**
	Choose: **Internet Resources	Network Information Services**

Networking Glossary of Terms

An Internet RFC (1208).

| **How to access** | Anonymous FTP: **ds.internic.net** |
| | Path: **rfc/rfc1208.txt** |

NICOL

Access to Internet resources, Electronic Publishing Service.

| **How to access** | Telnet: **nisc.jvnc.net** |
| | Login: **nicol** |

NICOLAS

Network Information Center On-Line Aid System

| **How to access** | Telnet: **dftnic.gsfc.nasa.gov** |
| | Login: **dftnic** |

People on the Internet

Collection of services to help you find that lost comrade on the Internet. It includes access to Knowbot, Netfind, PSI White Pages, UK searches, Whois, and X.500 searches.

How to access	Gopher: Yale University
	Address: **yaleinfo.yale.edu**
	Choose: **People on the Internet**

Popular FTP Archives

Browse some of the most popular FTP sites on the Internet from this menu. Just select the subject or system you wish to look at, and you will be connected.

How to access	Gopher: NASA Goddard Space Flight Center
	Address: **gopher.gsfc.nasa.gov**
	Choose: **FTP Archives**

Porter List

High Weirdness by E-mail.

| **How to access** | Anonymous FTP: **ftp.uu.net** |
| | Path: **/doc/political/umich-poli/Resources/weirdness.Z** |

Real-life on the Internet

Examples of specific uses of the Internet. Real people doing real things.

How to access   Anonymous FTP: **wuarchive.wustl.edu**
Path: **/doc/internet-info/user.profiles**

Responsibilities of Host and Network Managers

An Internet RFC (1173).

How to access   Anonymous FTP: **ds.internic.net**
Path: **rfc/rfc1173.txt**

RFC Archive

Complete set of the Internet "Request for Comments" technical documents.

How to access   Anonymous FTP: **wuarchive.wustl.edu**
Path: **/doc/rfc/***

RFC - List of Internet Repositories

A current list of repositories and instructions about how to obtain RFCs from each of the major U.S. sites.

How to access   Anonymous FTP: **isi.edu**
Path: **/in-notes/rfc-retrieval.txt**

RFC - List of RFCs

An index of all the RFCs. Lists each RFC, starting with the most recent, and for each RFC provides the number, title, author, issue date, and number of hardcopy pages. In addition, it lists the online formats for each RFC and the number of bytes of each version online.

How to access   1. Anonymous FTP: **isi.edu**
Path: **/in-notes/rfc-retrieval.txt**

2. Wais: **internet-rfcs**

Searching the Internet

A collection of all the most useful tools to search the Internet for that needed resource. It includes a FAQ about searches and archives, and offers all the important searching tools including Archie, Hytelnet, Veronica, Wais, and the World Wide Web.

How to access   Gopher: Yale University
Address: **yaleinfo.yale.edu**
Choose: **Internet Resources | Searching the Internet with...**

Subject Trees

A collection of different subject trees maintained all over the Internet. Each tree acts like a Gopher road map allowing you to find the subjects that you wish to investigate in the vastness of Gopherspace.

How to access   Gopher: Yale University
Address: **yaleinfo.yale.edu**
Choose: **Internet Resources | Information Organization...**

Talk Radio

Internet Talk Radio is a new type of publication: a news and information service about the Internet, distributed on the Internet. Internet Talk Radio is modeled on National Public Radio and has a goal of providing in-depth technical information to the Internet community.

How to access   1. Mail: **announce-request@radio.com**

2. Mail: **questions@radio.com**

3. Mail: **radio.ora.com**

4. Usenet: **alt.internet.services**

Trickle Server Documentation

All about TRICKLE servers. What they are, how to use them, technical information, and user tips.

How to access   Anonymous FTP: **oak.oakland.edu**
Path: **/pub/misc/trickle**

Using Internet Libraries

Multitude of software, documents, resource guides, including those for Archie, Wais, FTP, e-mail, Listserv, and resource lists, for librarians and library users on the Internet.

How to access   Anonymous FTP: **nic.funet.fi**
Path: **/pub/doc/library/***

Washington University Services

Washington University in St. Louis provides an easy-to-use menu interface to university library systems and Internet resources all over the world.

How to access   Telnet: **wugate.wustl.edu**
Login: **services**

Yanoff List

The Internet Services List, a comprehensive list of Internet resources.

How to access Anonymous FTP: **pit-manager.mit.edu**
Path: **/pub/usenet/news.answers/internet-services/list**

Zen and the Art of the Internet

The first edition of the booklet by Brendan Kehoe, which covers all the basics of the Internet, including e-mail, FTP, Usenet, Telnet, tools, and much more.

How to access 1. Gopher: University of Michigan
Address: **gopher.citi.umich.edu**
Choose: **Internet Documents | zen**

2. Anonymous FTP: **Fpickup coltp.cs.widener.edu**
Path: **/pub/zen/***

Intrigue

J.F.K. Conspiracy

John F. Kennedy assassination conspiracy material, and archives from the **alt.conspiracy.jfk** Usenet newsgroup.

How to access 1. Telnet: **grind.isca.uiowa.edu**
Login: **iscabbs**

2. Anonymous FTP: **grind.isca.uiowa.edu**
Path: **/info/jfk/***

Taylorology

In 1922, one of Hollywood's top movie directors, William Desmond Taylor, was shot to death in his home. The killing was never solved and remains Hollywood's most fascinating murder mystery. "Taylorology" is a lengthy newsletter devoted to analyzing and reprinting source material pertaining to the crime and the press coverage of it.

How to access 1. Usenet: **alt.true-crime**

2. Anonymous FTP: **uglymouse.css.itd.umich.edu**
Path: **/pub/Zines/Taylorology**

3. Gopher: University of Michigan
Address: **uglymouse.css.itd.umich.edu**
Choose: **Zines | Taylorology**

Language

Acronyms

Thousands of acronyms, defined.

How to access   1. Anonymous FTP: **ucselx.sdsu.edu**
Path: **/pub/doc/general/acronyms.txt**

    2. Gopher: Manchester Computing Centre
Address: **info.mcc.ac.uk**
Choose: **Miscellaneous items | Acronym dictionary**

    3. Wais: **acronyms**

Online Reference

Online dictionary and thesaurus. Also a database of famous (and not so famous) quotes.

How to access   Telnet: **info.rutgers.edu**
Choose: **Library | Reference**

Quotations Archive

A large selection of quotes from all walks of life. "Truth is more of a stranger than fiction."
—Mark Twain.

How to access   1. Anonymous FTP: **wilma.cs.brown.edu**
Path: **pub/alt.quotations/Archive/***

    2. Usenet: **alt.quotations**

Roget's Thesaurus

The complete reference. Provided by Project Gutenberg. Search and consult the original Roget's 1911 Thesaurus.

How to access   1. Gopher: University of Minnesota
Address: **consultant.micro.umn.edu**
Choose: **Libraries | Electronic Books | By Title**

    2. Gopher: Manchester Computing Centre, University of Manchester
Address: **uts.mcc.ac.uk**
Choose: **Experimental & New Services
| Roget's 1911 Thesaurus**

Shorter Oxford Dictionary

The complete shorter Oxford Dictionary word list, including part-of-speech information.

L

How to access   Anonymous FTP: **ftp.white.toronto.edu**
Path: **/pub/words/sodict.Z**

Webster's Dictionary Servers

Online Webster's dictionary and spelling reference. This service repeatedly prompts you for a word. If it misspells a word, the system will prompt you to choose between a number of similar words. Gives spelling, pronunciation, and definitions.

How to access   1. Telnet: **cs.indiana.edu 2627**
Login: **webster**

2. Telnet: **chem.ucsd.edu**
Login: **webster**

3. Gopher: University of California San Diego
Address: **infopath.ucsd.edu**
Choose: **Reference Services** | **Webster's Online Dictionary**

Word Lists

Word lists for several languages, including Dutch, English, German, Italian, Norwegian, and Swedish. Useful for linguistic research and text-processing applications such as spelling checkers.

How to access   Anonymous FTP: **nic.funet.fi**
Path: **/pub/doc/dictionaries/***

Languages

Chinese Text Viewers

Programs, documents, and fonts for viewing Chinese text on IBM-PC and Unix systems.

How to access   Anonymous FTP: **nic.funet.fi**
Path: **/pub/culture/chinese/***

Law

Computer Laws

Collection of information about laws involving computers, including computer crime laws for many states and countries.

How to access   Anonymous FTP: **ftp.eff.org**
Path: **/pub/cud/law/***

CU-LawNet

Legal reference resource and information about Columbia University and the Columbia University Law School by the Columbia Law School Public Information Service. Also offers Columbia University and Law School catalogs.

How to access   Telnet: **sparc-1.law.columbia.edu**
Login: **lawnet**

Law Server

Legal discussions and reference material for United States, foreign, and international law. Library resources, government agencies, periodicals, and lists of other legal resources on the Internet.

How to access   Gopher: Cornell University Law School
Address: **fatty.law.cornell.edu**

League for Programming Freedom

The League for Programming Freedom is a grass-roots organization of professors, students, businesspeople, programmers, and users dedicated to reversing the current trend toward copyright and patent laws covering software.

How to access   Anonymous FTP: **ftp.cs.widener.edu**
Path: **/pub/lpf/***

Net Law

Discussion of the various legal issues affecting the Net and the community therein.

How to access   Gopher: Electronic Frontier Foundation
Address: **gopher.eff.org**
Choose: **Electronic Frontier.. | Discussion of legal...**

Privacy Forum Digest

A moderated digest for the discussion and analysis of privacy in the information age.

How to access   1. Gopher: Vortex Technology
Address: **cv.vortex.com**
Choose: **Privacy Forum**

2. Anonymous FTP: **ftp.vortex.com**
Path: **/privacy/***

Steve Jackson Games

Collection of all the information surrounding the Secret Service raid on the Steve Jackson Games company, and the following court case.

L

How to access   Anonymous FTP: **ftp.eff.org**
Path: **/pub/SJG/***

Supreme Court Rulings

With Project Hermes, the United States Supreme Court makes its opinions and rulings available in electronic format within minutes of their release. Case Western Reserve University is one of the sites the Supreme Court supplies with this information. Get and read the files **INFO** and **README.FIRST**.

How to access   1. Gopher: Colorado State University
Address: **lobo.colorado.edu**
Choose: **Other Information Services
| Supreme Court Rulings**

2. Anonymous FTP: **ftp.cwru.edu**
Path: **/hermes/***

3. Wais: **supreme-court**

Washington & Lee University Law Library

Gateways to an amazing volume of Internet law resources, including many universities in the United States and around the world. Use an easy menu interface to navigate through the law libraries all over the nation. Search for key words, read the case law, or search for books, articles, and publications.

How to access   Telnet: **liberty.uc.wlu.edu**
Login: **lawlib**

Libraries

Accessible Library Catalogs & Databases

A large document with detailed instructions on how to access the computerized library systems of many universities around the world.

How to access   1. Anonymous FTP: **ariel.unm.edu**
Path: **/library/internet.library**

2. Anonymous FTP: **ftp.unt.edu**
Path: **/library/libraries.txt**

The Carl System

A computerized network of library systems. Search for key words from any of five databases (library catalogs, current articles, information databases, other library systems, library and system news).

How to access   Telnet: **pac.carl.org**

Eureka

Eureka is an easy-to-use search service. Any individual or institution can search the online resources of the Research Libraries Group (RLG), including the RLIN bibliographic files and the CitaDel article-citation and document-delivery service. Eureka contains information about more than 20 million books, serials, sound recordings, musical scores, archival collections, and other materials.

How to access	Telnet: **eureka-info.stanford.edu**

Hytelnet

This program assists people using library resources by automating the process. Hytelnet is a program that presents library resources on an easy menu interface. When you choose a resource, Hytelnet will show you how to access the resource, or even connect to it automatically. Executables for various machines, as well as source, are available.

How to access	Hytelnet is widely available on Gopher systems. Use Archie or Veronica for locations.
See also	Software

Launchpad

This system provides access to many library systems across the country. You can perform searches, download files, find other users, and even connect to the local Gopher client. Try the CIA World Fact Book in the experimental Gopherspace section.

How to access	Telnet: **launchpad.unc.edu** Login: **launch**

Library of Congress

The Library of Congress maintains millions of records of publications in the United States.

How to access	Telnet: **locis.loc.gov** Monday - Friday 6:30AM-9:30PM EST Saturday 8AM-5PM, Sunday 1PM-5PM EST

Library Newsletters

Selection of newsletters about cataloging, indexing, collecting, and preserving books.

How to access	Anonymous FTP: **nigel.msen.com** Path: **/pub/newsletters/Libraries/***

L

Library Resources on the Internet

This file contains strategies for selection and use of library resources on the Internet. A 40+ page document for download via FTP that includes how to get started, road maps, travel guides, search strategies, and information about other library resources.

How to access   Anonymous FTP: **dla.ucop.edu**
Path: **/pub/internet/libcat_guide**

University of Maryland Information Database

Information and documents on many subjects of interest.

How to access   Telnet: **info.umd.edu**
Login: **info**

Literature: Authors

L. Frank Baum

Two complete texts by L. Frank Baum, creator of the Wizard of Oz.

How to access   Anonymous FTP: **nic.funet.fi**
Paths: **/pub/doc/etext/ozland.txt; /pub/doc/etext/wizoz.txt**

Lewis Carroll

Several full texts by Lewis Carroll.

How to access   Anonymous FTP: **nic.funet.fi**
Paths: **/pub/doc/etext/carroll/*;**
/pub/doc/etext/alice.txt; /pub/doc/etext/jabber.txt;
/pub/doc/etext/looking.txt; /pub/doc/etext/snark.txt

Charles Dickens

Several complete novels by Charles Dickens.

How to access   Anonymous FTP: **nic.funet.fi**
Paths: **/pub/doc/etext/carol.txt; /pub/doc/etext/chimes.txt;**
/pub/doc/etext/cricket.txt

Sir Arthur Conan Doyle

Several complete texts and stories by Sir Arthur Conan Doyle, author of the Sherlock Holmes series.

How to access   Anonymous FTP: **nic.funet.fi**
Paths: **/pub/doc/etext/doyle/***; **/pub/doc/etext/hound.dyl**;
/pub/doc/etext/adventures.dyl; **/pub/doc/etext/lastbow.dyl**;
/pub/doc/etext/magicdoor.dyl; **/pub/doc/etext/memoirs.dyl**;
/pub/doc/etext/return.dyl; **/pub/doc/etext/signfour.dyl**;
/pub/doc/etext/valley.dyl; **/pub/doc/etext/casebook.dyl**;
/pub/doc/etext/study.dyl

Mansfield

Collection of works by K. Mansfield.

How to access   Anonymous FTP: **nic.funet.fi**
Path: **/pub/doc/etext/mansfield/***

John Milton

Several works by John Milton.

How to access   Anonymous FTP: **nic.funet.fi**
Paths: **/pub/doc/etext/pargain.txt**; **/pub/doc/etext/parlost.txt**

Edgar Allan Poe

Collection of works by Edgar Allan Poe.

How to access   Anonymous FTP: **nic.funct.fi**
Paths: **/pub/doc/etext/Poe/***; **/pub/doc/etext/cask.poe**;
/pub/doc/etext/pit.poe; **/pub/doc/etext/telltale.poe**

Terry Pratchett

Archive of information about the well-known author of humorous fantasy-based science
fiction novels, and his discworld creations.

How to access   Anonymous FTP: **ftp.cs.pdx.edu**
Path: **/pub/pratchett/***

Saki

Two complete texts, including *Reginald*.

How to access   Anonymous FTP: **nic.funet.fi**
Paths: **/pub/doc/etext/reginald.hh**; **/pub/doc/etext/russia.hh**

Shakespeare

The complete works of William Shakespeare.

L

How to access   1. Anonymous FTP: **nic.funet.fi**
Path: **/pub/doc/etext/shakespeare/***

2. Gopher: University of Minnesota
Address: **consultant.micro.umn.edu**
Choose: **Libraries | Electronic Books | By Title**
| Complete..

Elf Sternburg

The complete set of the journal entry stories of Elf Sternburg, tales of an erotic nature.

How to access   Anonymous FTP: **seismo.soar.cs.cmu.edu**
Path: **stories/journal/***

Mark Twain

Several complete texts by Mark Twain, including Tom Sawyer, Extracts from Adam's Diary, A Ghost Story, Niagara, and others.

How to access   Anonymous FTP: **nic.funet.fi**
Paths: **/pub/doc/etext/abroad.mt;**
/pub/doc/etext/adam.mt; /pub/doc/etext/detective.mt;
/pub/doc/etext/ghost.mt; /pub/doc/etext/niagara.mt;
/pub/doc/etext/pitcairn.mt; /pub/doc/etext/sawyer.mt;
/pub/doc/etext/yankee.mt

Virgil

Collection of works in Latin by Virgil.

How to access   Anonymous FTP: **nic.funet.fi**
Paths: **/pub/doc/etext/Virgil/***

H.G. Wells

Complete texts by H.G. Wells.

How to access   Anonymous FTP: **nic.funet.fi**
Path: **/pub/doc/etext/invisman.txt; /pub/doc/etext/timemach.txt**

W.B. Yeats

The complete works of the Irish poet William Butler Yeats.

How to access   Anonymous FTP: **nic.funet.fi**
Path: **/pub/doc/etext/yeats/***

Literature: Collection

Complete Books in ZIP Format

Many of the works mentioned elsewhere in this catalog in ZIP format. Other titles include *The Book of Mormon*, *Zen and the Art of the Internet*.

How to access   Anonymous FTP: **oak.oakland.edu**
Path: **/pub/misc/books**

Electronic Books

Access to the Library of Congress records and a number of books in electronic form available to be read or downloaded. Titles include *Aesop's Fables*, *Agrippa*, *Aladdin and the Wonderful Lamp*, *Alice's Adventures in Wonderland*, *CIA World Factbook*. Search by author, call letter, title, or by specific strings of text. Entries are arranged alphabetically.

How to access   Gopher: University of Minnesota
Address: **consultant.micro.umn.edu**
Choose: **Libraries I Electronic Books**

Electronic Books in ASCII Text

ASCII text versions of more of your favorite novels than you could ever have imagined, all in one place.

How to access   Anonymous FTP: **nic.funet.fi**
Path: **/pub/doc/etext/***

Fiction

A collection of interesting contemporary works of fiction.

How to access   Gopher: University of Michigan
Address: **uglymouse.css.itd.umich.edu**
Choose: **Fiction**

Narrative of the Life of Frederick Douglass

The full text of this narrative of the life of an American abolitionist.

How to access   Gopher: University of Minnesota
Address: **consultant.micro.umn.edu**
Choose: **Libraries I Electronic Books I By Title**

Poetry Assortments

A large collection of poetry arranged by author. Includes works of Housman, Jeffers, Millay, O'Shaugnessy, Russell, Whitman, and Yeats.

L

How to access
1. Anonymous FTP: **ocf.berkeley.edu**
   Path: **/pub/Library/Poetry**

2. Gopher: North Dakota State University
   Address: **chiphead.cc.ndsu.nodak.edu**
   Choose: **Other Stuff | Fun Stuff | Poems**

Project Gutenberg

Project Gutenberg is planned as a storage- and clearing-house for making books available very cheaply. Much of the work so far has focused on classic literature (for which the copyright has expired). They have books by many authors, including Mark Twain, H.G. Wells, and F. Scott Fitzgerald. They also have the Bible, Book of Mormon, and Koran in ASCII format. Also available from **info.umd.edu** is a collection of economic time series data from the federal government, as well as daily and long-term weather forecasts.

How to access
1. Anonymous FTP: **oes.orst.edu**
   Path: **/pub/etext**

2. Anonymous FTP: **mrcnext.cso.uiuc.edu**
   Path: **/pub/etext**

3. Anonymous FTP: **info.umd.edu**
   Path: **/info/ReadingRoom/Fiction**

4. Telnet: **info.umd.ude**
   Login: **info**

Short Stories

An interesting collection of enjoyable essays and short stories, covering different walks of life.

How to access
Gopher: Whole Earth Lectronic Link
Address: **gopher.well.sf.ca.us**
Choose: **Art and Culture | Essays and Short Stories**

Literature: Titles

A Christmas Carol

The full heartwarming tale by Charles Dickens.

How to access
Anonymous FTP: **nic.funet.fi**
Path: **/pub/doc/etext/carol.txt**

A Connecticut Yankee in King Arthur's Court

The complete text by Mark Twain.

How to access	Anonymous FTP: **nic.funet.fi** Path: **/pub/doc/etext/yankee.mt**

The Aeneid

The complete text, in Latin, by Virgil.

How to access	Anonymous FTP: **nic.funet.fi** Path: **/pub/doc/etext/Virgil/aeneid.tex**

Aesop's Fables

The complete text from Project Gutenberg.

How to access	Gopher: University of Minnesota Address: **consultant.micro.umn.edu** Choose: **Libraries I Electronic Books**

Agrippa

The complete text from Project Gutenberg.

How to access	Gopher: University of Minnesota Address: **consultant.micro.umn.edu** Choose: **Libraries I Electronic Books**

Aladdin and the Wonderful Lamp

The complete text from Project Gutenberg.

How to access	Gopher: University of Minnesota Address: **consultant.micro.umn.edu** Choose: **Libraries I Electronic Books**

Alice in Wonderland

The complete text by Lewis Carroll.

How to access	Anonymous FTP: **ucselx.sdsu.edu** Path: **/pub/doc/etext/alice26a.txt**

Anne of Green Gables

The complete text by Lucy Montgomery.

How to access	Anonymous FTP: **nic.funet.fi** Path: **/pub/doc/etext/anne.txt**

L

As a Man Thinketh

The complete text by James Allen.

How to access   Anonymous FTP: **nic.funet.fi**
Path: **/pub/doc/etext/thinketh.txt**

The Call of the Wild

The complete text by Jack London.

How to access   Anonymous FTP: **nic.funet.fi**
Path: **/pub/doc/etext/callwild.txt**

The Communist Manifesto

The document by Marx and Engels.

How to access   Anonymous FTP: **nic.funet.fi**
Path: **/pub/doc/etext/manifesto.txt**

Discourse on Reason

The complete essay by Descartes.

How to access   Anonymous FTP: **nic.funet.fi**
Path: **/pub/doc/etext/reason.txt**

Dracula

Collection of texts by Bram Stoker.

How to access   Anonymous FTP: **nic.funet.fi**
Paths: **/pub/doc/etext/dracula.txt**; **/pub/doc/etext/dracgst**

Essays in Radical Empiricism

The complete text by William James.

How to access   Anonymous FTP: **nic.funet.fi**
Path: **/pub/doc/etext/empiricism.txt**

Fairy Tales

Wonderful collection of childhood magic. Tales of puffing wolves, little pigs, ugly ducklings, dwarves, princesses, and thieves. Text of numerous fairy tales, including The Little Mermaid, Snow White, The Adventures of Aladdin, Beauty and the Beast, Hansel and Gretel, Jack and the Beanstalk, The Three Little Pigs, The Tin Soldier, and many more.

How to access   1. Anonymous FTP: **world.std.com**
Path: **obi/Fairy.Tales/Grimm/***

2. Anonymous FTP: **nic.funet.fi**
Path: **/pub/doc/etext/fairy-tale/***

See also   Movies, Television

Far from the Madding Crowd

The full text from Project Gutenberg.

How to access   Gopher: University of Minnesota
Address: **consultant.micro.umn.edu**
Choose: **Libraries | Electronic Books**

The Gift of the Magi

The full text from Project Gutenberg.

How to access   Gopher: University of Minnesota
Address: **consultant.micro.umn.edu**
Choose: **Libraries | Electronic Books | By Title**

Herland

The full text from Project Gutenberg.

How to access   Gopher: University of Minnesota
Address: **consultant.micro.umn.edu**
Choose: **Libraries | Electronic Books | By Title**

Hippocratic Oath and Law

The text by Hippocrates.

How to access   Anonymous FTP: **nic.funet.fi**
Path: **/pub/doc/etext/hippocrat.txt**

The Hunting of the Snark

A complete text.

How to access   Anonymous FTP: **ucselx.sdsu.edu**
Path: **/pub/doc/etext/snark11.txt**

The Invisible Man

The full text by H.G. Wells.

How to access   Anonymous FTP: **nic.funet.fi**
Path: **/pub/doc/etext/invisman.txt**

L

Jabberwocky

The full text by Lewis Carroll.

> **How to access**   Anonymous FTP: **nic.funet.fi**
> Path: **/pub/doc/etext/jabber.txt**

The Japan That Can Say No

A complete text.

> **How to access**   Anonymous FTP: **ucselx.sdsu.edu**
> Path: **/pub/doc/etext/japan-that-can-say-no.txt**

Keepsake Stories

The complete text by Walter Scott.

> **How to access**   Anonymous FTP: **nic.funet.fi**
> Path: **/pub/doc/etext/keepsake.txt**

The Legend of Sleepy Hollow

The complete story by Washington Irving.

> **How to access**   1. Anonymous FTP: **nic.funet.fi**
> Path: **/pub/doc/etext/sleepy.txt**
>
> 2. Gopher: University of Minnesota
> Address: **consultant.micro.umn.edu**
> Choose: **Libraries | Electronic Books | By Title**

Moby Dick

The full text from Project Gutenberg.

> **How to access**   Gopher: University of Minnesota
> Address: **consultant.micro.umn.edu**
> Choose: **Libraries | Electronic Books | By Title**

O Pioneers!

The full text by Willa Cather. Provided by Project Gutenberg.

> **How to access**   Gopher: University of Minnesota
> Address: **consultant.micro.umn.edu**
> Choose: **Libraries | Electronic Books | By Title**

Oedipus Trilogy

The full text, provided by Project Gutenberg.

How to access   Gopher: University of Minnesota
Address: **consultant.micro.umn.edu**
Choose: **Libraries** | **Electronic Books** | **By Title**

Paradise Lost

The complete text, provided by Project Gutenberg.

How to access   Gopher: University of Minnesota
Address: **consultant.micro.umn.edu**
Choose: **Libraries** | **Electronic Books** | **By Title**

Peter Pan

The complete text, provided by Project Gutenberg.

How to access   Gopher: University of Minnesota
Address: **consultant.micro.umn.edu**
Choose: **Libraries** | **Electronic Books** | **By Title**

The Pit and the Pendulum

The full text by Edgar Allan Poe.

How to access   Anonymous FTP: **nic.funet.fi**
Path: **/pub/doc/etext/pit.poe**

The Scarlet Letter

The complete text provided by Project Gutenberg.

How to access   Gopher: University of Minnesota
Address: **consultant.micro.umn.edu**
Choose: **Libraries** | **Electronic Books** | **By Title**

The Scarlet Pimpernel

The complete text by Baroness Orezy.

How to access   Anonymous FTP: **nic.funet.fi**
Path: **/pub/doc/etext/pimpernel.txt**

Scientific Secrets, 1861

The complete text of Daniel Young.

L

How to access	Anonymous FTP: **nic.funet.fi** Path: **/pub/doc/etext/science.txt**

Shakespeare

The full text of Shakespeare's plays and poems.

How to access	Anonymous FTP: **ocf.berkeley.edu** Path: **pub/Library/Shakeseare**

Sherlock Holmes Novels

Collection of texts by Sir Arthur Conan Doyle.

How to access	Anonymous FTP: **nic.funet.fi** Paths: **/pub/doc/etext/adventures.dyl;** **/pub/doc/etext/memoirs.dyl; /pub/doc/etext/lastbow.dyl;** **/pub/doc/etext/casebook.dyl; /pub/doc/etext/return.dyl;** **/pub/doc/etext/signfour.dyl; /pub/doc/etext/study.dyl;** **/pub/doc/etext/hound.dyl; /pub/doc/etext/magicdoor.dyl;** **/pub/doc/etext/valley.dyl**

Song of Hiawatha

The complete text, provided by Project Gutenberg.

How to access	Gopher: University of Minnesota Address: **consultant.micro.umn.edu** Choose: **Libraries** I **Electronic Books** I **By Title**

The Strange Case of Dr. Jekyll and Mr. Hyde

The full text, provided by Project Gutenberg.

How to access	Gopher: University of Minnesota Address: **consultant.micro.umn.edu** Choose: **Libraries** I **Electronic Books** I **By Title**

Through the Looking Glass

The complete text by Lewis Carroll.

How to access	Anonymous FTP: **ucselx.sdsu.edu** Path: **/pub/doc/etext/lglass15.txt**

Time Machine

The full text, provided by Project Gutenberg.

> **How to access**   Gopher: University of Minnesota
> Address: **consultant.micro.umn.edu**
> Choose: **Libraries** | **Electronic Books** | **By Title**

Tom Sawyer

Collection of complete Tom Sawyer texts by Mark Twain.

> **How to access**   Anonymous FTP: **nic.funet.fi**
> Paths: **/pub/doc/etext/abroad.mt**; **/pub/doc/etext/sawyer.mt**;
> **/pub/doc/etext/detective.mt**

United Nations Declaration of Human Rights

A complete text.

> **How to access**   Anonymous FTP: **ucselx.sdsu.edu**
> Path: **/pub/doc/etext/un_declaration.txt**

United We Stand

The full text, provided by Project Gutenberg.

> **How to access**   Gopher: University of Minnesota
> Address: **consultant.micro.umn.edu**
> Choose: **Libraries** | **Electronic Books** | **By Title**

Voices

The complete novella, by "Scotto". A science fiction/philosophy piece describing the author's so-called "brush with the Tremendum".

> **How to access**   Anonymous FTP: **penguin.gatech.edu**
> Path: **/pub/leri/text/stories/***

The War of the Worlds

The full novel by H.G. Wells.

> **How to access**   Anonymous FTP: **nic.funet.fi**
> Path: **/pub/doc/etext/warworld.txt**

The Wonderful Wizard of Oz

The complete text by L. Frank Baum.

> **How to access**   Anonymous FTP: **nic.funet.fi**
> Path: **/pub/doc/etext/wizos.txt**

L

Wuthering Heights

The complete text by Emily Bronte.

How to access   Anonymous FTP: **nic.funet.fi**
                         Path: **/pub/doc/etext/wuther.txt**

Locks and Keys

Lock Picking

Material covering lock pick sets, skeleton keys, Kryptonite locks, automatic pickers, related books, legal issues, and more.

How to access   Anonymous FTP: **rtfm.mit.edu**
                         Path: **/pub/usenet/alt.locksmithing/***

Locksmithing

Technical information and pictures related to locksmithing.

How to access   Anonymous FTP: **ftp.rahul.net**
                         Path: **/pub/atman/UTLCD-preview/locksmithing/***

Magazines

Byte Magazine

Back issues of Byte magazine in arc format.

How to access   Anonymous FTP: **oak.oakland.edu**
                         Path: **/pub/misc/byte**

Chips Online

Back issues of Chips and Chips Online magazine.

How to access   Anonymous FTP: **oak.oakland.edu**
                         Path: **/pub/misc/nardac**

PC Magazine

Programs and source code from the popular Ziff-Davis publication online.

How to access   Anonymous FTP: **ftp.cco.caltech.edu**
                         Path: **/pub/ibmpc/pcmag/**

Phrack

An electronic magazine devoted to hackers. Complete collection of the popular publication from the computer underground, containing technical and legal information relevant to hacking, phreaking, and other underground activities.

How to access   1. Anonymous FTP: **ftp.netsys.com**
Path: **/pub/phrack**

2. Anonymous FTP: **ftp.eff.org**
Path: **/pub/cud/phrack/***

Sound Site Newsletter

A monthly newsletter for PC sound enthusiasts. Tips on configuration, sound files, programming with sound, and much more.

How to access   Anonymous FTP: **oak.oakland.edu**
Path: **/pub/misc/sound**

Voices from the Net

An electronic magazine filled with interviews and essays presenting the voices of folks from a wide variety of online environments: entertaining and useful, net-literature and net-ethnography combined. An exploration of the odd corners of cyberspace. To subscribe, send mail to the address below. To download a back issue, use one of the FTP sites listed below.

How to access   1. Mail: **voices-request@andy.bgsu.edu**
Subject: **Voices from the Net**
Body: **subscribe**

2. Anonymous FTP: **aql.gatech.edu**
Path: **/pub/Zines/Voices_from_the_Net**

3. Anonymous FTP: **uglymouse.css.itd.umich.edu**
Path: **/pub/Zines/Voices**

4. Anonymous FTP: **wiretap.spies.com**
Path: **/Library/Zines**

Zines

A collection of many popular zines, including *The Unplastic News, Cyberspace, Vanguard,* and much more.

How to access   Gopher: Electronic Frontier Foundation
Address: **gopher.eff.org**
Choose: **Instant Karma Zine Stand**

M

Mathematics

E-Math

The BBS of the American Mathematics Society. Offers conversation, software, and software reviews.

How to access   Telnet: **e-math.ams.com**
Login: **e-math**

Math and Calculus Programs

Interesting and useful mathematics and calculus programs. See Chapter 17.

How to access   Anonymous FTP: **wuarchive.wustl.edu**
Path: **/edu/math/msdos/calculus/***

NetLib

Math software programs via e-mail.

How to access   1. Mail: **netlib@ornl.gov**
Body: **send index**

2. Mail: **netlib@uunet.uu.net**
Body: **send index**

Pi to 1 Million Digits

One million digits of the constant pi. Provided by Project Gutenberg.

How to access   Gopher: University of Minnesota
Address: **consultant.micro.umn.edu**
Choose: **Libraries | Electronic Books | By Title**

Pi to 1.25 Million Digits

The first 1.25 million digits of pi, that mathematical wonder!

How to access   Anonymous FTP: **wuarchive.wustl.edu**
Path: **/doc/misc/pi/***

Square Root of 2

The square root of 2 to many decimal places.

How to access   Gopher: University of Minnesota
Address: **consultant.micro.umn.edu**
Choose: **Libraries | Electronic Books | By Title**

StatLib Server

Programs, datasets, instructions, and help for statisticians.

How to access   Mail: **statlib@lib.stat.cmu.edu**
Body: **send index**

Medicine

AIDS Information via CHAT Database

CHAT (Conversational Hypertext Access Technology) database system. You can ask questions of this database with natural English questions. It correctly interprets an amazing number of questions and instantly provides answers about AIDS—especially in Canada.

How to access   Telnet: **debra.dgbt.doc.ca**
Login: **chat**

CancerNet

Health and clinical information about cancer and cancer research.

How to access   1. Gopher: National Institute of Health
Address: **gopher.nih.gov**

2. Mail: **cancernet@icicb.nci.nih.gov**

Drugs Information

Read the latest reports and journals from the Food and Drug Administration's database. Called "Medline", the database has the latest information.

How to access   1. Telnet: **library.umdnj.edu**
Login: **library**

2. Telnet: **utmem1.utmem.edu**
Login: **harvey**

E.T.Net

An information system run by the National Library of Medicine. It features conferences on computer aided education in health sciences, hypermedia, expert systems, patient simulations, nursing care, computer hardware and software—including medical shareware.

How to access   Telnet: **etnet.nlm.nih.gov**
Login: **etnet**

M

Epilepsy Information via CHAT Database

CHAT (Conversational Hypertext Access Technology) database system. You can ask questions of this database with natural English questions. It correctly interprets an amazing number of questions and instantly provides answers about epilepsy.

How to access   Telnet: **debra.dgbt.doc.ca**
                     Login: **chat**

Genetics Bank

A genetics database, including nucleic acid and protein sequence provided by the National Center of Biotechnology Information (part of the National Library of Medicine). Query the database by e-mail. Send a mail message for instructions.

How to access   1. Mail: **gene-server@bchs.uh.edu**
                        Body: **help**

                     2. Mail: **retrieve@ncbi.nlm.nih.gov**
                        Body: **help**

                     3. Mail: **blast@ncbi.nlm.nih.gov**
                        Body: **help**

Medical Resources

A guide to and large list of medical resources on the Internet.

How to access   Anonymous FTP: **nic.funet.fi**
                     Path: **/pub/doc/library/medical_resources.txt.Z**

Medical/Health Information

Offers access to a large number of medical resources from a single menu, including Camis, CancerNet, Medinfo, and many more.

How to access   Gopher: Albert Einstein College of Medicine
                     Address: **gopher.aecom.yu.edu**
                     Choose: **Internet Resources | Medical | Health Information**

National Cancer Center Database (Japan)

Database interfaces for physicians, patients, researchers, and other interested people. Other information includes design of clinical trials information, supportive care, cancer screening guidelines, news, and general information. Mailing list available.

How to access   Gopher: National Cancer Center
                     Address: **gan.ncc.go.jp**
                     Choose: **CancerNet service**

National Library of Medicine (NLM) Locator

The locator searches the book holdings database (CATLINE), the audiovisual holdings database (AVLINE), and the journal holdings database (SERLINE) of the U.S. National Library of Medicine. Information is also available on library hours, policies, interlibrary loan and the National Network of Libraries of Medicine (NN/LM)

How to access   Telnet: **locator.nlm.nih.gov**
Login: **locator**

Military

Defense Conversion Subcommittee

Forums for associations, institutes, think tanks, and consultants for defense conversion. Includes weekly updates, government contacts, Russian conversion issues, and success stories.

How to access   Gopher: University of California San Diego
Address: **infopath.ucsd.edu**
Choose: **News & Services | Economic.. | Defense Conversion..**

The Military

Collection of papers and viewpoints about the military, its people, policies, and practices, including book reviews and intelligence information.

How to access   Gopher: Whole Earth Lectronic Link
Address: **gopher.well.sf.ca.us**
Choose: **The Military, its People, Policies, and Practices**

Navy News Service (NAVNEWS)

An electronic magazine with articles of interest to those in, or interested in, the Navy. Includes good articles on the Internet by famous authors and such items as the air show schedule for the Blue Angels. Mail for information on how to get on the mailing list, and where to FTP files.

How to access   Mail: **navnews@nctamslant.navy.mil**

M

Movies

Film Database

Search a database of synopses, cast lists, and other information on thousands of films released before 1987.

> **How to access**   Gopher: Manchester Computing Centre
> Address: **info.mcc.ac.uk**
> Choose: **Miscellaneous items** | **Film database**

Reviews

Thousands of reviews of all the popular movies, from a variety of different critics.

> **How to access**   Anonymous FTP: **nic.funet.fi**
> Path: **/pub/culture/movies/***

MUDs

Actuator MUD

Actuator is about building cyberspace. It is for researching and designing drivers, clients, graphics, MUDlibs, worlds, networked objects, and social interaction.

> **How to access**   Telnet: **actlab.rtf.utexas.edu 4000**

Actuator MUD Gopher

Browse Actuator MUD, list its users, connect to other MUD Gopher servers, or simply connect to Actuator MUD itself.

> **How to access**   Gopher: University of Texas at Austin
> Address: **actlab.rtf.utexas.edu 3452**

AlexMUD

The oldest DikuMUD on the Internet, started on March 9th, 1991. Based in Sweden, it has its own distinctive style and depth, which gives it its popularity.

> **How to access**   1. Telnet: **mud.stacken.kth.se 4000**
>
>   2. Telnet: **marcel.stacken.kth.se 4000**

Apocalypse III (MUD)

A very popular DikuMUD, with lots of extras. Seven different races, nine different character classes, chit-chat channels, and even color! Check it out.

How to access   Telnet: **peabrain.humgen.upenn.edu 4000**

Black Knight's Realm

A growing MUD, mostly based on Copper Diku II.

How to access   Telnet: **sun1.cstore.ucf.edu 4000**

BurningDIKU (MUD)

An expanding DikuMUD, with lots of new areas, spells, skills, levels, and smart, tougher monsters.

How to access   Telnet: **next5.cas.muohio.edu 4000**

Chomestoru

A very changed DikuMUD, with eight races to choose from including faeries, mutants, and asuras. Over 130 spells, primary and tertiary skills, and stat preference.

How to access   Telnet: **mccool.cbi.msstate.edu 4000**

Copper Diku II

A friendly modified DikuMUD with selection of hometown, special city for killers, battle arena, new areas, and even jail for law-breaking players.

How to access   Telnet: **copper.denver.colorado.edu 4000**

Dark Shadow's DikuMUD

Based on the Dragonlance books (Weis & Hickman), with eight races including gnomish and kender, ten classes including druids and monks, guilds, and a Realm of Souls for those who met their fate.

How to access   Telnet: **hobbes.linfield.edu 7777**

Discworld MUD

An LPMUD based on the colorful Discworld books by the legendary Terry Prachett. Discworld is, as the Wombles and Blues will tell you, where all your dreams can't come true.

How to access   Telnet: **cix.compulink.co.uk 4242**

M

Discworld MUD Gopher

Allows one to view and find information about players of DiscWorld MUD, and offers easy access to other MUD and entertainment related Gophers.

How to access   Gopher: **cix.compulink.co.uk 3450**

Frontier MUD

This enhanced LPMUD has a medieval fantasy theme. Features include a very real weather system and 17 magnificent quests which will take you from one end of the land to the other, encountering many strange creatures and magical items on your way.

How to access   Telnet: **benford.cc.umanitoba.ca 9165**

Game Server

Choose from a multitude of exciting online games, including bucks, moria, tetris, sokoban, reversi, nethack, and many adventure games, including MUDs.

How to access   1. Telnet: **castor.tat.physik.uni-tuebingen.de**
Login: **GAMES**

2. Gopher: University of Tuebingen
Address: **rusinfo.rus.uni-stuttgart.de**
Choose: **Fun & Game | GamerServer in Tuebingen**
Login: **GAMES**

MUD Access via Gopher

Access all your favorite MUDs through Gopher. Simply select the MUD you wish to play from the massive selection available and you will be instantly connected. No more messing with lengthy MUD lists.

How to access   Gopher: Technische Universitaet Clausthal
Address: **solaris.rz.tu-clausthal.de**
Choose: **Studenten | Mud-Servers**

MUD Documents

An interesting selection of information about MUDs, including a history on MUD, inter-MUD communication, a MUD survey, and a paper on social virtual reality in the real world.

How to access   Gopher: University of Tuebingen
Address: **nova.tat.physik.uni-tuebingen.de 4242**
Choose: **Documents and papers about MUDs**

MUD Information

A large selection of information all about MUDs and the culture surrounding them, including research articles, clients, FTP sites, MUD lists, and descriptions of the various MUDs. It also categorizes MUDs into their different types and allows you to connect directly to them from the Gopher.

How to access   Gopher: University of Texas at Austin
                                Address: **actlab.rtf.utexas.edu**
                                Choose: **Virtual Spaces: MUD**

MUD List

List of all the Internet MUDS you will ever want to play. Classifies each MUD into its specific type, and provides both numeric and name addresses, status, and any further information such as how to register for each MUD. An updated list is released every Friday.

How to access   1. Anonymous FTP: **caisr2.caisr.edu**
                                Path: **/pub/mud**

                            2. Usenet: **rec.games.mud.announce** (every Friday)

MUDWHO Server

Shows the current players on various MUDs. Each MUDWHO Server will know who is using various MUDs, so check the different servers to see which one keeps track of players on your favorite MUD.

How to access   1. Telnet: **actlab.rtf.utexas.edu 6889**

                            2. Telnet: **af.itd.com 6889**

                            3. Telnet: **amber.ecst.csuchico.edu 6889**

                            4. Telnet: **dancer.athz.ch 6889**

                            5. Telnet: **nova.tat.physik.uni-tuebingen.de 6889**

                            6. Telnet: **riemann.math.okstate.edu 6889**

Multi-User Dimensions (MUDs)

Fight your way through a "Dungeons" type nerve teaser battling not a computer, but other intelligent players. Games include "Jay's House of MOO" and "Virtual Spaces". Explore monster-infested areas, solve puzzles, and do battle.

How to access   Gopher: University of Minnesota
                                Address: **consultant.micro.umn.edu**
                                Choose: **Fun & Games** | **Games** | **MUDs**

Nemesis MUD

Enter the ancient fantasy world of Nemesis, and explore its lands, finding the secrets that lay hidden in its depths.

How to access   Telnet: **dszenger9.informatik.tu-muenchen.de 4242**

M

Nemesis MUD Gopher

Offers lots of information and news about Nemesis MUD, including information about current players, maps, popular Nemesis bulletin boards, and Nemesis programming documentation. Also allows easy access to other LPMUD Gophers.

How to access   Gopher: **dszenger9.informatik.tu-muenchen.de 7000**

Nightfall MUD

Nightfall is an interactive, text-based, social virtual reality. It is an LPMUD allowing you to adventure through strange lands solving puzzles, killing monsters, and selling treasures on your way.

How to access   1. Telnet: **nova.tat.physik.uni-tuebingen.de 4242**

2. Gopher: **nova.tat.physik.uni-tuebingen.de 4242**
   Choose: **Enter Nightfall**

Nightfall MUD Information

This Gopher offers information about Nightfall MUD, and the MUD culture in general. It has access to the Nightfall MUD statistics and status, and it also allows you to connect to the MUD itself, or check who is currently playing.

How to access   Gopher: University of Tuebingen
Address: **nova.tat.physik.uni-tuebingen.de 4242**

Music

Acoustic Guitar Digest

Check out this electronic magazine for acoustic guitar buffs.

How to access   Anonymous FTP: **casbah.acns.nwu.edu**
Path: **/pub/acoustic-guitar**

Articles of Music Composition

A number of articles about composing music. Hints, tips, tricks, and ideas of all sorts.

How to access   Gopher: University of Wisconsin Parkside
Address: **cs.uwp.edu**
Choose: **Music Archives | Articles of Music Composition**

Billboard Magazine's Top 10 billboards

Top 10 charts for pop singles, pop albums, adult contemporary, and rhythm and blues.

How to access   Finger: **buckmr@rpi.edu**

Bob Dylan

Large archive of information about Bob Dylan, including interviews, details of his life and events for over 30 years, and CD and book lists.

How to access   Anonymous FTP: **ftp.cs.pdx.edu**
Path: **/pub/dylan/***

Folk Music

Selection of folk music, country blues, and fingerstyle guitarists, discographies and lyrics. Also offers lists of folk music societies, radio programs, publications, and other FTP sites.

How to access   1. Gopher: University of Wisconsin Parkside
Address: **gopher.uwp.edu**
Choose: **Music Archives | Folk Music Files and pointers**

2. Anonymous FTP: **ftp.uwp.edu**
Path: **/pub/music/folk/***

Grateful Dead Archives

Files of interest to fans of the Grateful Dead.

How to access   1. Anonymous FTP: **gdead.berkeley.edu**
Path: **/pub/gdead**

2. Gopher: University of California Berkeley
Address: **gdead.berkeley.edu**

Guitar Archive

Large collection of guitar Tab files, covering thousands of artists and groups, all organized in alphabetical order.

How to access   1. Gopher: University of Wisconsin Parkside
Address: **gopher.uwp.edu**
Choose: **Music Archives | Guitar TAB files**

2. Anonymous FTP: **ftp.nevada.edu**
Path: **/pub/guitar/***

Guitar Chords for Popular Songs

Song lyrics and guitar chords for many popular songs. Songs are categorized by group or artist.

How to access   Anonymous FTP: **ftp.nevada.edu**
Path: **/pub/guitar**

M

Lyrics Archive

Massive collection of song lyrics from thousands of artists and groups, with a variety of index methods.

How to access   1. Gopher: University of Wisconsin Parkside
Address: **gopher.uwp.edu**
Choose: **Music Archives | Lyrics Archives**

2. Anonymous FTP: **ftp.uwp.edu**
Path: **/pub/music/lyrics/***

3. Anonymous FTP: **ocf.berkeley.edu**
Path: **/pub/Library/Lyrics**

Music Archives

A massive collection of information about music, including artists, buying guides, picture files, lyrics, FTP site lists, and much more.

How to access   1. Gopher: University of Wisconsin Parkside
Address: **gopher.uwp.edu**
Choose: **Music Archives**

2. Anonymous FTP: **ftp.uwp.edu**
Path: **/pub/music/***

Music List of Lists

The master list for music subjects.

How to access   Mail: **mlol-request@wariat.org**

Music Server

This site has just about anything a music fan might be interested in. Archives by artist name, music databases, classical, folk music, guitar TAB files, lyrics, MIDI files, picture files, release listings, mailing lists, and on and on.

How to access   Anonymous FTP: **ftp.uwp.edu**
Path: **/pub/music**

New Music

Discover music with a difference, as you browse the online catalogs and playlists offering music and sound works that stretch the mind with their limitless boundaries.

How to access   Gopher: Whole Earth Lectronic Link
Address: **gopher.well.sf.ca.us**
Choose: **Art and Culture | New Music**

Pipe Organ

A programmable color organ program for the IBM-PC. Allows you to interchange music and graphics.

How to access   Anonymous FTP: **ftp.cs.pdx.edu**
Path: **/pub/music/ravel/pip.tar.Z**

Rock/Heavy Metal Lyric Quiz

Get a random quiz question each time you finger here. It's sort of an electronic version of "Name That Tune".

How to access   Finger: **gim@139.133.202.141**

Sid's Music Server

Lists of rare live recordings and CDs for sale. There is also a mailing list available. Send mail for more information.

How to access   Mail: **mwilkenf@silver.ucs.indiana.edu**
Subject: **BOOTHELP**

Song Lyrics

Read the lyrics to popular (and not so popular) songs. This Gopher server has an amazing volume of artists and songs. Directories are arranged alphabetically. Pick a directory for the band or artist (i.e., choose **J** for Elton John), then choose an album, then the song.

How to access   Gopher: University of Minnesota
Address: **consultant.micro.umn.edu**
Choose: **Fun & Games** | **Music** | **Music Archives**
| **Lyrics Archives**

Update Electronic Music Newsletter

An electronic newsletter for those interested in electronic music.

How to access   Listserv mailing list: **upnews@vm.marist.edu**

Used Music Server

Buy, sell, or trade CDs, tapes, and LP albums. You can also subscribe to the mailing list.

How to access   Mail: **used-music-server@cs.ucsb.edu**
Subject: **help**

M

News

Clarinet

Many Usenet-like newsgroups devoted to real news (see Chapter 9). In order to access Clarinet, your news site must subscribe to it.

How to access   Usenet: **clarinet.***

EFFector Online and EFF News

The complete set of the EFFector Online magazine and EFF News publications, which tackle issues relating to computers, the law, and privacy.

How to access   1. Gopher: Electronic Frontier Foundation
Address: **gopher.eff.org**
Choose: **Electronic Frontier..** I **Back issues of EFFector..**

                     2. Anonymous FTP: **ftp.eff.org**
Path: **/pub/EFF/newsletters/***

Freenet

USA Today, Sports, etc...

How to access   1. Telnet: **freenet-in-[a,b,c].cwru.edu**

                     2. Telnet: **yfn.ysu.edu**
Login: **visitor**

French Language Press Review

French summaries of news reported in the French language press, updated on an almost daily basis.

How to access   Gopher: Michigan State University
Address: **gopher.msu.edu**
Choose: **News & Weather** I **Electronic Newspapers**
I **General Newspapers** I **French Language Press Review**

French News

Daily newsbriefs in French, about events in France, including a general discussion of the main articles published in the French press on that day, often quoting the articles extensively.

How to access   Gopher: Yale University
Address: **yaleinfo.yale.edu**
Choose: **Internet Resources** I **News and weather** I **France**

Moscow News

Sample of English articles from the Moscow News newspaper.

How to access   Gopher: Michigan State University
Address: **gopher.msu.edu**
Choose: **News & Weather** | **Electronic Newspapers**
| **Sample Newspapers** | **Moscow News**

News Mail Servers

Post to Usenet news via mail.

How to access   Mail: **[newsgroup]@cs.utexas.edu**

Newsletters and Journals Available through Gopher

A large collection of online newsletters and journals, covering art, computing, education, humanities, languages, law, medicine, politics, religion, and more. They are all neatly categorized and available through Gopher.

How to access   1. Gopher: University of North Texas
Address: **gopher.unt.edu**
Choose: **Newsletters & Journals**

2. Gopher: Swedish University Network
Address: **sunic.sunet.se**
Choose: **Newsletters & Journals**

USA Today

Sample articles from USA Today newspaper.

How to access   Gopher: Michigan State University
Address: **gopher.msu.edu**
Choose: **News & Weather** | **Electronic Newspapers**
| **Sample Newspapers** | **USA Today**

Washington Post

Sample articles from the Washington Post newspaper.

How to access   Gopher: Michigan State University
Address: **gopher.msu.edu**
Choose: **News & Weather** | **Electronic Newspapers**
| **Sample Newspapers** | **Washington Post**

N

Oceanography

Oceanic (The Ocean Network Information Center)

Oceanic data sets, research ship schedules and information, science and program information.

How to access   Telnet: **delocn.udel.edu**
Login: **info**

Oceanography Information

Exchange information with oceanographers and oceanography buffs. From fishery science to the effects of natural and man-made disasters on the ocean's ecology, this is the place to find the information.

How to access   Anonymous FTP: **biome.bio.dfo.ca**
Path: **/pub**

Operating Systems

See also   Unix, Software

Mach

Source code and documentation for the Mach operating system.

How to access   Anonymous FTP: **gatekeeper.dec.com**
Path: **/.0/Mach**

Minix Operating System

Sources and discussion of the popular Unix-like operating system.

How to access   Anonymous FTP: **oak.oakland.edu**
Path: **/pub/misc/minix**

Organizations

Association for Computing Machinery

The Association for Computing Machinery (ACM) is the largest and oldest educational and scientific computing organization in the world today, and access to current information about the Association's activities is readily available here, including many technical computing areas.

How to access   Gopher: Association for Computing Machinery
Address: **gopher.acm.org**

Electronic Frontier Foundations Gopher

The Electronic Frontier Foundation's purpose is to ensure that the new communications technology era is available to all, and that individual's constitutional rights are preserved therein. Plenty of legal information, EFF publications, and many related articles and zines are available here.

How to access   1. Gopher: Electronic Frontier Foundation
Address: **gopher.eff.org**

    2. Anonymous FTP: **ftp.eff.org**
Path: **pub/EFF/***

SEA Gopher

The Society for Electronic Access (SEA) works to educate the populace about computer networks and how to use them to communicate and find information. Their Gopher includes a list of U.S. government BBSs, articles, archives, and telecom law information.

How to access   Gopher: Panix
Address: **gopher.panix.com**
Choose: **Society for Electronic Access (SEA)**

Pets

Aquaria

Buyer guides, filter information, magazine list, plant and water quality basics, and much more related to aquarium and fish keeping.

How to access   Anonymous FTP: **caldera.usc.edu**
Path: **/pub/aquaria/***

Bird Keeping

Bird magazines, books, terminology, buying guides, cage and toy reviews, diet and feeding information, training help, and much more.

How to access   Anonymous FTP: **rtfm.mit.edu**
Path: **/pub/usenet/news.answers/pets-birds-faq/***

Cats

Basic cat care, guide to getting a cat, medical information, problem behaviors, entertainment, and much more related material all about cats.

How to access   Anonymous FTP: **rtfm.mit.edu**
Path: **/pub/usenet/news.answers/cats-faq/***

Dogs

Owner guides, puppy needs, health care issues, training tips, behavior understanding, kennel clubs, publications, resources, and much more material for man's best friend.

How to access   Anonymous FTP: **rtfm.mit.edu**
Path: **/pub/usenet/news.answers/dogs-faq/***

Fleas and Ticks

Learn how to rid your pet or home of fleas, and how to deal with ticks.

How to access   Anonymous FTP: **rtfm.mit.edu**
Path: **/pub/usenet/news.answers/fleas-ticks**

Pet Discussions

Discussion groups for pet and animal enthusiasts. Training, feeding, caring for, and generally enjoying pets.

How to access   Usenet: **rec.pets.birds, rec.pets.cats, rec.pets.dogs, rec.pets.herp** [reptiles, etc.]

Physics

National Nuclear Data Center Online Data Service

All the data you could possibly want regarding nuclear physics and statistical measurements, including radiation levels and other information for the U.S.

How to access   Telnet: **bnlnd2.dne.bnl.gov**
Login: **nndc**

Physics Gopher

Access to lots of physics resources and information, including the areas of astrophysics, general relativity and quantum cosmology, high energy physics, and nuclear theory.

How to access   Gopher: University of Chicago
Address: **granta.uchicago.edu**

Physics Mailing List

Covers current developments in theoretical and experimental physics, including plasmaphysics, particle physics, and astrophysics.

How to access   Mail: **physics-request@qedqcd.rye.ny.us**
Body: **SUBSCRIBE** [your name]

Theoretical Physics Pre-print List

Papers on general relativity and quantum cosmology, and high energy physics.

How to access   1. Mail: **gr-qc@xxx.lanl.gov**
Subject: **help**

2. Mail: **hep-th@xxx.lanl.gov**
Subject: **help**

3. Anonymous FTP: **xxx.lanl.gov**
Path: **/hep-th/*; /gr-qc/***

Pictures and Sound

Acid Warp

Much sought-after graphics program, with a wonderful psychedelic graphics display.

How to access   Anonymous FTP: **ftp.rahul.net**
Path: **/pub/atman/UTLCD-preview/mind-candy/acidwarp.arj**

Image File Formats

See   Programing

Next Sounds

Hundreds of megabytes of sounds for the Next machine, including lots of theme songs and samples from well-known artists.

How to access   Anonymous FTP: **wuarchive.wustl.edu**
Path: **/pub/NeXT-Music/***

Picture Related Files Anonymous FTP Site List

Large list of Anonymous FTP sites that contain files related to viewing, extracting, encoding, compressing, archiving, converting, and anything else you can do to pictures of all format types.

How to access   Anonymous FTP: **bongo.cc.utexas.edu**
Path: **/gifstuff/ftpsites**

Pictures

Explicit scanned photos.

> **How to access**   Usenet: **alt.sex.pictures**

Shuttle and Satellite Images

Photographs in electronic formats of spacecraft and spectacular views from Earth and space.

> **How to access**   1. Anonymous FTP: **ames.arc.nasa.gov**
>   Path: **/pub/GIF /pub/SPACE/GIF /pub/space/CDROM**
>
> 2. Anonymous FTP: **sseop.jsc.nasa.gov**
>
> 3. Anonymous FTP: **sanddunes.scd.ucar.edu**
>
> 4. Mail: **kelley@sanddunes.scd.ucar.edu**
>
> 5. Anonymous FTP: **pioneer.unm.edu**
>   Path: **/pub/info**
>
> 6. Anonymous FTP: **iris1.ucis.dal.ca**
>   Path: **/pub/GIF**
>
> 7. Gopher: NASA Goddard Space Flight Center
>   Address: **gopher.gsfc.nasa.gov**
>   Choose: **NASA information | Space images and information**
>
> 8. Anonymous FTP: **explorer.ar**

Sun Sparcstation Sound Files

A large collection of songs and tunes, and miscellaneous sounds to play on your Sun Sparcstation. Includes the Sparctracker software.

> **How to access**   Anonymous FTP: **toybox.gsfc.nasa.gov**
>   Path: **/pub/sounds/***

Politics

Coalition for Networked Information (CNI)

Search databases for interesting publications, transcripts of Congressional sessions, and other political events.

> **How to access**   Telnet: **a.cni.org**
>   Login: **brsuser**

Iowa Political Stock Market

Buy and sell shares in political candidates. This is a non-profit research project.

How to access   Telnet: **ipsm.biz.uiowa.edu**

National Research and Education Network Bill

The full text of the bill for an information superhighway, including editorial comments.

How to access   Gopher: University of Minnesota
Address: **consultant.micro.umn.edu**
Choose: **Libraries** I **Electronic Books** I **By Title** I **NREN**

Politics and the Network Community

Information concerning several areas of politics, especially those issues concerning politics, computers, and the network communities.

How to access   Gopher: Whole Earth Lectronic Link
Address: **gopher.well.sf.ca.us**
Choose: **Politics**

Weird Politics and Conspiracy

Lots of documents and archives from some of the more unusual political movements. Includes Arm the Spirit, The Disability Rag, NativeNet archives, Workers World, and more.

How to access   Anonymous FTP: **red.css.itd.umich.edu**
Paths: **/pub/poli/*; /pub/zines/***

Programming

Ada

Public library containing compilers, tools, documentation, FAQs and other software for the Ada programming language, as used by the U.S. Department of Defense.

How to access   Anonymous FTP: **wuarchive.wustl.edu**
Path: **/languages/ada/***

C Programs

Some interesting little C programs for the Unix gurus amongst you, although some of them require you to have root privileges to run them.

| **How to access** | Anonymous FTP: **ftp.cs.widener.edu** |
| | Path: **/pub/brendan/*** |

C++ Frequently Asked Questions

Answers to hundreds of the most frequently asked questions about the C++ programming language. Nearly one hundred pages of densely packed information for programmers interested in C++.

| **How to access** | Anonymous FTP: **sun.soe.clarkson.edu** |
| | Path: **/pub/C++/FAQ** |

CCMD Source Code in C

The complete source code and makefiles to a user-interface program based on the COMND jsys from tops20. This program gives escape completion on many different types of data (filenames, users, groups, key words, etc.). Columbia's MM (Mail Manager) program is written using it.

| **How to access** | Anonymous FTP: **oak.oakland.edu** |
| | Path: **/pub/misc/ccmd/*** |

CompuServe B File Transfer Protocol

The complete C source and documentation for implementing the CompuServer B file transfer protocol.

| **How to access** | Anonymous FTP: **oak.oakland.edu** |
| | Path: **/pub/misc/cis/*** |

Data File Formats

Documents and descriptions of image and data file formats.

How to access	Anonymous FTP: **wuarchive.wustl.edu**
	Path: **/doc/graphic-formats/***
See also	Pictures and Sound

Eyes X Window Environment

Eyes is a visual programming environment for X Window, and is useful in the areas of image processing, document recognition, database management, dataflow programming, and network analysis.

| **How to access** | Anonymous FTP: **ftp.ulowell.edu** |
| | Path: **/Eyes/*** |

Free Language Tools

Extensive list of free language tools with source code, including compilers and interpreters.

How to access   1. Gopher: Pacific Systems Group
  Address: **gopher.psg.com**
  Choose: **Programming Languages | List of Free Compilers...**

2. Anonymous FTP: **csd4.csd.uwm.edu**
  Path: **/pub/compilers/list**

FSP

Information, utilities, and the latest Unix FSP software, the new file transfer protocol alternative to FTP.

How to access   Anonymous FTP: **seismo.soar.cs.cmu.edu**
  Path: **/fsp/***

Gnuplot Tutorial

A postscript file containing a tutorial to teach you the nuances of gnuplot, the graph plotting program.

How to access   Anonymous FTP: **ccosun.caltech.edu**
  Path: **/pub/documents/gnuplot-tutorial.ps**

Gopher and Utilities for VMS

Collection of files and programs for VMS, which can be browsed or downloaded. Includes latest Gopher client, compression and archive utilities, and more.

How to access   Gopher: Sam Houston State University
  Address: **niord.shsu.edu**
  Choose: **VMS Gopher-related file library**

Hello World!

The classic neophyte program in many different languages.

How to access   Anonymous FTP: **ocf.berkeley.edu**
  Path: **/pub/Library/Hello_World**

Interactive Fiction

Interactive fiction games, development tools, game solutions, and programming examples, from the Usenet newsgroups **rec.arts.int-fiction** and **rec.games.int-fiction**.

How to access   Anonymous FTP: **wuarchive.wustl.edu**
  Path: **/doc/misc/if-archive/***

Language List

An extensive list of collected information on more than 2,000 computer languages, past and present.

How to access   1. Gopher: Pacific Systems Group
Address: **gopher.psg.com**
Choose: **Programming Languages**
| **List of 'All' Programming...**

2. Anonymous FTP: **rzsun2.informatik.uni-hamburg.de**
Path: **/pub/doc/news.answers/language-list/***

League for Programming Freedom

See   Law.

Make Tutorial

An easy to use introduction to make, the C program maintenance utility.

How to access   Gopher: Manchester Computing Centre
Address: **uts.mcc.ac.uk**
Choose: **Experimental and New Services**
| **Make: A Tutorial**

Obfuscated C Code

Entries for the International Obfuscated C Code Contest, which asked people to write, in 512 bytes or less, the worst complete C program.

How to access   Anonymous FTP: **ftp.white.toronto.edu**
Path: **/pub/obfuscated/***

Pascal to C translator

This Unix program translates Pascal programs to C programs, just like that!

How to access   Anonymous FTP: **ccosun.caltech.edu**
Path: **/pub/misc/p2c-1.18.tar.Z**

Programming in Ada

Tips, tricks, utilities, source code, and complete programs for users of this programming language.

How to access   Anonymous FTP: **oak.oakland.edu**
Path: **/pub/ada/***

Programming Languages

Information on lots of computer programming languages, including C++, Modula 2, Oberon, and Pascal.

How to access   Gopher: Pacific Systems Group
Address: **gopher.psg.com**
Choose: **Programming Languages**

Ravel

A C-like interpreted programming language for the IBM-PC, that directly supports MIDI music constructs. The package includes music files and source code.

How to access   Anonymous FTP: **ftp.cs.pdx.edu**
Path: **/pub/music/ravel/***

TCP/IP development tools

This directory contains instructions and C source code for a mail program, and for implementing the TCP/IP protocol on PCs and Macintosh computers.

How to access   Anonymous FTP: **oak.oakland.edu**
Path: **/pub/misc/ka9q-tcpip**

X Window Software Index

An index of public domain software to exploit or enhance the X Window system. It lets you peruse archives, and tells you what software exists, with a brief description for each item, and where to find it.

How to access   Gopher: University of Edinburgh
Address: **gopher.ed.ac.uk**
Choose: **Index to public domain X sources**

Publications

Dartmouth College Library Online System

Search and browse a wealth of database files. Get information about titles, authors, and their publications. You can even search books and periodicals for key words and view the text.

How to access   Telnet: **library.dartmouth.edu**

Journalism

This Wais site indexes over 10,000 published journals and periodicals.

How to access   Wais: **journalism.periodicals**

Journals with a Difference

Selection of journals covering some of the more unusual topics in life. Includes such publications as The Unplastic News, Athene, Scream Baby, and Quanta.

How to access   Anonymous FTP: **ftp.eff.org**
Path: **/pub/journals/***

Newsletters, Electronic Journals, Zines

A wide selection of online publications covering biology, computing, health, kids, libraries, networks, politics, and space.

How to access   Anonymous FTP: **nigel.msen.com**
Path: **/pub/newsletters/***

Online Publications

Multitudes of online zines, essays, and articles covering lots of varied and unusual subjects. Includes articles from such publications as Mondo 2000, Whole Earth Review, Locus, and The Unplastic News.

How to access   Gopher: Whole Earth Lectronic Link
Address: **gopher.well.sf.ca.us**
Choose: **Publications**

Whole Earth Review Articles

A selection of articles from the Whole Earth Review, which is dedicated to demystification, to self-teaching, and to encouraging people to think for themselves.

How to access   Gopher: Whole Earth Lectronic Link
Address: **gopher.well.sf.ca.us**
Choose: **Whole Earth Review, the Magazine**

Zine Reviews

A large collection of reviews, from the Factsheet Five-Electric zine, covering various and esoteric zines, those opinionated publications with a press run of between 50 and 5,000.

How to access   1. Gopher: Whole Earth Lectronic Link
Address: **gopher.well.sf.ca.us**
Choose: **Publications | Factsheet Five, Electric**

    2. Anonymous FTP: **nigel.msen.com**
Path: **/pub/newsletters/Zine-Reviews/F5-E**

Zines

Collection of zines, those opinionated publications, covering such topics as science fiction, computers, short fiction, poetry, and cyberpunk.

How to access   Gopher: Electronic Frontier Foundation
Address: **gopher.eff.org**
Choose: **Instant Karma Zine Stand**

Religion

Bible Browser for Unix

A program for browsing the Bible.

How to access   Anonymous FTP: **cs.arizona.edu**
Path: **/icon/contrib/bibleref-2.1.tar.Z**

Bible Online

The complete Bible online with a Gopher interface. There actually are many places where you can read the Bible online. Use Veronica to find other sites.

How to access   Gopher: University of Minnesota
Address: **joeboy.micro.umn.edu**
Choose: **Ebooks | By Title | King James Bible**

Bible Program

The complete online Bible program, with instructions for installation on an IBM-PC or Mac.

How to access   Anonymous FTP: **wuarchive.wustl.edu**
Path: **/doc/bible/***

Bible Promises Macintosh Hypercard Stack

A Hypercard program for the Macintosh.

How to access   Anonymous FTP: **f.ms.uky.edu**
Path: **/pub/mac/hypercard/stack.cpt.bin**

Bible Quiz Game

Over a thousand questions.

How to access   Anonymous FTP: **oak.oakland.edu**
Path: **/pub/bible/bibleq.arc**

R

Bible Retrieval System for Unix

A Bible query system for Unix systems.

How to access   Anonymous FTP: **ftp.uu.net**
                     Path: **/doc/literary/obi/Religion/Bible.Retrieval.System/***

Bible Search Program

More Bible search utilities.

How to access   Anonymous FTP: **oak.oakland.edu**
                     Path: **/pub/bible/bible14.arc**

Bible Search Tools

Assorted Bible search utilities.

How to access   Anonymous FTP: **oak.oakland.edu**
                     Path: **/pub/bible/kjv-tool.arc**

Bible Text

The complete text of the Bible.

How to access   1. Anonymous FTP: **oak.oakland.edu**
                        Path: **/doc/bible/journey.arc**

                     2. Anonymous FTP: **ocf.berkeley.edu**
                        Path: **/pub/Library/Religion**

                     3. Wais: **Bible**

Bible Verses

A memory resident pop-up TSR for DOS computers.

How to access   Anonymous FTP: **oak.oakland.edu**
                     Path: **/pub/bible/biblepop.arc**

Bible Word and Phrase Counts of King James Version

Word counts, phrase counts, and other statistics for the King James Bible.

How to access   Anonymous FTP: **oak.oakland.edu**
                     Path: **/pub/bible/kjvcount.txt**

Biblical Search and Extraction Tool

More Bible search utilities.

How to access	Anonymous FTP: **oak.oakland.edu**
	Path: **/pub/bible/refrkjv1.zip**

The Book of Mormon

The complete text.

How to access	1. Anonymous FTP: **ocf.berkeley.edu**
	Path: **/pub/Library/Religion/Book_of_Mormon**
	2. Anonymous FTP: **gatekeeper.dec.com**
	Path: **/.2/micro/msdos/simtel20/books/mormon13.zip**
	3. Wais: **Book_of_Mormon**

Buddhism

Lots of documents about the different types of Buddhism, Shamanism, and Taoism, including the e-mail addresses of individuals and organizations interested in Zen around the world.

How to access	Anonymous FTP: **coombs.anu.edu.au**
	Path:
	/coombspapers/otherarchives/electronic-buddhist-archives/*

The Electric Mystics Guide to the Internet

A complete bibliography of electronic documents, online conferences, serials, software, and archives relevant to religious studies.

How to access	Anonymous FTP: **panda1.uottawa.ca**
	Path: **/pub/religion/electric-mystics-guide**

Genesis

A study aid and reference for the King James Bible.

How to access	Anonymous FTP: **oak.oakland.edu**
	Path: **/pub/bible/genaidc.zip**

GRASS Mailing List

The Generic Religions and Secret Societies (GRASS) mailing list is a forum for the development of religions and secret societies for use in role-playing games.

How to access	Mail: **grass-server@wharton.upenn.edu**
	Body: **SUBSCRIBE** [your name]

Hebrew Quiz

A biblical Hebrew language tutor.

How to access   Anonymous FTP: **oak.oakland.edu**
Path: **/pub/msdos/hebrew**

King James Bible

The full text from Project Gutenberg.

How to access   Gopher: University of Minnesota
Address: **consultant.micro.umn.edu**
Choose: **Libraries | Electronic Books | By Title
| King James Bible**

The Koran (or Quran)

The complete text.

How to access   1. Anonymous FTP: **ocf.berkeley.edu**
Path: **/pub/Library/Religion/Koran**

2. Anonymous FTP: **oes.orst.edu**
Path: **/pub/data/etext/koran/koran**

2. Anonymous FTP: **quake.think.com**
Path: **/pub/etext/koran**

4. Wais: **Quran**

Pagan Mailing List

Discusses the religions, philosophy, and related material of paganism.

How to access   Mail: **pagan-request@drycas.club.cc.cmu.edu**
Body: **SUBSCRIBE** [your name]

Religious Studies Publications Journal - CONTENTS

CONTENTS is an electronic publication which examines new and recent print
publications of relevance to religious studies.

How to access   Listserv mailing list: **contents@acadvm1.uottawa.ca**

Science

National Science Foundation Publications

Access the publications of the National Science Foundation, including award information, letters, program guidelines, and reports for each directorate.

How to access   Gopher: National Science Foundation
Address: **stis.nsf.gov**

Science and Technology Information System

STIS offers discussion, help, and a number of interesting topics for people interested in science and technology.

How to access   Telnet: **stis.nsf.gov**
Login: **public**

Scientific Articles

Contains some interesting online science articles and book reviews, such as a scientific article about forecasting eclipses.

How to access   Gopher: **Whole Earth Lectronic Link**
Address: **gopher.well.sf.ca.us**
Choose: **Science**

Scientist's Workbench

The Scientist's Workbench is an X Window based application designed to bring together a set of tools for enhancing the development, testing, and execution of scientific codes.

How to access   1. Gopher: Cornell University
Address: **gopher.tc.cornell.edu**
Choose: **Anonymous FTP | Scientist's Workbench**

2. Anonymous FTP: **info.tc.cornell.edu**
Path: **/pub/swb/***

Sex

Bestiality

Of bestiality.

How to access   Usenet: **alt.sex.bestiality**

Bondage

Of bondage.

How to access   Usenet: **alt.sex.bondage**

Discussion of Sex Stories

Discussion of the stories of sexual encounters in **alt.sex.stories**.

How to access   Usenet: **alt.sex.stories.d**

Fetishes

Of anatomical obsessions.

How to access   Usenet: **alt.sex.fetish.feet**

Homosexuality

Of alternative lifestyles, politics, and homosexuality, and encounters with members of the same sex (MOTSS).

How to access   Usenet: **alt.homosexual; alt.politics.homosexuality; alt.sex.motss**

Law, Crime, and Sex

Discussion of sex crimes, trials, and the law.

How to access   Usenet: **clari.news.law.crime.sex**

Masturbation

Of masturbation.

How to access   Usenet: **alt.sex.masturbation**

Movies

Sex and the movies.

How to access   Usenet: **alt.sex.movies**

Queer Resource Directory

FTP site for gay interests, including rights issues and AIDS information.

How to access   Anonymous FTP: **nifty.andrew.cmu.edu**

Questions and Answers

All your wildest questions answered, including a special list of FAQs for sex wizards. Includes sex terms and acronyms, purity test guides, and truly much more.

How to access　　Anonymous FTP: **rtfm.mit.edu**
　　　　　　　　　　Path: **/pub/usenet/news.answers/alt-sex/***

Sex

Sexual issues, sex-related political stories.

How to access　　Usenet: **clari.news.sex**

Sex Abuse Recovery

Helping others deal with traumatic experiences.

How to access　　Usenet: **alt.sexual.abuse.recovery**

Sounds

Audio excitement.

How to access　　Usenet: **alt.sex.sounds**

Stories

Stories about sexual encounters and experiences.

How to access　　Usenet: **alt.sex.stories**

Wanted

Requests for erotica, either literary or in the flesh.

How to access　　Usenet: **alt.sex.wanted**

Wizards

Questions only for true sex wizards.

How to access　　Usenet: **alt.sex.wizards**

S

Software

ASK-Software-Informationssystem

Bilingual BBS offering database searches for software, news, and information. Choose option 3 on the main menu to change from German to English.

How to access   Telnet: **askhp.ask.uni-karlsruhe.de**
Login: **ask**
Password: **ask**

Encryption

Encryption routines and software. Information, algorithms, hardware, and other related material.

How to access   Anonymous FTP: **nic.funet.fi**
Path: **/pub/crypt/***

GNU

Large, multifaceted software project to create and distribute "free" software.

How to access   Usenet: **gnu.***

GNU for PCs

GNU style programs and utilities for the PC, including emacs, make, grep, sed, and sort.

How to access   Anonymous FTP: **wuarchive.wustl.edu**
Path: **/systems/ibmpc/gnuish/***

Info und Softserver

Journals, Unix programs and utilities, recipes, online cookbook. A bilingual (but mostly German) BBS.

How to access   Telnet: **rusinfo.rus.uni-stuttgart.de**
Login: **info**

Microsoft Windows

Applications, tips, utilities, drivers, and bitmaps for Windows.

How to access   1. Anonymous FTP: **ftp.cica.indiana.edu**
Path: **/pub/pc/win3/***

2. Anonymous FTP: **wuarchive.wustl.edu**
Path: **/systems/ibmpc/win3/***

NCSA Telnet

A TCP/IP package for PCs.

    How to access   Anonymous FTP: **ftp.ncsa.uiuc.edu**
                              Path: **/HDF/contrib**

SIMTEL20 Software Archives

A huge repository of software of all types for all types of computers. These archives are mirrored at many sites around the world.

    How to access   Anonymous FTP: **wsmr-simtel20.army.mil**

Software Sites

Some of the most well-known software archive sites. They contain far more than we could ever print in this book. Surely these are good for weeks of exploration.

    How to access   1. Anonymous FTP: **bongo.cc.utexas.edu**

                                  2. Anonymous FTP: **ftp.rahul.net**

                                  3. Anonymous FTP: **nic.funet.fi**

                                  4. Anonymous FTP: **export.lcs.mit.edu**

                                  5. Anonymous FTP: **wsmr-simtel20.army.mil**

                                  6. Anonymous FTP: **wuarchive.wustl.edu**

                                  7. Anonymous FTP: **oak.oakland.edu**

                                  8. Anonymous FTP: **garbo.uwasa.fi**

System Software

Large software and documentation archives for numerous systems, including Xenix, Unix, Sinclair, Novell, MS-DOS, HPUX, Apple2, and VAX-VMS.

    How to access   Anonymous FTP: **wuarchive.wustl.edu**
                                Path: **/systems/***

TeX Text Typesetter

Bundles of information about TeX, the software system written by Donald Knuth to typeset text, including access to archives, font information, FAQ lists, and LaTeX information.

    How to access   Gopher: Sam Houston State University
                                Address: **niord.shsu.edu**
                                Choose: **TeX-related Materials**

S

Thesaurus Construction Program

Tim Craven's thesaurus construction program for the IBM-PC. Assists in creating, modifying, viewing, and printing out a small thesaurus. Includes source code and is in self-extracting archive form.

How to access   Anonymous FTP: **nic.funet.fi**
Path: **/pub/doc/library/thesauri.exe**

X-10 Protocol

Technical information about the X-10 remote control standard.

How to access   Anonymous FTP: **oak.oakland.edu**
Path: **/pub/misc/x-10.info**

ZIB Electronic Library

Software, hot links to NetLib, archives, and catalogs.

How to access   Telnet: **elib.zib-berlin.de**
Login: **elib**

ZMODEM

Complete C source, including makefiles, for the sz and rz Unix programs for sending and receiving with the XMODEM, YMODEM, and ZMODEM protocols. You can download these programs and type make to produce the executables.

How to access   Anonymous FTP: **oak.oakland.edu**
Path: **/pub/misc/zmodem**

Software: Archives

Amiga

Demos, games, utilities, programming tools, mailing list information, and documentation for the Commodore Amiga computer range.

How to access   Anonymous FTP: **wuarchive.wustl.edu**
Path: **/systems/amiga/***

Apple II Archive

Large archive of games, demos, utilities, source code, and other material for the Apple II range of computers.

How to access Anonymous FTP: **ccosun.caltech.edu**
Path: **/pub/apple2/***

Atari

Bundles of programs, source code, graphics, sounds, magazines, and documentation for the Atari ST range of computers.

How to access Anonymous FTP: **wuarchive.wustl.edu**
Path: **/systems/atari/***

CPM Archives

Tips, tricks, utilities, source code, and complete programs for users of CPM computers.

How to access Anonymous FTP: **oak.oakland.edu**
Path: **/pub/cpm/***

Commodore 64/128 Archive

Large archive of information, graphics, games, utilities and other material for the Commodore 64 and Commodore 128 computers.

How to access Anonymous FTP: **ccosun.caltech.edu**
Path: **/pub/rknop/***

Garbo (University of Vaasa, Finland)

Software, software, software! This Anonymous FTP site has gobs of software for DOS, Windows, Macs, and Unix machines. Demos, utilities, screen savers, fonts, bitmaps, icons, games, patches, programs from astronomy and business to virus interceptors, more games, and more.

How to access Anonymous FTP: **garbo.uwasa.fi**
Paths: **/pc**; **/win3**; **/win31**; **/mac**; **/next**; **/unix**

Macintosh Archives

Tips, tricks, utilities, source code, and complete programs for users of Macintosh computers.

How to access Anonymous FTP: **oak.oakland.edu**
Path: **/pub2/macintosh/***

The PC Archives

Probably one of the largest collections of IBM-PC related material to ever exist in one place at the same time. Mirrors from several other large sites also. You name it, it's here.

How to access   Anonymous FTP: **wuarchive.wustl.edu**
                     Path: **/systems/ibmpc/***

PC Game Archives

Central repository for the PC gaming community, containing all manner of freely
distributable games and accessories that run under MS-DOS or MS-Windows.

How to access   1. Anonymous FTP: **ftp.ulowell.edu**
                        Path: **/msdos/Games/***

                     2. Anonymous FTP: **wuarchive.wustl.edu**
                        Path: **/systems/ibmpc/msdos-games/***

Sinclair Archive

Dedicated to those wonderful creations of Clive Sinclair; if you're looking for information or
specs for a ZX-81, QL, or Cambridge Z88, this is the place to find it.

How to access   Anonymous FTP: **wuarchive.wustl.edu**
                     Path: **/systems/sinclair/***

TeX Archive

Large repository of TeX related material, accumulated by the Comprehensive TeX Archive
Network (CTAN). It offers TeX digests, documentation, fonts, graphics, listings, macros,
and much more.

How to access   Anonymous FTP: **ftp.shsu.edu**
                     Path: **/tex-archive/***

Unix C Archive

More serious software for Unix than you could ever imagine, much of it written in C or
related tools and complete with source. Included are database utilities, editors, graphics,
languages, and much more.

How to access   Anonymous FTP: **wuarchive.wustl.edu**
                     Path: **/systems/unix/unix-c/***

Software: Internet

Archie Clients

Client programs for accessing Archie servers (see Chapter 17).

How to access	Anonymous FTP: **ftp.cs.widener.edu**
	Path: **/pub/archie/***

Gopher Clients

A Gopher client is a program used to access a Gopher server (see Chapter 21). Information about downloading Gopher clients can be found in the Gopher frequently asked question (FAQ) list.

How to access	Anonymous FTP: **rtfm.mit.edu**
	Path: **/pub/usenet/new.answers/gopher-faq**

Wais Clients

A Wais client is a program used to access a Wais server (see Chapter 23). Information about downloading Wais clients can be found in the Wais frequently asked question (FAQ) list.

How to access	Anonymous FTP: **rtfm.mit.edu**
	Path: **/pub/usenet/new.answers/wais-faq**

World Wide Web Browsers

A browser is a program used to access the World Wide Web (see Chapter 24). You can download a browser via Anonymous FTP. The **xmosaic** browser, used with X Window, can be found at the first address listed below. Browsers for all other systems can be found at the second address (look for a file named **README.txt**).

How to access:	1. Anonymous FTP: **ftp.ncsa.uiuc.edu**
	Path: **/Web/xmosaic-binaries**
	2. Anonymous FTP: **cern.info.ch**
	Path: **/pub/www**

Software: Macintosh

Apple Macintosh System 7 Operating System

Not System 7.1, but System 7.0 (as well as other software, drivers, patches, and utilites) is available by Anonymous FTP.

How to access	Anonymous FTP: **ftp.apple.com**
	Path: **/dts/mac/sys.soft/***

Columbia University's AppleTalk Package

A program for Macs.

How to access	Anonymous FTP: **rutgers.edu**
	Path: **/pub/***

Software for Macintosh

Archives of Macintosh software.

> **How to access**   Anonymous FTP: **sumex-aim.stanford.edu**

Software: Utilities

ARJ Utilities

Utilities to un-ARJ those .ARJ files, for Unix, VMS, IBM-PC, Mac, and Amiga.

> **How to access**   Anonymous FTP: **ftp.rahul.net**
> Path: **/pub/atman/UTLCD-preview/unarjers/***

ASCII Table

The complete ASCII code table, including control character abbreviations. ASCII stands for the American Standard Code for Information Interchange.

> **How to access**   Gopher: University of Surrey
> Address: **gopher.cpe.surrey.ac.uk**
> Choose: **Misc Technical notes | ASCII table**

Automatic Login Executor (ALEX)

A slick C-Shell program that you can run on a Unix system to automate your Internet explorations. This amazing collection of shell scripts was written by a high school student in 10 weeks.

> **How to access**   Anonymous FTP: **dftsrv.gsfc.nasa.gov**
> Path: **/alex/csh-alex/***

OS/2 Utilities

Utility programs for OS/2, including replacements for CMD.EXE, Phil Katz' PKZIP for OS/2, and other useful programs.

> **How to access**   Anonymous FTP: **oak.oakland.edu**
> Path: **/pub/misc/os2**

PC Archiving Utilities

Collection of the ARJ, LHA, ZIP, and ZOO common PC archive programs for MS-DOS.

| **How to access** | Anonymous FTP: **ftp.ulowell.edu** |
| | Path: **/msdos/Archivers/*** |

PC Utilities

Lots of PC utilities and files from PC Magazine's Interactive Reader Service.

| **How to access** | Anonymous FTP: **ftp.cso.uiuc.edu** |
| | Choose: **/pc/pcmag/*** |

Source Code for Unix Utilities

Source code for Unix and X Window utilities such as uuencode, uudecode, xxencode, xxdecode, whois, uucat, and more.

| **How to access** | Anonymous FTP: **oak.oakland.edu** |
| | Path: **/pub/misc/unix** |

VMS Utilities

Collection of various utilities for VMS, including diff, grep, more, and others.

| **How to access** | Anonymous FTP: **ftp.cs.widener.edu** |
| | Path: **/pub/vms/*** |

Space

European Space Agency

The Data Dissemination Network of the European Space Agency (ESA). Offers information retrieval services, prototype international directory, and other information and services.

| **How to access** | Telnet: **esrin.esa.it** |

FIFE Information System

Scientific data from satellites, space flights, and other databases.

| **How to access** | Telnet: **pldsg3.gsfc.nasa.gov** |
| | Login: **FIFEUSER** |

Frequently Asked Questions about Space

Get answers to the most frequently asked questions (FAQs) regarding NASA, spaceflight, and astrophysics.

> **How to access**   Anonymous FTP: **ames.arc.nasa.gov**
> Path: **/pub/SPACE/FAQ**

Goddard Space Flight Center

Contains lots of news, images, pictures, information, and other resources connected with NASA and the Goddard Space Flight Center.

> **How to access**   Gopher: NASA Goddard Space Flight Center
> Address: **gopher.gsfc.nasa.gov**

Hubble Space Telescope

The Space Telescope Electronic Information System (STEIS) contains information for Hubble Space Telescope proposers and observers. It offers documents, status reports, plans, and weekly summaries. Get the daily update on the scheduled events and the outcome of experiments with the Hubble Space Telescope.

> **How to access**   1. Telnet: **stinfo.hq.eso.org**
> Login: **stinfo**
>
> 2. Gopher: Space Telescope Electronic Information System
> Address: **stsci.edu**
>
> 3. Anonymous FTP: **stsci.edu**

Lunar and Planetary Institute

The Institute is near the NASA Johnson Space Center and includes a computing center, extensive collections of lunar and planetary data, an image processing facility, an extensive library, a publishing facility, and facilities for workshops and conferences. Current topics include the origin and early evolution of the solar system; studies of the moon, meteorites, and the Earth; the outer solar system with emphasis on studies of icy satellites.

> **How to access**   Telnet: **lpi.jsc.nasa.gov**
> Login: **lpi**

NASA Extragalactic Database (NED)

At present NED contains extensive cross-identifications for over 200,000 objects—galaxies, quasars, infrared and radio sources, etc. NED provides positions, names, and basic data (e.g., magnitudes, redshifts), as well as bibliographic references, abstracts, and notes.

> **How to access**   Telnet: **ned.ipac.caltech.edu**
> Login: **ned**

NASA Information

Allows connection to several important Nasa information resources, including Nasa Spacelink, the Nasa online database, the Lunar and Planetary Institute, and the Space Physics Analysis Network Information Center.

How to access   Gopher: Yale University
Address: **yaleinfo.yale.edu**
Choose: **Internet Resources** I **Miscellaneous resources**
I **NASA**

NASA News

Up-to-date information of the status of current spacecraft in space and other NASA happenings. Find out about discoveries made with the space-based Hubble telescope and unmanned probes launched toward distant planets and galaxies.

How to access   Finger: **nasanews@space.mit.edu**

NASA Spacelink

History, current events, project and plans at NASA.

How to access   Telnet: **spacelink.msfc.nasa.gov**

NSSDC's Online Data & Information Service

NODIS provides menu-based access to data and information services provided by NSSDC and by the Space Physics Data Facility. Primarily through the multi-disciplinary Master Directory, NODIS also describes and enables electronic access (E-links) to data/information/systems remote to NSSDC.

How to access   Telnet: **nssdc.gsfc.nasa.gov**
Login: **nodis**

Southwest Data Display and Analysis System

SDDAS offers scientific data and analysis services. It also provides high resolution graphic servers, including X Window servers for data displays.

How to access   Telnet: **espsun.space.swri.edu 540**

Space Newsletter

Selection of newsletters about the final frontier, including Space News, Satscan, and Biosphere.

How to access   Anonymous FTP: **nigel.msen.com**
Path: **/pub/newsletters/Space/***

Sports and Athletics

Baseball Schedule

Get the day's game schedule for your favorite major league baseball teams. Enter "**help**" for help. Full schedules are also available.

How to access   Telnet: **culine.colorado.edu 862**

Baseball Scores

Get major league baseball scores by mail.

How to access   Mail: **jtchern@ocf.berkeley.edu**
Subject: **MLB**

Climbing

Open discussion about climbing techniques, specific climbs, and competition announcements.

How to access   Usenet: **rec.climbing**

Cycling

FAQs, interesting information, racing details, bike guides, technical hints, general tips, and more from the world of bicycles.

How to access   1. Anonymous FTP: **rtfm.mit.edu**
Path: **/pub/usenet/news.answers/bicycles-faq/***

2. Anonymous FTP: **ugle.unit.no**
Path: **/local/biking/***

3. Anonymous FTP: **draco.acs.uciedu**
Path: **/pub/rec.bicycles/***

Cycling in Canada

Information, discussion groups, pictures, programs, and more for biking enthusiasts. This FTP site is in Canada and has great data about routes, rules, and routines for biking in Canada (especially Nova Scotia).

How to access   Anonymous FTP: **biome.bio.dfo.ca**
Path: **/pub/biking**

Martial Arts

Archive of aikido-related material, including dojo addresses around the world.

How to access   Anonymous FTP: **cs.ucsd.edu**
Path: **/pub/aikido/***

NBA Schedule

Get the day's game schedule for your favorite NBA basketball teams. Enter "**help**" for help. Full schedules are also available.

How to access   Telnet: **culine.colorado.edu 859**

NFL Schedule

Get the day's game schedule for your favorite NFL football teams. Enter "**help**" for help. Full schedules are also available.

How to access   Telnet: **culine.colorado.edu 862**

NHL Schedule

Get the day's game schedule for your favorite NHL hockey teams. Enter "**help**" for help. Full schedules are also available.

How to access   Telnet: **culine.colorado.edu 860**

Paddling

Terminology, river ratings, equipment information, safety guides, books, and addresses for canoeing, kayaking, and rafting.

How to access   Anonymous FTP: **rtfm.mit.edu**
Path: **/pub/usenet/news.answers/paddling-faq**

Scouting

Scouting material for both scouts and leaders, including campfire songs, games, history, world news, unit administration information, and official policies.

How to access   1. Anonymous FTP: **rtfm.mit.edu**
Path: **/pub/usenet/news.answers/scouting/***

2. Anonymous FTP: **ftp.ethz.ch**
Path: **/rec.scouting/***

Scuba Diving

Equipment reviews, magazine list, buying guide, huge archives, and lots of related material about scuba diving, snorkeling, dive travel, and other underwater activities.

How to access   1. Anonymous FTP: **rtfm.mit.edu**
Path: **/pub/usenet/rec.answers/scuba-faq**

S

2. Anonymous FTP: **ames.arc.nasa.gov**
   Path: **/pub/SCUBA/***

Skating

Origins, equipment reviews, technique instructions, maintenance advice, FAQs, location lists, and much more for in-line (rollerblading), roller figure, and speed skating.

How to access   Anonymous FTP: **rtfm.mit.edu**
                       Path: **/pub/usenet/news.answers/rec-skate-faq/***

Skiing

Get ski conditions and read frequently asked questions about skiing in Utah, Wyoming, and Idaho.

How to access   Anonymous FTP: **ski.utah.edu**
                       Path: **/pub/skiing**

See also   Weather

Skydiving

FAQs, and their answers, about skydiving, learning to skydive, and the newsgroup **rec.skydiving**.

How to access   Anonymous FTP: **rtfm.mit.edu**
                       Path: **/pub/usenet/news.answers/skydiving-faq**

Sports Schedules

Easy access to professional football, hockey, basketball, and baseball schedules from this single menu.

How to access   Gopher: Ball State University
                       Address: **gopher.bsu.edu**
                       Choose: **x-Professional Sports Schedules**

Sports Statistics

Statistics for baseball (MLB), basketball (NBA), football (NFL), and hockey (NHL).

How to access   Anonymous FTP: **wuarchive.wustl.edu**
                       Path: **/doc/misc/sports/***

Windsurfing

Information on windsurfing at various areas in the United States. Get the scoop on windsurf shops, launch sites, and conditions. You can also participate in lively discussion topics and download cool windsurfing bitmaps.

How to access	Anonymous FTP: **bears.ece.ucsb.edu** Path: **/pub/windsurf**
See also	Weather

Technology

Artificial Intelligence

Technical papers, journals, and surveys about artificial intelligence, robotics, and neural networks.

How to access	1. Anonymous FTP: **ftp.uu.net** Path: **/pub/ai/***
	2. Anonymous FTP: **gargoyle.uchicago.edu** Path: **/pub/artificial_intelligence/***
	3. Anonymous FTP: **solaria.cc.gatech.edu** Path: **/pub/ai/***
	4. Anonymous FTP: **flash.bellcore.com** Path: **/pub/ai/***

Audio

Mail order surveys, lists, and related information for the art and science of audio.

How to access	Anonymous FTP: **ssesco.com** Path: **/pub/rec.audio/***

Canada Department of Communications

CHAT (Conversational Hypertext Access Technology) database system. You can ask questions of this database with natural English questions. It correctly interprets an amazing number of questions and instantly provides answers about the Canadian Department of Communications. The Canadian DOC manages radio in Canada, participates in high-technology ventures, and promotes Canada's cultural infrastructure through communications.

How to access	Telnet: **debra.dgbt.doc.ca** Login: **chat**

Video Laserdiscs

Introduction to video laserdisc technology, media quality reports, how-to reports, care, repair, and retail sources—everything you could want to know about this technology.

How to access	Telnet: **panda.uiowa.edu**
	Choose: **General Information**
	| Information on Video Laserdiscs

Virtual Reality

Papers, FAQs, and bibliographies concerning virtual reality.

How to access	Anonymous FTP: **ftp.u.washington.edu**
	Path: **/public/VirtualReality/***

Virtual Reality Mailing List

Open discussion and news about everything to do with virtual reality.

How to access	Listserv mailing list: **virtu-l@uiucvmd.bitnet**

Telephones

Phone Number to Word Translator

Unix C source code to an interesting program that converts phone numbers into words.

How to access	Anonymous FTP: **gatekeeper.dec.com**
	Path: **/.0/usenet/comp.sources.misc/volume12/telewords**

Phonecards

Lists of phone card collections, manufacturer lists, and chip diagram pictures of phone cards, those cards you use in public phones instead of coins.

How to access	Anonymous FTP: **nic.funet.fi**
	Path: **/pub/doc/telecom/phonecard/***

Telefax Country Codes

List of the international telephone and fax dialing codes for different countries of the world.

How to access	Anonymous FTP: **nic.funet.fi**
	Path: **/pub/doc/telecom/telefax_country_codes**

Toll-free Phone Numbers for Computer Companies

This file contains 800 phone numbers for many companies in the computer industry.

How to access	Anonymous FTP: **oak.oakland.edu**
	Path: **/pub/misc/telephone/tollfree.num**

Toll-free Phone Numbers for Non-Profit Organizations

This file contains a list of 800 numbers for non-profit organizations and crisis hotlines.

How to access   Anonymous FTP: **oak.oakland.edu**
Path: **/pub/misc/telephone/1800help.inf**

U.S. Telephone Area Codes

Find a specific area code or list all of the area code districts in the United States.

How to access   Gopher: NASA Goddard Space Flight Center
Address: **gopher.gsfc.nasa.gov**
Choose: **Other Resources | US telephone areacodes**

Television

Cheers

Episode guides, theme music, Norm sayings, and other trivia from the classic *Cheers* television series.

How to access   Anonymous FTP: **nic.funet.fi**
Path: **/pub/culture/tv+film/Cheers/***

Nielsen TV Ratings

Weekly TV ratings according to the Nielsen rating system.

How to access   Finger: **normg@halcyon.halcyon.com**

Parker Lewis

All about the *Parker Lewis* television series, including scripts, pictures, sounds, digests, cast lists, and much more.

How to access   Anonymous FTP: **ftp.cs.pdx.edu**
Path: **/pub/flamingos/***

Series and Sitcoms

Episode guides, lists, FAQs, scripts, and much more to do with films and television series, including *Star Trek, Cheers, Blade Runner, Twin Peaks, Monty Python*, and many more.

How to access   Anonymous FTP: **nic.funet.fi**
Path: **/pub/culture/tv+film/***

Simpsons

Get the lowdown on the popular animated TV show *The Simpsons*. Bone up on *Simpsons* trivia. This FTP site maintains summaries of each episode, bibliographies of the characters, and series schedules.

| **How to access** | Anonymous FTP: **tfp.cs.widener.edu** |
| | Path: **/pub/simpsons** |

Simpsons Archive

Large collection of information about *The Simpsons*, including pictures, bibliography, cast list, lyrics, episode guides, and quotes.

| **How to access** | Anonymous FTP: **ftp.cs.widener.edu** |
| | Path: **/pub/simpsons/*** |

Star Trek Archives

Quotes, parodies, episodes, reviews, and more.

| **How to access** | Anonymous FTP: **ftp.uu.net** |
| | Path: **/usenet/rec.arts.startrek** |

Star Trek Reviews and Synopses

Synopses and reviews for every episode of *Star Trek: The Next Generation* and *Deep Space Nine*, by Timothy Lynch.

| **How to access** | Telnet: **panda.uiowa.edu** |
| | Choose: **General Information** I **Star Trek Reviews** |

UK Sitcom List

A complete list of all situation comedies that have ever been made in the UK and broadcast on UK terrestrial TV.

How to access	Gopher: Manchester Computing Centre
	Address: **info.mcc.ac.uk**
	Choose: **Miscellaneous** I **The definitive list of UK sitcoms**

Travel

Travel Resources

Guides to where to find everything available on the Internet about travel. Consists of a large list of mailing lists, FTP sites and Listservs of interest to travellers.

How to access   Anonymous FTP: **rtfm.mit.edu**
Paths: **/pub/usenet/rec.answers/travel/online-info**;
/pub/usenet/rec.answers/travel/ftp-archive

Travel information

Get travel information and personal accounts about a number of destinations. Travelers post detailed accounts of their vacations, including places to go, places you should pass up, where and what to eat, reviews of hotels, and much more.

How to access   Anonymous FTP: **ftp.cc.umanitoba.ca**
Path: **/pub/rec-travel/**

Trivia

Coke Servers

Interesting places to finger for trivial information.

How to access   1. Finger: **coke@cmu.edu**

2. Finger: **@coke.elab.cs.edu**

3. Finger: **mnm@coke.elab.cs.cmu.edu**

4. Finger: **bargraph@coke.elab.cs.cmu.edu**

5. Finger: **drink@drink.csh.rit.edu**

6. Finger: **graph@drink.csh.rit.edu**

7. Finger: **coke@gu.uwa.edu.au**

8. Finger: **coke@cs.wisc.edu**

Star Trek Quotes

Quotes from the popular TV show.

How to access   Finger: **franklin.ug.cs.dal.ca**

Today's Events in History

Important historical events that happened on today's date in years gone by.

How to access   Gopher: Manchester Computing Centre
Address: **uts.mcc.ac.uk**
Choose: **Today's events in history**

T

The Unofficial Smiley Dictionary

This smiley dictionary is one of the most complete lists of smileys available on the Internet. Do you feel ;-) or :-(or even just :-I ?

How to access   Gopher: Universitaet des Saarlandes
Address: **pfsparc02.phil15.uni-sb.de**
Choose: **INFO-SYSTEM BENUTZEN | Fun | Cartoons
| Smilies :-)**

Unix

Linux Operating System

A free Unix operating system. Source code, make files, and documentation. Get the FAQ files first from the Usenet archives on **rtfm.mit.edu (comp.os.linux)**.

How to access   1. Anonymous FTP: **nic.funet.fi**
Path: **/pub/OS/Linux/***

2. Anonymous FTP: **ftp.informatik.tu-muenchen.de**
Path: **/pub/Linux/***

3. Anonymous FTP: **tsx-11.mit.edu**
Path: **/pub/linux**

4. Anonymous FTP: **sunsite.unc.edu**
Path: **/pub/Linux**

See also   Operating Systems

NQS Unix Batch System

A collection of available information about NQS, the most commonly used batch software on Unix. The information is grouped into two areas: that relevant to users of NQS, and that required by administrators of systems running NQS.

How to access   Gopher: Manchester Computing Centre
Address: **uts.mcc.ac.uk**
Choose: **NQS...The Unix Batch System**

Source Code for Berkeley Unix

Volumes of source and makefile for the BSD version of Unix. Includes source for Intel platforms, networking code, and Unix manuals.

How to access   Anonymous FTP: **gatekeeper.dec.com**
Path: **/.0/BSD**

Unix Dial-up Site List

A long listing of open access dial-up Unix sites, including both fee and no fee hosts. Contains detailed information on what each site has to offer.

How to access   1. Usenet: **comp.misc, comp.bbs.misc, alt.bbs**

2. Anonymous FTP: **gvl.unisys.com**
Path: **/pub/nixpub/***

3. Mail: **nixpub@access.digex.net**

4. Mail: **mail-server@bts.com**
Body: **get PUB nixpub**

5. Mail: **archive-server@cs.widener.edu**
Body: **send nixpub long**

Unix Manual

Access the Unix manual by searching for key words.

How to access   Wais: **unix-manual**

Unix Software and Source Code

Large general purpose archive of systems software, sources, and documentation for Unix.

How to access   Anonymous FTP: **ftp.uunet.uu.net**
Paths: **/systems/unix/***; **/systems/unix-today/***;
/systems/unix-world/*

Unix Text Formatter

A scaled-down nroff/troff style text formatter for Unix systems.

How to access   1. Usenet: **comp.sources.unix**

2. Anonymous FTP: **world.std.docm**
Path: **/src/print/awf/***

Vi Reference Card

A complete list of all the commands for the popular Unix **vi** text editor. The commands are suitably grouped into topics, so it's easy to locate the one you need.

How to access   Gopher: Manchester Computing Centre
Address: **uts.mcc.ac.uk**
Choose: **Experimental and New Services**
| **VI Reference Card**

U

Vi Tutorial

An interactive tutor teaching you all the necessities of the popular Unix **vi** text editor. The file needs to be captured or downloaded and then used with **vi**.

How to access   Gopher: Manchester Computing Centre
Address: **uts.mcc.ac.uk**
Choose: **Experimental and New Gopher Services**
| **VI Tutorial**

Usenet

Anonymous Posting Service

Post messages to Usenet groups anonymously. To find out how, send mail to the address below.

How to access   Mail: **help@anon.penet.fi**

Frequently Asked Question (FAQ) Lists

Collections of frequently asked questions from many Usenet newsgroups.

How to access   Anonymous FTP: **rtfm.mit.edu**
Path: **/pub/usenet/news.answers**

Newsgroup Descriptions

Detailed descriptions of most of the Usenet groups and categories.

How to access   Telnet: **kufacts.cc.ukans.edu**
Login: **kufacts** | **Usenet News**
Select: **Reference Shelf** | **Usenet News**

NNTP News Servers

Post to Usenet news via telnet. Type **help** after connecting.

How to access   1. Telnet: **vaxc.cc.monash.edu.au 119**

2. Telnet: **munnari.oz.au 119**

3. Telnet: **etl.go.jp 119**

4. Telnet: **news.fu-berlin.de 119**

Oracle

The Usenet Oracle answers any question you have about Usenet.

How to access	Mail: **oracle@cs.indiana.edu**
	Subject: **help**

Usenet Groups with a Difference

A list of Usenet groups that are connected with interaction, interfaces and agency, for those of you looking for the real Net News.

How to access	Gopher: University of Texas at Austin
	Address: **actlab.rtf.utexas.edu**
	Choose: **USENET	USENET Groups related to...**

Usenet News via Gopher

Allows you to read Usenet news through Gopher.

How to access	Gopher: Michigan State University
	Address: **gopher.msu.edu**
	Choose: **News & Weather	USENET News**

Usenet Olympics

Parodies of Usenet and its inhabitants, combined into a single file covering each year.

How to access	Anonymous FTP: **ocf.berkeley.edu**
	Path: **/pub/Usenet_Olympics/***

W

Weather

Gray's Atlantic Seasonal Hurricane Forecast

Provides 1944 to date seasonal means and current year forecast for the number of named storms, named storm days, hurricanes, hurricane days, major hurricanes, destruction potential, and so on.

How to access	Finger: **forecast@typhoon.atmos.colostate.edu**

Hourly Auroral Activity Status Report

Get the latest reports on the activities of the Aurora Borealis (northern lights). Reports watches and warnings. This information is updated hourly.

How to access	Finger: **aurora@xi.uleth.ca**

NOAA Earth Systems Data Directory

The NESDD is an information resource for identification, location, and overview descriptions of Earth Science Data Sets (data from satellites).

How to access   1. Telnet: **esdim1.nodc.noaa.gov**
Login: **noaadir**

2. Telnet: **nodc.nodc.noaa.gov**
Login: **noaadir**

Solar and Geophysical Reports

Get the latest reports on solar activities. Reports watches and warnings. Information is extensive and includes numerous text graphs and tabular data. This information is updated every three hours.

How to access   Finger: **solar@xi.uleth.ca** (updated every 3 hours)
Finger: **daily@xi.uleth.ca** (daily summary)

Surface Analysis and Weather Maps

This Anonymous FTP site carries current surface analysis and infrared weather maps. They are updated often, and the old ones are not archived. Maps are mostly in GIF format.

How to access   Anonymous FTP: **vmd.cso.uiuc.edu**
Path: **WX/***

Weather Reports

Get up-to-date weather reports for any location on the planet. An easy-to-use interface guides you through the process of selecting a city or location, then viewing the weather report on-screen or downloading it to your computer.

How to access   Telnet: **downwind.sprl.umich.edu 3000**

Weather Reports (Australia)

Current weather, forecasts, recent trends, and future developments, including boating and coastal conditions, river and solar reports for all parts of Australia.

How to access   Telnet: **vicbeta.vic.bom.gov.au 55555**

APPENDIX A

Public Access to the Internet

The Purpose of This Appendix

Suppose you have your own PC or Macintosh. What do you need to use the Internet? You need two things. First, you need the proper hardware and software: a computer, a modem, and a communications program. Second, you need a connection to the Internet. (We cover all of this in Chapter 3.)

The general scenario is that your communications program uses the modem and dials another computer, one that is on the Internet. Once the connection is made, your program emulates a terminal (usually a VT-100). Your keyboard and screen are now connected to the remote Internet computer.

The thing is, in order for all this to work, you need to know the phone number of an Internet computer, and you need an account on that computer. That is the purpose of this appendix.

Note to Internet access providers:

If you are not listed in this appendix but would like to be, send us a message with the appropriate information and we will include you in the next edition of this book.

If you are listed in this appendix and would like to make changes to your entry, just let us know.

Please send all changes and comments that relate to this appendix to **access@rain.org**.

Special Offer for Our Readers

One of our goals in writing this book is to help you get started using the Internet as easily as possible. To this end, we have arranged for certain Internet access providers to offer our readers a special trial offer of one month's free Internet usage. We have made such arrangements with Internet providers in virtually every large metropolitan area in the United States, Canada, and Europe. In addition, there are also such providers in Australia and New Zealand. To take advantage of this offer, see the coupon at the back of this book.

Bear in mind that you will find variations from one organization to another. Many of the providers listed in this appendix have promised to participate in this offer. However, some of them may make the reasonable request that, before they start up your account, you should register as a regular user. Some may also ask you to sign up for extra time past the first month. These conditions do not have anything to do with money—the providers who are honoring our arrangement really do want you to be able to try the Internet for free—it is a matter of security. Unfortunately, there are too many people who will take advantage of free service merely to cause as much trouble as they can (for a month anyway). That is why some of the Internet providers will insist that you do not remain anonymous, and that you show that you are serious about using (or at least about trying out) their service.

You can take our word for it that such conditions are fair and reasonable. Moreover, as an Internet user yourself, you will want to have access to a secure, well-run system.

Hint

Some of the Internet access providers on our list could not participate in our one-free-month offer, but did want to give a special discount to our readers. For example, some of them offered to waive a registration fee, or provide two week's free access.

Thus, whenever you call any provider listed in this appendix, mention that you read about them in this book and ask if they can offer you a discount of some kind. Many of them will do so.

Choosing an Internet Access Provider

We have collected the names of many companies and organizations that provide Internet access to the public. This appendix contains the list of names. We have carefully screened our list to ensure that each provider (1) offers a full set of Internet services, and (2) offers accounts to individuals.

As a general rule, "a full set of Internet services" means at least electronic mail, Usenet, FTP, and Telnet. There are many public providers that do not have an Internet connection. Such providers typically offer only mail and Usenet. In our opinion, it is important for you to have a real Internet connection, and, with one exception, providers who are not on the Internet did not make our list.

The exception is the Freenets (see Chapter 3). These are community-based organizations that provide free public computing services. Although all Freenets have an Internet connection, they do not, as a rule, offer full Internet services. They will usually provide mail, Usenet, and a Gopher, but you may not get Telnet and FTP.

However, we have included the Freenets for three reasons. First, they are free. Second, for many people, they are the only way to get at least some Internet services. Third, we like what they are doing and we want to encourage them.

How to Use This Appendix

The first step in using this appendix is to find the names of the Internet providers in your area. To do so, use Table A-1 if you are in the United States or Canada. Use Table A-2 if you are in Europe, Australia, or New Zealand. Please bear in mind that the details about Internet providers are changing all the time. To the best of our knowledge, the information in this appendix was accurate at the time this book was printed.

Table A-1 is organized by area code. Our goal is to make it easy for you to find a computer that you can use by making a local phone call. If a provider is listed for a particular area code, it means that they have a phone number that you can call in that area. Each such facility is called a *point of presence*. Although many providers have only a single point of presence, the larger organizations have more than one area code from which you can make local calls.

Some providers have a toll-free 800 number that you can reach from anywhere in the country. However, if you check carefully, you will usually find that there is an extra charge for using this number, and that the charge is larger than you would have to pay for a long distance call.

Similarly, some American providers advertise that they serve "the entire country". But when you ask about the details, you find that they offer local phone access in only a few places. Sure, you can call from anywhere—if you are willing to pay the long distance costs or a special access charge—but that is true for any of the services. We have carefully weeded out all of this in Table A-1.

Once you have found the Internet providers in your area, look up the details in the master list. Contact each provider and ask about their rates and services.

Remember, you can use the coupon at the back of this book to take advantage of the free trial offer.

If you have a choice of providers in your area, there are two important things you should check out. First, look for full Internet service (mail, Usenet, FTP, and Telnet).

Second, look for a provider that offers a flat rate, allowing you to use the Internet as much as you want for a fixed amount of money per month. Unless you have no alternative, stay away from providers that charge by the hour. No matter how reasonable they look, your bill will be larger than you expect. What is worse, you will tend to use the service sparingly to conserve money. We want you to use the Internet a lot.

The master list contains the basic information about each provider and is arranged in alphabetical order. We have purposely not detailed prices or services as these are subject to change and, anyway, you will want to check for yourself.

The first line of each description shows the name of the organization. In brackets, you will see the address of their main computer. (We explain Internet addresses in Chapter 4.)

Next, you will see the area codes (or country) served by that provider, followed by a phone number you can call to ask for more information.

The next line contains a mail address. This address is handy if you have some way to send and receive mail.

Finally, many providers allow you to connect to their system as a guest, to try out the services or to sign up for an account. There are two ways you can do this. First, if you have some type of Internet service that you can use, even temporarily, you can telnet to the provider. Second, you can use your PC, modem, and communications program and dial in. After all, since you are buying a service that you will use by calling their computer, there is no reason why you shouldn't be able to sign up the same way.

Just so this all makes sense, let's take a look at a typical entry:

Rain [**rain.org**]

Area Code:	805
Voice Phone:	(805) 899-8610
Mail:	**rain@rain.org**
Guest Access:	(805) 899-8600, log in as **guest**

This is the description of Rain, the Regional Access Information Network in Santa Barbara, California. We see that they offer a single point of presence (the 805 area). If you want to talk to a person, you can call (805) 899-8610. If you have access to mail, you can send a message to **rain@rain.org**.

If you want to try out the system as a guest and you have some type of Internet access, you can telnet to **rain.org** (the name on the top line). The actual command would be:

L A.1
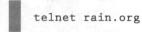

```
telnet rain.org
```

Otherwise, you can connect by dialing (805) 899-8600. Regardless of how you connect, you should log in with a userid of **guest**.

Hint

When you dial a remote host, set the parameters in your communications program to: No parity, 8 data bits, 1 stop bit (N 8 1). If this doesn't work, try: Even parity, 7 data bits, 1 stop bit (E 7 1).

Area Code Reference List

Area Code	Internet Access Providers
201	JVNCnet, PSI, Panix
202	Clark, Digex, PSI
203	JVNCnet, PSI
204	MBnet
206	Cyberlink, Eskimo, Netcom, Northwest Nexus, PSI, Seattle
208	PSI, Westnet
212	MindVOX, NYsernet, PSI, Panix
213	Cerfnet, Los Nettos, PSI

Table A-1.   *Area Code Reference List*

Area Code	Internet Access Providers
214	Netcom, PSI, Texas
215	DSC/Voicenet, Digital, JVNCnet, PSI
216	Akron, Cleveland, Lorain, Medina, OARnet, Youngstown
217	Prairienet
301	Cap-Access, Clark, Digital, Grebyn, SURAnet
303	Colorado, Community, Denver, Netcom, Nyx, PSI, Westnet
305	PSI
307	NorthWestNet, Westnet
308	MIDnet
309	Heartland
310	Cerfnet, Netcom
312	Cicnet, Ddsw1, MCSNet, Netcom, NetIllinois, PSI
313	Advanced, Arbornet, Huron, Merit, Msen, Netcom, PSI
314	Coin, MIDnet
315	NYsernet, PSI
316	MIDnet
319	MIDnet
401	Anomaly, IDS, JVNCnet
402	MIDnet
403	Alberta
404	CR, Netcom, PSI
405	MIDnet
406	Big Sky, NorthWestNet
407	PSI
408	Duck, Netcom, PSI, Portal, TSoft, a2i
410	Clark, Digital
412	PSCnet, PSI, Telerama
414	Mix, WiscNet
415	Barrnet, CR, Cerfnet, IGC, Netcom, Netsys, PSI, Portal, TSoft, Well
416	Hookup, Io, Uunorth
417	Coin, MIDnet
419	OARnet
501	MIDnet

Table A-1. *Area Code Reference List* (continued)

Area Code	Internet Access Providers
503	Agora, Netcom, PSI, Techbook
505	New Mexico Technet, Westnet
508	Anomaly, NEARnet, PSI
509	NorthWestNet, PSI
510	Cerfnet, CR, Netcom
512	PSI
513	Dayton, OARnet, TSO
514	Communications
515	MIDnet, NetIowa
516	JVNCnet, NYsernet, PSI, Panix
517	Merit, Msen
518	NYsernet, PSI
519	Hookup
602	Aztec, CR, PSI, Westnet
603	NEARnet
604	Ciao, Mind, Victoria, Wimsey
605	MIDnet
607	NYsernet, PSI
608	WiscNet
609	Digex, JVNCnet, PSI
612	MRNet
613	National
614	OARnet, PSI
615	Vnet
616	BCA, Merit, Msen
617	Delphi, NEARnet, Netcom, PSI, World
619	CTS, Cerfnet, Cyberspace, Netcom
703	Clark, Digex, Netcom, PSI, Vernet
704	Concert, Vnet
707	CR

Table A-1. *Area Code Reference List* (continued)

Area Code	Internet Access Providers
708	Cicnet, Ddsw1, InterAccess, MCSNet
712	MIDnet
713	Neosoft, PSI
714	Cerfnet, Digex, Netcom, Orange
715	Chippewa, WiscNet
716	Buffalo, NYsernet, PSI
717	Circuit
718	MindVOX, NYsernet, Panix
719	CSN, Colorado, Old Colorado, Westnet
801	Westnet
804	Vernet
805	Rain, SLONET
808	Pacific
813	PSI, Suncoast
815	Ddsw1
816	Coin, MIDnet
817	Texas
818	Cerfnet, Netcom
902	NSTN
904	NE Regional, PSI, Tallahassee
906	Merit
908	Digex, JVNCnet
909	Cerfnet
913	MIDnet
914	NYsernet, PSI, Panix
916	Netcom
917	NYsernet
918	MIDnet
919	Concert, Vnet

Table A-1. *Area Code Reference List* (continued)

Country Reference List

Country	Internet Access Providers
Australia	Aarnet, Connect.com.au
Belgium	S-Team
Denmark	DKnet
Europe	EUnet
Germany	Bayreuth, Contributed, Erlangen, INS, Individual, Odb
Greece	FORTHnet
Ireland	Ireland on-line
Israel	Public
Italy	IUnet
The Netherlands	NLnet
New Zealand	Actrix, Wellington
Norway	Telepost
Switzerland	Switch
UK	Almac, Cix, Demon, Direct, Exnet, Pipex, RJB, UKnet, UK PC

Table A-2. *Country Reference List*

List of Public Internet Providers

a2i [**rahul.net**]

Area Code:	408, 415
Voice Phone:	None
Mail:	**info@rahul.net**
Guest Access:	(408) 293-9010, log in as **guest**

Aarnet [**aarnet.edu.au**]

Country:	Australia
Voice Phone:	+61 6-249-3385
Mail:	**aarnet@aarnet.edu.au**

Actrix [**actrix.gen.nz**]

Country:	New Zealand
Voice Phone:	(04) 389-6316
Mail:	john@actrix.gen.nz

Advanced Networks and Services [**nis.ans.net**]

Area Code:	313
Voice Phone:	(313) 663-7610
Mail:	maloff@nis.ans.net

Agora [**agora.rain.com**]

Area Code:	503
Voice Phone:	None
Mail:	batie@agora.rain.com
Guest Access:	(503) 293-1772, log in as **apply**

Akron Free-Net [**vm1.cc.uakron.edu**]

Area Code:	216
Voice Phone:	(216) 972-6352
Mail:	r1asm@vm1.cc.uakron.edu

Alberta Research Council [**arc.ab.ca**]

Area Code:	403
Voice Phone:	(403) 450-5189
Mail:	penno@arc.ab.ca

Almac [**almac.co.uk**]

Country:	UK
Voice Phone:	+44 0324-665371
Mail:	alastair.mcintyre@almac.co.uk

Anomaly [**anomaly.sbs.risc.net**]

Area Codes:	401, 508
Voice Phone:	(401) 273-4669
Mail:	info@anomaly.sbs.risc.net

Arbornet [**m-net.ann-arbor.mi.us**]

Area Code:	313
Voice Phone:	(313) 594-1769, (313) 973-7528
Mail:	help@m-net.ann-arbor.mi.us
Guest Access:	(313) 996-4644, type **G** for **guest** when prompted

Aztec [**asu.edu**]

Area Code:	602
Voice Phone:	(602) 965-5985
Mail:	joe.askins@asu.edu

Barrnet [**forsythe.stanford.edu**]

Area Code:	415
Voice Phone:	(415) 723-3104
Mail:	gd.why@forsythe.stanford.edu

Bayreuth Free-Net [**uni-bayreuth.de**]

Country:	Germany
Voice Phone:	+49 0921-553134
Mail:	wolfgang.kiessling@uni-bayreuth.de

BCA Free-Net [**aol.com**]

Area Code:	616
Voice Phone:	(616) 961-3676
Mail:	fredj@aol.com

Big Sky Telegraph [**bigsky.bigsky.dillon.mt.us**]

Area Code:	406
Voice Phone:	None
Mail:	Franko@bigsky.bigsky.dillon.mt.us
Guest Access:	(406) 683-7680, log in as **bbs**

Buffalo Free-Net [**freenet.buffalo.edu**]

Area Code:	716
Voice Phone:	None
Mail:	None
Guest Access:	(716) 645-6128, log in as **freeport**

CapAccess [**capital.com**]

Area Code:	301
Voice Phone:	(301) 942-1262
Mail:	taylor@capital.com

Cerfnet [**cerf.net**]

Area Codes:	213, 310, 415, 510, 619, 714, 818, 909
Voice Phone:	(800) 879-2373, (619) 455-3990
Mail:	help@cerf.net

Chippewa Valley Free-Net [**uwec.edu**]

Area Code:	715
Voice Phone:	(715) 836-3715
Mail:	smarquar@uwec.edu

Ciao [**first.etc.bc.ca**]

Area Code:	604
Voice Phone:	(604) 368-6434
Mail:	kmcclean@first.etc.bc.ca

Cicnet [**cic.net**]

Area Codes:	312, 708
Voice Phone:	(313) 998-6103
Mail:	info@cic.net

Circuit Board [**ka2qhd.de.com**]

Area Code:	717
Voice Phone:	None
Mail:	jimd%cktbd@ka2qhd.de.com

Cix [**cix.compulink.co.uk**]

Country:	UK
Voice Phone:	+44 49 2641 961
Mail:	cixadmin@cix.compulink.co.uk

Clark [**clark.net**]

Area Codes:	202, 301, 410, 703
Voice Phone:	(800) 735-2258: and then ask for 410-730-9764
Mail:	info@clark.net

Cleveland Free-Net [**cleveland.freenet.edu**]

Area Code:	216
Voice Phone:	None
Mail:	None
Guest Access:	(216) 368-3888, select #2 at the **first menu**

CNS (Community News Service) [**cscns.com**]

Area Codes:	303, 719
Voice Phone:	(800) 748-1200
Mail:	klaus@cscns.com
Guest Access:	(719) 520-1700, log in as **new**, password **newuser**

Coin [**bigcat.missouri.edu**]

Area Codes:	314, 417, 816
Voice Phone:	(314) 882-2000
Mail:	ccwam@mizzoui.missouri.edu
Guest Access:	(314) 884-7000, log in as **guest**

Colorado SuperNet [**csn.org**]

Area Codes:	303, 719
Voice Phone:	(303) 273-3471
Mail:	info@csn.org

Communications Accessibles Montreal [**cam.org**]

Area Code:	514
Voice Phone:	(514) 923-2102
Mail:	info@cam.org
Guest Access:	(514) 671-6723, log in as **new**

Concert [**rock.concert.net**]

Area Codes:	704, 919
Voice Phone:	(919) 248-1404
Mail:	info@concert.net

Connect.com.au [**yarrina.connect.com.au**]

Country:	Australia
Voice Phone:	+61 3-528-2239
Mail:	connect@connect.com.au

Contributed Software [**uropax.contrib.de**]

Country:	Germany
Voice Phone:	+49 30-694-69-07
Mail:	info@contrib.de
Guest Access:	+49 30-694-60-55, log in as **guest** or **gast**

CR Laboratories [**crl.com**]

Area Codes:	404, 415, 510, 602, 707
Voice Phone:	(415) 381-2800
Mail:	info@crl.com
Guest Access:	(415) 389-8649, log in as **guest**, password **guest**

CTS [**crash.cts.com**]

Area Code:	619
Voice Phone:	(619) 593-9597
Mail:	bblue@crash.cts.com
Guest Access:	(619) 593-6400, log in as **help**

Cyberlink [**cyberspace.com**]

Area Code:	206
Voice Phone:	(206) 282-4919
Mail:	brc@cyberspace.com

Cyberspace Station [**cyber.net**]

Area Code:	619
Voice Phone:	(619) 944-9498 x626
Mail:	info@cyber.net or help@cyber.net
Guest Access:	(619) 634-1376, log in as **guest**

Dayton Free-Net [**desire.wright.edu**]

Area Code:	513
Voice Phone:	(513) 873-4035
Mail:	pvendt@desire.wright.edu

Ddsw1 [**ddsw1.mcs.com**]

Area Codes:	312, 708, 815
Voice Phone:	None
Mail:	info@ddsw1.mcs.com or karl@ddsw1.mcs.com
Guest Access:	(312) 248-0900, follow the instructions

Delphi [**delphi.com**]

Area Code:	617
Voice Phone:	(800) 544-4005
Mail:	walthowe@delphi.com
Guest Access:	(800) 365-4636, log in as **JOINDELPHI** with password **INTERNETSIG**

Demon [**gate.demon.co.uk**]

Country:	UK
Voice Phone:	+44 81-349-0063
Mail:	internet@demon.co.uk

Denver Free-Net [**freenet.hsc.colorado.edu**]

Area Code:	303
Voice Phone:	(303) 270-4300
Mail:	info@freenet.hsc.colorado.edu
Guest Access:	(303) 270-4865, log in as **guest**

Digex [**access.digex.net**]

Area Codes:	202, 215, 301, 410, 609, 703, 714, 908
Voice Phone:	(800) 969-9090
Mail:	info@digex.com
Guest Access:	(301) 220-0462 or (410) 766-1855, log in as **new**

Direct Connection [**dircon.co.uk**]

Country:	UK
Voice Phone:	+44 081-317-0100
Mail:	helpdesk@dircon.co.uk
Guest Access:	telnet, log in as **help** or **demo**

DKnet [**dknet.dk**]

Country:	Denmark
Voice Phone:	+45 39 179900
Mail:	netpasser@dknet.dk

DSC/Voicenet [**voicenet.com**]

Area Code:	215
Voice Phone:	(215) 957-4060
Mail:	ron@ubos.voicenet.com

Duck Pond [**quack.kfu.com**]

Area Code:	408
Voice Phone:	None
Mail:	postmaster@quack.kfu.com
Guest Access:	(408) 249-9630, log in as **guest**

Erlangen Free-Net [**cnve.rrze.uni-erlangen.dbp.de**]

Country:	Germany
Voice Phone:	+49 09131-85-4735
Mail:	fimpsy@cnve.rrze.uni-erlangen.dbp.de

Eskimo North [**eskimo.com**]

Area Code:	206
Voice Phone:	(206) 367-7457
Mail:	**nanook@eskimo.com**
Guest Access:	telnet, log in as **new**

EUnet [**eu.net**]

Country:	Europe
Voice Phone:	+31 20-59-25-12-4
Mail:	**info@eu.net**
Guest Access:	send e-mail to **eunet@**_country_**.eu.net** (where _country_ is **uk**, **it**, **gr**, **de**, **fr**, etc.)

Exnet [**exnet.co.uk**]

Country:	UK
Voice Phone:	+44 081-755-0077
Mail:	**helpex@exnet.co.uk**

FORTHnet [**csi.forth.gr**]

Country:	Greece
Voice Phone:	+30 81 221171, + 30 81 229302
Mail:	**stelios@ics.forth.gr**

Heartland Free-Net [**heartland.bradley.edu**]

Area Code:	309
Voice Phone:	(309) 677-2544
Mail:	**xxadm@heartland.bradley.edu**
Guest Access:	(309) 674-1100, log in as **bbguest**

Hookup [**nic.hookup.net**]

Area Codes:	416, 519
Voice Phone:	(519) 747-4110
Mail:	**info@nic.hookup.net**
Guest Access:	(519) 883-4665, log in as **guest**

Huron Valley Free-Net [**umich.edu**]

Area Code:	313
Voice Phone:	(313) 662-8374
Mail:	michael.todd.glazier@umich.edu

IDS World Network [**ids.net**]

Area Code:	401
Voice Phone:	(401) 884-7856
Mail:	sysadmin@ids.net
Guest Access:	telnet, log in as **guest**

IGC Networks [**igc.apc.org**]

Area Code:	415
Voice Phone:	(415) 442-0220
Mail:	support@igc.apc.org

Individual Network [**individual.net**]

Country:	Germany
Voice Phone:	+49 69-331461
Mail:	in-info@individual.net
Guest Access:	+49 69-39048414, log in as **gast**

Individual Network-Rhein-Main [**rhein-main.de**]

Country:	Germany
Voice Phone:	+49 69 39048413, +49 69 6312083
Mail:	ip-admins@rhein-main.de

INS [**ins.net**]

Country:	Germany
Voice Phone:	+49 2305-356505
Mail:	info@ins.net

InterAccess [**interaccess.com**]

Area Code:	708
Voice Phone:	(800) 967-1580
Mail:	info@interaccess.com
Guest Access:	(708) 671-0237, log in a **guest**

InterAddcesInternex [**io.org**]

Area Code:	416
Voice Phone:	(416) 363-8676
Mail:	**info@io.org**
Guest Access:	(416) 363-3783, log in as **new**

Ireland On-line [**iol.ie**]

Country:	Ireland
Voice Phone:	+353 (0) 91 92727
Mail:	**info@iol.i**

IUnet [**iunet.it**]

Country:	Italy
Voice Phone:	+39 10 3532747
Mail:	**abi@iunet.it**

JVNCnet [**jvnc.net**]

Area Codes:	201, 203, 215, 401, 516, 609, 908
Voice Phone:	(800) 358-4437, (609) 897-7300
Mail:	**info@jvnc.net**

Lorain County Free-Net [**freenet.lorain.oberlin.edu**]

Area Code:	216
Voice Phone:	None
Mail:	None
Guest Access:	(216) 366-9721, log in as **guest**

Los Nettos [**isi.edu**]

Area Code:	213
Voice Phone:	(310) 822-1511
Mail:	**los-nettos-request@isi.edu**

MBnet [**mbnet.mb.ca**]

Area Code:	204
Voice Phone:	(204) 474-8230
Mail:	**info@mbnet.mb.ca**
Guest Access:	(204) 275-6132, log in as **guest**

MCSNet [**genesis.mcs.com**]

Area Codes:	312, 708
Voice Phone:	(312) 248-8649
Mail:	info@mcs.com
Guest Access:	(312) 248-0900

Medina County Freenet [**medina.freenet.edu**]

Area Code:	216
Voice Phone:	(216) 725-1000
Mail:	aa001@medina.freenet.edu

Merit [**merit.edu**]

Area Codes:	313, 517, 616, 906
Voice Phone:	(313) 764-9430
Mail:	info@merit.edu
Guest Access:	telnet **hermes.merit.edu**, type **help** at **Which host?**

MIDnet [**westie.unl.edu**]

Area Codes:	308, 314, 316, 319, 402, 405, 417, 501, 515, 605, 712, 816, 913, 918
Voice Phone:	(402) 472-7600
Mail:	dmf@westie.unl.edu

Mind Link [**mindlink.bc.ca**]

Area Code:	604
Voice Phone:	(604) 534-5663
Mail:	info@mindlink.bc.ca
Guest Access:	(604) 576-1683, log in as **guest**

MindVOX [**phantom.com**]

Area Codes:	212, 718
Voice Phone:	(212) 989-2481, (800) 646-3869
Mail:	info@phantom.com
Guest Access:	(212) 989-4141, log in as **mindvox** or **guest**

Mix [**mixcom.com**]

Area Code:	414
Voice Phone:	(414) 962-8172
Mail:	sysop@mixcom.com
Guest Access:	(414) 241-5469, log in as **newuser**

MRNet [**mr.net**]

Area Code:	612
Voice Phone:	(612) 342-2570
Mail:	info@mr.net

Msen [**Msen.com**]

Area Codes:	313, 517, 616
Voice Phone:	(313) 998-4562
Mail:	info@mail.msen.com
Guest Access:	telnet, log in as **newuser**

National Capital Free-Net [**freenet.carleton.ca**]

Area Code:	613
Voice Phone:	(613) 788-3947
Mail:	ncf@freenet.carleton.ca
Guest Access:	(613) 780-3733, log in as **guest**

NE Regional Data Center [**ufl.edu**]

Area Code:	904
Voice Phone:	(904) 392-2061
Mail:	poke@ufl.edu

NEARnet [**nic.near.net**]

Area Codes:	508, 603, 617
Voice Phone:	(617) 873-8730
Mail:	nearnet-join@nic.near.net

Neosoft [**sugar.neosoft.com**]

Area Code:	713
Voice Phone:	(713) 438-4964
Mail:	info@neosoft.com
Guest Access:	(713) 684-5969, log in as **new**

Netcom [**netcom.com**]

Area Codes:	202, 206, 213, 214, 303, 310, 312, 313, 404, 408, 415, 503, 510, 512, 617, 619, 703, 713, 714, 818, 916
Voice Phone:	(408) 554-8649
Mail:	info@netcom.com
Guest Access:	telnet, log in as **guest**

NetIllinois [**bradley.edu**]

Area Code:	312
Voice Phone:	(309) 677-3100
Mail:	p-roll@nwu.edu

NetIowa [**ins.infonet.net**]

Area Code:	515
Voice Phone:	(800) 546-6587
Mail:	info@ins.infonet.net
Guest Access:	(800) 469-9990, log in as **guest**

Netsys [**netsys.com**]

Area Code:	415
Voice Phone:	(415) 424-0384
Mail:	netsys@netsys.com

New Mexico Technet [**technet.nm.org**]

Area Code:	505
Voice Phone:	(505) 345-6555
Mail:	granoff@technet.nm.org

NLnet [**nl.net**]

Country:	The Netherlands
Voice Phone:	+31 20 5924245
Mail:	nl.net@nl.net

Northwest Nexus [**halcyon.com**]

Area Code:	206
Voice Phone:	(206) 455-3505, (800) 539-3505
Mail:	info@nwnexus.wa.com
Guest Access:	(206) 382-6245, log in as **guest** or **new**

NSTN [**nstn.ns.ca**]

Area Code:	902
Voice Phone:	(902) 468-6786
Mail:	parsons@hawk.nstn.ns.ca

NYsernet [**nysernet.org**]

Area Codes:	212, 315, 516, 518, 607, 716, 718, 914, 917
Voice Phone:	(315) 453-2912
Mail:	luckett@nysernet.org

Nyx [**nyx.cs.du.edu**]

Area Code:	303
Voice Phone:	None
Mail:	aburt@nyx.cs.du.edu
Guest Access:	(303) 871-3324, log in as **guest**

OARnet [**oar.net**]

Area Codes:	216, 419, 513, 614
Voice Phone:	(614) 292-8100, (614) 292-9248
Mail:	demetris@oar.net or nic@oar.net

Odb [**odb.rhein-main.de**]

Country:	Germany
Voice Phone:	+49 69-331461
Mail:	oli@odb.rhein-main.de

Old Colorado [**oldcolo.com**]

Area Code:	719
Voice Phone:	(719) 632-4848, (719) 593-7575
Mail:	dave@oldcolo.com
Guest Access:	(719) 632-4111, log in as **newuser**

Orange County Free-Net [**world.std.com**]

Area Code:	714
Voice Phone:	(714) 762-8551
Mail:	palmer@world.std.com

Pacific Network Consortium [**freenet-in-a.cwru.edu**]

Area Code:	808
Voice Phone:	(808) 533-3969
Mail:	bm189@po.cwru.edu

Panix [**panix.com**]

Area Codes:	201, 212, 516, 718, 914
Voice Phone:	(212) 787-6160, (212) 877-4854
Mail:	info@panix.com
Guest Access:	(212) 787-3100, log in as **help**

Pipex [**pipex.net**]

Country:	UK
Voice Phone:	+44 0223-424616
Mail:	pipex@pipex.net

Portal [**portal.com**]

Area Codes:	408, 415
Voice Phone:	(408) 973-9111
Mail:	cs@cup.portal.com or info@portal.com
Guest Access:	(408) 725-0561, log in as **info**

Prairienet [**firefly.prairienet.org**]

Area Code:	217
Voice Phone:	(217) 244-3299
Mail:	prairienet@uiuc.edu or prairienet@prairienet.org
Guest Access:	(217) 255-9000, log in as **visitor**

PSCnet [**psc.edu**]

Area Code:	412
Voice Phone:	(412) 268-4960
Mail:	hastings@psc.edu

PSI [**psi.com**]

Area Codes:	201, 202, 203, 206, 208, 212, 213, 214, 215, 303, 305, 312, 313, 315, 404, 407, 408, 412, 415, 503, 508, 509, 512, 516, 518, 602, 607, 609, 614, 617, 703, 713, 716, 813, 904, 914
Voice Phone:	(703) 620-6651
Mail:	world-dial-info@psi.com

Public InterNet Provider [**datasrv.co.il**]

Country:	Israel
Voice Phone:	+972 3-574-6640
Mail:	gil@datasrv.co.il

Rain [**rain.org**]

Area Code:	805
Voice Phone:	(805) 899-8610
Mail:	rain@rain.org
Guest Access:	(805) 899-8600, log in as **guest**

RJB Communications [**sound.demon.co.uk**]

Country:	UK
Voice Phone:	+44 0932-253131
Mail:	rob@sound.demon.co.uk

Seattle Community Network [**atc.boeing.com**]

Area Code:	206
Voice Phone:	(206) 865-3832
Mail:	douglas@atc.boeing.com

SLONET [**callamer.com**]

Area Code:	805
Voice Phone:	(805) 545-5002
Mail:	slonet@callamer.com

S-Team [**k12.be**]

Country:	Belgium
Voice Phone:	+32 3-455-1653
Mail:	lvg@k12.be

Suncoast Free-Net [**firnvx.firn.edu**]

Area Code:	813
Voice Phone:	(813) 273-3714
Mail:	mullam@firnvx.firn.edu

SURAnet [**sura.net**]

Area Code:	301
Voice Phone:	(301) 982-4600
Mail:	**info@sura.net**

Switch (Swiss Academic & Research Network) [**switch**]

Country:	Switzerland
Voice Phone:	+41 1 268 1515
Mail:	**postmaster@switch.ch**

Tallahassee Free-Net [**freenet.fsu.edu**]

Area Code:	904
Voice Phone:	None
Mail:	None
Guest Access:	(904) 488-5056, log in as **visitor**

Techbook [**techbook.com**]

Area Code:	503
Voice Phone:	(503) 223-4245
Mail:	**info@techbook.com**
Guest Access:	telnet, log in as **new**

Telepost [**telepost.telemax.no**]

Country:	Norway
Voice Phone:	+47 22-73-37-00
Mail:	**firmapost@telepost.telemax.no**

Telerama [**telerama.pgh.pa.us**]

Area Code:	412
Voice Phone:	(412) 481-3505
Mail:	**info@telerama.pgh.pa.us**
Guest Access:	(412) 481-5302, log in as **new**

Texas Metronet [**metronet.com**]

Area Codes:	214, 817
Voice Phone:	(214) 705-2900, (214) 401-2800
Mail:	**srl@metronet.com** or **info@metronet.com**
Guest Access:	(214) 705-2902, log in as **info**, password **info**

TSO [**cbos.uc.edu**]

Area Code:	513
Voice Phone:	None
Mail:	None
Guest Access:	(513) 579-1990, choose **cbos**, then follow instructions

TSoft [**tsoft.net**]

Area Codes:	408, 415
Voice Phone:	None
Mail:	**info@tsoft.net**
Guest Access:	telnet, log in as **bbs**

UKnet [**uknet.ac.uk**]

Country:	UK
Voice Phone:	+44 227-475497
Mail:	**postmaster@uknet.ac.uk**

UK PC User Group [**ibmpcug.co.uk**]

Country:	UK
Voice Phone:	+44 81-863-6646, +44 81-863-1191
Mail:	**info@ibmpcug.co.uk** or **adrian@ibmpcug.co.uk**
Guest Access:	telnet, follow the directions

Uunorth [**uunorth.north.net**]

Area Code:	416
Voice Phone:	(416) 225-8649
Mail:	**info@uunorth.north.net**

Vernet [**virginia.edu**]

Area Codes:	703, 804
Voice Phone:	(804) 924-0616
Mail:	**jaj@virginia.edu**

Victoria Free-Net [**freenet.victoria.bc.ca**]

Area Code:	604
Voice Phone:	(604) 389-6026
Mail:	**vifa@freenet.victoria.bc.ca**
Guest Access:	(604) 595-2300, log in as **guest**

Vnet [**char.vnet.net**]

Area Codes:	615, 704, 919
Voice Phone:	(704) 374-0779
Mail:	**brat@char.vnet.net**
Guest Access:	telnet, log in as **new** or **newuser**

Well [**well.sf.ca.us**]

Area Code:	415
Voice Phone:	(415) 332-4335
Mail:	info@well.sf.ca.us
Guest Access:	(415) 332-6106, log in as **guest** or **newuser**

Wellington Citynet [**kosmos.wcc.govt.nz**]

Country:	New Zealand
Voice Phone:	+64 4-801-3303
Mail:	rich@tosh.wcc.govt.nz

Westnet [**yuma.acns.colostate.edu**]

Area Codes:	208, 303, 307, 505, 602, 719, 801
Voice Phone:	(303) 491-7260
Mail:	pburns@yuma.acns.colostate.edu

Wimsey [**van-bc.wimsey.com**]

Area Code:	604
Voice Phone:	(604) 936-8649
Mail:	admin@wimsey.com
Guest Access:	(604) 937-7411, log in as **guest**

WiscNet [**macc.wisc.edu**]

Area Codes:	414, 608, 715
Voice Phone:	(608) 262-8874
Mail:	dorl@macc.wisc.edu

World [**world.std.com**]

Area Code:	617
Voice Phone:	(617) 739-0202
Mail:	office@world.std.com
Guest Access:	(617) 739-9753, log in as **new**

Youngstown Free-Net [**yfn.ysu.edu**]

Area Code:	216
Voice Phone:	(216) 742-3075
Mail:	lou@yfn.ysu.edu
Guest Access:	(216) 742-3072, log in as **visitor**

APPENDIX B

Public Archie Servers

Archie servers are used to search for files that are available by Anonymous FTP. We explain how to use an Archie server in Chapter 17. As we discuss in that chapter, it is usually best to use an Archie client program on your own computer. Your client will contact one of the Archie servers, issue your request, and return the output to you.

If you do not have an Archie client on your computer, you can telnet (see Chapter 7) to one of the Archie servers shown in the table and log in using a userid of **archie**. You can then enter commands to perform your search. Alternatively, you can send requests to Archie through the mail and have the output mailed back to you. Again, see Chapter 17 for the details.

Location	Internet Address	IP Address
Austria	archie.edvz.uni-linz.ac.at	140.78.3.8
Austria	archie.univie.ac.at	131.130.1.23
Australia	archie.au	139.130.4.6
Canada	archie.uqam.ca	132.208.250.10
England	archie.doc.ic.ac.uk	146.169.11.3
Finland	archie.funet.fi	128.214.6.102
Germany	archie.th-darmstadt.de	130.83.22.60
Israel	archie.cs.huji.ac.il	132.65.6.15
Japan	archie.wide.ad.jp	133.4.3.6
South Korea	archie.sogang.ac.kr	163.239.1.11
Spain	archie.rediris.es	130.206.1.2
Sweden	archie.luth.se	130.240.18.4
Switzerland	archie.switch.ch	130.59.1.40
Taiwan	archie.ncu.edu.tw	140.115.19.24
USA: Maryland	archie.sura.net	128.167.254.179
USA: Nebraska	archie.unl.edu	129.93.1.14
USA: New Jersey	archie.internic.net	198.49.45.10
USA: New Jersey	archie.rutgers.edu	128.6.18.15
USA: New York	archie.ans.net	147.225.1.10

Log in with a userid of **archie**.

Table B-1.   *Public Archie Servers*

APPENDIX C

Public Gopher Clients

Gophers are menu-driven systems that allow you to access information and remote computers all over the Internet. We discuss Gophers in Chapter 21. The best way to use Gopher is to run a Gopher client program on your own computer. From there, you can connect to any other Gopher you need.

If your computer does not have a Gopher client, you can use one of the public Gopher clients listed in the table. Simply telnet (see Chapter 7) to one of the computers and log in as indicated. To get an up-to-date list of public Gopher clients, look in the Gopher frequently asked question list. (FAQ lists are discussed in Chapter 9.)

The Gopher FAQ list is available by Anonymous FTP from the computer **rtfm.mit.edu**. The path is **/pub/usenet/news.answers/gopher-faq**. (Anonymous FTP is explained in Chapter 16.)

Hint

If you use a remote public Gopher client, you cannot take advantage of local services. For example, when you use a Gopher client that runs on your own computer, you can save information to a file. When you use a public Gopher client on a remote computer, you will not be able to save information in this way, because you will not have permission to create files on the remote computer.

Location	Internet Address	IP Address	Log in as...
Australia	info.anu.edu.au	150.203.84.20	info
Chile	gopher.puc.cl	146.155.1.16	gopher
Denmark	gopher.denet.dk	129.142.6.66	gopher
Ecuador	ecnet.ec	157.100.45.2	gopher
England	gopher.brad.ac.uk	143.53.2.5	info
Germany	gopher.th-darmstadt.de	130.83.55.75	gopher
Japan	gopher.ncc.go.jp	160.190.10.1	gopher
Spain	gopher.uv.es	147.156.1.12	gopher
Sweden	gopher.chalmers.se	129.16.221.40	gopher
Sweden	gopher.sunet.se	192.36.125.2	gopher
USA: California	infopath.ucsd.edu	132.239.50.100	infopath
USA: California	scilibx.ucsc.edu	128.114.143.4	gopher
USA: Georgia	grits.valdosta.peachnet.edu	131.144.8.206	gopher
USA: Illinois	gopher.uiuc.edu	128.174.5.61	gopher
USA: Iowa	panda.uiowa.edu	128.255.40.201	—
USA: Michigan	gopher.msu.edu	35.8.2.61	gopher
USA: Minnesota	consultant.micro.umn.edu	134.84.132.4	gopher
USA: N. Carolina	gopher.unc.edu	152.2.22.81	gopher
USA: N. Carolina	twosocks.ces.ncsu.edu	152.1.45.21	gopher
USA: Ohio	gopher.ohiolink.edu	130.108.120.25	gopher
USA: Virginia	ecosys.drdr.virginia.edu	128.143.96.10	gopher
USA: Virginia	gopher.virginia.edu	128.143.22.36	gwis
USA: Washington	wsuaix.csc.wsu.edu	134.121.1.40	wsuinfo

Table C-1. *Public Gopher Clients*

APPENDIX D

Public Wais Clients

Wais (Wide Area Information Service) is an Internet service that allows you to search a large number of specially indexed databases. We discuss Wais in Chapter 23.

The best way to use Wais is to run a Wais client program on your own computer. There are a number of different Wais clients—for different types of computers—available via Anonymous FTP. If you do not have a Wais client on your computer, you can use telnet (see Chapter 7) to connect to one of the public Wais clients listed in the table.

Location	Internet Address	IP Address	Log in as...
Finland	info.funet.fi	128.214.6.102	wais
USA: California	swais.cwis.uci.edu	128.200.15.2	wais
USA: Massachusetts	nnsc.nsf.net	128.89.1.178	wais
USA: Massachusetts	quake.think.com	192.31.181.1	wais
USA: North Carolina	kudzu.cnidr.org	128.109.130.57	wais
USA: North Carolina	sunsite.unc.edu	152.2.22.81	swais

Table D-1. *Public Wais Clients*

Public World Wide Web Browsers

The *World Wide Web*—also known as the *Web* and *W3*—is a hypertext-based service that allows you to search for and retrieve data based on keyword searches.

We discuss the Web in Chapter 24. Hypertext is data that contains links to other data. You use a client program, called a browser, to access a Web server. Using your browser, you can search for and display information. As you read the information, you can follow the hypertext links to other information.

There are a number of different browsers available that you can obtain via Anonymous FTP and run on your own computer. If you will be using the Web a lot, this is the best way to go. If you do not have your own browser, you can telnet (see Chapter 7) to a public browser. These are shown in Table E-1.

Hint

The public Web databases and browsers listed below are not all the same. Your experience using the Web will vary, depending on which system you use. You might want to take a look at all of them. (The Slovakian Web, however, is in Czech. The Israeli Web has both English and Hebrew.)

Location	Internet Address	IP Address
Finland	info.funet.fi	128.214.6.100
Israel	vms.huji.ac.il	128.139.4.3
Slovakia	sun.uakom.cs	192.108.131.11
Switzerland	info.cern.ch	128.141.201.74
USA: (Kansas)	ukanaix.cc.ukans.edu	129.237.1.30
USA: (Illinois)	www.bradley.edu	136.176.5.252
USA: (New Jersey)	www.njit.edu	128.235.163.2

If you are not given information as to how to log in, use a userid of **www**.

Table E-1.   *Public World Wide Web Browsers*

APPENDIX F

List of Top-Level Internet Domains

733

In Chapter 4, we explained that Internet addresses have the form:

> *userid@domain*

Here are two examples:

> `harley@nipper.ucsb.edu`
> `michael@music.tuwien.ac.at`

The domain consists of a number of sub-domains, separated by **.** characters. The rightmost sub-domain is called the top-level domain. Top-level domains are standardized across the Internet.

There are two types of top-level domains: organizational and geographical. For reference, Tables F-1 and F-2 show the top-level domains that were in effect at the time we wrote this book. For more information about Internet addressing, see Chapter 4.

Domain	Meaning
com	commercial organization
edu	educational institution
gov	government
int	international organization
mil	U.S. military
net	networking organization
org	non-profit organization

Table F-1. *Organizational Top-Level Domains*

Domain	Meaning
aq	Antarctica
ar	Argentina
at	Austria
au	Australia
be	Belgium
bg	Bulgaria
br	Brazil
ca	Canada
ch	Switzerland ("Confoederatio Helvetia")
cl	Chile
cn	China
cr	Costa Rica
cs	Czech Republic and Slovakia
de	Germany ("Deutschland")
dk	Denmark
ec	Ecuador
ee	Estonia
eg	Egypt
es	Spain ("España")
fi	Finland
fr	France
gb	Great Britain (same as uk)
gr	Greece
hk	Hong Kong
hr	Croatia
hu	Hungary
ie	Republic of Ireland
il	Israel
in	India
is	Iceland
it	Italy
jp	Japan
kr	Korea (South)
kw	Kuwait

Table F-2.   *Geographical Top-Level Domains* (continued)

Domain	Meaning
li	Liechtenstein
lt	Lithuania
lu	Luxembourg
lv	Latvia
mx	Mexico
my	Malaysia
nl	Netherlands
no	Norway
nz	New Zealand
pl	Poland
pr	Puerto Rico (U.S.)
pt	Portugal
re	Reunion (Fr.)
se	Sweden
sg	Singapore
si	Slovenia
su	Soviet Union
th	Thailand
tn	Tunisia
tw	Taiwan
uk	United Kingdom (England, Scotland, Wales, N. Ireland)
us	United States
ve	Venezuela
yu	Yugoslavia
za	South Africa

Table F-2.   *Geographical Top-Level Domains* (continued)

APPENDIX G

List of Usenet Discussion Groups

This appendix contains a listing of all the non-local Usenet newsgroups as of May 1993. They are divided into three categories: mainstream, alternative, and Clarinet.

New groups are constantly being created and removed. However, most of the newsgroups, at least in the mainstream hierarchies, will not change. The alternative hierarchies are always adding new groups, and any master list will always be incomplete.

The master lists of current newsgroups are posted regularly to **news.lists**. (The newsgroup descriptions won't be as good as in this appendix, however, as we spent a lot of time rewriting to make them better.) For a discussion of Usenet and how to use it, see Chapters 9 through 15.

Mainstream Hierarchies

`comp.admin.policy`	Site administration policies
`comp.ai`	Artificial intelligence
`comp.ai.edu`	Applications of artificial intelligence to education
`comp.ai.fuzzy`	Fuzzy set theory, aka fuzzy logic
`comp.ai.genetic`	Using artificial intelligence for genetic research
`comp.ai.nat-lang`	Natural language processing by computers
`comp.ai.neural-nets`	All aspects of neural networks
`comp.ai.nlang-know-rep`	Natural language/knowledge representation (Moderated)
`comp.ai.philosophy`	Philosophical aspects of artificial intelligence
`comp.ai.shells`	Artificial intelligence applied to shells
`comp.ai.vision`	Artificial intelligence vision research (Moderated)
`comp.answers`	Repository for periodic Usenet articles (Moderated)
`comp.apps.spreadsheets`	Spreadsheets on various platforms
`comp.arch`	Computer architecture
`comp.arch.bus.vmebus`	Hardware and software for VMEbus systems
`comp.arch.storage`	Storage system issues, both hardware and software
`comp.archives`	Public access archives (Moderated)
`comp.archives.admin`	Computer archive administration
`comp.archives.msdos.announce`	Announcements about DOS archives (Moderated)
`comp.archives.msdos.d`	Discussion of materials available in DOS archives
`comp.bbs.misc`	General discussion of computer bulletin board systems
`comp.bbs.waffle`	Waffle BBS and Usenet system on all platforms
`comp.benchmarks`	Benchmarking techniques and results
`comp.binaries.acorn`	Binary-only postings for Acorn machines (Moderated)
`comp.binaries.amiga`	Encoded public domain programs in binary (Moderated)
`comp.binaries.apple2`	Binary-only postings for the Apple II computer

`comp.binaries.atari.st`	Binary-only postings for the Atari ST (Moderated)
`comp.binaries.ibm.pc`	Binary-only postings for IBM PC/DOS (Moderated)
`comp.binaries.ibm.pc.d`	Discussions of IBM PC binary postings
`comp.binaries.ibm.pc.wanted`	Requests for IBM PC and compatible programs
`comp.binaries.mac`	Encoded Macintosh programs in binary (Moderated)
`comp.binaries.ms-windows`	Binary programs for Microsoft Windows (Moderated)
`comp.binaries.os2`	Binaries for use under the OS/2 ABI (Moderated)
`comp.bugs.2bsd`	Reports of Unix version 2BSD-related bugs
`comp.bugs.4bsd`	Reports of Unix version 4BSD-related bugs
`comp.bugs.4bsd.ucb-fixes`	Bug reports/fixes for BSD Unix (Moderated)
`comp.bugs.misc`	General Unix bug reports and fixes (incl V7, Uucp)
`comp.bugs.sys5`	Reports of USG (System III, V, etc) bugs
`comp.cad.cadence`	Users of Cadence Design Systems products
`comp.client-server`	Client/server technology
`comp.cog-eng`	Cognitive engineering
`comp.compilers`	Compiler construction, theory, etc (Moderated)
`comp.compression`	Data compression algorithms and theory
`comp.compression.research`	Data compression research
`comp.databases`	Database and data management issues and theory
`comp.databases.informix`	Informix database management software discussions
`comp.databases.ingres`	Ingres products
`comp.databases.object`	Object-oriented paradigms in databases systems
`comp.databases.oracle`	SQL database products of the Oracle Corporation
`comp.databases.sybase`	Implementations of the SQL Server
`comp.databases.theory`	Advances in database technology
`comp.dcom.cell-relay`	Cell relay-based products
`comp.dcom.fax`	Fax hardware, software, and protocols
`comp.dcom.isdn`	The Integrated Services Digital Network (ISDN)
`comp.dcom.lans.ethernet`	Ethernet/IEEE 802.3 protocols
`comp.dcom.lans.fddi`	FDDI protocol suite
`comp.dcom.lans.hyperchannel`	Hyperchannel networks within an IP network
`comp.dcom.lans.misc`	Local area network hardware and software
`comp.dcom.modems`	Data communications hardware and software
`comp.dcom.servers`	Selecting and operating data communications servers
`comp.dcom.sys.cisco`	Cisco routers and bridges
`comp.dcom.sys.wellfleet`	Wellfleet bridge and router systems
`comp.dcom.telecom`	Telecommunications digest (Moderated)
`comp.doc`	Archived public-domain documentation (Moderated)
`comp.doc.techreports`	Lists of technical reports (Moderated)
`comp.dsp`	Digital signal processing using computers
`comp.editors`	Computerized text editing
`comp.edu`	Computer science education

`comp.edu.composition`	Writing instruction in computer-based classrooms
`comp.emacs`	Emacs editors of different flavors
`comp.fonts`	Typefonts: design, conversion, use, etc
`comp.graphics`	Computer graphics, art, animation, image processing
`comp.graphics.animation`	Technical aspects of computer animation
`comp.graphics.avs`	Application Visualization System
`comp.graphics.explorer`	Explorer Modular Visualization Environment (MVE)
`comp.graphics.gnuplot`	Gnuplot interactive function plotter
`comp.graphics.opengl`	OpenGL 3D application programming interface
`comp.graphics.research`	Highly technical computer graphics (Moderated)
`comp.graphics.visualization`	Scientific visualization
`comp.groupware`	Shared interactive environments
`comp.human-factors`	Human-computer interaction (HCI)
`comp.infosystems`	General discussion of about information systems
`comp.infosystems.gis`	Geographic Information Systems
`comp.infosystems.gopher`	The Internet Gopher (menu-driven information service)
`comp.infosystems.wais`	Wais full-text search service
`comp.internet.library`	Electronic libraries (Moderated)
`comp.ivideodisc`	Interactive videodiscs: uses, potential, etc
`comp.lang.ada`	Ada programming language
`comp.lang.apl`	APL programming language
`comp.lang.asm370`	IBM System/370 Assembly language
`comp.lang.c`	C programming language
`comp.lang.c++`	Object-oriented C++ programming language
`comp.lang.clos`	Common Lisp Object System
`comp.lang.clu`	CLU language and related topics
`comp.lang.dylan`	Dylan language
`comp.lang.eiffel`	Object-oriented Eiffel language
`comp.lang.forth`	Forth programming language
`comp.lang.forth.mac`	CSI MacForth programming environment
`comp.lang.fortran`	Fortran programming language
`comp.lang.functional`	Functional programming languages
`comp.lang.hermes`	Hermes language for distributed applications
`comp.lang.icon`	ICON programming language
`comp.lang.idl`	IDL (Interface Description Language)
`comp.lang.idl-pvwave`	IDL and PV-Wave language
`comp.lang.lisp`	Lisp programming language
`comp.lang.lisp.franz`	Franz Lisp programming language
`comp.lang.lisp.mcl`	Apple's Macintosh Common Lisp
`comp.lang.lisp.x`	XLISP language system
`comp.lang.logo`	Logo teaching and learning language
`comp.lang.misc`	Computer languages not specifically listed

comp.lang.ml	ML languages: ML, CAML, Lazy ML, etc (Moderated)
comp.lang.modula2	Modula-2 programming language
comp.lang.modula3	Modula-3 programming language
comp.lang.oberon	Oberon language and system
comp.lang.objective-c	Objective-C language and environment
comp.lang.pascal	Pascal programming language
comp.lang.perl	Perl scripting language
comp.lang.pop	Pop11 and the Plug user group
comp.lang.postscript	PostScript Page Description Language
comp.lang.prolog	Prolog programming language
comp.lang.rexx	IBM's REXX command-scripting language
comp.lang.scheme	Scheme programming language
comp.lang.scheme.c	Scheme language environment
comp.lang.sigplan	Info and announcements from ACM SIGPLAN (Moderated)
comp.lang.smalltalk	Smalltalk 80 programming language
comp.lang.tcl	TCL programming language and related tools
comp.lang.verilog	Verilog and PLI programming languages
comp.lang.vhdl	VHSIC Hardware Description Language, IEEE 1076/87
comp.lang.visual	Visual programming languages
comp.laser-printers	Laser printers, hardware and software (Moderated)
comp.lsi	Large scale integrated circuits
comp.lsi.cad	Computer-aided design
comp.lsi.testing	Testing of electronic circuits
comp.mail.elm	Discussion and fixes for Elm mail system
comp.mail.headers	Gatewayed from the Internet header-people list
comp.mail.maps	Various maps, including Uucp maps (Moderated)
comp.mail.mh	UCI version of the Rand Message Handling system
comp.mail.mime	Multipurpose Internet Mail Extensions of RFC 1341
comp.mail.misc	General discussions of computer mail
comp.mail.multi-media	Multimedia mail
comp.mail.mush	Mail User's Shell (MUSH)
comp.mail.sendmail	Configuring and using the BSD sendmail agent
comp.mail.uucp	Mail in the Uucp network environment
comp.misc	General computer topics not covered elsewhere
comp.multimedia	Interactive multimedia technologies of all kinds
comp.music	Applications of computers in music research
comp.newprod	Announcements of new products (Moderated)
comp.object	Object-oriented programming and languages
comp.org.acm	Association for Computing Machinery
comp.org.decus	Digital Equipment Computer Users' Society
comp.org.eff.news	Electronic Frontiers Foundation (Moderated)
comp.org.eff.talk	Discussion of EFF goals, strategies, etc

`comp.org.fidonet`	Official digest of FidoNet Assoc (Moderated)
`comp.org.ieee`	Issues and announcements about the IEEE
`comp.org.issnnet`	International Student Society for Neural Networks
`comp.org.sug`	Sun User's Group
`comp.org.usenix`	Usenix Association events and announcements
`comp.org.usenix.roomshare`	Finding lodging during Usenix conferences
`comp.os.386bsd.announce`	386bsd operating system (Moderated)
`comp.os.386bsd.apps`	Applications which run under 386bsd OS
`comp.os.386bsd.bugs`	Bugs and fixes for the 386bsd OS and its clients
`comp.os.386bsd.development`	Working on 386bsd OS internals
`comp.os.386bsd.misc`	386bsd OS topics not covered elsewhere
`comp.os.386bsd.questions`	General questions about 386bsd
`comp.os.aos`	Data General's AOS/VS
`comp.os.coherent`	Coherent operating system
`comp.os.cpm`	CP/M operating system
`comp.os.cpm.amethyst`	Amethyst, CP/M-80 software package
`comp.os.linux`	Linux, the free Unix clone for the 386/486
`comp.os.linux.announce`	Announcements for the Linux community (Moderated)
`comp.os.mach`	Mach OS from CMU and other places
`comp.os.minix`	Andy Tanenbaum's Minix system
`comp.os.misc`	OS-oriented discussion, not carried elsewhere
`comp.os.ms-windows.advocacy`	Debate about Microsoft Windows
`comp.os.ms-windows.announce`	Announcements relating to MS Windows (Moderated)
`comp.os.ms-windows.apps`	Applications in the MS Windows environment
`comp.os.ms-windows.misc`	General discussion of MS Windows issues
`comp.os.ms-windows.programmer.misc`	Programming MS Windows
`comp.os.ms-windows.programmer.tools`	Development tools for MS Windows
`comp.os.ms-windows.programmer.win32`	32-bit MS Windows programming interfaces
`comp.os.ms-windows.setup`	Installing and configuring MS Windows
`comp.os.msdos.4dos`	The 4DOS command processor for DOS
`comp.os.msdos.apps`	DOS applications
`comp.os.msdos.desqview`	QuarterDeck's Desqview and related products
`comp.os.msdos.misc`	DOS topics not covered elsewhere
`comp.os.msdos.pcgeos`	GeoWorks PC/GEOS and PC/GEOS-based packages
`comp.os.msdos.programmer`	DOS programming
`comp.os.os2.advocacy`	Debate about IBM's OS/2 operating system
`comp.os.os2.announce`	Announcements related to OS/2 (Moderated)
`comp.os.os2.apps`	OS/2 applications
`comp.os.os2.beta`	Beta releases of OS/2
`comp.os.os2.bugs`	Bug reports and fixes for OS/2
`comp.os.os2.misc`	OS/2 system topics not covered elsewhere
`comp.os.os2.multimedia`	Multimedia products and implementation on OS/2

`comp.os.os2.networking`	Networking in OS/2 environments
`comp.os.os2.programmer`	OS/2 programming
`comp.os.os2.programmer.misc`	More OS/2 programming
`comp.os.os2.programmer.porting`	Porting software to OS/2
`comp.os.os2.setup`	Setting up and configuring OS/2
`comp.os.os2.ver1x`	OS/2 versions 1.0 through 1.3
`comp.os.os9`	OS9 operating system
`comp.os.research`	Operating systems and related areas (Moderated)
`comp.os.rsts`	PDP-11 RSTS/E operating system
`comp.os.v`	The V distributed operating system from Stanford
`comp.os.vms`	DEC's VAX line of computers and VMS operating system
`comp.os.vxworks`	VxWorks real-time operating system
`comp.os.xinu`	Xinu operating system from Purdue (Doug Comer)
`comp.parallel`	Massively parallel hardware/software (Moderated)
`comp.parallel.pvm`	PVM system of multi-computer parallelization
`comp.patents`	Patents of computer technology (Moderated)
`comp.periphs`	Peripheral devices
`comp.periphs.printers`	Printers
`comp.periphs.scsi`	SCSI-based peripheral devices
`comp.programming`	Programming issues that transcend languages and OSs
`comp.protocols.appletalk`	Applebus hardware and software
`comp.protocols.ibm`	Networking with IBM mainframes
`comp.protocols.iso`	ISO protocol stack
`comp.protocols.iso.dev-environ`	ISO development environment
`comp.protocols.iso.x400`	X400 mail protocol
`comp.protocols.iso.x400.gateway`	X400 mail gateway (Moderated)
`comp.protocols.kerberos`	Kerberos authentication server
`comp.protocols.kermit`	Kermit communications package (Moderated)
`comp.protocols.misc`	Protocol topics not covered elsewhere
`comp.protocols.nfs`	Network File System protocol
`comp.protocols.pcnet`	PCNET (a personal computer network)
`comp.protocols.ppp`	Point to Point protocol
`comp.protocols.snmp`	Simple Network Management potocol
`comp.protocols.tcp-ip`	TCP and IP network protocols
`comp.protocols.tcp-ip.domains`	Domain-style names
`comp.protocols.tcp-ip.ibmpc`	TCP/IP for IBM-compatible computers
`comp.protocols.time.ntp`	Network Time protocol
`comp.realtime`	Real-time computing
`comp.research.japan`	Research in Japan (Moderated)
`comp.risks`	Risks to public from computers and users (Moderated)
`comp.robotics`	Robots and their applications
`comp.security.announce`	Announcements from CERT about security (Moderated)

`comp.security.misc`	Security issues about computers and networks
`comp.simulation`	Simulation methods, problems, uses (Moderated)
`comp.society`	Impact of technology on society (Moderated)
`comp.society.cu-digest`	The Computer Underground Digest (Moderated)
`comp.society.development`	Computer technology in developing countries
`comp.society.folklore`	Computer folklore, past and present (Moderated)
`comp.society.futures`	Events in technology affecting future computing
`comp.society.privacy`	Effects of technology on privacy (Moderated)
`comp.soft-sys.andrew`	Andrew file system from Carnegie Mellon University
`comp.soft-sys.khoros`	Khoros X11 visualization system
`comp.soft-sys.matlab`	MathWorks calculation and visualization package
`comp.soft-sys.nextstep`	Nextstep computing environment
`comp.soft-sys.shazam`	Shazam software
`comp.software-eng`	Software engineering and related topics
`comp.software.licensing`	Software licensing technology
`comp.sources.3b1`	Source code-only postings for AT&T 3b1 (Moderated)
`comp.sources.acorn`	Source code-only postings for Acorn (Moderated)
`comp.sources.amiga`	Source code-only postings for Amiga (Moderated)
`comp.sources.apple2`	Source code and discussion for AppleII (Moderated)
`comp.sources.atari.st`	Source code-only postings for Atari ST (Moderated)
`comp.sources.bugs`	Bug reports, fixes, discussion for posted sources
`comp.sources.d`	Discussion of source postings
`comp.sources.games`	Postings of recreational software (Moderated)
`comp.sources.games.bugs`	Bug reports and fixes for posted game software
`comp.sources.hp48`	Programs for HP48 and HP28 calculators (Moderated)
`comp.sources.mac`	Software for Apple Macintosh (Moderated)
`comp.sources.misc`	Software postings not covered elsewhere (Moderated)
`comp.sources.postscript`	Source code for Postscript programs (Moderated)
`comp.sources.reviewed`	Source code evaluated by peer review (Moderated)
`comp.sources.sun`	Software for Sun workstations (Moderated)
`comp.sources.testers`	Finding people to test software
`comp.sources.unix`	Complete, Unix-oriented sources (Moderated)
`comp.sources.wanted`	Requests for software and fixes
`comp.sources.x`	Software for the X Window system (Moderated)
`comp.specification`	Languages and methodologies for formal specification
`comp.specification.z`	Formal specification notation Z
`comp.speech`	Research/applications in speech science
`comp.std.announce`	Announcements about standards activities (Moderated)
`comp.std.c`	C language standards
`comp.std.c++`	C++ language standards
`comp.std.internat`	International standards
`comp.std.misc`	Standards topics not covered elsewhere

`comp.std.mumps`	X11.1 committee on Mumps (Moderated)
`comp.std.unix`	P1003 committee on Unix (Moderated)
`comp.sw.components`	Software components and related technology
`comp.sys.3b1`	AT&T 7300/3B1/UnixPC
`comp.sys.acorn`	Acorn and ARM-based computers
`comp.sys.acorn.advocacy`	Debate about Acorn computers
`comp.sys.acorn.announce`	Announcements for Acorn and ARM users (Moderated)
`comp.sys.acorn.tech`	Technical aspects of Acorn and ARM products
`comp.sys.alliant`	Alliant computers
`comp.sys.amiga.advocacy`	Debate about Amiga computers
`comp.sys.amiga.announce`	Announcements about Amiga (Moderated)
`comp.sys.amiga.applications`	Amiga applications not covered elsewhere
`comp.sys.amiga.audio`	Amiga music, MIDI, speech, other sounds
`comp.sys.amiga.datacomm`	Amiga data communications
`comp.sys.amiga.emulations`	Amiga hardware and software emulators
`comp.sys.amiga.games`	Amiga games
`comp.sys.amiga.graphics`	Amiga charts, graphs, pictures, etc
`comp.sys.amiga.hardware`	Amiga computer hardware
`comp.sys.amiga.introduction`	Newcomers to Amigas
`comp.sys.amiga.marketplace`	Buying and selling Amigas
`comp.sys.amiga.misc`	Amiga topics not covered elsewhere
`comp.sys.amiga.multimedia`	Animations, video, and multimedia
`comp.sys.amiga.programmer`	Developers and hobbyists
`comp.sys.amiga.reviews`	Reviews of Amiga software and hardware (Moderated)
`comp.sys.apollo`	Apollo computer systems
`comp.sys.apple2`	Apple II computers
`comp.sys.apple2.gno`	Apple IIgs GNO multitasking environment
`comp.sys.atari.8bit`	8-bit Atari micros
`comp.sys.atari.advocacy`	Debate about Atari computers
`comp.sys.atari.st`	16-bit Atari micros
`comp.sys.atari.st.tech`	Technical discussions of Atari ST
`comp.sys.att`	AT&T microcomputers
`comp.sys.cbm`	Commodore computers
`comp.sys.cdc`	Control Data Corporation computers
`comp.sys.concurrent`	Concurrent/Masscomp computers (Moderated)
`comp.sys.convex`	Convex computers
`comp.sys.dec`	DEC computers
`comp.sys.dec.micro`	DEC micros (Rainbow, Professional 350/380)
`comp.sys.encore`	Encore's MultiMax computers
`comp.sys.handhelds`	Handheld computers and programmable calculators
`comp.sys.hp`	Hewlett-Packard equipment
`comp.sys.hp48`	Hewlett-Packard's HP48 and HP28 calculators

`comp.sys.ibm.pc.demos`	Demo programs that showcase programmer skill
`comp.sys.ibm.pc.digest`	IBM-compatible PCs (Moderated)
`comp.sys.ibm.pc.games`	Games for PCs
`comp.sys.ibm.pc.games.action`	Action games
`comp.sys.ibm.pc.games.adventure`	Adventure games
`comp.sys.ibm.pc.games.announce`	Announcements about PC game groups
`comp.sys.ibm.pc.games.flight-sim`	PC flight simulators
`comp.sys.ibm.pc.games.misc`	Games not covered elsewhere
`comp.sys.ibm.pc.games.rpg`	Roleplaying games on PCs
`comp.sys.ibm.pc.games.strategic`	Strategy games
`comp.sys.ibm.pc.hardware`	PC hardware, any vendor
`comp.sys.ibm.pc.misc`	PC topics not covered elsewhere
`comp.sys.ibm.pc.rt`	IBM's RT computer
`comp.sys.ibm.pc.soundcard`	Hardware and software aspects of PC sound cards
`comp.sys.ibm.ps2.hardware`	PS/2 and Microchannel hardware, any vendor
`comp.sys.intel`	Intel systems and parts
`comp.sys.intel.ipsc310`	Intel 310
`comp.sys.isis`	ISIS distributed system from Cornell
`comp.sys.laptops`	Laptop computers
`comp.sys.m6809`	6809 processors
`comp.sys.m68k`	68000 processors
`comp.sys.m68k.pc`	68000-based computers (Moderated)
`comp.sys.m88k`	88000-based computers
`comp.sys.mac.advocacy`	Debate about Macintosh computers
`comp.sys.mac.announce`	Important notices for Macintosh users (Moderated)
`comp.sys.mac.apps`	Macintosh applications
`comp.sys.mac.comm`	Macintosh communications
`comp.sys.mac.databases`	Macintosh database systems
`comp.sys.mac.digest`	Macintosh general talk, no programs (Moderated)
`comp.sys.mac.games`	Macintosh games
`comp.sys.mac.hardware`	Macintosh hardware
`comp.sys.mac.hypercard`	Macintosh Hypercard
`comp.sys.mac.misc`	Macintosh topics not covered elsewhere
`comp.sys.mac.oop.macapp3`	Version 3 of the MacApp object-oriented system
`comp.sys.mac.oop.misc`	Macintosh object-oriented programming
`comp.sys.mac.oop.tcl`	Programming the Macintosh with Think Class Libraries
`comp.sys.mac.portables`	Laptop Macintoshes
`comp.sys.mac.programmer`	Programming the Macintosh
`comp.sys.mac.system`	Macintosh system software
`comp.sys.mac.wanted`	Requests for Macintosh-related software/hardware
`comp.sys.mentor`	Mentor Graphics products and Silicon Compiler System
`comp.sys.mips`	Systems based on MIPS chips

`comp.sys.misc`	Computer topics not covered elsewhere
`comp.sys.ncr`	NCR computers
`comp.sys.next.advocacy`	Debate about Next computers
`comp.sys.next.announce`	Announcements about Next computer system (Moderated)
`comp.sys.next.bugs`	Discussion and solutions of known Next bugs
`comp.sys.next.hardware`	The physical aspects of Next computers
`comp.sys.next.marketplace`	Next hardware, software and jobs
`comp.sys.next.misc`	Next topics not covered elsewhere
`comp.sys.next.programmer`	Next-related programming issues
`comp.sys.next.software`	Next computer programs
`comp.sys.next.sysadmin`	Next system administration
`comp.sys.northstar`	Northstar microcomputer users
`comp.sys.novell`	Novell Netware products
`comp.sys.nsc.32k`	National Semiconductor 32000 series chips
`comp.sys.palmtops`	Super-powered calculators for the palm of your hand
`comp.sys.pen`	Interacting with computers through pen gestures
`comp.sys.prime`	Prime Computer products
`comp.sys.proteon`	Proteon gateway products
`comp.sys.pyramid`	Pyramid 90x computers
`comp.sys.ridge`	Ridge 32 computers and ROS
`comp.sys.sequent`	Sequent systems (Balance and Symmetry)
`comp.sys.sgi.admin`	Silicon Graphics's Irises system administration
`comp.sys.sgi.announce`	Announcements for the SGI community (Moderated)
`comp.sys.sgi.apps`	Iris applications
`comp.sys.sgi.bugs`	Bugs in the IRIX operating system
`comp.sys.sgi.graphics`	Graphics packages on SGI machines
`comp.sys.sgi.hardware`	Base systems and peripherals for Iris computers
`comp.sys.sgi.misc`	Silicon Graphics topics not covered elsewhere
`comp.sys.stratus`	Stratus products (System/88, CPS-32, VOS and FTX)
`comp.sys.sun.admin`	Sun system administration
`comp.sys.sun.announce`	Sun announcements and Sunergy mailings (Moderated)
`comp.sys.sun.apps`	Sun applications
`comp.sys.sun.hardware`	Sun hardware
`comp.sys.sun.misc`	Sun topics not covered elsewhere
`comp.sys.sun.wanted`	Requests for Sun products and support
`comp.sys.super`	Supercomputers
`comp.sys.tahoe`	CCI 6/32, Harris HCX/7, and Sperry 7000 computers
`comp.sys.tandy`	Tandy computers, new and old
`comp.sys.ti`	Texas Instruments products
`comp.sys.ti.explorer`	Texas Instruments Explorer
`comp.sys.transputer`	Transputer computer and OCCAM language
`comp.sys.unisys`	Sperry, Burroughs, Convergent and Unisys systems

`comp.sys.xerox`	Xerox 1100 workstations and protocols
`comp.sys.zenith`	Heath terminals and related Zenith products
`comp.sys.zenith.z100`	Zenith Z-100 (Heath H-100) family of computers
`comp.terminals`	All sorts of terminals
`comp.terminals.bitgraph`	BB&N BitGraph terminal
`comp.terminals.tty5620`	AT&T dot mapped display terminals (5620 and BLIT)
`comp.text`	Text processing
`comp.text.desktop`	Desktop publishing
`comp.text.frame`	Desktop publishing with FrameMaker
`comp.text.interleaf`	Interleaf software
`comp.text.sgml`	ISO 8879 SGML, structured documents, markup lang.
`comp.text.tex`	TeX and LaTeX systems and macros
`comp.theory`	Theoretical computer science
`comp.theory.cell-automata`	Cellular automata
`comp.theory.dynamic-sys`	Ergodic theory and dynamical systems
`comp.theory.info-retrieval`	Information retrieval topics (Moderated)
`comp.theory.self-org-sys`	Self organization topics
`comp.unix.admin`	Administering a Unix-based system
`comp.unix.aix`	IBM's version of Unix
`comp.unix.amiga`	Minix, SYSV4 and other Unix on an Amiga
`comp.unix.aux`	Unix for Macintosh computers
`comp.unix.bsd`	Berkeley Software Distribution Unix
`comp.unix.cray`	Cray computers and their operating systems
`comp.unix.dos-under-unix`	DOS running under Unix by whatever means
`comp.unix.internals`	Hacking Unix internals
`comp.unix.large`	Unix on mainframes and in large networks
`comp.unix.misc`	Unix topics not covered elsewhere
`comp.unix.osf.misc`	Open Software Foundation
`comp.unix.osf.osf1`	Open Software Foundation's OSF/1 operating system
`comp.unix.pc-clone.16bit`	Unix on 16-bit PC architectures
`comp.unix.pc-clone.32bit`	Unix on 23-bit PC architectures
`comp.unix.programmer`	Questions and answers regarding Unix programming
`comp.unix.questions`	Question and answer forum for Unix beginners
`comp.unix.shell`	Unix shells
`comp.unix.solaris`	Solaris operating system
`comp.unix.sys3`	System III Unix
`comp.unix.sys5.misc`	Versions of System V which predate Release 3
`comp.unix.sys5.r3`	System V Release 3
`comp.unix.sys5.r4`	System V Release 4
`comp.unix.ultrix`	DEC's Ultrix
`comp.unix.wizards`	Questions for true Unix wizards only
`comp.unix.xenix.misc`	Non-SCO Xenix

comp.unix.xenix.sco	Santa Cruz Operation (SCO) Xenix
comp.virus	Computer viruses and security (Moderated)
comp.windows.interviews	InterViews object-oriented windowing system
comp.windows.misc	Windowing systems topics not covered elsewhere
comp.windows.news	Sun Microsystems' NeWS window system
comp.windows.open-look	Open Look graphical user interface
comp.windows.x	General discussion of X Window
comp.windows.x.announce	X Consortium announcements (Moderated)
comp.windows.x.apps	Getting and using (not programming) X applications
comp.windows.x.i386unix	XFree86 window system and others
comp.windows.x.intrinsics	X toolkit
comp.windows.x.motif	Motif graphical user interface
comp.windows.x.pex	PHIGS extension of the X Window System
misc.activism.progressive	Progressive activism (Moderated)
misc.answers	FAQ lists and other periodic postings (Moderated)
misc.books.technical	Books about technical topics (including computers)
misc.consumers	Consumer interests, product reviews, etc
misc.consumers.house	Owning and maintaining a house
misc.education	The educational system
misc.emerg-services	Paramedics and other first responders
misc.entrepreneurs	Operating your own business
misc.fitness	Physical fitness, exercise
misc.forsale	Short postings about items for sale
misc.forsale.computers.d	Discussion of misc.forsale.computers.* topics
misc.forsale.computers.mac	Macintosh-related computer items (no discussion)
misc.forsale.computers.other	Selling miscellaneous computer items (no discussion)
misc.forsale.computers.pc-clone	PC-related computer items (no discussion)
misc.forsale.computers.workstation	Workstation-related computer items (no discussion)
misc.handicap	Issues about the handicapped (Moderated)
misc.headlines	Current events
misc.health.alternative	Alternative health care
misc.health.diabetes	Diabetes
misc.int-property	Intellectual property rights
misc.invest	Investments and the handling of money
misc.invest.real-estate	Property investments
misc.invest.technical	Highly technical discussion of investment strategy
misc.jobs.contract	Contract labor
misc.jobs.misc	Employment, workplaces, careers
misc.jobs.offered	Announcements of positions available
misc.jobs.offered.entry	Job listings, entry-level positions only
misc.jobs.resumes	Postings of resumes and jobs wanted

`misc.kids`	Children, their behavior and activities
`misc.kids.computer`	Use of computers by children
`misc.legal`	Legalities and the ethics of law
`misc.legal.computing`	Legal climate of the computing world
`misc.misc`	Discussions of topics not covered elsewhere
`misc.news.east-europe.rferl`	Radio Free Europe/Radio Liberty (Moderated)
`misc.news.southasia`	News from Southeast Asia (Moderated)
`misc.rural`	Rural living
`misc.taxes`	Tax laws and advice
`misc.test`	Place to send test articles
`misc.wanted`	Requests for things (NOT software)
`misc.writing`	Discussion of writing in all its forms
`news.admin.misc`	General network news administration
`news.admin.policy`	Policy issues of Usenet
`news.admin.technical`	Technical aspects of Usenet (Moderated)
`news.announce.conferences`	Calls for papers and conference notices (Moderated)
`news.announce.important`	General announcements of interest to all (Moderated)
`news.announce.newgroups`	Calls for new groups (Moderated)
`news.announce.newusers`	Explanatory postings for new users (Moderated)
`news.answers`	Repository for periodic Usenet articles (Moderated)
`news.config`	Postings of system down times and interruptions
`news.future`	Future technology of network news systems
`news.groups`	Discussions and lists of newsgroups
`news.lists`	Usenet statistics and lists (Moderated)
`news.lists.ps-maps`	Maps relating to Usenet traffic flows (Moderated)
`news.misc`	Discussions of Usenet itself
`news.newsites`	Postings of new site announcements
`news.newusers.questions`	Questions and answers for new Usenet users
`news.software.anu-news`	VMS B-news software from Australian National Univ
`news.software.b`	B-news-compatible software
`news.software.nn`	The **nn** newsreader
`news.software.nntp`	Network News Transfer Protocol
`news.software.notes`	Notesfile software from the Univ of Illinois
`news.software.readers`	Usenet newsreader programs topics
`rec.answers`	Repository for periodic Usenet articles (Moderated)
`rec.antiques`	Antiques and vintage items
`rec.aquaria`	Fish and aquariums as a hobby
`rec.arts.animation`	Various kinds of animation
`rec.arts.anime`	Japanese animation
`rec.arts.anime.info`	The art of animation
`rec.arts.anime.marketplace`	Making money with animation

`rec.arts.anime.stories`	Animated stories
`rec.arts.bodyart`	Tattoos and body decoration
`rec.arts.bonsai`	Miniature trees and shrubbery
`rec.arts.books`	Books of all genres, and the publishing industry
`rec.arts.books.tolkien`	Works of J.R.R. Tolkien
`rec.arts.cinema`	Art of the cinema (Moderated)
`rec.arts.comics.info`	Reviews, conventions, other comics news (Moderated)
`rec.arts.comics.marketplace`	Exchange of comics and related items
`rec.arts.comics.misc`	Comic books, graphic novels, sequential art
`rec.arts.comics.strips`	Comic strips
`rec.arts.comics.xbooks`	Mutant Universe of Marvel Comics
`rec.arts.dance`	Dance topics not covered elsewhere
`rec.arts.disney`	Any Disney-related subjects
`rec.arts.drwho`	Dr. Who
`rec.arts.erotica`	Erotic fiction and verse (Moderated)
`rec.arts.fine`	Fine arts and artists
`rec.arts.int-fiction`	Interactive fiction
`rec.arts.manga`	Japanese storytelling art form
`rec.arts.marching.drumcorps`	Drum and bugle corps
`rec.arts.marching.misc`	Marching-related performance activities
`rec.arts.misc`	Arts topics not covered elsewhere
`rec.arts.movies`	Movies and movie making
`rec.arts.movies.reviews`	Movie reviews (Moderated)
`rec.arts.poems`	For the posting of poems
`rec.arts.sf.announce`	Major science fiction announcements (Moderated)
`rec.arts.sf.fandom`	SF fan activities
`rec.arts.sf.marketplace`	Personal for sale notices of SF materials
`rec.arts.sf.misc`	SF topics not covered elsewhere
`rec.arts.sf.movies`	SF motion pictures
`rec.arts.sf.reviews`	Reviews of SF/fantasy/horror works (Moderated)
`rec.arts.sf.science`	Real and speculative aspects of SF science
`rec.arts.sf.starwars`	Star Wars universe (not SDI)
`rec.arts.sf.tv`	General television SF
`rec.arts.sf.written`	Written science fiction and fantasy
`rec.arts.startrek.current`	New Star Trek shows, movies and books
`rec.arts.startrek.fandom`	Star Trek conventions and memorabilia
`rec.arts.startrek.info`	The universe of Star Trek (Moderated)
`rec.arts.startrek.misc`	Star Trek topics not covered elsewhere
`rec.arts.startrek.reviews`	Reviews of Star Trek books, shows, films (Moderated)
`rec.arts.startrek.tech`	Star Trek's depiction of future technologies
`rec.arts.theatre`	All aspects of stage work and theatre
`rec.arts.tv`	Television: history, past and current shows

`rec.arts.tv.soaps`	Soap operas
`rec.arts.tv.uk`	Telly shows from the UK
`rec.arts.wobegon`	Literary and music esoterica
`rec.audio`	High fidelity audio
`rec.audio.car`	Automobile audio systems
`rec.audio.high-end`	High-end audio systems (Moderated)
`rec.audio.pro`	Professional audio recording and studio engineering
`rec.autos`	Automobiles, automotive products and laws
`rec.autos.antique`	Automobiles over 25 years old
`rec.autos.driving`	Driving automobiles
`rec.autos.sport`	Organized, legal auto competitions
`rec.autos.tech`	Technical aspects of automobiles
`rec.autos.vw`	Volkswagen products
`rec.aviation.announce`	Events for the aviation community (Moderated)
`rec.aviation.answers`	Questions and answers about aviation (Moderated)
`rec.aviation.homebuilt`	Selecting, designing, building, restoring aircraft
`rec.aviation.ifr`	Flying under Instrument Flight Rules
`rec.aviation.military`	Military aircraft of the past, present and future
`rec.aviation.misc`	Aviation topics not covered elsewhere
`rec.aviation.owning`	Owning airplanes
`rec.aviation.piloting`	General discussion for pilots
`rec.aviation.products`	Products useful to pilots
`rec.aviation.simulators`	Flight simulation on all levels
`rec.aviation.soaring`	Sailplanes and hang-gliders
`rec.aviation.stories`	Anecdotes of flight experiences (Moderated)
`rec.aviation.student`	Learning to fly
`rec.backcountry`	Outdoor activities
`rec.bicycles.marketplace`	Buying, selling and reviewing cycling items
`rec.bicycles.misc`	Bicycling topics not covered elsewhere
`rec.bicycles.racing`	Bicycle racing techniques, rules and results
`rec.bicycles.rides`	Cycling tours, training, commuting routes
`rec.bicycles.soc`	Societal issues of bicycling
`rec.bicycles.tech`	Cycling product design, construction, maintenance
`rec.birds`	Bird watching
`rec.boats`	Boating
`rec.boats.paddle`	Any boats with oars, paddles, etc
`rec.climbing`	Climbing techniques, announcements, etc
`rec.collecting`	Collecting topics not covered elsewhere
`rec.collecting.cards`	Collecting sport and non-sport cards
`rec.crafts.brewing`	Making beers and meads
`rec.crafts.metalworking`	Working with metal
`rec.crafts.misc`	Craft topics not covered elsewhere

`rec.crafts.textiles`	Sewing, weaving, knitting and other fiber arts
`rec.equestrian`	Horses and riding
`rec.folk-dancing`	Folk dances, dancers, and dancing
`rec.food.cooking`	Food, cooking, cookbooks, and recipes
`rec.food.drink`	Wines and spirits
`rec.food.historic`	History of food preparation
`rec.food.recipes`	Interesting recipes for food and drink (Moderated)
`rec.food.restaurants`	Dining out
`rec.food.sourdough`	Baking with sourdough
`rec.food.veg`	Vegetarian food and recipes
`rec.gambling`	Games of chance and betting
`rec.games.abstract`	Perfect information, pure strategy games
`rec.games.backgammon`	Backgammon
`rec.games.board`	Discussion and hints on board games
`rec.games.board.ce`	Cosmic Encounter board game
`rec.games.bridge`	Bridge card game
`rec.games.chess`	Chess and computer chess
`rec.games.corewar`	Core War computer game
`rec.games.design`	Game design-related issues
`rec.games.diplomacy`	Diplomacy conquest game
`rec.games.empire`	Empire game
`rec.games.frp.advocacy`	Debate about various roleplaying systems
`rec.games.frp.announce`	Announcements in the roleplaying world (Moderated)
`rec.games.frp.archives`	Archivable fantasy stories, etc (Moderated)
`rec.games.frp.cyber`	Cyberpunk roleplaying games
`rec.games.frp.dnd`	Fantasy roleplaying with TSR's Dungeons and Dragons
`rec.games.frp.marketplace`	Roleplaying game materials wanted and for sale
`rec.games.frp.misc`	Roleplaying game topics not covered elsewhere
`rec.games.go`	The Go game
`rec.games.hack`	The Hack game
`rec.games.int-fiction`	Interactive fiction games
`rec.games.mecha`	Giant robot games
`rec.games.miniatures`	Tabletop wargaming
`rec.games.misc`	Games and computer games
`rec.games.moria`	The Moria game
`rec.games.mud.admin`	Administrative issues of multiuser dimensions
`rec.games.mud.announce`	Announcements about multiuser dimensions (Moderated)
`rec.games.mud.diku`	DikuMuds
`rec.games.mud.lp`	LPMUD computer roleplaying game
`rec.games.mud.misc`	Mud topics not covered elsewhere
`rec.games.mud.tiny`	Tiny muds, like MUSH, MUSE and MOO
`rec.games.netrek`	X window system game Netrek (XtrekII)

`rec.games.pbm`	Play by mail games
`rec.games.pinball`	Pinball games
`rec.games.programmer`	Adventure game programming
`rec.games.rogue`	Rogue game
`rec.games.trivia`	Trivia
`rec.games.vectrex`	Vectrex game system
`rec.games.video`	Video games
`rec.games.video.arcade`	Coin-operated video games
`rec.games.video.classic`	Older home video entertainment systems
`rec.games.video.marketplace`	Home video game stuff for sale or trade
`rec.games.video.misc`	Home video game topics not covered elsewhere
`rec.games.video.nintendo`	Nintendo video game systems and software
`rec.games.video.sega`	Sega video game systems and software
`rec.games.xtank.play`	Distributed game Xtank
`rec.games.xtank.programmer`	Coding the Xtank game and its robots
`rec.gardens`	Gardening, methods and results
`rec.guns`	Firearms (Moderated)
`rec.heraldry`	Coats of arms
`rec.humor`	Jokes (may be offensive)
`rec.humor.d`	Discussions on the content of rec.humor articles
`rec.humor.funny`	Jokes that a moderator thinks are funny (Moderated)
`rec.humor.oracle`	Sagacious advice from the Usenet Oracle (Moderated)
`rec.humor.oracle.d`	Comments about the Usenet Oracle
`rec.hunting`	Hunting (Moderated)
`rec.juggling`	Juggling techniques, equipment and events
`rec.kites`	Kites and kiting
`rec.mag`	Magazine summaries, tables of contents, etc
`rec.mag.fsfnet`	A science fiction fanzine (Moderated)
`rec.martial-arts`	Martial arts
`rec.misc`	General topics about recreational activities
`rec.models.railroad`	Model railroads of all scales
`rec.models.rc`	Radio-controlled models
`rec.models.rockets`	Model rockets
`rec.models.scale`	Construction of models
`rec.motorcycles`	Motorcycles, related products and laws
`rec.motorcycles.dirt`	Motorcycles and ATVs off-road
`rec.motorcycles.harley`	Harley-Davidson motorcycles
`rec.motorcycles.racing`	Racing motorcycles
`rec.music.afro-latin`	Music with African and Latin influences
`rec.music.beatles`	The Beatles
`rec.music.bluenote`	Jazz, blues, and related types of music
`rec.music.cd`	Music and CDs

`rec.music.christian`	Christian music, both contemporary and traditional
`rec.music.classical`	Classical music
`rec.music.classical.guitar`	Classical guitar music
`rec.music.compose`	Creating musical and lyrical works
`rec.music.country.western`	Country and western music
`rec.music.dementia`	Comedy and novelty music
`rec.music.dylan`	Music of Bob Dylan
`rec.music.early`	Pre-classical European music
`rec.music.folk`	Folk music
`rec.music.funky`	Funk, rap, hip-hop, house, soul, R&B, etc
`rec.music.gaffa`	Kate Bush and other alternative music (Moderated)
`rec.music.gdead`	Music of the Grateful Dead
`rec.music.indian.classical`	Hindustani and Carnatic Indian classical music
`rec.music.indian.misc`	Indian music in general
`rec.music.industrial`	Industrial-related music styles
`rec.music.info`	News on musical topics (Moderated)
`rec.music.makers`	Performers and their discussions
`rec.music.makers.bass`	Upright bass and bass guitar techniques, equipment
`rec.music.makers.guitar`	Electric and acoustic guitar techniques, equipment
`rec.music.makers.guitar.tablature`	Guitar tablature and chords
`rec.music.makers.percussion`	Drum, other percussion techniques and equipment
`rec.music.makers.synth`	Synthesizers and computer music
`rec.music.marketplace`	Records, tapes, and CDs: wanted, for sale, etc
`rec.music.misc`	Music topics not covered elsewhere
`rec.music.newage`	New Age music
`rec.music.phish`	Music of Phish
`rec.music.reggae`	Reggae music
`rec.music.reviews`	Reviews of all types of music (Moderated)
`rec.music.video`	Music videos and music video software
`rec.nude`	Naturist and nudist activities
`rec.org.mensa`	Mensa high IQ society
`rec.org.sca`	Society for Creative Anachronism
`rec.outdoors.fishing`	Sport and commercial fishing
`rec.pets`	Pets, pet care, and household animals in general
`rec.pets.birds`	Indoor birds
`rec.pets.cats`	Domestic cats
`rec.pets.dogs`	Dogs as pets
`rec.pets.herp`	Reptiles, amphibians and other exotic vivarium pets
`rec.photo`	Photography
`rec.puzzles`	Puzzles, problems, and quizzes
`rec.puzzles.crosswords`	Making and playing gridded word puzzles
`rec.pyrotechnics`	Fireworks, rocketry, safety, and other topics

rec.radio.amateur.misc	Amateur radio practices, contests, events, etc
rec.radio.amateur.packet	Packet radio setups
rec.radio.amateur.policy	Radio use and regulation policy
rec.radio.broadcasting	Local area broadcast radio (Moderated)
rec.radio.cb	Citizen band radio
rec.radio.info	Informative postings related to radio (Moderated)
rec.radio.noncomm	Noncommercial radio
rec.radio.shortwave	Shortwave radio
rec.radio.swap	Trading and swapping radio equipment
rec.railroad	Real and model trains
rec.roller-coaster	Roller coasters and other amusement park rides
rec.running	Running for enjoyment, sport, exercise, etc
rec.scouting	Scouting youth organizations worldwide
rec.scuba	Scuba diving
rec.skate	Ice skating and roller skating
rec.skiing	Snow skiing
rec.skydiving	Skydiving
rec.sport.baseball	Baseball
rec.sport.baseball.college	Baseball on the collegiate level
rec.sport.baseball.fantasy	Rotisserie (fantasy) baseball play
rec.sport.basketball.college	Basketball on the collegiate level
rec.sport.basketball.misc	Basketball topics not covered elsewhere
rec.sport.basketball.pro	Professional basketball
rec.sport.cricket	Cricket
rec.sport.cricket.scores	Scores from cricket matches (Moderated)
rec.sport.disc	Flying disc-based sports
rec.sport.fencing	All aspects of swordplay
rec.sport.football.australian	Australian rules football
rec.sport.football.college	American-style college football
rec.sport.football.misc	General discussion of American-style football
rec.sport.football.pro	American-style professional football
rec.sport.golf	Golf
rec.sport.hockey	Ice hockey
rec.sport.hockey.field	Field hockey
rec.sport.misc	Spectator sports
rec.sport.olympics	Olympic Games
rec.sport.paintball	Survival game paintball
rec.sport.pro-wrestling	Professional wrestling
rec.sport.rowing	Crew for competition or fitness
rec.sport.rugby	Rugby
rec.sport.soccer	Soccer (Association football)
rec.sport.swimming	Training for and competing in swimming events

`rec.sport.table-tennis`	Ping-Pong
`rec.sport.tennis`	Tennis
`rec.sport.triathlon`	Multi-event sports
`rec.sport.volleyball`	Volleyball
`rec.travel`	Traveling anywhere in the world
`rec.travel.air`	Airline travel anywhere in the world
`rec.travel.marketplace`	Tickets and accommodations wanted and for sale
`rec.video`	Video and video components
`rec.video.cable-tv`	Technical and regulatory issues of cable television
`rec.video.production`	Professional-quality video productions
`rec.video.releases`	Prerecorded video releases, laserdisc and videotape
`rec.video.satellite`	Receiving video via satellite
`rec.windsurfing`	Wind surfing
`rec.woodworking`	Woodworking
`sci.aeronautics`	Aeronautics and related technology
`sci.aeronautics.airliners`	Airliner technology (Moderated)
`sci.answers`	Repository for periodic Usenet articles (Moderated)
`sci.anthropology`	All aspects of studying humankind
`sci.aquaria`	Scientifically oriented postings about aquaria
`sci.archaeology`	Studying antiquities of the world
`sci.astro`	Astronomy
`sci.astro.fits`	Flexible Image Transport System
`sci.astro.hubble`	Hubble space telescope data (Moderated)
`sci.bio`	Biology and related sciences
`sci.bio.technology`	Biotechnology
`sci.chem`	Chemistry and related sciences
`sci.chem.organomet`	Organometallic chemistry
`sci.classics`	Classical history, languages, art and more
`sci.cognitive`	Perception, memory, judgment and reasoning
`sci.comp-aided`	Computers as tools in scientific research
`sci.cryonics`	Biostasis, suspended animation, etc
`sci.crypt`	Data encryption and decryption
`sci.econ`	Economics
`sci.edu`	Education
`sci.electronics`	Circuits, theory, electrons, etc
`sci.energy`	Energy, science and technology
`sci.engr`	Technical discussions about engineering tasks
`sci.engr.biomed`	Biomedical engineering
`sci.engr.chem`	Chemical engineering
`sci.engr.civil`	Civil engineering
`sci.engr.control`	Control systems

`sci.engr.mech`	Mechanical engineering
`sci.environment`	Environment and ecology
`sci.fractals`	Objects of non-integral dimension and other chaos
`sci.geo.fluids`	Geophysical fluid dynamics
`sci.geo.geology`	Solid earth sciences
`sci.geo.meteorology`	Meteorology and related topics
`sci.image.processing`	Scientific image processing and analysis
`sci.lang`	Natural languages, communication, etc
`sci.lang.japan`	Japanese language, both spoken and written
`sci.life-extension`	Discussions about living longer
`sci.logic`	Logic: math, philosophy and computational aspects
`sci.materials`	Materials engineering
`sci.math`	Mathematics topics not covered elsewhere
`sci.math.num-analysis`	Numerical analysis
`sci.math.research`	Current mathematical research (Moderated)
`sci.math.stat`	Statistics
`sci.math.symbolic`	Symbolic algebra
`sci.med`	Medicine, related products and regulations
`sci.med.aids`	AIDS and HIV virus (Moderated)
`sci.med.dentistry`	Dentistry and teeth
`sci.med.nutrition`	Physiological aspects of diet and eating
`sci.med.occupational`	Occupational injuries
`sci.med.physics`	Physics in medical testing and care
`sci.med.telemedicine`	Clinical consulting through computer networks
`sci.military`	Science and the military (Moderated)
`sci.misc`	Short-lived discussions on subjects in the sciences
`sci.nanotech`	Self-reproducing molecular-size machines (Moderated)
`sci.optics`	Optics
`sci.philosophy.meta`	Metaphilosophy
`sci.philosophy.tech`	Technical philosophy: math, science, logic, etc
`sci.physics`	Physics
`sci.physics.fusion`	Fusion
`sci.physics.research`	Current physics research (Moderated)
`sci.psychology`	Psychology
`sci.psychology.digest`	Psycoloquy, a refereed psychology journal (Moderated)
`sci.research`	Research methods, funding, ethics, etc
`sci.research.careers`	Careers in scientific research
`sci.skeptic`	Skeptics discussing pseudo-science
`sci.space`	Space, space programs, space-related research, etc
`sci.space.news`	Space-related news items (Moderated)
`sci.space.shuttle`	Space shuttle and the STS program
`sci.systems`	Theory and application of systems science

`sci.virtual-worlds`	Modelling the universe (Moderated)
`sci.virtual-worlds.apps`	Virtual worlds technology (Moderated)
`soc.answers`	Repository for periodic Usenet articles (Moderated)
`soc.bi`	Bisexuality
`soc.college`	College activities
`soc.college.grad`	Graduate schools
`soc.college.gradinfo`	Information about graduate schools
`soc.couples`	Discussions about couples
`soc.culture.afghanistan`	Afghanistan
`soc.culture.african`	Africa
`soc.culture.african.american`	African-American issues
`soc.culture.arabic`	Technological and cultural issues (not politics)
`soc.culture.asean`	Countries of the Association of SE Asian Nations
`soc.culture.asian.american`	Asian-American issues
`soc.culture.australian`	Australia
`soc.culture.baltics`	Baltic states
`soc.culture.bangladesh`	Bangladesh
`soc.culture.bosna-herzgvna`	Bosnia and Herzegovina
`soc.culture.brazil`	Brazil
`soc.culture.british`	Great Britain
`soc.culture.bulgaria`	Bulgaria
`soc.culture.canada`	Canada
`soc.culture.caribbean`	The Caribbean
`soc.culture.celtic`	Irish, Scottish, Breton, Cornish, Manx and Welsh
`soc.culture.china`	China
`soc.culture.croatia`	Croatia
`soc.culture.czecho-slovak`	Bohemian, Slovak, Moravian and Silesian life
`soc.culture.esperanto`	Esperanto, the neutral international language
`soc.culture.europe`	All-European society
`soc.culture.filipino`	The Philippines
`soc.culture.french`	France
`soc.culture.german`	Germany
`soc.culture.greek`	Greece
`soc.culture.hongkong`	Hong Kong
`soc.culture.indian`	India
`soc.culture.indian.telugu`	Telugu people of India
`soc.culture.indonesia`	Indonesia
`soc.culture.iranian`	Iran and things Persian
`soc.culture.italian`	Italy
`soc.culture.japan`	Japan, except the Japanese language
`soc.culture.jewish`	Jewish culture and religion

`soc.culture.korean`	Korea
`soc.culture.latin-america`	Latin America
`soc.culture.lebanon`	Lebanon
`soc.culture.magyar`	Hungary
`soc.culture.malaysia`	Malaysia
`soc.culture.mexican`	Mexico
`soc.culture.misc`	Cultural topics not covered elsewhere
`soc.culture.native`	Indigenous people around the world
`soc.culture.nepal`	Nepal
`soc.culture.netherlands`	The Netherlands and Belgium
`soc.culture.new-zealand`	New Zealand
`soc.culture.nordic`	Scandinavia and Nordic topics
`soc.culture.pakistan`	Pakistan
`soc.culture.polish`	Poland
`soc.culture.portuguese`	Portugal
`soc.culture.romanian`	Romania and Moldavia
`soc.culture.singapore`	Singapore
`soc.culture.soviet`	Russia and the former Soviet Union
`soc.culture.spain`	Spain
`soc.culture.sri-lanka`	Sri Lanka
`soc.culture.taiwan`	Taiwan
`soc.culture.tamil`	Tamil (northern Sri Lanka)
`soc.culture.thai`	Thailand
`soc.culture.turkish`	Turkey
`soc.culture.usa`	United States of America
`soc.culture.vietnamese`	Vietnam
`soc.culture.yugoslavia`	The former Yugoslavia
`soc.feminism`	Feminism and feminist issues (Moderated)
`soc.history`	History
`soc.libraries.talk`	Libraries
`soc.men`	Men: their problems and relationships
`soc.misc`	Socially oriented topics not covered elsewhere
`soc.motss`	Homosexuality (members of the same sex)
`soc.net-people`	Announcements, requests, about people on the Net
`soc.penpals`	Penpals
`soc.politics`	Political problems, systems, solutions (Moderated)
`soc.politics.arms-d`	Arms discussion digest (Moderated)
`soc.religion.bahai`	Baha'i (Moderated)
`soc.religion.christian`	Christianity and related topics (Moderated)
`soc.religion.christian.bible-study`	Bible studies (Moderated)
`soc.religion.eastern`	Eastern religions (Moderated)
`soc.religion.islam`	Islam (Moderated)

`soc.religion.quaker`	Quakers (Religious Society of Friends)
`soc.rights.human`	Human rights and activism
`soc.roots`	Genealogy, tracing your roots
`soc.singles`	Single people, their activities, etc
`soc.veterans`	Military veterans
`soc.women`	Women: their problems and relationships
`talk.abortion`	Discussions and arguments on abortion
`talk.answers`	Repository for periodic Usenet articles (Moderated)
`talk.bizarre`	The unusual, bizarre, curious, and often stupid
`talk.environment`	The environment
`talk.origins`	Evolution versus creationism
`talk.philosophy.misc`	Philosophical musings on all topics
`talk.politics.animals`	Use and abuse of animals
`talk.politics.china`	Politics of China
`talk.politics.drugs`	Politics of drug issues
`talk.politics.guns`	Politics of firearm ownership
`talk.politics.medicine`	Politics of health care
`talk.politics.mideast`	Politics of the Middle East
`talk.politics.misc`	Political topics not covered elsewhere
`talk.politics.soviet`	Politics of the former Soviet Union
`talk.politics.space`	Non-technical issues affecting space exploration
`talk.politics.theory`	Theory of politics and political systems
`talk.rape`	Rape (not to be crossposted)
`talk.religion.misc`	Religious, ethical, moral topics not covered elsewhere
`talk.religion.newage`	New Age religions and philosophies
`talk.rumors`	Rumors

Alternative Hierarchies

`alt.1d`	One-dimensional imaging
`alt.3d`	Three-dimensional imaging
`alt.abortion.inequity`	The inequity of abortion
`alt.abuse-recovery`	Helping victims of abuse recover (Moderated)
`alt.abuse.offender.recovery`	Helping offenders recover
`alt.abuse.recovery`	Helping victims of abuse recover (Moderated)
`alt.activism`	Activities for activists (no discussion)
`alt.activism.d`	Discussion of issues in alt.activism
`alt.adoption`	Adoption
`alt.aeffle.und.pferdle`	German TV cartoon characters
`alt.agriculture.fruit`	Fruit farming and agriculture
`alt.agriculture.misc`	Agriculture and farming
`alt.aldus.freehand`	Aldus Freehand software
`alt.aldus.misc`	Other Aldus software products
`alt.aldus.pagemaker`	Aldus PageMaker
`alt.alien.visitors`	Space aliens on Earth! Abduction! Gov't cover-up!
`alt.amateur-comp`	The amateur computerist
`alt.angst`	Anxiety in the modern world
`alt.answers`	Frequently asked questions about alt newsgroups
`alt.appalachian`	Appalachian regional issues
`alt.aquaria`	Aquariums and tropical fish
`alt.archery`	Archery
`alt.architecture`	Building design/construction and related topics
`alt.artcom`	Artistic community, arts and communication
`alt.astrology`	Astrology
`alt.atheism`	Atheism
`alt.atheism.moderated`	Atheism (Moderated)
`alt.autos.rod-n-custom`	Souped-up and customized autos
`alt.bacchus`	The nonprofit Bacchus organization
`alt.backrubs`	Massage and back rubs
`alt.bbs`	Computer BBS systems and software
`alt.bbs.ads`	Ads for various computer BBSs
`alt.bbs.allsysop`	A forum for BBS system operators (sysops)
`alt.bbs.internet`	BBS systems accessible via the Internet
`alt.bbs.lists`	Postings of regional BBS listings (no discussion)
`alt.bbs.lists.d`	Discussion about regional BBS listings
`alt.bbs.metal`	Metal telecomm environment
`alt.bbs.pcbuucp`	Uucp for DOS systems

`alt.bbs.searchlight`	Searchlight BBS systems
`alt.bbs.unixbbs`	Unix BBS systems
`alt.beer`	Beer and related beverages
`alt.binaries.multimedia`	Sound, text, graphics and so on
`alt.binaries.pictures`	Pictures (suitable for displaying on a computer)
`alt.binaries.pictures.d`	Discussion about pictures
`alt.binaries.pictures.erotica`	Pictures: erotic
`alt.binaries.pictures.erotica.blondes`	Pictures: erotic blonds
`alt.binaries.pictures.erotica.d`	Discussion about erotic blond pictures
`alt.binaries.pictures.erotica.female`	Pictures: female erotic
`alt.binaries.pictures.erotica.male`	Pictures: male erotic
`alt.binaries.pictures.fine-art.d`	Discussion about fine art pictures (Moderated)
`alt.binaries.pictures.fine-art.digitized`	Pictures: fine art pictures (Moderated)
`alt.binaries.pictures.fine-art.graphics`	Pictures: graphics (Moderated)
`alt.binaries.pictures.fractals`	Pictures: fractals
`alt.binaries.pictures.misc`	Pictures: miscellaneous
`alt.binaries.pictures.supermodels`	Pictures: supermodels
`alt.binaries.pictures.tasteless`	Pictures: tasteless
`alt.binaries.pictures.utilities`	Programs for scanning and viewing picture files
`alt.binaries.sounds.d`	Discussion about digitized sounds
`alt.binaries.sounds.misc`	Miscellaneous digitized sounds
`alt.birthright`	Birthright Party propaganda
`alt.bonsai`	The art of bonsai
`alt.books.deryni`	Katherine Kurtz's books, esp. the Deryni series
`alt.books.technical`	Technical books
`alt.boomerang`	Technology and use of the boomerang
`alt.brother-jed`	Born-again minister touring U.S. campuses
`alt.business.multi-level`	Multi-level marketing businesses
`alt.cable-tv.re-regulate`	Regulation of TV networks and cable
`alt.cad`	Computer-aided design
`alt.cad.autocad`	Autodesk CAD products
`alt.california`	All about California
`alt.callahans`	Callahan's bar for puns and fellowship
`alt.cascade`	Long, silly followup articles (cascades)
`alt.cd-rom`	Optical storage devices (specifically CD-ROM)
`alt.censorship`	Restricting speech and press
`alt.cesium`	Trivia relating to the element cesium
`alt.chess.ics`	Internet chess server
`alt.child-support`	Child support
`alt.chinchilla`	Chinchilla farming and cultivation
`alt.chinese.text`	Postings in Chinese; Chinese language software
`alt.chinese.text.big5`	Posting in Chinese [BIG 5]

`alt.clearing.technology`	Clearer and healthier skin
`alt.co-evolution`	Whole Earth Review and associated lifestyles
`alt.co-ops`	Cooperatives
`alt.cobol`	Programming in Cobol
`alt.collecting.autographs`	Autograph collectors and enthusiasts
`alt.college.college-bowl`	College Bowl competition
`alt.college.food`	Dining halls, cafeterias, mystery meat, and more
`alt.colorguard`	Marching bands, etc
`alt.comedy.british`	British humour
`alt.comics.buffalo-roam`	A postscript comic strip
`alt.comics.lnh`	Interactive net.madness in the superhero genre
`alt.comp.acad-freedom.news`	Academic freedom related to computers (Moderated)
`alt.comp.acad-freedom.talk`	Academic freedom issues related to computers
`alt.comp.fsp`	FSP file transport protocol
`alt.config`	Discussion about forming new alternative newsgroups
`alt.consciousness`	Topics about consciousness
`alt.conspiracy`	Conspiracy theories
`alt.conspiracy.jfk`	Kennedy assassination conspiracy theories
`alt.cosuard`	Council of Sysops, Users Against Rate Discrimination
`alt.cult-movies`	Movies with a cult following
`alt.cult-movies.rocky-horror`	Rocky Horror Picture Show
`alt.culture.alaska`	Alaska
`alt.culture.argentina`	Argentina
`alt.culture.austrian`	Austria
`alt.culture.electric-midget`	Midgets
`alt.culture.indonesia`	Indonesia
`alt.culture.karnataka`	Indian state of Karnataka
`alt.culture.kerala`	People of Keralite origin and the Malayalam language
`alt.culture.ny-upstate`	Upstate New York
`alt.culture.oregon`	Oregon
`alt.culture.theory`	Cultural theory and current practical problems
`alt.culture.tuva`	Republic of Tannu Tuva
`alt.culture.us.asian-indian`	Asian Indians in the U.S. and Canada
`alt.culture.usenet`	Usenet culture
`alt.current-events.somalia`	Somalia
`alt.cyb-sys`	Cybernetics and systems
`alt.cyberpunk`	Computer-mediated high-tech lifestyle
`alt.cyberpunk.chatsubo`	Literary virtual reality in a cyberpunk hangout
`alt.cyberpunk.movement`	Cybernizing the universe
`alt.cyberpunk.tech`	Cyberspace and cyberpunk technology
`alt.cyberspace`	Cyberspace and how it should work
`alt.dads-rights`	Rights of fathers trying to win custody in court

`alt.dcom.catv`	Data communications and equipment
`alt.dcom.telecom`	Telecommunications technology
`alt.decathena`	DEC Athena
`alt.desert-storm`	The war against Iraq in Kuwait
`alt.desert-storm.facts`	Factual information on the Gulf War
`alt.desert-thekurds`	Kurds in Iraq
`alt.destroy.the.earth`	General destruction
`alt.discordia`	Social discord
`alt.discrimination`	Quotas, affirmative action, bigotry, persecution
`alt.divination`	Divination techniques (I Ching, Tarot, runes, etc)
`alt.dreams`	Dreams interpretation and related topics
`alt.drugs`	Recreational pharmaceuticals and related flames
`alt.drugs.caffeine`	Caffeine
`alt.drumcorps`	Drum and bugle corps
`alt.education.bangkok`	Education issues in Thailand
`alt.education.bangkok.cmc`	Education issues in Thailand
`alt.education.bangkok.databases`	Education issues in Thailand
`alt.education.bangkok.planning`	Education issues in Thailand
`alt.education.bangkok.research`	Education issues in Thailand
`alt.education.bangkok.student`	Education issues in Thailand
`alt.education.bangkok.theory`	Education issues in Thailand
`alt.education.disabled`	Learning experiences for the disabled
`alt.education.distance`	Learning over networks, etc
`alt.emulators.ibmpc.apple2`	Emulating a PC with an Apple II
`alt.emusic`	Electronic music
`alt.etext`	Electronic text
`alt.evil`	Tales from the dark side
`alt.exotic-music`	Exotic and esoteric music
`alt.exploding.kibo`	People who like to blow things up
`alt.fan.alok.vijayvargia`	Alok Vijayvargia
`alt.fan.amy-fisher`	Amy Fisher and the famous trial
`alt.fan.anne-rice`	Anne Rice, writer
`alt.fan.asprin`	Robert Lynn Asprin
`alt.fan.bill-gates`	Bill Gates of Microsoft
`alt.fan.chris-elliott`	Chris Elliott, comic actor
`alt.fan.dale-bass`	Dale Bass
`alt.fan.dan-quayle`	Dan Quayle, former U.S. Vice President
`alt.fan.dan-wang`	Dan Wang
`alt.fan.dave_barry`	Dave Barry, humorist
`alt.fan.dice-man`	Andrew Dice Clay, comedian
`alt.fan.dick-depew`	Dick Depew
`alt.fan.disney.afternoon`	Disney afternoon characters and shows

`alt.fan.douglas-adams`	Douglas Adams, writer
`alt.fan.eddings`	David Eddings, writer
`alt.fan.enya`	Enya music
`alt.fan.frank-zappa`	Frank Zappa, musician
`alt.fan.furry`	Furry animals, anthropomorphized & stuffed
`alt.fan.greaseman`	Doug Tracht, disc jockey
`alt.fan.holmes`	Sherlock Holmes
`alt.fan.howard-stern`	Howard Stern, radio and TV personality
`alt.fan.itchy-n-scratchy`	Bart Simpson's favorite TV cartoon
`alt.fan.james-bond`	James Bond
`alt.fan.jimmy-buffett`	Jimmy Buffett, country singer
`alt.fan.joel-furr`	Joel Furr
`alt.fan.john-palmer`	John Palmer
`alt.fan.kent-montana`	Kent Montana book series
`alt.fan.kevin-darcy`	Kevin Darcy
`alt.fan.lemurs`	Lemur, monkey-like mammal from Madagascar
`alt.fan.lemurs.cooked`	Weird talk, lemur and non-lemur
`alt.fan.letterman`	David Letterman, TV personality
`alt.fan.madonna`	Madonna, singer, etc
`alt.fan.marla-thrift`	Marla Thrift
`alt.fan.mike-jittlov`	Mike Jittlov, animator
`alt.fan.monty-python`	Monty Python, humor group
`alt.fan.naked-guy`	Andrew Martinez, the Naked Guy
`alt.fan.nathan.brazil`	Hero of a Jack Chalker novel
`alt.fan.pern`	Anne McCaffery's science fiction oeuvre
`alt.fan.piers-anthony`	Piers Anthony, science fiction author
`alt.fan.pratchett`	Terry Pratchett, science fiction humor writer
`alt.fan.q`	Star Trek: The Next Generation's Q
`alt.fan.ren-and-stimpy`	Ren and Stimpy, cartoon characters
`alt.fan.rumpole`	Rumpole of the Bailey, British TV show
`alt.fan.rush-limbaugh`	Rush Limbaugh, conservative talk show host
`alt.fan.rush-limbaugh.transcripts`	Transcripts of the Rush Limbaugh show
`alt.fan.rush-limbaugh.tv-show`	Rush Limbaugh TV show
`alt.fan.serdar-argic`	Armenia, Turkey, genocide
`alt.fan.shostakovich`	Shostakovitch, classical music composer
`alt.fan.tank-girl`	Tank girl
`alt.fan.tolkien`	J.R.R. Tolkien, fantasy writer
`alt.fan.tom-robbins`	Tom Robbins, novelist
`alt.fan.tom_peterson`	Tom Peterson from Portland, Oregon
`alt.fan.wal-greenslade`	Wal Greenslade, the BBC Home Service
`alt.fan.warlord`	War Lord of the West Preservation Fan Club

`alt.fan.wodehouse`	P.G. Wodehouse (pronounced "Woodhouse"), humorist
`alt.fandom.cons`	Conventions, science fiction and others
`alt.fandom.misc`	Other topics for fans of various kinds
`alt.fashion`	Fashion industry
`alt.feminism`	Like soc.feminism, only different
`alt.fishing`	Fishing
`alt.flame`	Complaints, and complaints about complaints
`alt.flame.sean-ryan`	Detractors of Sean Ryan
`alt.folklore.college`	Collegiate humor
`alt.folklore.computers`	Stories and anecdotes about computers (some true!)
`alt.folklore.ghost-stories`	Ghost stories
`alt.folklore.science`	The folklore of science, not the science of folklore
`alt.folklore.urban`	Urban legends, ala Jan Harold Brunvand
`alt.food.sugar-cereals`	Sugary breakfast cereals
`alt.forever.linette`	Andrew K. Heller's first (and possibly last) love
`alt.forgery`	One place for all forgeries, crossposting encouraged
`alt.fractals.pictures`	Pictures of fractals
`alt.galactic-guide`	Entries for actual Hitchhiker's Guide to the Galaxy
`alt.games.frp.live-action`	Live-action gaming
`alt.games.gb`	Galactic Bloodshed conquest game
`alt.games.lynx`	Atari Lynx
`alt.games.mornington.cresent`	Fans of a game
`alt.games.omega`	Computer game Omega
`alt.games.sf2`	Video game Street Fighter 2
`alt.games.tiddlywinks`	Game of Tiddlywinks
`alt.games.torg`	Gateway for TORG mailing list
`alt.games.vga-planets`	Tim Wisseman's VGA Planets
`alt.games.video.classic`	From early TV remote controls to Space Invaders, etc
`alt.games.xtrek`	Star Trek game for X Window
`alt.good.news`	Good news (really!)
`alt.gopher`	The Internet Gopher (menu-driven information)
`alt.gorets`	Imaginary and not so imaginary silliness
`alt.gothic`	The gothic movement, things mournful and dark
`alt.gourmand`	Recipes and cooking information (Moderated)
`alt.graphics.pixutils`	Pixmap utilities
`alt.great-lakes`	The Great Lakes and adjacent places
`alt.guitar`	Guitar
`alt.guitar.bass`	Bass guitar
`alt.guitar.tab`	Music and lyrics (tablature) for guitar fans
`alt.hackers`	Hacking (Moderated)
`alt.heraldry.sca`	Heraldry in the Society of Creative Anachronism
`alt.hindu`	Hindu religion (Moderated)

`alt.history.living`	Living history
`alt.history.what-if`	Historical conjecture
`alt.homosexual`	Homosexuality
`alt.horror`	Horror genre
`alt.horror.cthulhu`	Mythos/role playing based on H.P.Lovecrafts' Cthulhu
`alt.horror.werewolves`	Werewolves
`alt.hotrod`	High speed automobiles (Moderated)
`alt.housing.nontrad`	Non-traditional housing concepts
`alt.hurricane.andrew`	1992 Florida hurricane disaster
`alt.hypertext`	Hypertext
`alt.hypnosis`	Hypnosis
`alt.illuminati`	Conspiracy, political and financial intrigue
`alt.individualism`	Philosophies where individual rights are paramount
`alt.industrial`	Industrial Computing Society
`alt.info-science`	Library science
`alt.info-theory`	Information theory
`alt.internet.access.wanted`	People looking for Internet access information
`alt.internet.services`	Various Internet services
`alt.internet.talk-radio`	Internet Talk Radio program
`alt.irc`	Internet Relay Chat
`alt.irc.ircii`	IRC II client program
`alt.irc.recovery`	Recovery from too much IRC
`alt.journalism`	Topics about and by journalists
`alt.journalism.criticism`	Criticism of journalists and the media
`alt.journalism.music`	Music journalism
`alt.kids-talk`	Discussion among children
`alt.lang.asm`	Various types of Assembly languages
`alt.lang.awk`	Unix awk scripting language
`alt.lang.basic`	Basic programming language
`alt.lang.cfutures`	Future of the C programming language
`alt.lang.intercal`	Intercal programming language
`alt.lang.ml`	ML and SML symbolic programming languages
`alt.lang.teco`	TECO editor language
`alt.locksmithing`	Locksmithing: locks, keys, etc
`alt.lucid-emacs.bug`	Bug reports about Lucid Emacs
`alt.lucid-emacs.help`	Question and answer forum for Lucid Emacs
`alt.lycra`	Clothes made of Lycra, Spandex, etc
`alt.magic`	Stage magic
`alt.magick`	Supernatural arts
`alt.manga`	Non-Western comics
`alt.materials.simulation`	Computer modeling of materials (Moderated)
`alt.med.cfs`	Chronic fatigue syndrome

`alt.meditation`	Meditation
`alt.meditation.transcendental`	Transcendental meditation
`alt.mensa.boston`	Mensa members in Boston
`alt.messianic`	Messianic traditions
`alt.military.cadet`	Military school cadets
`alt.mindcontrol`	Mind control (You will buy 10 copies of this book!)
`alt.misanthropy`	Hatred and distrust toward all
`alt.missing-kids`	Locating missing children
`alt.models`	Model building, design, etc
`alt.msdos.programmer`	Tips, questions and answers for DOS programming
`alt.mud`	Multiuser dimension games
`alt.mud.bsx`	Mud systems on BSX VR
`alt.mud.chupchups`	Mud game called chupcups
`alt.mud.german`	German-speaking Mud-ers
`alt.mud.lp`	Help setting up a mud
`alt.mud.tiny`	Fans of small mud sites
`alt.music.a-cappella`	Unaccompanied singing
`alt.music.alternative`	Alternative music
`alt.music.bela-fleck`	Bela and the Flecktones
`alt.music.canada`	Canadian music
`alt.music.enya`	Enya, Gaelic set to spacey music
`alt.music.filk`	SciFi/fantasy-related folk music
`alt.music.hardcore`	Hard core music fans
`alt.music.marillion`	Marillion, pop music group
`alt.music.progressive`	Music such as Yes, Marillion, Asia, King Crimson
`alt.music.queen`	Queen, pop music group
`alt.music.rush`	Rush, pop music group
`alt.music.ska`	Ska (skank) music, bands, and so on
`alt.music.tmbg`	They Might Be Giants
`alt.mythology`	Mythology
`alt.national.enquirer`	National Enquirer newspaper
`alt.native`	Indigenous peoples of the world
`alt.necromicon`	Bizarre political talk
`alt.netgames.bolo`	Game of bolo
`alt.news-media`	News media
`alt.news.macedonia`	News and current affairs in Macedonia
`alt.non.sequitur`	Interesting bits of imaginative writing + much junk
`alt.online-service`	Large commercial online services
`alt.org.food-not-bombs`	Food Not Bombs, an anti-hunger organization
`alt.org.pugwash`	Technological issues from a social stance
`alt.os.bsdi`	BSD (Berkeley) Unix
`alt.os.multics`	Multics operating system

`alt.out-of-body`	Out of body experiences
`alt.overlords`	Office of the Omnipotent Overlords of the Omniverse
`alt.pagan`	Paganism
`alt.pantyhose`	Talk about pantyhose
`alt.paranet.abduct`	Stories of abductions by aliens
`alt.paranet.paranormal`	Paranormal experiences
`alt.paranet.science`	Theories behind paranormal events
`alt.paranet.skeptic`	Skeptics explain and refute paranormal stories
`alt.paranet.ufo`	UFO sightings
`alt.paranormal`	Phenomena that are not scientifically explicable
`alt.parents-teens`	Parent-teenager relationships
`alt.party`	Parties, celebration and general debauchery
`alt.pcnews`	PCNews software
`alt.peeves`	Pet peeves and related topics
`alt.periphs.pcmcia`	Credit card-sized plug-in peripherals for PCs
`alt.personals`	General personal ads
`alt.personals.ads`	More personal ads
`alt.personals.bondage`	Personals: bondage
`alt.personals.misc`	Personals: miscellaneous
`alt.personals.poly`	Personals: multiple people
`alt.philosophy.objectivism`	Ayn Rand-derived philosophy of Objectivism
`alt.planning.urban`	Urban planning
`alt.politics.bush`	George Bush, former U.S. President
`alt.politics.clinton`	Bill Clinton, U.S. President
`alt.politics.correct`	Political correctness
`alt.politics.democrats`	Politics: Democrats
`alt.politics.democrats.clinton`	Politics: Democrats and Bill Clinton
`alt.politics.democrats.d`	Politics: Discussion about Democrats
`alt.politics.democrats.governors`	Politics: Democrats and governors
`alt.politics.democrats.house`	Politics: House of Representatives
`alt.politics.democrats.senate`	Politics: Senate
`alt.politics.ec`	The European Community
`alt.politics.economics`	Economics
`alt.politics.elections`	Elections
`alt.politics.equality`	Political equality
`alt.politics.europe.misc`	European politics
`alt.politics.greens`	Green party politics and activities worldwide
`alt.politics.homosexuality`	Politics and homosexuality
`alt.politics.india.communist`	Communism in India
`alt.politics.india.progressive`	Progressive Indian politics
`alt.politics.libertarian`	Libertarian ideology
`alt.politics.media`	Politics and the communications media

`alt.politics.org.misc`	Political organizations
`alt.politics.org.un`	Politics at the United Nations
`alt.politics.radical-left`	The radical left
`alt.politics.reform`	Political reform
`alt.politics.sex`	Politics and sex
`alt.politics.usa.constitution`	U.S. Constitutional politics
`alt.politics.usa.misc`	U.S. political topics not covered elsewhere
`alt.politics.usa.republican`	U.S. Republican party
`alt.politics.vietnamese`	Politics and Vietnam
`alt.polyamory`	Multiple-love relationships
`alt.postmodern`	Postmodernism, semiotics, deconstruction, etc
`alt.president.clinton`	U.S. President Bill Clinton
`alt.prisons`	Prisons and prison life
`alt.privacy`	Privacy issues in cyberspace
`alt.privacy.anon-server`	Anonymous Usenet postings
`alt.privacy.clipper`	Clipper encoding chip
`alt.prophecies.nostradamus`	Nostradamus' predictions and score card
`alt.prose`	Original writings, fiction and otherwise
`alt.prose.d`	Discussion of alt.prose.d articles
`alt.psychoactives`	Psychoactive drugs
`alt.psychology.personality`	Personality taxonomy, such as Myers-Briggs
`alt.pub.cloven-shield`	Roleplaying stories
`alt.pub.dragons-inn`	Computer fantasy environment
`alt.pub.havens-rest`	Roleplaying stories
`alt.pulp`	Paperback fiction
`alt.quotations`	Famous (and not so famous) quotes
`alt.radio.pirate`	Pirate radio stations
`alt.radio.scanner`	Scanning radio receivers
`alt.rap`	Rap music
`alt.rap-gdead`	Grateful Dead and Rap (music)
`alt.rave`	Techno-culture: music, dancing, drugs, dancing
`alt.recovery`	Recovery programs (AA, ACA, GA, etc)
`alt.recovery.codependency`	Codependency
`alt.religion.all-worlds`	The Church of All Worlds from Heinlein's book
`alt.religion.computers`	Computing as a Way of Life
`alt.religion.emacs`	Emacs as a Way of Life
`alt.religion.kibology`	Silliness based on the mythical Kibo
`alt.religion.monica`	Net-venus Monica and her works
`alt.religion.sabaean`	Sabaean religious order
`alt.religion.santaism`	Christmas and Santa Claus
`alt.religion.scientology`	Scientology and Dianetics
`alt.restaurants`	Restaurant industry

`alt.revisionism`	Changing interpretations of history
`alt.revolution.counter`	Counter-revolutionary issues
`alt.rhode_island`	The U.S. state of Rhode Island
`alt.rmgroup`	For people who like to rmgroup/newgroup things
`alt.rock-n-roll`	General rock and roll
`alt.rock-n-roll.acdc`	AC/DC, music group
`alt.rock-n-roll.classic`	Classic rock music
`alt.rock-n-roll.hard`	Hard rock
`alt.rock-n-roll.metal`	General metal rock music
`alt.rock-n-roll.metal.gnr`	Heavy metal rock
`alt.rock-n-roll.metal.heavy`	More heavy metal rock music
`alt.rock-n-roll.metal.ironmaiden`	Iron Maiden, music group
`alt.rock-n-roll.metal.metallica`	Metallica, music group
`alt.rock-n-roll.metal.progressive`	Progressive metal rock
`alt.rock-n-roll.stones`	The Rolling Stones, music group
`alt.rock-n-roll.symphonic`	Symphonic rock and roll
`alt.rodney-king`	The L.A. riots and the aftermath
`alt.romance`	Romantic side of love
`alt.romance.chat`	Romantic talk
`alt.rush-limbaugh`	Rush Limbaugh, conservative radio talk show host
`alt.satanism`	Satanism
`alt.save.the.earth`	Environmentalist causes
`alt.sb.programmer`	Programming the Sound Blaster card for PCs
`alt.sci.astro.aips`	Astronomical Image Processing System
`alt.sci.astro.figaro`	Figaro data reduction package
`alt.sci.astro.fits`	Programming with FITS
`alt.sci.physics.acoustics`	Acoustics
`alt.sci.physics.new-theories`	New scientific theories you won't find in journals
`alt.sci.planetary`	Planetary science
`alt.sect.ahmadiyya`	Ahmadiyyat sect of Islam
`alt.security`	Building and automobile security systems
`alt.security.index`	Good references to {alt,misc}.security (Moderated)
`alt.security.keydist`	Exchange of keys for public key encryption systems
`alt.security.pgp`	The Pretty Good Privacy encryption package
`alt.security.ripem`	A secure electronic mail system
`alt.sega.genesis`	Genesis video game
`alt.self-improve`	Self-improvement
`alt.sewing`	Sewing
`alt.sex`	General discussion about sex
`alt.sex.bestiality`	Sex: bestiality
`alt.sex.bondage`	Sex: bondage
`alt.sex.fetish.feet`	Sex: foot fetishes

`alt.sex.masturbation`	Sex: masturbation
`alt.sex.motss`	Sex: homosexuality (members of the same sex)
`alt.sex.movies`	Sex and the movies
`alt.sex.pictures`	Erotic pictures (no discussion)
`alt.sex.pictures.d`	Discussion about erotic pictures
`alt.sex.pictures.female`	Erotic pictures of women
`alt.sex.pictures.male`	Erotic pictures of men
`alt.sex.sounds`	Erotic sounds
`alt.sex.stories`	Stories about sex (no discussion)
`alt.sex.stories.d`	Discussion about sexual stories
`alt.sex.wanted`	Requests for erotica, literary or in the flesh
`alt.sex.wizards`	Questions for sex experts
`alt.sexual.abuse.recovery`	Helping others deal with traumatic experiences
`alt.shenanigans`	Practical jokes, pranks, etc
`alt.sigma2.height`	Very short or tall people (> 2 standard deviations)
`alt.skate`	Skating
`alt.skate-board`	Skateboarding
`alt.skinheads`	Skinhead culture/anti-culture
`alt.slack`	The Church of the Subgenius
`alt.snowmobiles`	Snowmobiles
`alt.society.anarchy`	Anarchy
`alt.society.ati`	The Activist Times Digest (Moderated)
`alt.society.civil-disob`	Civil disobedience
`alt.society.civil-liberties`	Individual rights
`alt.society.civil-liberty`	Civil libertarians
`alt.society.foia`	U.S. Freedom of Information Act
`alt.society.futures`	Future of society
`alt.society.resistance`	Political talk: resistance
`alt.society.revolution`	Political talk: revolution
`alt.society.sovereign`	Political talk: sovereignty
`alt.soft-sys.tooltalk`	ToolTalk
`alt.sources`	Free computer programs, unmoderated, caveat emptor
`alt.sources.amiga`	Technically oriented Amiga PC programs
`alt.sources.d`	Discussion about programs that have been posted
`alt.sources.index`	Good references to free programs
`alt.sources.patches`	Corrections to programs, from non-bugs groups
`alt.sources.wanted`	Requests for programs
`alt.sport.bowling`	Bowling
`alt.sport.bungee`	Bungee cord jumping
`alt.sport.darts`	Darts
`alt.sport.lasertag`	Laser tag
`alt.sport.paintball`	Paintball combat game

`alt.stagecraft`	Technical aspects of the theatre
`alt.startrek.creative`	Stories and parodies related to Star Trek
`alt.stupidity`	Stupid topics
`alt.suburbs`	The suburbs
`alt.suicide.holiday`	Why suicides increase at holidays
`alt.suit.att-bsdi`	The AT&T vs BSDI lawsuit
`alt.super.nes`	Super Nintendo video game
`alt.supermodels`	Famous and beautiful models
`alt.support`	General support group
`alt.support.big-folks`	Support group for very tall people
`alt.support.cancer`	Support group for cancer patients
`alt.support.diet`	Support group for dieters
`alt.support.step-parents`	Support group for step-parents
`alt.sustainable.agriculture`	Politics and agriculture
`alt.sys.amiga.demos`	Demo programs for Amiga computers
`alt.sys.amiga.uucp`	Uucp for the Amiga
`alt.sys.amiga.uucp.patches`	Corrections for Amiga Uucp software
`alt.sys.intergraph`	Intergraph computers
`alt.sys.pdp8`	DEC PDP-8 computers
`alt.sys.perq`	Antique computers
`alt.sys.sun`	Sun computers
`alt.tasteless`	Tasteless and disgusting topics
`alt.tasteless.jokes`	Tasteless jokes (offensive to everybody)
`alt.tasteless.pictures`	Tasteless pictures
`alt.test`	Place to send test articles
`alt.text.dwb`	AT&T Documenter's WorkBench
`alt.thrash`	Thrash music (an acquired taste...)
`alt.tla`	Palindromic fun, Three-Letter-Acronyms
`alt.toolkits.xview`	X windows XView toolkit
`alt.toon-pics`	Cartoon pictures
`alt.toys.lego`	Lego building blocks
`alt.transgendered`	Changing one's sex
`alt.true-crime`	Famous crimes
`alt.tv.antagonists`	The Antagonists, TV show
`alt.tv.babylon-5`	Babylon 5, TV show
`alt.tv.beakmans-world`	Beakman's World, TV show
`alt.tv.bh90210`	Beverly Hills 90210, TV show
`alt.tv.dinosaurs`	Dinosaurs and television
`alt.tv.infomercials`	Infomercials
`alt.tv.la-law`	L.A. Law, TV show
`alt.tv.liquid-tv`	BBC/MTV Liquid TV series (weird...)
`alt.tv.mash`	MASH, TV show

alt.tv.melrose-place	Melrose Place, TV show
alt.tv.mst3k	Mystery Science Theatre 3000, TV show
alt.tv.muppets	The Muppets
alt.tv.mwc	More TV shows
alt.tv.northern-exp	Northern Exposure, TV show
alt.tv.prisoner	The Prisoner, TV show
alt.tv.red-dwarf	Red Dwarf (British SciFi/comedy), TV show
alt.tv.ren-n-stimpy	Ren and Stimpy, TV show
alt.tv.saved-bell	Saved by the Bell, TV Show
alt.tv.seinfeld	Seinfeld, TV show
alt.tv.simpsons	The Simpsons, TV show
alt.tv.simpsons.itchy-scratchy	Itchy and Scratchy (Bart Simpson's cartoon friends)
alt.tv.tiny-toon	Tiny Toons, TV Show
alt.tv.twin-peaks	Twin Peaks, TV show
alt.usage.english	English grammar, word usages, and related topics
alt.usenet.offline-reader	Reading the news off-line
alt.usenet.recovery	Support group for Usenet junkies
alt.uu.announce	Announcements of Usenet University (UU)
alt.uu.comp.misc	Computer department of Usenet University
alt.uu.comp.os.linux.questions	UU Linux learning group, questions and answers
alt.uu.future	Planning the future of Usenet University
alt.uu.lang.esperanto.misc	Usenet University: Esperanto
alt.uu.lang.misc	Usenet University: language department
alt.uu.lang.russian.misc	Usenet University: Russian
alt.uu.math.misc	Usenet University: math
alt.uu.misc.misc	Usenet University: miscellaneous departments
alt.uu.tools	Usenet University: tools for education
alt.uu.virtual-worlds.misc	Usenet University: study of virtual worlds
alt.vampyres	Vampires
alt.video.laserdisc	Laser disk movies and games
alt.visa.us	U.S. immigration visas
alt.wais	WAIS (Wide Area Information Service)
alt.war	War
alt.war.civil.usa	U.S. Civil War
alt.war.vietnam	Vietnam War
alt.wedding	First meetings, dates, romance, weddings, etc
alt.whine	Whining and complaining
alt.windows.text	Text-based (non-graphical) window systems
alt.winsock	Socket implementations for Microsoft Windows
alt.wolves	Wolves and wolf-mix dogs
alt.world.taeis	Shared World project
alt.zines	Small magazines, mostly noncommercial

`alt.znet.aeo`	Atari computer enthusiasts
`alt.znet.pc`	Z*NET International PC, an electronic magazine
`bionet.agroforestry`	Agroforestry
`bionet.announce`	Announcements for biologists (Moderated)
`bionet.biology.computational`	Computer and mathematical biology (Moderated)
`bionet.biology.n2-fixation`	Biological nitrogen fixation
`bionet.biology.tropical`	Tropical biology
`bionet.drosophila`	Biology of Drosophila (fruit flies)
`bionet.general`	General forum for the biological sciences
`bionet.genome.arabidopsis`	The Arabidopsis project
`bionet.genome.chrom22`	Chromosome 22
`bionet.immunology`	Immunology
`bionet.info-theory`	Biological information theory
`bionet.jobs`	Job opportunities in biology
`bionet.journals.contents`	Contents of biology journals
`bionet.journals.note`	Advice on dealing with biology journals
`bionet.molbio.ageing`	Cellular and organismal aging
`bionet.molbio.bio-matrix`	Computer applications to biological databases
`bionet.molbio.embldatabank`	EMBL nucleic acid database
`bionet.molbio.evolution`	Evolution of genes and proteins
`bionet.molbio.gdb`	Messages to and from the GDB database staff
`bionet.molbio.genbank`	GenBank nucleic acid database
`bionet.molbio.genbank.updates`	News about GenBank (Moderated)
`bionet.molbio.gene-linkage`	Genetic linkage analysis
`bionet.molbio.genome-program`	Human Genome Project
`bionet.molbio.hiv`	Molecular biology of HIV (AIDS virus)
`bionet.molbio.methds-reagnts`	Requests for information and lab reagents
`bionet.molbio.proteins`	Proteins and protein databases
`bionet.molbio.rapd`	Randomly amplified polymorphic DNA
`bionet.molbio.yeast`	Molecular biology and genetics of yeast
`bionet.n2-fixation`	Nitrogen fixation
`bionet.neuroscience`	Neuroscience
`bionet.photosynthesis`	Photosynthesis
`bionet.plants`	All aspects of plant biology
`bionet.population-bio`	Population biology
`bionet.sci-resources`	Funding agencies, research grants, etc
`bionet.software`	Software for biology
`bionet.software.gcg`	GCG software
`bionet.software.sources`	Free programs relating to biology (Moderated)
`bionet.users.addresses`	Names and addresses in the world of biology
`bionet.virology`	Virology

`bionet.women-in-bio`	Women in biology
`bionet.xtallography`	Protein crystallography
`bit.admin`	Administrating Bitnet newsgroups
`bit.general`	General information on Bitnet and Usenet
`bit.listserv.3com-1`	3Com Products
`bit.listserv.9370-1`	IBM's 9370 and VM/IS operating system
`bit.listserv.ada-law`	ADA Law
`bit.listserv.advanc-1`	Geac Advanced Integrated Library System
`bit.listserv.advise-1`	User services
`bit.listserv.aix-1`	IBM's AIX (Unix) operating system
`bit.listserv.allmusic`	All forms of music
`bit.listserv.appc-1`	IBM's APPC
`bit.listserv.apple2-1`	Apple II
`bit.listserv.applicat`	Applications under Bitnet
`bit.listserv.arie-1`	RLG Ariel Document Transmission Group
`bit.listserv.ashe-1`	Higher education policy and research
`bit.listserv.asm370`	IBM 370 Assembly language programming
`bit.listserv.autism`	Autism and developmental disability
`bit.listserv.banyan-1`	Banyan Vines network software
`bit.listserv.big-lan`	Campus-size LANs (Moderated)
`bit.listserv.billing`	Chargeback of computer resources
`bit.listserv.biosph-1`	Biosphere, ecology
`bit.listserv.bitnews`	News about Bitnet
`bit.listserv.blindnws`	Blindness (Moderated)
`bit.listserv.buslib-1`	Business libraries
`bit.listserv.c+health`	Computer and health
`bit.listserv.c18-1`	18th century interdisciplinary forum
`bit.listserv.c370-1`	IBM's C/370 programming language
`bit.listserv.candle-1`	Candle products
`bit.listserv.catholic`	Free Catholics
`bit.listserv.cdromlan`	CD-ROM on local area networks
`bit.listserv.cfs.newsletter`	Chronic Fatigue Syndrome newsletter (Moderated)
`bit.listserv.christia`	Practical Christian life
`bit.listserv.cics-1`	IBM's CICS (transaction processing)
`bit.listserv.cinema-1`	The cinema
`bit.listserv.circplus`	Circulation reserve and related library topics
`bit.listserv.cmspip-1`	IBM's VM/SP CMS Pipelines
`bit.listserv.commed`	Communication education
`bit.listserv.csg-1`	Control System Group Network
`bit.listserv.cumrec-1`	Computer use, college and university administration
`bit.listserv.cw-email`	Campus-wide electronic mail

`bit.listserv.cwis-1`	Campus-wide information systems
`bit.listserv.cyber-1`	CDC computers
`bit.listserv.dasig`	Database administration
`bit.listserv.db2-1`	IBM's DB2 (relational database)
`bit.listserv.dbase-1`	dBase IV (PC database)
`bit.listserv.deaf-1`	Deafness
`bit.listserv.decnews`	Digital Equipment Corporation news
`bit.listserv.dectei-1`	DECUS Education Software Library
`bit.listserv.devel-1`	Technology transfer in international development
`bit.listserv.disarm-1`	Disarmament
`bit.listserv.domain-1`	Domain listings
`bit.listserv.earntech`	EARN technical group
`bit.listserv.ecolog-1`	Ecological Society of America
`bit.listserv.edi-1`	Electronic data interchange
`bit.listserv.edpolyan`	Professionals and students and education
`bit.listserv.edstat-1`	Statistics education
`bit.listserv.edtech`	Educational technology (Moderated)
`bit.listserv.edusig-1`	Education special interest group
`bit.listserv.emusic-1`	Electronic music
`bit.listserv.endnote`	Bibsoft Endnote
`bit.listserv.envbeh-1`	Environment and human behavior
`bit.listserv.erl-1`	Educational research
`bit.listserv.ethics-1`	Ethics in computing
`bit.listserv.ethology`	Ethology
`bit.listserv.euearn-1`	Eastern Europe
`bit.listserv.film-1`	Film making and reviews
`bit.listserv.fnord-1`	New ways of thinking
`bit.listserv.frac-1`	Fractals
`bit.listserv.free-1`	Fathers' rights
`bit.listserv.games-1`	Computer games
`bit.listserv.gaynet`	GayNet (Moderated)
`bit.listserv.gddm-1`	GDDM
`bit.listserv.geodesic`	Buckminster Fuller
`bit.listserv.gguide`	Bitnic Gguide
`bit.listserv.govdoc-1`	Government documents
`bit.listserv.gutnberg`	Project Gutenberg
`bit.listserv.hellas`	Hellenic topics (Moderated)
`bit.listserv.help-net`	Help on Bitnet and the Internet
`bit.listserv.hindu-d`	Hindu digest (Moderated)
`bit.listserv.history`	History
`bit.listserv.hytel-1`	Hytelnet (Moderated)
`bit.listserv.i-amiga`	Amiga computers

bit.listserv.ibm-hesc	IBM higher education consortium
bit.listserv.ibm-main	IBM mainframe computers
bit.listserv.ibm-nets	IBM mainframes and networking
bit.listserv.ibm7171	IBM's 7171 Protocol Converter
bit.listserv.ibmtcp-l	IBM's TCP/IP products
bit.listserv.india-d	India (Moderated)
bit.listserv.info-gcg	GCG genetics software
bit.listserv.ingrafx	Information graphics
bit.listserv.innopac	Innovative Interfaces Online Public Access
bit.listserv.ioob-l	Industrial psychology
bit.listserv.ipct-l	Interpersonal computing and technology (Moderated)
bit.listserv.isn	ISN Data Switch
bit.listserv.jes2-l	IBM's JES2
bit.listserv.jnet-l	Jnet running under VMS
bit.listserv.l-hcap	Handicaps (Moderated)
bit.listserv.l-vmctr	VMCENTER Components
bit.listserv.lawsch-l	Law schools
bit.listserv.liaison	Bitnic liaison
bit.listserv.libref-l	Library reference issues (Moderated)
bit.listserv.libres	Library and information science research
bit.listserv.license	Software licensing
bit.listserv.linkfail	Link failure announcements
bit.listserv.literary	Literature
bit.listserv.lstsrv-l	Listserv
bit.listserv.mail-l	Mail
bit.listserv.mailbook	Mail/Mailbook
bit.listserv.mba-l	MBA student curricula
bit.listserv.mbu-l	Megabyte University: computers and writing
bit.listserv.mdphd-l	Dual degree programs
bit.listserv.medforum	Medical students (Moderated)
bit.listserv.medlib-l	Medical libraries
bit.listserv.mednews	Mednews: Health Info-Com Network newsletter
bit.listserv.mideur-l	Middle Europe
bit.listserv.netnws-l	Netnews (Usenet)
bit.listserv.nettrain	Network trainers
bit.listserv.new-list	Announcements about new Bitnet lists (Moderated)
bit.listserv.next-l	Next computers
bit.listserv.nodmgt-l	Node management
bit.listserv.notabene	Nota Bene
bit.listserv.notis-l	Notis/Dobis
bit.listserv.novell	Novell LANs
bit.listserv.omrscan	OMR scanners

`bit.listserv.os2-1`	IBM's OS/2 operating system
`bit.listserv.ozone`	OZONE
`bit.listserv.pacs-1`	Public Access computer systems (Moderated)
`bit.listserv.page-1`	IBM's 3812/3820 tips and problems
`bit.listserv.pagemakr`	PageMaker for desktop publishers
`bit.listserv.physhare`	Physics
`bit.listserv.pmdf-1`	PMDF
`bit.listserv.politics`	Politics
`bit.listserv.postcard`	Postcard collectors
`bit.listserv.power-1`	IBM's RS/6000 computers (the Power architecture)
`bit.listserv.powerh-1`	PowerHouse
`bit.listserv.psycgrad`	Psychology grad students
`bit.listserv.qualrs-1`	Qualitative research of the human sciences
`bit.listserv.relusr-1`	Relay users
`bit.listserv.rhetoric`	Rhetoric, social movements, persuasion
`bit.listserv.rscs-1`	IBM's VM/RSCS
`bit.listserv.rscsmods`	RSCS modifications
`bit.listserv.s-comput`	Supercomputers
`bit.listserv.sas-1`	SAS
`bit.listserv.script-1`	IBM vs Waterloo SCRIPT
`bit.listserv.scuba-1`	Scuba diving
`bit.listserv.seasia-1`	Southeast Asia
`bit.listserv.seds-1`	Interchapter SEDS communications
`bit.listserv.sfs-1`	IBM's VM shared file system
`bit.listserv.sganet`	Student government global mail network
`bit.listserv.simula`	SIMULA programming (simulation) language
`bit.listserv.slart-1`	SLA research and teaching
`bit.listserv.slovak-1`	Slovaks
`bit.listserv.snamgt-1`	IBM's SNA network management
`bit.listserv.sos-data`	Social science
`bit.listserv.spires-1`	Spires conference
`bit.listserv.sportpsy`	Exercise and sports psychology
`bit.listserv.spssx-1`	SPSSX
`bit.listserv.sqlinfo`	IBM's SQL/DS (database language)
`bit.listserv.stat-1`	Statistical consulting
`bit.listserv.tech-1`	Bitnic TECH-L mailing list
`bit.listserv.techwr-1`	Technical writing
`bit.listserv.test`	Place to send test articles
`bit.listserv.tex-1`	TeX typesetting system
`bit.listserv.tn3270-1`	tn3270 protocol
`bit.listserv.toolb-1`	Asymetrix Toolbook
`bit.listserv.trans-1`	Bitnic TRANS-L mailing list

`bit.listserv.travel-l`	Travel and tourism
`bit.listserv.ucp-l`	University Computing Project
`bit.listserv.ug-l`	Usage guidelines
`bit.listserv.uigis-l`	User interface for Geographical Information Systems
`bit.listserv.urep-l`	UREP software
`bit.listserv.usrdir-l`	User directory
`bit.listserv.uus-l`	Unitarian-Universalist church
`bit.listserv.valert-l`	Virus alerts
`bit.listserv.vfort-l`	IBM's VS-Fortran programming language
`bit.listserv.vm-util`	IBM's VM Utilities
`bit.listserv.vmesa-l`	IBM's VM/ESA operating system
`bit.listserv.vmslsv-l`	DEC's VAX/VMS operating system
`bit.listserv.vmxa-l`	IBM's VM/XA operating system
`bit.listserv.vnews-l`	VNEWS
`bit.listserv.vpiej-l`	Electronic publishing
`bit.listserv.win3-l`	Microsoft Windows 3.x
`bit.listserv.words-l`	The English language
`bit.listserv.wpcorp-l`	WordPerfect products
`bit.listserv.wpwin-l`	WordPerfect for MS Windows
`bit.listserv.x400-l`	X.400 protocol
`bit.listserv.xcult-l`	International Intercultural Newsletter
`bit.listserv.xedit-l`	VM System Editor
`bit.listserv.xerox-l`	Xerox products
`bit.listserv.xmailer`	Crosswell mailer
`bit.listserv.xtropy-l`	Extopian topics
`bit.mailserv.word-mac`	Word processing on the Macintosh
`bit.mailserv.word-pc`	Word processing on PCs
`biz.americast`	The Americast company
`biz.americast.samples`	Americast samples
`biz.books.technical`	Selling and buying books
`biz.clarinet`	Announcements about Clarinet
`biz.clarinet.sample`	Free samples from Clarinet newsgroups
`biz.comp.hardware`	Generic commercial hardware
`biz.comp.services`	Generic commercial service
`biz.comp.software`	Generic commercial software
`biz.comp.telebit`	Telebit modems
`biz.comp.telebit.netblazer`	Telebit Netblazer
`biz.config`	Configuration and administration of biz newsgroups
`biz.control`	Usenet control information for biz newsgroups
`biz.dec`	DEC equipment and software
`biz.dec.decathena`	DEC Athena

`biz.dec.decnews`	DEC news releases
`biz.dec.ip`	IP networking on DEC machines
`biz.dec.workstations`	DEC workstations
`biz.digex.announce`	Digex
`biz.jobs.offered`	Jobs available
`biz.misc`	Miscellaneous postings of a commercial nature
`biz.next.newprod`	New products from the Next Company
`biz.sco.announce`	Santa Cruz Operation (SCO) news (Moderated)
`biz.sco.binaries`	Programs for SCO Xenix, Unix, or ODT (Open Desktop)
`biz.sco.general`	General discussion about SCO products
`biz.sco.magazine`	SCO magazine
`biz.sco.opendesktop`	Open Desktop
`biz.sco.sources`	Free programs to run under an SCO environment
`biz.stolen`	Stolen items
`biz.tadpole.sparcbook`	Sparcbook portable computer
`biz.test`	Place to send test articles
`biz.zeos.announce`	Announcements by Zeos International
`biz.zeos.general`	Zeos computers
`ddn.mgt-bulletin`	Defense Data Network Management Bulletin (Moderated)
`ddn.newsletter`	The DDN newsletter (Moderated)
`gnu.announce`	News about the GNU project (Moderated)
`gnu.bash.bug`	Bugs and fixes: Bash/Bourne Again shell (Moderated)
`gnu.chess`	GNU chess program
`gnu.emacs.announce`	GNU Emacs (Moderated)
`gnu.emacs.bug`	Bugs and fixes: GNU Emacs (Moderated)
`gnu.emacs.gnews`	News reading under GNU Emacs using Weemba's Gnews
`gnu.emacs.gnus`	News reading under GNU Emacs using GNUS
`gnu.emacs.help`	Questions and answers about Emacs
`gnu.emacs.sources`	Free C and Lisp programs for GNU Emacs (no discussion)
`gnu.emacs.vm.bug`	Bugs and fixes: Emacs VM mail package
`gnu.emacs.vm.info`	Emacs VM mail package
`gnu.emacs.vms`	Port of GNU Emacs to VMS
`gnu.epoch.misc`	Epoch X11 extensions to Emacs
`gnu.g++.announce`	News about g++ (the GNU C++ compiler) (Moderated)
`gnu.g++.bug`	Bugs and fixes: g++ (Moderated)
`gnu.g++.help`	Questions and answers about g++
`gnu.g++.lib.bug`	Bugs and fixes: g++ library (Moderated)
`gnu.gcc.announce`	News about gcc (the GNU C compiler) (Moderated)
`gnu.gcc.bug`	Bugs and fixes: gcc (Moderated)
`gnu.gcc.help`	GNU C Compiler (gcc) user queries and answers

`gnu.gdb.bug`	Bugs and fixes: gcc/g++ debugger (Moderated)
`gnu.ghostscript.bug`	Bugs and fixes: Ghostscript interpreter (Moderated)
`gnu.gnusenet.config`	GNU's Not Usenet administration and configuration
`gnu.gnusenet.test`	Place to send test articles
`gnu.groff.bug`	Bugs and fixes: GNU roff programs (Moderated)
`gnu.misc.discuss`	General talk about the GNU (Gnu's Not Unix) project
`gnu.smalltalk.bug`	Bugs and fixes: GNU Smalltalk (Moderated)
`gnu.utils.bug`	Bugs and fixes: GNU utilities: gawk, etc (Moderated)
`ieee.announce`	News for the IEEE community
`ieee.config`	Managing the ieee.* newsgroups
`ieee.general`	IEEE: general discussion
`ieee.pcnfs`	Tips on PC-NFS
`ieee.rab.announce`	Regional Activities Board: announcements
`ieee.rab.general`	Regional Activities Board: general discussion
`ieee.region1`	Region 1 announcements
`ieee.tab.announce`	Technical Activities Board: announcements
`ieee.tab.general`	Technical Activities Board: general discussion
`ieee.tcos`	TCOS newsletter and discussion (Moderated)
`ieee.usab.announce`	USAB: announcements
`ieee.usab.general`	USAB: general discussion
`info.admin`	Managing the info.* newsgroups
`info.big-internet`	Issues facing a huge Internet
`info.bind`	Berkeley BIND server (Moderated)
`info.brl-cad`	BRL's Solid modeling CAD system (Moderated)
`info.bytecounters`	NSstat network analysis
`info.convex`	Convex Company computers (Moderated)
`info.firearms`	Firearms: non-political (Moderated)
`info.firearms.politics`	Firearms: political
`info.gated`	Cornell's GATED program
`info.grass.programmer`	Programming: GRASS geographic information system
`info.grass.user`	Using: GRASS geographic information system
`info.ietf`	Internet Engineering Task Force (IETF)
`info.ietf.hosts`	IETF host requirements
`info.ietf.isoc`	The Internet society
`info.ietf.njm`	Joint Monitoring Access, Adjacent Nets (Moderated)
`info.ietf.smtp`	IETF SMTP extension
`info.isode`	ISO Development Environment package
`info.jethro-tull`	Jethro Tull, pop music group
`info.labmgr`	Computer lab managers (Moderated)
`info.mach`	Mach operating system (Moderated)

`info.mh.workers`	MH development (Moderated)
`info.nets`	Inter-network connectivity
`info.nsf.grants`	NSF grants (Moderated)
`info.nsfnet.cert`	Computer Emergency Response Team
`info.nsfnet.status`	Status of NSFnet
`info.nupop`	Northwestern University's POP for PCs
`info.nysersnmp`	SNMP software distributed by PSI
`info.osf`	OSF's electronic bulletin (Moderated)
`info.pem-dev`	IETF: privacy enhanced mail (Moderated)
`info.ph`	Qi, ph, sendmail/phquery
`info.rfc`	Announcements of newly released RFCs
`info.slug`	Symbolics Lisp machines
`info.snmp`	SNMP (Simple Gateway/Network Monitoring Protocol)
`info.solbourne`	Solbourne computers
`info.sun-managers`	Sun Managers digest (Moderated)
`info.sun-nets`	Sun Nets digest
`info.theorynt`	Theory (Moderated)
`info.unix-sw`	Unix software available by Anonymous FTP
`k12.chat.elementary`	Elementary students forum: grades K-5
`k12.chat.junior`	Elementary students forum: grades 6-8
`k12.chat.senior`	High school students forum
`k12.chat.teacher`	Teachers forum
`k12.ed.art`	Art curriculum
`k12.ed.business`	Business education curriculum
`k12.ed.comp.literacy`	Computer literacy
`k12.ed.health-pe`	Health and physical education
`k12.ed.life-skills`	Home economics and career education
`k12.ed.math`	Mathematics
`k12.ed.music`	Music and performing arts
`k12.ed.science`	Science
`k12.ed.soc-studies`	Social studies and history
`k12.ed.special`	Students with handicaps or special needs
`k12.ed.tag`	Talented and gifted students
`k12.ed.tech`	Industrial arts and vocational education
`k12.lang.art`	Language arts
`k12.lang.deutsch-eng`	German/English practice with native speakers
`k12.lang.esp-eng`	Spanish/English practice with native speakers
`k12.lang.francais`	French/English practice with native speakers
`k12.lang.russian`	Russian/English practice with native speakers
`k12.library`	Libraries and librarians
`k12.sys.channel0`	Forum for teachers

`k12.sys.channel1`	Forum for teachers
`k12.sys.channel2`	Forum for teachers
`k12.sys.channel3`	Forum for teachers
`k12.sys.channel4`	Forum for teachers
`k12.sys.channel5`	Forum for teachers
`k12.sys.channel6`	Forum for teachers
`k12.sys.channel7`	Forum for teachers
`k12.sys.channel8`	Forum for teachers
`k12.sys.channel9`	Forum for teachers
`k12.sys.channel10`	Forum for teachers
`k12.sys.channel11`	Forum for teachers
`k12.sys.channel12`	Forum for teachers
`k12.sys.projects`	Teaching projects
`u3b.config`	3B distribution configuration
`u3b.misc`	AT&T 3B computers
`u3b.sources`	Free programs for AT&T 3B systems
`u3b.tech`	3B technical topics
`u3b.test`	Place to send test articles
`vmsnet.admin`	Managing VMS internals, MACRO-32, Bliss, etc
`vmsnet.alpha`	Alpha AXP architecture, systems, porting, etc
`vmsnet.announce`	General announcements (Moderated)
`vmsnet.announce.newusers`	Orientation information for new users (Moderated)
`vmsnet.decus.journal`	DECUServe Journal (Moderated)
`vmsnet.decus.lugs`	DECUS Local User Groups and related issues
`vmsnet.employment`	Jobs sought and offered, employment-related issues
`vmsnet.internals`	VMS internals, MACRO-32, Bliss, etc
`vmsnet.mail.misc`	Other electronic mail software
`vmsnet.mail.mx`	MX email system from RPI
`vmsnet.mail.pmdf`	PMDF email system
`vmsnet.misc`	General VMS topics not covered elsewhere
`vmsnet.networks.desktop.misc`	Other desktop integration software
`vmsnet.networks.desktop.pathworks`	DEC Pathworks desktop integration software
`vmsnet.networks.management.decmcc`	DECmcc and related software
`vmsnet.networks.management.misc`	Other network management software
`vmsnet.networks.misc`	General networking topics not covered elsewhere
`vmsnet.networks.tcp-ip.cmu-tek`	CMU-TEK TCP/IP package
`vmsnet.networks.tcp-ip.misc`	Other TCP/IP software for VMS
`vmsnet.networks.tcp-ip.multinet`	TGV's Multinet TCP/IP, gatewayed to info-multinet
`vmsnet.networks.tcp-ip.tcpware`	Process Software's TCPWARE TCP/IP software
`vmsnet.networks.tcp-ip.ucx`	DEC's VMS/Ultrix Connection, TCP/IP services for VMS

`vmsnet.networks.tcp-ip.wintcp`	The Wollongong Group's WIN-TCP TCP/IP software
`vmsnet.pdp-11`	PDP-11 hardware and software
`vmsnet.sources`	Free programs (no discussion) (Moderated)
`vmsnet.sources.d`	Discussion about and requests for free programs
`vmsnet.sources.games`	Recreational software
`vmsnet.sysmgt`	VMS system management
`vmsnet.test`	Place to send test articles
`vmsnet.tpu`	TPU language and applications
`vmsnet.uucp`	DECUS Uucp software
`vmsnet.vms-posix`	VMS Posix

Clarinet Hierarchies

clari.biz.commodity	Commodity news and price reports (Moderated)
clari.biz.courts	Lawsuits, business-related legal matters (Moderated)
clari.biz.economy	Economic news and indicators (Moderated)
clari.biz.economy.world	Economy stories for non-US countries (Moderated)
clari.biz.features	Business feature stories (Moderated)
clari.biz.finance	Finance, currency, corporate finance (Moderated)
clari.biz.finance.earnings	Earnings and dividend reports (Moderated)
clari.biz.finance.personal	Personal investing and finance (Moderated)
clari.biz.finance.services	Banks and financial industries (Moderated)
clari.biz.invest	News for investors (Moderated)
clari.biz.labor	Strikes, unions and labor relations (Moderated)
clari.biz.market	General stock market news (Moderated)
clari.biz.market.amex	American Stock Exchange reports and news (Moderated)
clari.biz.market.dow	Dow Jones NYSE reports (Moderated)
clari.biz.market.ny	NYSE reports (Moderated)
clari.biz.market.otc	NASDAQ reports (Moderated)
clari.biz.market.report	General market reports, S&P, etc (Moderated)
clari.biz.mergers	Mergers and acquisitions (Moderated)
clari.biz.misc	Other business news (Moderated)
clari.biz.products	Important new products and services (Moderated)
clari.biz.top	Top business news (Moderated)
clari.biz.urgent	Breaking business news (Moderated)
clari.canada.biz	Canadian business summaries (Moderated)
clari.canada.briefs	Canadian news briefs
clari.canada.briefs.ont	Canadian news briefs (Ontario)
clari.canada.briefs.west	Canadian news briefs (Western)
clari.canada.features	Almanac, Ottawa Special, Arts (Moderated)
clari.canada.general	Short items on Canadian news stories (Moderated)
clari.canada.gov	Government-related news (all levels) (Moderated)
clari.canada.law	Crimes, the courts and the law. (Moderated)
clari.canada.newscast	Regular newscast for Canadians (Moderated)
clari.canada.politics	Political and election items (Moderated)
clari.canada.trouble	Mishaps, accidents and serious problems (Moderated)
clari.feature.dave_barry	Columns of humorist Dave Barry (Moderated)
clari.feature.kinsey	Sex Q&A and advice from Kinsey Institute (Moderated)
clari.feature.lederer	Richard Lederer's "Looking at Language" (Moderated)
clari.feature.mike_royko	Chicago opinion columnist Mike Royko (Moderated)
clari.feature.miss_manners	Judith Martin's etiquette advice (Moderated)
clari.feature.movies	Discussion and reviews of feature movies

`clari.local.alberta.briefs`	Local news briefs (Moderated)
`clari.local.arizona`	Local news (Moderated)
`clari.local.arizona.briefs`	Local news briefs (Moderated)
`clari.local.bc.briefs`	Local news briefs (Moderated)
`clari.local.california`	Local news (Moderated)
`clari.local.california.briefs`	Local news briefs (Moderated)
`clari.local.chicago`	Local news (Moderated)
`clari.local.chicago.briefs`	Local news briefs (Moderated)
`clari.local.florida`	Local news (Moderated)
`clari.local.florida.briefs`	Local news briefs (Moderated)
`clari.local.georgia`	Local news (Moderated)
`clari.local.georgia.briefs`	Local news briefs (Moderated)
`clari.local.headlines`	Various local headline summaries (Moderated)
`clari.local.illinois`	Local news (Moderated)
`clari.local.illinois.briefs`	Local news briefs (Moderated)
`clari.local.indiana`	Local news (Moderated)
`clari.local.indiana.briefs`	Local news briefs (Moderated)
`clari.local.iowa`	Local news (Moderated)
`clari.local.iowa.briefs`	Local news briefs (Moderated)
`clari.local.los_angeles`	Local news (Moderated)
`clari.local.los_angeles.briefs`	Local news briefs (Moderated)
`clari.local.louisiana`	Local news (Moderated)
`clari.local.manitoba.briefs`	Local news briefs (Moderated)
`clari.local.maritimes.briefs`	Local news briefs (Moderated)
`clari.local.maryland`	Local news (Moderated)
`clari.local.maryland.briefs`	Local news briefs (Moderated)
`clari.local.massachusetts`	Local news (Moderated)
`clari.local.massachusetts.briefs`	Local news briefs (Moderated)
`clari.local.michigan`	Local news (Moderated)
`clari.local.michigan.briefs`	Local news briefs (Moderated)
`clari.local.minnesota`	Local news (Moderated)
`clari.local.minnesota.briefs`	Local news briefs (Moderated)
`clari.local.missouri`	Local news (Moderated)
`clari.local.missouri.briefs`	Local news briefs (Moderated)
`clari.local.nebraska`	Local news (Moderated)
`clari.local.nebraska.briefs`	Local news briefs (Moderated)
`clari.local.nevada`	Local news (Moderated)
`clari.local.nevada.briefs`	Local news briefs (Moderated)
`clari.local.new_england`	Local news (Moderated)
`clari.local.new_hampshire`	Local news (Moderated)
`clari.local.new_jersey`	Local news (Moderated)
`clari.local.new_jersey.briefs`	Local news briefs (Moderated)

`clari.local.new_york`	Local news (Moderated)
`clari.local.new_york.briefs`	Local news briefs (Moderated)
`clari.local.nyc`	Local news (New York City) (Moderated)
`clari.local.nyc.briefs`	Local news briefs (Moderated)
`clari.local.ohio`	Local news (Moderated)
`clari.local.ohio.briefs`	Local news briefs (Moderated)
`clari.local.ontario.briefs`	Local news briefs (Moderated)
`clari.local.oregon`	Local news (Moderated)
`clari.local.oregon.briefs`	Local news briefs (Moderated)
`clari.local.pennsylvania`	Local news (Moderated)
`clari.local.pennsylvania.briefs`	Local news briefs (Moderated)
`clari.local.saskatchewan.briefs`	Local news briefs (Moderated)
`clari.local.sfbay`	Stories datelined San Francisco Bay Area (Moderated)
`clari.local.texas`	Local news (Moderated)
`clari.local.texas.briefs`	Local news briefs (Moderated)
`clari.local.utah`	Local news (Moderated)
`clari.local.utah.briefs`	Local news briefs (Moderated)
`clari.local.virginia+dc`	Local news (Moderated)
`clari.local.virginia+dc.briefs`	Local news briefs (Moderated)
`clari.local.washington`	Local news (Moderated)
`clari.local.washington.briefs`	Local news briefs (Moderated)
`clari.local.wisconsin`	Local news (Moderated)
`clari.local.wisconsin.briefs`	Local news briefs (Moderated)
`clari.nb.apple`	Newsbytes Apple/Macintosh news (Moderated)
`clari.nb.business`	Newsbytes business and industry news (Moderated)
`clari.nb.general`	Newsbytes general computer news (Moderated)
`clari.nb.govt`	Newsbytes legal and gov't computer news (Moderated)
`clari.nb.ibm`	Newsbytes IBM PC World coverage (Moderated)
`clari.nb.index`	Newsbytes index (Moderated)
`clari.nb.review`	Newsbytes new product reviews (Moderated)
`clari.nb.telecom`	Newsbytes telecom & online industry news (Moderated)
`clari.nb.trends`	Newsbytes new developments and trends (Moderated)
`clari.nb.unix`	Newsbytes Unix news (Moderated)
`clari.net.admin`	Announcements admins at Clarinet sites (Moderated)
`clari.net.announce`	Announcements for all Clarinet readers (Moderated)
`clari.net.newusers`	All about Clarinet (Moderated)
`clari.net.products`	New Clarinet products (Moderated)
`clari.net.talk`	Discussion of Clarinet: only unmoderated group
`clari.news.almanac`	Daily almanac: "This date in history" (Moderated)
`clari.news.arts`	Stage, drama and other fine arts (Moderated)
`clari.news.aviation`	Aviation industry and mishaps (Moderated)
`clari.news.books`	Books and publishing (Moderated)

`clari.news.briefs`	Regular news summaries (Moderated)
`clari.news.bulletin`	Major breaking stories of the week (Moderated)
`clari.news.canada`	News related to Canada (Moderated)
`clari.news.cast`	Regular U.S. news summary (Moderated)
`clari.news.children`	Children and parenting (Moderated)
`clari.news.consumer`	Consumer news, car reviews, etc (Moderated)
`clari.news.demonstration`	Demonstrations around the world (Moderated)
`clari.news.disaster`	Major problems, accidents, disasters (Moderated)
`clari.news.economy`	General economic news (Moderated)
`clari.news.election`	U.S. and international elections (Moderated)
`clari.news.entertain`	Entertainment industry news and features (Moderated)
`clari.news.europe`	News related to Europe (Moderated)
`clari.news.features`	Unclassified feature stories (Moderated)
`clari.news.fighting`	Clashes around the world (Moderated)
`clari.news.flash`	Ultra-important once-a-year news flashes (Moderated)
`clari.news.goodnews`	Stories of success and survival (Moderated)
`clari.news.gov`	General government-related stories (Moderated)
`clari.news.gov.agency`	Government agencies, FBI, etc (Moderated)
`clari.news.gov.budget`	Budgets at all levels (Moderated)
`clari.news.gov.corrupt`	Government corruption, kickbacks, etc (Moderated)
`clari.news.gov.international`	International government-related stories (Moderated)
`clari.news.gov.officials`	Government officials and their problems (Moderated)
`clari.news.gov.state`	State gov't news of national importance (Moderated)
`clari.news.gov.taxes`	Tax laws, trials, etc (Moderated)
`clari.news.gov.usa`	U.S. federal government news [high volume] (Moderated)
`clari.news.group`	Special interest groups (Moderated)
`clari.news.group.blacks`	News of interest to Blacks (Moderated)
`clari.news.group.gays`	Homosexuality and Gay rights (Moderated)
`clari.news.group.jews`	Jews and Jewish interests (Moderated)
`clari.news.group.women`	Women's issues and abortion (Moderated)
`clari.news.headlines`	Hourly list of top U.S./World headlines (Moderated)
`clari.news.hot.east_europe`	News from Eastern Europe (Moderated)
`clari.news.hot.iraq`	Persian Gulf crisis news (Moderated)
`clari.news.hot.somalia`	News from Somalia (Moderated)
`clari.news.hot.ussr`	News from the Soviet Union (Moderated)
`clari.news.interest`	Human interest stories (Moderated)
`clari.news.interest.animals`	Animals in the news (Moderated)
`clari.news.interest.history`	Human interest/history in the making (Moderated)
`clari.news.interest.people`	Famous people in the news (Moderated)
`clari.news.interest.people.column`	Famous people in the news dailies (Moderated)
`clari.news.interest.quirks`	Unusual or funny news stories (Moderated)
`clari.news.issues`	Major issues not covered elsewhere (Moderated)

`clari.news.issues.civil_rights`	Freedom, racism, civil rights issues (Moderated)
`clari.news.issues.conflict`	Conflict between groups around the world (Moderated)
`clari.news.issues.family`	Family, child abuse, etc (Moderated)
`clari.news.labor`	Unions, strikes (Moderated)
`clari.news.labor.strike`	Strikes (Moderated)
`clari.news.law`	General group for law-related issues (Moderated)
`clari.news.law.ciwil`	Civil trials and litigation (Moderated)
`clari.news.law.crime`	Major crimes (Moderated)
`clari.news.law.crime.sex`	Sex crimes and trials (Moderated)
`clari.news.law.crime.trial`	Trials for criminal actions (Moderated)
`clari.news.law.crime.violent`	Violent crime and criminals (Moderated)
`clari.news.law.drugs`	Drug-related crimes and stories (Moderated)
`clari.news.law.investigation`	Investigation of crimes (Moderated)
`clari.news.law.police`	Police and law enforcement (Moderated)
`clari.news.law.prison`	Prisons, prisoners and escapes (Moderated)
`clari.news.law.profession`	Lawyers, judges, etc (Moderated)
`clari.news.law.supreme`	U.S. Supreme Court rulings and news (Moderated)
`clari.news.lifestyle`	Fashion, leisure, etc (Moderated)
`clari.news.military`	Military equipment, people and issues (Moderated)
`clari.news.movies`	Reviews, news and stories on movie stars (Moderated)
`clari.news.music`	Reviews and issues concerning music (Moderated
`clari.news.politics`	Politicians and politics (Moderated)
`clari.news.politics.people`	Politicians and political personalities (Moderated)
`clari.news.religion`	Religion and religious leaders (Moderated)
`clari.news.sex`	Sexual issues and sex-related politics (Moderated)
`clari.news.terrorism`	Terrorist actions around the world (Moderated)
`clari.news.top`	Top U.S. news stories (Moderated)
`clari.news.top.world`	Top international news stories (Moderated)
`clari.news.trends`	Surveys and trends. (Moderated)
`clari.news.trouble`	Less major accidents, problems, mishaps (Moderated)
`clari.news.tv`	TV schedules, news, reviews and stars (Moderated)
`clari.news.urgent`	Major breaking stories of the day (Moderated)
`clari.news.weather`	Weather and temperature reports (Moderated)
`clari.sfbay.briefs`	S.F. Bay Area: breaking news briefs (Moderated)
`clari.sfbay.entertain`	S.F. Bay Area: entertainment and reviews (Moderated)
`clari.sfbay.fire`	S.F. Bay Area: current fire information (Moderated)
`clari.sfbay.general`	S.F. Bay Area: general announcements (Moderated)
`clari.sfbay.misc`	S.F. Bay Area: miscellaneous information (Moderated)
`clari.sfbay.police`	S.F. Bay Area: police activities/reports (Moderated)
`clari.sfbay.roads`	S.F. Bay Area: road/traffic conditions (Moderated)
`clari.sfbay.short`	S.F. Bay Area: headlines (Moderated)
`clari.sfbay.weather`	S.F. Bay Area: weather reports/forecasts (Moderated)

`clari.sports.baseball`	Baseball scores, stories, games, stats (Moderated)
`clari.sports.basketball`	Basketball coverage (Moderated)
`clari.sports.basketball.college`	College basketball coverage (Moderated)
`clari.sports.features`	Sports feature stories (Moderated)
`clari.sports.football`	Pro football coverage (Moderated)
`clari.sports.football.college`	College football coverage (Moderated)
`clari.sports.hockey`	NHL coverage (Moderated)
`clari.sports.misc`	Other sports, plus general sports news (Moderated)
`clari.sports.motor`	Racing and motor sports (Moderated)
`clari.sports.olympic`	Olympic Games (Moderated)
`clari.sports.tennis`	Tennis news and scores (Moderated)
`clari.sports.top`	Top sports news (Moderated)
`clari.streetprice`	Direct buyer prices: computer equipment (Moderated)
`clari.tw.aerospace`	Aerospace industry and companies (Moderated)
`clari.tw.computers`	Computer industry (Moderated)
`clari.tw.defense`	Defense industry issues (Moderated)
`clari.tw.education`	Universities and colleges (Moderated)
`clari.tw.electronics`	Electronics makers and sellers (Moderated)
`clari.tw.environment`	Environmental news, hazardous waste, etc (Moderated)
`clari.tw.health`	Health care and medicine (Moderated)
`clari.tw.health.aids`	AIDS stories, research, political issues (Moderated)
`clari.tw.misc`	General technical industry stories (Moderated)
`clari.tw.nuclear`	Nuclear power and waste (Moderated)
`clari.tw.science`	General science stories (Moderated)
`clari.tw.space`	NASA, astronomy and spaceflight (Moderated)
`clari.tw.stocks`	Computer and technology stock prices (Moderated)
`clari.tw.telecom`	Phones, satellites, media, telecom (Moderated)

Index

Bold page numbers indicate information found in Catalog of Internet Resources

Bold page numbers indicate information found in Catalog of Internet Resources

---

Bold page numbers indicate information found in Catalog of Internet Resources

Bold page numbers indicate information found in Catalog of Internet Resources

Bold page numbers indicate information found in Catalog of Internet Resources

Bold page numbers indicate information found in Catalog of Internet Resources

Bold page numbers indicate information found in Catalog of Internet Resources

Bold page numbers indicate information found in Catalog of Internet Resources

Bold page numbers indicate information found in Catalog of Internet Resources

Bold page numbers indicate information found in Catalog of Internet Resources

Bold page numbers indicate information found in Catalog of Internet Resources

Bold page numbers indicate information found in Catalog of Internet Resources

About the Authors...

Harley Hahn is an internationally recognized author, analyst and consultant. He is the author of more than 10 books, including **A Student's Guide to Unix**, and the well known **Peter Norton's Guide to Unix**. He is also the author of Osborne's highly regarded **Assembler Inside & Out**. Hahn has a degree in Mathematics and Computer Science from the University of Waterloo, Canada, and a graduate degree in Computer Science from the University of California at San Diego.

Rick Stout, CPA, holds degrees in business administration and accounting. Formerly a consulting manager at the accounting firm of Ernst & Young, Stout is a specialist in business computer and accounting systems. He is the author of several books and numerous articles for computer magazines. Stout lives in the San Diego area where he works on literary projects, develops software and consults.

Free Internet Access Offer on Following Page

Free Internet Access Offer

To help new users gain Internet access, we have made arrangements with more than 50 access providers to give our readers a special trial offer of one month's free Internet usage. This special deal includes Internet providers in virtually every major metropolitan area in the United States, Canada, and Europe, as well as providers in Australia and New Zealand.

Note: The various providers have different terms and conditions and this offer may vary from one provider to another. Some providers may charge a start-up fee while others may ask you to sign up for more than one month. All have agreed, however, to give one month's free access to new users.

To receive information on how to obtain your free month's access please fill out the form below and send it along with a proof of purchase to:

Special Internet Deal
Osborne/McGraw-Hill
2600 Tenth Street
Berkeley, CA 94710

- -

Please send me information on how I can obtain a free month's access on the Internet.

_____
Name

_____
Street Address

_____
City   **State** (Province)   **Zip** (Postal) **Code**

_____
Country

_____
Phone Number (necessary to determine nearest provider)